Heterocyclic Chemistry

Heterocyclic Chemistry

Second edition

T. L. Gilchrist
University of Liverpool

233 4

Longman
Scientific &
Technical

Copublished in the United States with
John Wiley & Sons, Inc., New York

Longman Scientific & Technical
Longman Group UK Limited,
Longman House,
Burnt Mill, Harlow,
Essex CM20 2JE, England
and Associated Companies throughout the world

Copublished in the United States of America with
John Wiley & Sons Inc., 605 Third Avenue, New York, NY 10158.

First published in Great Britain 1985 by Pitman Publishing Ltd

Reprinted by Longman Scientific & Technical 1987

Second Edition 1992

British Library Cataloguing in Publication Data

Gilchrist, Thomas Lonsdale
　Heterocyclic chemistry.—2nd ed.
　I. Title
　547

　ISBN 0-582-06420-1　PPR

Library of Congress Cataloging-in-Publication Data

Gilchrist, T. L. (Thomas Lonsdale)
　Heterocyclic chemistry/T. L. Gilchrist.—2nd ed.
　　p.　cm.
　Includes bibliographical references and index.
　1. Heterocyclic chemistry.　I. Title.
QD400.G55　1992
547′.59—dc20

Set in 10/12 Times roman by 6
Produced by Longman Group (FE) Limited
Printed in Hong Kong

Contents

Preface to the second edition

I have received several very helpful comments and criticisms of the first edition and I have tried to take these into account in producing a second edition. I have also taken the opportunity to update some of the original material. The most substantial additions have been made to the chapter on the synthesis of heterocycles (Chapter 4). There have been important advances in the use of 1,3-dipolar cycloaddition and Diels–Alder cycloaddition in heterocyclic synthesis; the sections covering these topics have therefore been enlarged and reorganized. Other general methods that have gained increased prominence, such as radical cyclization, are also discussed more fully. The introductory chapter and that on the properties of aromatic heterocycles (Chapter 2) have also been substantially revised. Elsewhere new material has been added to reflect the increasing importance of topics such as the use of organolithium reagents in heterocyclic chemistry and the value of heterocycles as intermediates in synthesis. Several new problems have also been added.

Another major change is a cosmetic one: the order of the later chapters has been altered. This has been done because the simpler six- and five-membered aromatic heterocycles seem to be taught first in most courses on heterocyclic chemistry. The chapters on the various classes of ring system (Chapters 5 to 10) are, however, largely self-contained and the material can be presented in any sequence.

There is rarely time to cover all the material presented here in a course on heterocyclic chemistry. The book is therefore intended also to serve as a reference source and as a guide to the enormous literature of heterocyclic chemistry. The reference list has been updated, mainly to include recent review articles. In doing this I have adopted the policy of making little specific reference to review chapters in *Comprehensive Heterocyclic Chemistry* (ed. A. R. Katritzky and C. W. Rees, Pergamon Press, Oxford, 1984). This work is nevertheless the best place to start a search for more detailed information on heterocyclic chemistry and its literature. The periodic guides to the literature provided in *Advances in Heterocyclic Chemistry* (ed. A. R. Katritzky, Academic Press, San Diego) are also invaluable: the latest of these, by L. I. Belen'kii, is in Volume 44 (1988).

I am most grateful to those who have made suggestions for changes to the original edition. In particular, I should like to thank Professor Wilhelm

Preface to the first edition

In this book I have aimed to present a description of the preparation and properties of heterocyclic compounds which can be used by anyone with a good background knowledge of organic chemistry. The number of different heterocyclic systems and their structural diversity make it difficult to gain a good understanding of the subject by studying individual ring systems in isolation. I have therefore tried to identify and emphasize general features of the properties of heterocyclic compounds throughout the book.

The concept of aromaticity has long proved useful in rationalizing the chemistry of many unsaturated heterocycles. The meaning of the term armoaticity, as applied to heterocyclic compounds, is discussed in Chapter 2. This is intended to provide a firm basis for understanding the chemistry of particular ring systems described later in the book. The chemistry of nonaromatic heterocycles is often regarded as a simple extrapolation of aliphatic and alicyclic chemistry. Although this undoubtedly has some justification there are some important special features of the properties of nonaromatic heterocycles which are discussed in Chapter 3. Chapter 4 is intended to provide a framework for rationalizing the vast literature on heterocyclic synthesis. The approach I have taken is to identify the reactions involved in the ring-forming processes and to classify the synthetic methods accordingly. There are many examples in this chapter, presented in tabular form, which are given to emphasize the common features involved in the synthesis of different ring systems. Many of these examples are referred to in later chapters.

Descriptive accounts of the more important ring systems are presented in Chapters 5 to 10. The ring systems are described in order of increasing ring size and complexity, but the material in this part of the book need not be read in the order in which it is presented. For example, the reader may prefer to study the chemistry of pyridines (Chapter 8) before that of pyrroles (Chapter 6). Imidazole and the other azoles have an interesting and diverse chemistry which is sometimes given little attention in textbooks. I have felt justified in describing these ring systems in more detail (Chapter 7) because of their apparently increasing importance as pharmaceuticals. In all of the chapters covering particular ring systems examples of currently important derivatives are given, and the diverse roles of the heterocycles as synthetic intermediates in organic chemistry are illustrated. A guide to the various nomenclature systems for heterocycles is presented in Chapter 11.

Flitsch, Professor Charles Rees, Dr John Boulton, and Dr Derek Hurst, who provided detailed written comments. If their suggestions have not all been incorporated it is because I have had to restrict the length of the book. I am also grateful to Dr Dick Storr, who read part of the new material, to Dr Frank King for his help with Chapter 1, to Dr Michael Rodgers for commissioning a second edition, and to the staff of Longman Higher Education and Reference for their expert guidance.

Each chapter is followed by a summary, listing the main points in the text, and by a set of problems. Most of these problems are taken from the literature, and a list of references to the original papers is given at the end of the book. Literature citations are also given throughout the text. These can be ignored by the reader who simply wants an introduction to the topics being discussed, but they are provided to allow access to the original work on much of the factual material. Where this is not possible, for reasons of space, references are given to reviews. Heterocyclic chemistry is well served by specialized reviews, notably those in the series *The Chemistry of Heterocyclic Compounds*, Edited by A. Weissberger and E. C. Taylor, and in the *Advances in Heterocyclic Chemistry* series edited by A. R. Katritzky and A. J. Boulton. An important additional source of specialized information, which has become available since this text was written, is *Comprehensive Heterocyclic Chemistry*, edited by A. R. Katritzky and C. W. Rees.

In writing the book I have been fortunate to have the expert guidance of the staff of Pitman Publishing Ltd. I should like to thank them, and particularly Mr Navin Sullivan, for their help. I am also indebted to the referees who read parts of the manuscript and made invaluable suggestions for its improvement, and to my colleagues at Liverpool who have read and commented upon parts of the text. My wife Lorraine typed most of the manuscript and was a constant source of advice, encouragement and practical assistance during its preparation.

T. L. Gilchrist. September 1984
University of Liverpool

Acknowledgements

We are grateful to the following for permission to reproduce copyright material:

Verlag Chemie GmbH and Professor Dr G. Schröder for Fig. 2.18 from a figure, p. 242 (Gilb & Schröder, 1982).

Whilst every effort has been made to trace the owners of copyright material, in a few cases this has proved impossible and we take this opportunity to offer our apologies to any copyright holders whose rights we may have unwittingly infringed.

1 Introduction

The known organic compounds have an enormous diversity of structure. Many of these structures contain ring systems. If the ring system is made up of atoms of carbon and at least one other element, the compound can be classed as *heterocyclic*. The elements that occur most commonly with carbon in ring systems are nitrogen, oxygen, and sulphur. About half of the known organic compounds have structures that incorporate at least one heterocyclic component.

Heterocyclic compounds have a wide range of applications: they are predominant among the types of compounds used as pharmaceuticals, as agrochemicals, and as veterinary products. They are used as optical brightening agents, as antioxidants, as corrosion inhibitors, and as additives with a variety of other functions. Many dyestuffs and pigments have heterocyclic structures.[1]

Heterocyclic compounds are also widely distributed in nature. Many are of fundamental importance to living systems: it is striking how often a heterocyclic compound is found as a key component in biological processes. We can, for example, identify the nucleic acid bases, which are derivatives of the pyrimidine and purine ring systems (Chapter 7), as being crucial to the mechanism of replication. Chlorophyll and heme, which are derivatives of the porphyrin ring system (Chapter 6), are the components required for photosynthesis and for oxygen transport in higher plants and in animals, respectively. Essential diet ingredients such as thiamin (vitamin B_1), riboflavin (vitamin B_2), pyridoxol (vitamin B_6), nicotinamide (vitamin B_3), and ascorbic acid (vitamin C) are heterocyclic compounds. Of the twenty amino acids commonly found in proteins, three, namely histidine, proline, and tryptophan, are heterocyclic. It is not surprising, therefore, that a great deal of current research work is concerned with methods of synthesis and properties of heterocyclic compounds.

One of the reasons for the widespread use of heterocyclic compounds is that their structures can be subtly manipulated to achieve a required modification in function. As we shall see in Chapter 2, many heterocycles can be fitted into one of a few broad groups of structures which have overall

1 Reviews of several applications of heterocyclic compounds can be found in *Comprehensive Heterocyclic Chemistry*, Vol. 1.

similarities in their properties but significant variations within the group. Such variations can include differences in acidity or basicity, different susceptibility to attack by electrophiles or nucleophiles, and different polarity. The possible structural variations include the change of one heteroatom for another in a ring and the different positioning of the same heteroatoms within the ring. Another important feature of the structures of many heterocyclic compounds is that it is possible to incorporate functional groups either as substituents or as part of the ring system itself. For example, basic nitrogen atoms can be incorporated both as amino substituents and as part of a ring. This means that the structures are particularly versatile as a means of providing, or of mimicking, a functional group. An example of the latter, which is discussed further in Chapter 8, is the use of the $1H$-tetrazole ring system as a mimic of a carboxylic acid function, because of its similarity in acidity and in steric requirement. One of the main purposes of the later chapters of this book is to provide a framework for understanding and predicting the effects of structure on properties. Armed with this under-standing, the heterocyclic chemist can 'tailor' a structure to meet a particular need by modifying the heterocyclic component.

Heterocyclic compounds are also finding an increasing use as intermediates in organic synthesis.[2] Very often this is because a relatively stable ring system can be carried through a number of synthetic steps and then cleaved at the required stage in a synthesis to reveal other functional groups. Many examples are given in later chapters, but the principle is illustrated by the common use of the furan ring system as a 'masked' γ-dicarbonyl compound (Fig. 1.1): the furan ring can be cleaved by acid to reveal this functionality.

A great deal of research in heterocyclic chemistry is concerned with discovering new methods of ring synthesis. Some of the classical methods of synthesis of heterocyclic compounds continue to be used widely, but there is a constant need for methods that can provide new derivatives, methods that are highly selective, and methods that can be carried out under particularly mild conditions. Examples of both the established methods of ring synthesis and recent methods are given in later chapters.

Many of the pharmaceuticals and most of the other heterocyclic compounds with practical applications are not extracted from natural sources but are manufactured. The origins of organic chemistry do, however, lie in the study of natural products. These formed the basis for the design of many

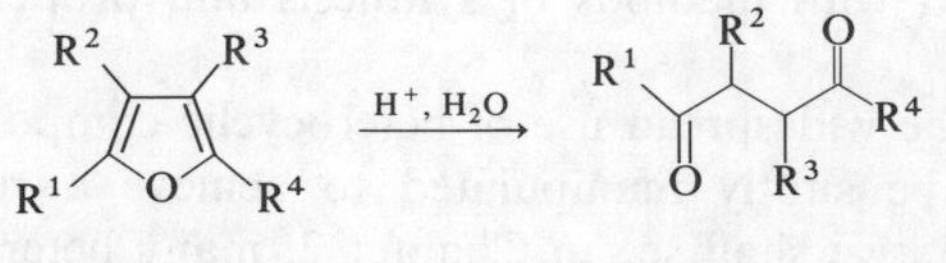

Fig. 1.1 Furans as masked γ-dicarbonyl compounds.

2 Reviews: A. P. Kozikowski, in *Comprehensive Heterocyclic Chemistry*, Vol. 1, p. 413; B. H. Lipshutz, *Chem. Rev.*, 1986, **86**, 795.

of the useful compounds developed subsequently: examples are the early development of vat dyes based on the structure of indigo and the continuing invention of new antibacterial agents based on the β-lactam structure of penicillin. Three different groups of pharmaceuticals with structures related to natural products are described briefly below as illustrations of the ways in which natural product chemistry interacts with synthetic heterocyclic chemistry.

1 *Compounds related to serotonin.*   Natural products often occur in very small quantities and are therefore difficult to investigate if the compounds have to be extracted from the natural source. Organic chemists can provide a solution to the problem by devising practicable laboratory syntheses. An example is provided by serotinin, **1** (also commonly called 5-hydroxy-tryptamine, or 5-HT). This compound is widely distributed in nature but occurs only in low concentration. It is derived in nature from the aminoacid tryptophan by hydroxylation at C-5 followed by decarboxylation. It was first isolated from natural sources in 1948 as the vasoconstricting agent present in serum after blood has clotted, and it was also detected soon afterwards in the gut and the brain. However, it was its laboratory synthesis a few years later that enormously widened the possibilities for investigating its mode of action. Serotonin is now known to have a wide and complex range of pharmacological actions, including contraction of blood vessels by stimulation of smooth muscle and blood platelet aggregation. Its acts as a constrictor of arteries in the brain and is implicated in migraine.[3] Changes in serotonin concentration in the brain are also connected with changes in mood and in appetite. However, serotonin is too rapidly metabolized to have any potential as a pharmaceutical agent.

There is a group of alkaloids with hallucinogenic properties (that is, they alter perception and mood) which are closely related in structure to serotonin, but which are more stable *in vivo*.[4] Psilocin, **2**, is one of the active constituents of Mexican mushrooms which were used since at least 1500 BC in Aztec and Mayan culture as hallucinogens. Bufotenine, **3**, is another hallucinogen which occurs in toadstools. These hallucinogenic compounds act as agonists (that is, as promotors of activity) at serotonin receptors in the brain. The simple indole derivative sumatriptan, **4**, a product of pharmaceutical research, also

1   **2**

3 This and other compounds potentially useful for relief of pain are described by R. Hill and K. Pittaway, *Chem. Br.*, 1987, 758.
4 For a discussion of these and other hallucinogenic compounds, see J. Mann, *Chem. Br.*, 1989, 478.

appears to act as a selective agonist at serotonin receptor sites in the brain and it may offer an effective treatment for migraine. The discovery of a useful pharmaceutical such as this requires the synthesis of a large number of other compounds on the way, in order to reveal which structural features provide the desired activity and selectivity. The ergot alkaloids have more complex structures (the structure of one of them, ergotamine, is shown on p. 224) but they are all based on indole with a β-aminoethyl side chain at the 3-position. Ergotamine in small doses has been shown to be useful in treating migraine but it is also highly toxic and its action is not sufficiently specific. The structurally related compound lysergic acid diethylamide (LSD), now notorious as a hallucinogen, is a product of laboratory synthesis; its properties were first discovered following accidental inhalation of the compound. The structure of LSD is shown on p. 224.

2 *Pharmaceuticals related to histamine.*   Histamine, **5**, a monosubstituted imidazole, is derived *in vivo* from the amino acid histidine by decarboxylation. It was not established unequivocally as a natural constituent of the body until 1927, but it had become available twenty years earlier as a product of laboratory synthesis. This had allowed its pharmacological effects to be investigated. It was shown to have powerful effects in contracting smooth muscle and in lowering blood pressure. Histamine is involved in allergic reactions and is liberated from skin cells on injury. It also participates in the regulation of gastric acid secretion.

From the 1940s onwards several synthetic drugs that act as histamine antagonists became available. An example of a widely used antihistamine drug of this type is pyrilamine, **6**, a pyridine derivative. These drugs inhibit several of the actions of histamine but they fail to block gastric acid secretion. Two different types of histamine receptor were therefore distinguished in the body; these were labelled H_1 and H_2 receptors. Pyrilamine and similar drugs act as blockers of the H_1 receptors.

The search for a specific histamine H_2 receptor antagonist which could potentially control gastric acid secretion, and thus provide the basis of a treatment for peptic ulcers, started in 1964. Having little information to guide their initial choice, the chemists used the structure of histamine as a starting point for the design and synthesis of new compounds. Although it took a long period to discover active compounds, the work eventually resulted in 1976 in the introduction of a new drug, cimetidine, **7**, for the treatment of peptic ulcers.[5] The design of cimetidine, an imidazole derivative, made use

5 Reviews: C. R. Ganellin, in *Medicinal Chemistry*, ed. S. M. Roberts and B. J. Price, Academic Press, London, 1985, p. 93; G. J. Durant, *Chem. Soc. Rev.*, 1985, **14**, 375.

of some of the fundamental chemistry of imidazoles, such as a knowledge of tautomeric equilibria and the synthesis and properties of methyl-substituted imidazoles. Cimetidine was one of the most important and successful drugs of the late 1970s and 1980s, being the first nonsurgical treatment for peptic ulcers.

The success of cimetidine has prompted the introduction of other drugs having structures which are closely related but with the imidazole ring replaced by a different heterocycle. Ranitidine, **8**, containing a furan ring, is also an important and successful drug for the treatment of peptic ulcers. The recognition that distinct types of histamine receptor exist has also led to efforts to develop new specific H_1 receptor antagonists as potential antiasthmatic and antiallergic agents. Drugs with structures based on a variety of heterocyclic ring systems have been introduced for such purposes. An example is azelastine, **9**, which is particularly useful as an antiasthmatic since it can be taken orally.

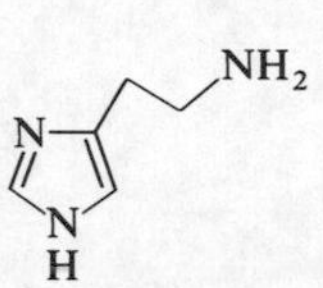

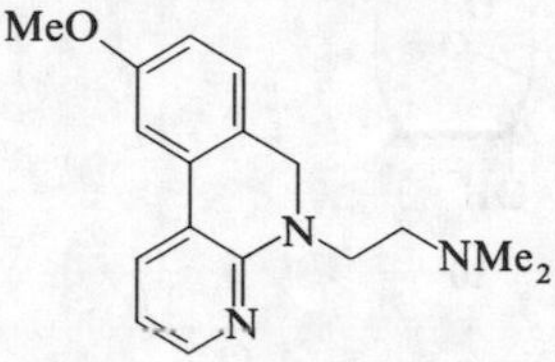

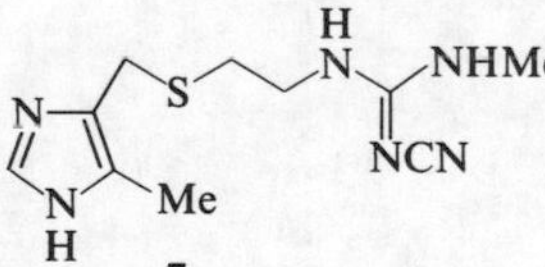

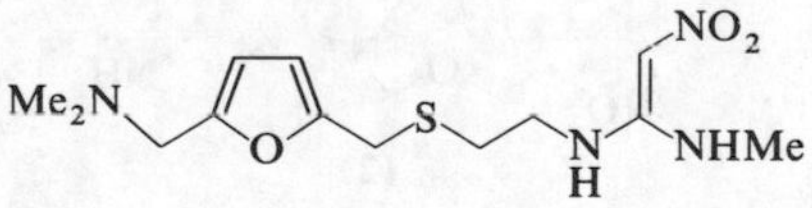

3 *Nucleoside analogues.*[6] A logical starting point in the search for drugs to combat cancer and viruses is the structure of DNA. One approach that has been intensively explored is to investigate the use of analogues of the DNA nucleosides. These DNA fragments consist of a heterocycle (a purine

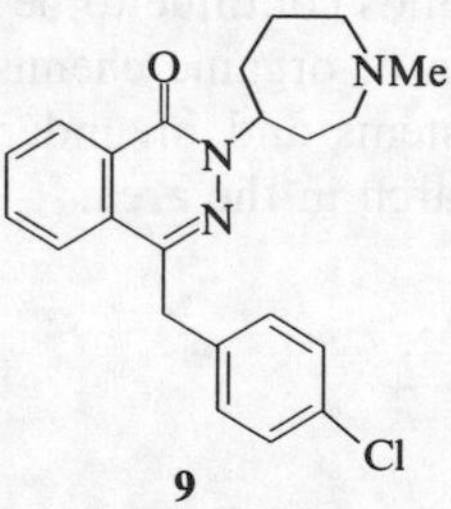

6 For reviews of useful drugs of this type see M. F. Jones, *Chem. Br.*, 1988, 1122; G. B. Elion, *Angew. Chem. Int. Edn Engl.*, 1989, **28**, 870.

or a pyrimidine) linked to a sugar; an example is thymidine, **10**. Analogues can be made in which the structure of the heterocycle is modified, or that of the sugar is modified, or both. Such a compound might interfere in the replicative cycle of a virus, for example, by being incorporated in place of the natural nucleoside. The main problem is one of selectivity, because the majority of such compounds are also likely to prove toxic to normal cells. Nevertheless, some important drugs of this type have been produced. Zidovudine (AZT), **11**, used in the treatment of AIDS, is a close analogue of thymidine. Acyclovir (ACV), **12**, is a guanine derivative of which the acyclic side chain mimics part of the structure of the sugar in the natural nucleoside. It is a useful drug for the treatment of herpes virus infections because of its high selectivity, and its mechanism of action has been elucidated.

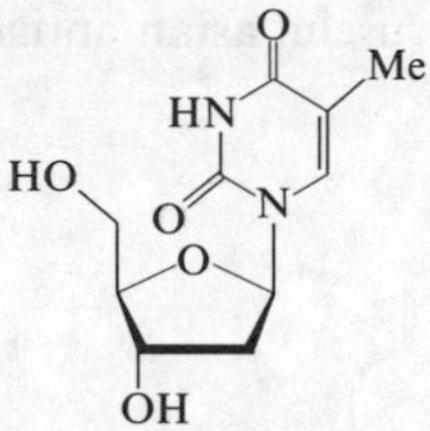

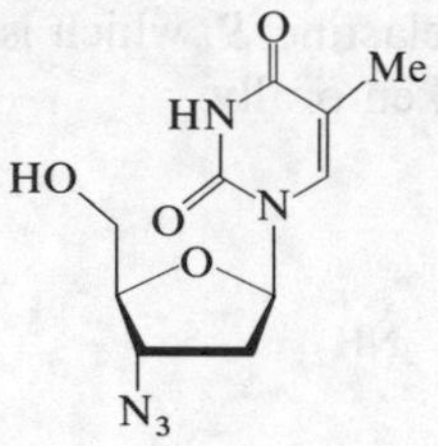

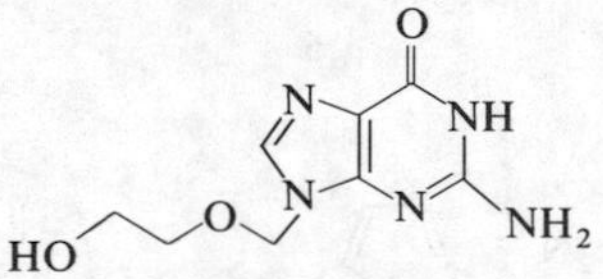

Later chapters contain the structures of many other compounds with important uses. The rate at which such compounds continue to be invented testifies to the strength and vitality of this area of organic chemistry. The challenges of discovering new heterocyclic systems and of understanding their properties also continue to stimulate research in the area.

2 Aromatic heterocycles

2.1 The common structural types

2.1.1 Six-atom, six-π-electron heterocycles

All six-membered fully unsaturated heterocycles are formally related to benzene in that one or more of the carbon atoms of the benzene ring is replaced by a heteroatom. *Pyridine* (azabenzene) is the simplest of such compounds, having one ring nitrogen atom. Like benzene, pyridine is planar, the ring system being a slightly distorted hexagon because the C—N bonds are shorter than the C—C bonds. It can be represented as shown in Fig. 2.1(a) by a cyclic structure made up of five sp^2 hybridized carbon atoms, to each of which is attached a hydrogen atom (not shown), and one sp^2 hybridized nitrogen atom. Each of the six atoms of the ring has one p-orbital orthogonal to the plane of the ring. This structure is similar to that of benzene in having a complete cycle of p-orbitals containing six electrons, but it differs in having a lone pair of electrons in the plane of the ring. Like benzene, pyridine, is conveniently represented by one of two equivalent Kekulé structures [Fig. 2.1(b)].

(a)

(b)

Fig. 2.1 Representations of the structure of pyridine.

Molecular orbital theory is widely used to interpret the structure of benzene, pyridine, and related cyclic organic compounds. In the Hückel approximation the electrons in the p-orbitals are treated separately from those forming the bonds in the plane of the ring. The six p-orbitals are combined to give six delocalized π-molecular orbitals, each of which can contain a maximum of two electrons. Calculations of the energies of these orbitals lead to the conclusion that three are of lower energy than the isolated p-orbitals, and three are of higher energy. The three lower energy orbitals, the bonding π-orbitals, are thus able to accommodate the six electrons, leaving the higher energy orbitals, the antibonding π*-orbitals, vacant. The *Hückel rule*, which is based on simple molecular orbital calculations, states that planar molecules with a complete and uninterrupted cycle of p-orbitals are stabilized, relative to acyclic counterparts, when they contain $(4n+2)$ π-electrons, n being zero or an integer. Benzene, pyridine, and other compounds with six π-electrons thus obey the Hückel rule for $n=1$. These are by far the most numerous and the most important group of structures that obey the Hückel rule and which we can refer to as 'aromatic'.[1]

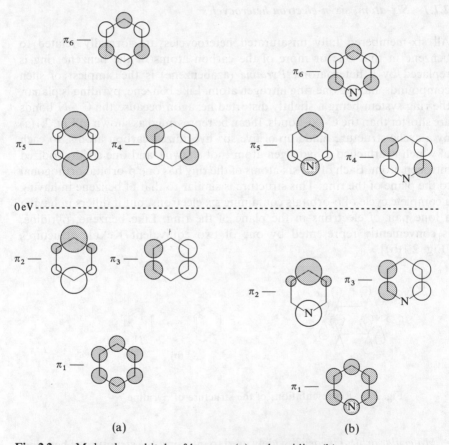

(a)   (b)

Fig. 2.2   π Molecular orbitals of benzene (a) and pyridine (b).

1 For a general review of the concept of aromaticity and the Hückel rule see P. J. Garratt, *Aromaticity*, Wiley-Interscience, New York, 1986. For applications of spin-coupled valence bond theory to heterocycles see D. L. Cooper, S. C. Wright, J. Gerratt, and M. Raimondi, *J. Chem. Soc., Perkin Trans.* **2**, 1989, 255, 263.

The π-molecular orbitals of benzene and pyridine are illustrated in Fig. 2.2. In these drawings the sizes of the circles give an approximate indication of the relative orbital coefficients and the open and shaded circles indicate opposite phases of the wave functions. The orbitals of pyridine are qualitatively very similar to those of benzene, but there are two important differences. The first is that the π-orbitals of pyridine are lower in energy than those of benzene. This is a consequence of the greater electronegativity of nitrogen compared with carbon. The second difference is that whereas the highest occupied molecular orbitals (HOMOs) of benzene, orbitals π_2 and π_3, are degenerate (equal in energy), the corresponding orbitals of pyridine have different energies. Orbital π_2, with a large coefficient on nitrogen, is lower in energy than π_3, with coefficients only on carbon: π_3 is the HOMO. The energies of the HOMOs are particularly significant because, as we shall see in Section 2.4, they can provide a basis for estimating chemical reactivity.

An alternative interpretation of the structures of these molecules is provided by *valence bond theory*. Wave functions for the molecules are constructed from combinations of the wave functions of the constituent atoms or fragments. For benzene and related molecules this results in a description of the structure as a weighted average of several classical bonded structures, called canonical forms. The theory has tended to be regarded by organic chemists as less useful than molecular orbital theory because the numerical results it gives for physical properties have been less good than those based on the best molecular orbital calculations. It does, however, have the great merit of providing a theoretical basis for structural representations such as those for pyridine in Fig. 2.1(b): pyridine is regarded as a *resonance hybrid* of these two structures, together with some others of less importance. Calculations based on recent developments in valence bond theory – so-called 'spin-coupled' valence bond theory – have provided much more accurate estimates than previously of physical properties. This theory gives a quite different picture of the structure of benzene and pyridine to that provided by molecular orbital theory. The six electrons orthogonal to the plane are localized in six overlapping π-type orbitals which are essentially distorted p-orbitals on each of the ring atoms. The characteristic properties of the aromatic systems arise by symmetric couplings of the electron spins around the ring. These calculations indicate that in the case of pyridine, the two Kekulé structures shown in Fig. 2.1(b) are the major structures resulting from such couplings and account for 80% of the total of spin couplings in pyridine. The theory provides a similar justification for the traditional valence bond structures of other common heterocycles.

The relationship between the structure of benzene and that of pyridine extends to other six-membered conjugated ring systems with more than one nitrogen atom in the ring. There are altogether eight heterocyclic systems known (including pyridine) which have one or more sp^2-hybridized nitrogen atoms in the ring. These are shown in Fig. 2.3.

The reduced π-electron density at the ring carbon atoms compared with those of benzene, particularly at the 2- and 4-positions, is illustrated for

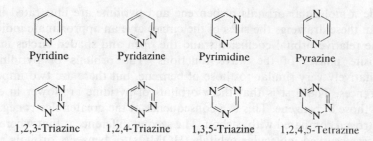

Pyridine   Pyridazine   Pyrimidine   Pyrazine

1,2,3-Triazine   1,2,4-Triazine   1,3,5-Triazine   1,2,4,5-Tetrazine

Fig. 2.3 The known aza derivatives of benzene.

pyridine by the calculated values shown in Fig. 2.4. As the number of nitrogen atoms in the ring increases, this reduction of electron density at carbon becomes more marked. For this reason the aromatic heterocycles of this type are often referred to as *π-deficient heterocyclic compounds.*

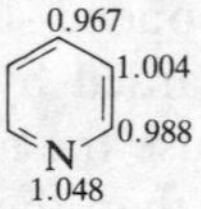

0.967
1.004
0.988
1.048

Fig. 2.4 Distribution of π-electron density in pyridine. (Calculated by an *ab initio* MO method: J. E. Del Bene, *J. Am. Chem. Soc.*, 1979, **101**, 6184.)

Nitrogen is the only atom from the first row of the Periodic Table which can replace a CH group of benzene and give an uncharged aromatic heterocycle. Oxygen can participate only by forming a positively charged ionic species, the pyrylium cation. Sulphur can participate in the same way as oxygen. Other elements from Groups IV and V of the Periodic Table can, however, form uncharged heterocycles of this type; some examples are shown in Fig. 2.5.[2]

All these ring systems are considered to have aromatic character: the criteria are discussed in Section 2.2.

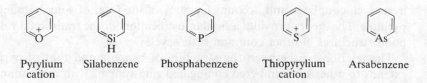

Pyrylium   Silabenzene   Phosphabenzene   Thiopyrylium   Arsabenzene
cation   cation

Fig. 2.5 Heterobenzenes containing atoms other than nitrogen.

2. For reviews of the properties of 'heterobenzenes' of Group V atoms see P. Jutzi, *Angew. Chem. Int. Edn Engl.*, 1975, **14**, 232; A. J. Ashe, *Accounts Chem. Res.*, 1978, **11**, 153.

2.1.2 Five-atom, six-π-electron heterocycles

Planar, unsaturated heterocycles containing five atoms can be considered as aromatic systems if they have an uninterrupted cycle of p-orbitals containing altogether six electrons. The carbocyclic analogue of these heterocycles is the cyclopentadienyl anion, which is a planar symmetrical pentagon with five sp^2-hybridized carbon atoms and a cyclic array of five p-orbitals containing six electrons [Fig. 2.6(a)]. *Pyrrole* can be considered as an example of the five-membered aromatic heterocycles. Pyrrole is a planar molecule, which indicates that the nitrogen atom is sp^2-hybridized. The three σ-bonds to nitrogen lie in the plane of the molecule, and the p-orbital, orthogonal to the plane, bears the remaining lone pair of electrons. This p-orbital on nitrogen interacts with the four p-orbitals on the carbon atoms, leading to a cyclic π-electron system derived from five p-orbitals but containing a total of six electrons [Fig. 2.6(b)].

The π molecular orbitals of pyrrole are shown in Fig. 2.7. The perturbation brought about by the introduction of the heteroatom is seen as a splitting

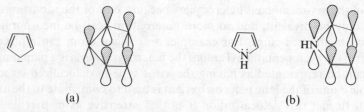

(a)   (b)

Fig. 2.6 Representations of the structure of the cyclopentadienyl anion (a) and of pyrrole (b).

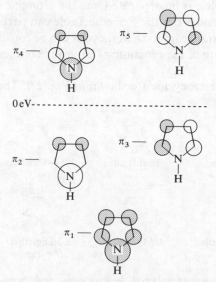

Fig. 2.7 π Molecular orbitals of pyrrole.

of the π_2 and π_3 levels, with the orbital π_2, which has a large coefficient on nitrogen, being of lower energy. This separation of the π_2 and π_3 levels can be envisaged as representing a partial localization of the lone pair on nitrogen. The calculated π-electron distribution in pyrrole (Fig. 2.8) shows that the ring system is electron rich (six π-electrons being distributed over five atoms) and that all four carbon atoms have greater π-electron density than in benzene, although the π-electron density is greatest at nitrogen.

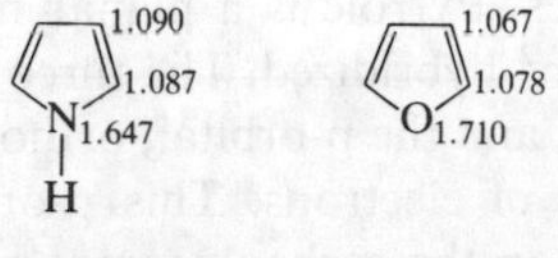

Fig. 2.8 Calculated π-electron densities in pyrrole[a] and in furan[b]. [a] J. Kao, A. L. Hinde, and L. Radom, *Nouv. J. Chim.*, 1979, **3**, 473. [b] I. G. John and L. Radom, *J. Am. Chem. Soc.*, 1978, **100**, 3981.

The group of five-membered aromatic heterocycles is much larger than that of the six-membered heterocycles because one of the atoms in the ring need only be divalent, and so more heteroatoms can be incorporated into the five-membered rings. For example, an oxygen atom can replace a CH group of the cyclopentadienyl anion; the heterocycle *furan* is a planar molecule and can be represented as having the same type of delocalized structure as pyrrole if one of the lone pairs on oxygen is used to contribute to the aromatic sextet. In fact this delocalization is not as extensive as in pyrrole because the oxygen atom is more electronegative and the oxygen lone pair is more tightly held by the heteroatom (Fig. 2.8). *Thiophene* can similarly be regarded as aromatic if a lone pair on sulphur is incorporated into the aromatic sextet. Other elements from Groups V and VI of the Periodic Table can participate in the same way. In addition, neutral aromatic heterocycles can be constructed by the replacement of one or more of the remaining CH groups of these ring systems by nitrogen.

Examples of five-membered heterocycles are shown in Fig. 2.9. Those on

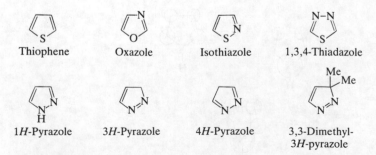

Thiophene   Oxazole   Isothiazole   1,3,4-Thiadazole

1*H*-Pyrazole   3*H*-Pyrazole   4*H*-Pyrazole   3,3-Dimethyl-3*H*-pyrazole

Fig. 2.9 Some aromatic five-membered heterocycles and nonaromatic pyrazoles.

the first row can all be regarded as aromatic systems if a lone pair on the oxygen or sulphur atom is incorporated into a delocalized π-electron sextet. The second row shows pyrazole in three tautomeric forms. Of these only 1H-pyrazole is aromatic; both the 3H- and the 4H-tautomers contain an sp^3-hybridized carbon atom in the ring and so cannot show cyclic delocalization. Such isomers are normally very unstable unless tautomerism is blocked, as in 3,3-dimethyl-3H-pyrazole.

The increased π-electron density at carbon compared with benzene, which is shown for pyrrole and furan in Fig. 2.8, has led to a general description of this group of heterocycles as π *excessive heterocyclic compounds*. Note, however, that for rings of this type which contain additional nitrogen atoms in the ring, the carbon atoms become increasingly electron deficient as the number of nitrogens in the ring increases.

2.1.3   Benzo-fused ring systems

The criterion of aromatic character provided by the Hückel rule strictly applies only to monocyclic compounds. It has long been recognized, however, that compounds with structures in which a benzene ring is fused to another aromatic ring system retain their aromatic properties, albeit in modified form. In naphthalene, for example, the 1,2-bond is shorter than the 2,3-bond so the hexagonal symmetry of benzene is lost. Many substituted naphthalenes also show significant deviations from planarity because of steric interactions across the 1- and 8-positions. Nevertheless, naphthalenes retain sufficient of the typical properties of a benzenoid system for them to be regarded as aromatic. In the same way, benzo-fused analogues of the six- and five-membered aromatic heterocycles are most conveniently classified as aromatic systems.

The benzo-fused six-membered aza heteroaromatics are shown in Fig. 2.10. These compounds, like naphthalene, show alternation in bond lengths consistent with the Kekulé structures shown. This leads to partial localization of the bonds, which is reflected in their chemistry, as is discussed in Chapter 5. The bond localization is more pronounced in dibenzo-fused hetero-aromatics such as phenanthridine, **1**.

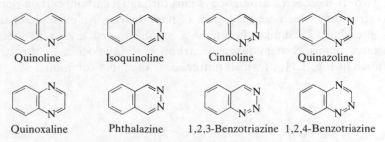

| Quinoline | Isoquinoline | Cinnoline | Quinazoline |

| Quinoxaline | Phthalazine | 1,2,3-Benzotriazine | 1,2,4-Benzotriazine |

Fig. 2.10 Benzo-fused six-membered nitrogen heteroaromatics.

1

Some benzo-fused five-membered heteroaromatics are shown in Fig. 2.11. Indole is the most important of these. It is a near-planar compound and the bond lengths show some alternation, as represented by the Kekulé structure in Fig. 2.11. The compounds which are annelated across the 'long' carbon–carbon bond of the heterocycle, such as isoindole and isobenzofuran, show a much lower degree of aromatic character than the others, as will be discussed in Section 2.2.

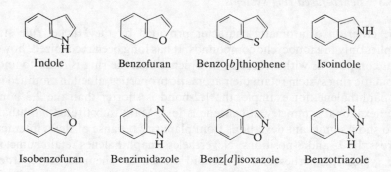

Indole	Benzofuran	Benzo[b]thiophene	Isoindole
Isobenzofuran	Benzimidazole	Benz[d]isoxazole	Benzotriazole

Fig. 2.11 Representative five-membered benzo-fused heterocycles.

2.1.4 *Other fused heterocycles*

An enormous variety of fused heterocycles exists which are formally constructed by the fusion of two monocyclic heteroaromatic systems. Fusion of two six-membered aromatic systems through a carbon–carbon bond leads to structures which are analogues of naphthalene. An example of this type is pteridine, **2**. Similar fusion of a six-membered and a five-membered heteroaromatic system through carbon gives analogues of the structures shown in Fig. 2.11, of which purine, **3**, is the most common.

2   **3**

4   5

Fusion across the C—N bond can also lead to structures that can be classed as aromatic. The quinolizinium cation, **4**, is of this type. The fusion of five- and six-membered rings in this way gives a series of neutral bicyclic structures of which indolizine, **5**, is the prototype. The classification of these compounds as aromatic is based on the facts that they are planar and contain a total of ten π-electrons, thus complying with the Hückel rule for $n = 2$. The status of these and some other conjugated heterocycles is discussed further in Section 2.3.

2.2 Some criteria of aromaticity in heterocycles[3]

2.2.1 Bond lengths

It is a general property of aromatic compounds that the lengths of the bonds in the rings are intermediate between the values for single and double bonds. In benzene the bond lengths are all equal (1.395 Å) whereas in a conjugated acyclic polyene the bond lengths alternate. In butadiene, for example, the C1—C2 bond length is 1.34 Å and the C2—C3 bond length is 1.48 Å. As a general rule, bond lengths tend to alternate in nonaromatic compounds but they tend not to do so in aromatic compounds. Typical values for the lengths of some common bonds in acyclic compounds are shown in Table 2.1. These values can be compared with the bond lengths in some heteroaromatic compounds which are shown in Fig. 2.12.

Table 2.1 Typical lengths of single and double bonds (Å) linking sp²-hybridized atoms.[a]

C—C	1.48	C=C	1.34
C—N	1.45	C=N	1.27
C—O	1.36	C=O	1.22
C—S	1.75	C=S	1.64
N—N	1.41	N=N	1.23

a M. Burke-Laing and M. Laing, *Acta Cryst.* (*B*), 1976, **32**, 3216; 'Tables of interatomic distances and configuration in molecules and ions', Chemical Society, London, 1965.

3 A. R. Katritzky, P. Barczynski, G. Musumarra, D. Pisano, and M. Szafran, *J. Am. Chem. Soc.*, 1989, **111**, 7.

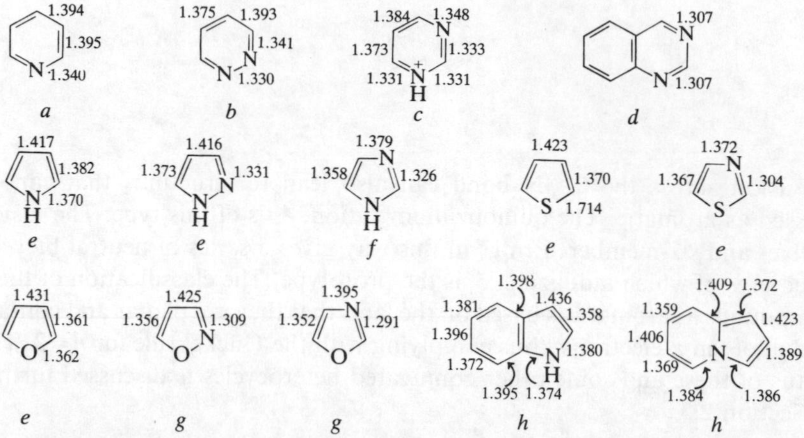

Fig. 2.12 Bond distances (Å) in some heteroaromatic compounds. [a] B. Bak, L. Hansen-Nygaard, and J. Rastrup-Andersen, *J. Mol. Spectroscopy*, 1958, **2**, 361. [b] A. Almenningen, G. Bjornsen, T. Ottersen, R. Seip, and T. G. Strand, *Acta Chem. Scand.* (A), 1977, **31**, 63. [c] S. Furberg and J. B. Aas, *Acta Chem. Scand.* (A), 1975, **29**, 713. [d] C. Huiszoon, *Acta Cryst.* (B), 1976, **32**, 998. [e] From J. Sheridan, in *Physical Methods in Heterocyclic Chemistry*, Academic Press, New York, Vol. VI, 1974, p. 53. [f] G. J. Visser and A. Voss, *Acta Cryst.* (B), 1971, **27**, 1802. [g] A. Kumar, J. Sheridan, and O. L. Stiefvater, *Z. Naturforsch.* (A), 1978, **33**, 145. [h] Average values from X-ray structures of derivatives.

The bond distances in the four six-membered heterocycles shown in Fig. 2.12 are all intermediate between the values for single and double bonds. The carbon–carbon lengths show little alternation in the three monocyclic six-membered rings, and are close to the value in benzene. We can conclude, on this basis, that there is substantial cyclic delocalization of the π-electrons in these compounds. The values for the five-membered heterocycles, on the other hand, show a considerable degree of alternation, consistent with the localized structures drawn, Because different heteroatoms are involved, it is not valid to compare the bond lengths directly, but the oxygen heterocycles generally show the greatest degree of bond localization. Even in the oxygen heterocycles, however, the bond lengths differ from those of 'pure' single and double bonds. All these five-membered ring systems show some evidence of cyclic delocalization but it is less than for the six-membered heterocycles. The two bicyclic five-membered ring systems shown (indole and indolizine) both have a higher degree of bond localization in the five-membered rings than does pyrrole; this is generally true for other fused heterocycles in comparison with the corresponding monocyclic compounds.

An indirect measure of bond order can sometimes be provided by vicinal coupling constants in ^1H NMR spectra. For example, the coupling constants J_{ab} and J_{bc} on adjacent carbon atoms a, b, and c remote from a heteroatom should be equal if the bonds C_a—C_b and C_b—C_c are equal in length, but not otherwise. The ratio $J_{ab}:J_{bc}$ can lie between 0.5 and 1.0 depending upon the degree of bond alternation. Comparison of values in a series of similar compounds can thus provide a measure of the degree of bond localization. For example, the variation in the ratio $J_{ab}:J_{ac}$ for the four compounds shown in Fig. 2.13 provides an indirect estimate of the degree of bond fixation in isoindole, for which X-ray data are not available.[4]

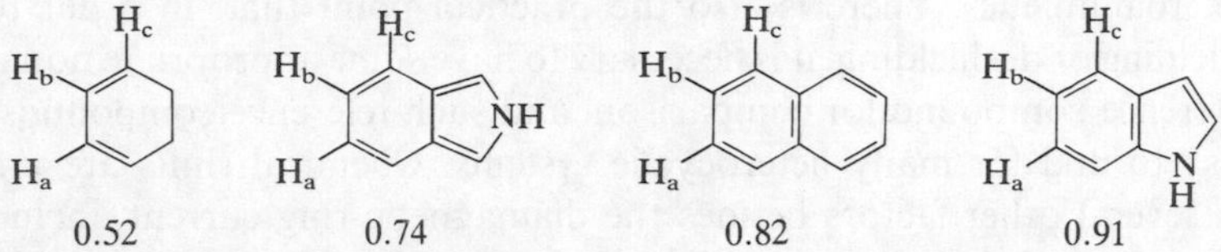

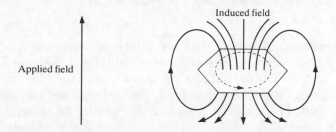

0.52   0.74   0.82   0.91

Fig. 2.13 Variation of $J_{ab}:J_{bc}$ with structure.

2.2.2   Ring current effects and chemical shifts in ^1H NMR spectra

^1H chemical shifts for benzenes are greater than for the analogous acyclic polyenes. This is ascribed, in part, to a 'diamagnetic ring current' effect.[5] When a solution of the benzenoid compound is placed in a magnetic field, the molecules are aligned at right angles to the field and a diamagnetic ring current is induced because of the presence of the delocalized π-electrons. This produces a secondary magnetic field which opposes the applied field within the ring but which reinforces it outside the ring (Fig. 2.14). Thus, hydrogen nuclei lying in an area above or below the centre of the ring are shielded, and those on the periphery of the ring are deshielded. The changes are more difficult to observe in ^{13}C NMR spectra because ^{13}C chemical shifts are much larger and the additional shielding or deshielding caused by the ring current is less significant in comparison.

Induced field

Applied field

Fig. 2.14 Field induced by a diamagnetic ring current.

4 B. A. Hess and L. J. Schaad, *Tetrahedron Lett.*, 1977, 535; E. Chacko, J. Bornstein, and D. J. Sardella, *Tetrahedron Lett.*, 1977, 1095.
5 Reviews: R. K. Harris, *Nuclear Magnetic Resonance Spectroscopy*, Pitman, London, 1983, p. 197; R. C. Haddon, V. R. Haddon, and L. M. Jackman, *Top. Curr. Chem.*, 1971, **16**, 103; F. Sondheimer, *Accounts Chem. Res.*, 1972, **5**, 81.

The existence of a diamagnetic ring current, as indicated by the appropriate shielding or deshielding effects on proton chemical shifts, has been proposed as a diagnostic test of aromatic character. This seems to have some justification in that a theoretical link has been established between diamagnetic susceptibility and resonance energy (Section 2.5).[6] The criterion needs to be applied with caution because ring current effects increase with the size of the ring and so are very marked in large annulenes and heteroannulenes. There is also the practical point that, in order to detect shielding or deshielding, it is necessary to have some appropriate nonaromatic reference compound for comparison, and such reference compounds are not easy to find for many heterocyclic systems. Chemical shifts are influenced by several other factors besides the diamagnetic ring current, including the distortion of the π-electron distribution by the heteroatoms, and solvent effects. Chemical shift values for many heterocycles are strongly solvent dependent. We can, however, see the qualitative effect of ring currents by comparing the ^{1}H NMR spectra of pyridine, furan and thiophene with those of dihydro analogues (Fig. 2.15).

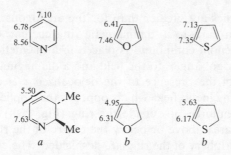

Fig. 2.15 Comparison of chemical shifts of aromatic and nonaromatic heterocycles. [a] Values in benzene solution: I. Hasan and F. W. Fowler, *J. Am. Chem. Soc.*, 1978, **100**, 6696. [b] R. J. Abraham, R. C. Sheppard, W. A. Thomas, and S. Turner, *Chem. Commun.*, 1965, 43.

Comparisons with nonaromatic systems of this type are open to criticism,[7] and indeed it is difficult to design suitable model systems for some simple heterocycles such as pyrrole. Indirect methods for estimating the influence of the diamagnetic ring current have been used: for example, the values of the chemical shifts of the methyl groups of the heterocycles shown in Fig. 2.16 are compared with values calculated for open-chain models.[7] The downfield shifts observed were taken as an indication of the relative aromaticities of the heterocycles. In general, however, the 'ring current effect' is best regarded as a qualitative indication of aromaticity rather than as a quantitative measure.

6 J. Aihara, *J. Am. Chem. Soc.*, 1981, **103**, 5704.
7 J. A. Elvidge, *Chem. Commun.*, 1965, 160.

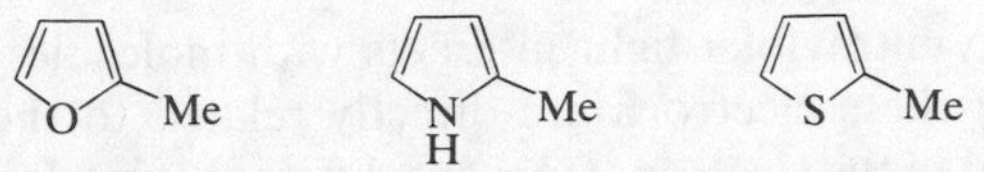

δ (Me) observed	2.30	2.18	2.48
calculated	2.14	1.97	2.19
downfield shift	0.16	0.21	0.29

Fig. 2.16 Ring current effects on methyl substituents.

2.2.3 Other physical probes of electronic structure

There are several experimental methods of determining electronic energy levels or electron distribution. These cannot be regarded as criteria of aromaticity, but they do provide independent experimental tests of the validity of molecular orbital calculations on the energies of heterocycles.

Ultraviolet absorption spectra have been used for many years in a qualitative way to indicate similarities in the bonding patterns of different compounds. The spectra of some heteroaromatic compounds are compared with those of benzene and of naphthalene in Table 2.2. The bands which are assigned to $\pi \rightarrow \pi^*$ absorptions are similar to those of the carbocyclic analogues, although many of the spectra of the heterocycles show additional bands which can be ascribed to $n \rightarrow \pi^*$ absorptions.

Energy levels of filled molecular orbitals can be obtained by *photoelectron spectroscopy*.[8] Electrons are ejected from occupied molecular orbitals when

Table 2.2 Ultraviolet absorption spectra of some heteroaromatic compounds (in aliphatic hydrocarbon solvents).[a]

	$\pi \rightarrow \pi^*$		$n \rightarrow \pi^*$	
	λ_{max} (nm)	$\log \varepsilon$	λ_{max} (nm)	$\log \varepsilon$
(Benzene)	256	2.40		
Pyridine	251	3.30	270	2.65
Pyridazine	246	3.11	340	2.50
Pyrimidine	243	3.31	298	2.51
Pyrazine	260	3.75	328	3.02
1,2,4,5-Tetrazine	252	3.33	542	2.92
(Naphthalene)	219, 275, 311	5.10, 3.75, 2.39	—	
Quinoline	225, 270, 313	4.48, 3.59, 3.37	*b*	
Isoquinoline	217, 266, 317	4.57, 3.61, 3.49	*b*	
Quinazoline	220, 267, 311	4.61, 3.45, 3.32	330	2.30
1,5-Naphthyridine	206, 257, 308	4.73, 3.69, 3.84	330	2.70
Indole	215, 226, 279	4.38, 3.70, 3.62	—	

a From S. F. Mason, in *Physical Methods in Heterocyclic Chemistry*, ed. A. R. Katritzky, Academic Press, New York, Vol. II, 1963, p. 1. *b* Masked by long wavelength $\pi \rightarrow \pi^*$ absorptions.

8 Review: H. Bock, *Pure Appl. Chem.*, 1975, **44**, 343.

high-energy ultraviolet light interacts with molecules in the gas phase. The energies of these electrons are directly related to the ionization potentials for removal of the electrons from the different orbitals. Analysis of the spectra involves the assignment of spectral bands to the electronic states of the molecular ions and hence the identification of the orbitals from which the electrons are ejected. The technique thus provides an experimental test of predicted changes in bonding levels in a series of heterocycles. For example, the analogues of pyridine in which the nitrogen atom is replaced by other Group V elements (P, As, Sb, Bi) have all been prepared,[2] and their photoelectron spectra provide clear evidence that the π-bonding is similar to that in benzene and pyridine. The π-bonding orbitals resemble those shown in Fig. 2.2. The ionization energies associated with the π_2 level (Fig. 2.2) decrease with increasing size of the heteroatom as it becomes more electropositive. The photoelectron spectrum of silabenzene is also in accord with that expected for a benzene analogue.[9]

A complementary technique, which enables electron affinities to be measured and which therefore provides an estimate of the energy levels of unoccupied orbitals, is known as *electron transmission spectroscopy*.[10] An electron from an electron beam is temporarily captured in an unoccupied orbital of a molecule to give an anion of very short lifetime (10^{-12}–10^{-15} s). The electron affinities are obtained by analysis of the change in the electron scattering spectrum that is produced. The technique has been applied to some of the common aromatic heterocycles and their electron affinities have been determined.[10] These techniques provide values that support those calculated from the energies of the π-orbitals in aromatic heterocycles, and which were represented in Figs 2.2 and 2.7.

2.2.4   Thermochemical estimates of aromaticity: empirical resonance energies[1]

Two types of thermochemical determination have commonly been used to estimate the stabilization of aromatic compounds: measurement of the standard enthalpy of combustion and the standard enthalpy of hydrogenation. The heat of combustion of pyridine, for example, is the enthalpy change associated with

$$C_5H_5N(g) + \tfrac{25}{4}O_2 \rightarrow 5\,CO_2(g) + \tfrac{5}{2}H_2O\,(l) + \tfrac{1}{2}N_2(g)$$

and the value can be determined experimentally by calorimetry. The method can also be used to obtain an experimental value for the heat of formation of the compound. The atomic heat of formation of pyridine is the enthalpy

9 G. Maier, G. Nihm, and H. P. Reisenauer, *Angew. Chem. Int. Edn. Engl.*, 1980, **19**, 52.
10 Review: K. D. Jordan and P. D. Burrow, *Accounts Chem. Res.*, 1978, **11**, 341.
11 An evaluation of empirical resonance energies is given by P. George, *Chem. Rev.*, 1975, **75**, 85.

change associated with

$$C_5H_5N(g) \rightarrow 5C(g) + 5H(g) + N(g).$$

The value can be obtained from the heat of combustion by making use of known values for the heats of combustion and of atomization of carbon, hydrogen, and nitrogen.[12]

The heat of formation can be calculated by adding together individual bond energy values for the molecule: for pyridine these would be the values appropriate to a localized (Kekulé) structure. The difference between the (numerically smaller) experimental value and the calculated value is a measure of the stabilization of the delocalized system and is called the *empirical resonance energy*. The values obtained depend upon the bond energy terms used in the calculation, and also upon the choice of the 'localized bond' model system.

Heats of hydrogenation of aromatic compounds can be used for calculating empirical resonance energies by comparing them with experimental values for appropriate model compounds. For example, the enthalpy of hydrogenation of benzene [$\Delta H = -49.7 \, \text{kcal mol}^{-1} \, (-208 \, \text{kJ mol}^{-1})$] can be compared with that for 3 mol of cyclohexene [$\Delta H = -28.4 \, \text{kcal mol}^{-1} \, (-119 \, \text{kJ mol}^{-1})$; $3\Delta H = -85.3 \, \text{kcal mol}^{-1} \, (-357 \, \text{kJ mol}^{-1})$]. The difference, $35.6 \, \text{kcal mol}^{-1}$ ($149 \, \text{kJ mol}^{-1}$), is the empirical resonance energy of benzene. A slightly different value is obtained if the model system is chosen in a different way. Thus, the values for the hydrogenation of the first and second double bonds of 1,3-cyclohexadiene can be extrapolated to give an estimate for the addition of a third mole of hydrogen to a hypothetical 'cyclohexatriene'. The total of the three values is then taken as that for the localized model, as shown (estimated by extrapolation):

$$C_6H_{10} + H_2 \rightarrow C_6H_{12} \qquad \Delta H = -28.4 \, \text{kcal mol}^{-1} \, (-119 \, \text{kJ mol}^{-1})$$

$$1,3\text{-}C_6H_8 + H_2 \rightarrow C_6H_{10} \qquad \Delta H = -26.5 \, \text{kcal mol}^{-1} \, (-111 \, \text{kJ mol}^{-1})$$

$$C_6H_6 + H_2 \rightarrow C_6H_8 \qquad \Delta H = -24.6 \, \text{kcal mol}^{-1} \, (-103 \, \text{kJ mol}^{-1})$$

The total heat of hydrogenation of this localized model is $-79.5 \, \text{kcal mol}^{-1}$ ($-333 \, \text{kJ mol}^{-1}$), giving an empirical resonance energy for benzene of $29.8 \, \text{kcal mol}^{-1}$ ($125 \, \text{kJ mol}^{-1}$).

It is clear, therefore, that not too much significance should be placed on absolute values of empirical resonance energies; all we can expect is that values obtained by similar methods for a series of compounds should give reasonable comparative estimates of the degrees of stabilization. Most of the values of heterocyclic compounds are based on combustion data because many of the model systems necessary for measuring heats of hydrogenation are not readily available. Literature values vary over quite a wide range, mainly because of differences in the choice of bond energy values. Some comparable values (obtained by similar methods) are given in Table 2.3.

12 Review: K. Schofield, *Hetero-aromatic Nitrogen Compounds*, Butterworths, London, 1967, p. 17.

Table 2.3 Empirical resonance energies.[a]

	kcal mol^{-1}	(kJ mol^{-1})
Benzene	35.9	(150)
Pyridine	27.9	(117)
Quinoline	48.4	(200)
Pyrrole	21.6	(90)
Indole	46.8	(196)
Thiophene	29.1	(122)
Furan	16.2	(68)

a F. Klages, *Chem. Ber.*, 1949, **82**, 358; G. W. Wheland, *Resonance in Organic Chemistry*, Wiley, New York, 1955.

2.2.5 Molecular orbitals and delocalization energy

We are concerned with heterocycles which are fully unsaturated and which are planar or nearly so, with a complete cycle of atoms with interacting p-orbitals. In the Hückel approximation the electrons in the π molecular orbitals are treated separately from those in the σ-bonds. The energies of the π molecular orbitals can be expressed in terms of two constants. The first, the *Coulomb integral*, denoted by the symbol α, is an approximate measure of the electron-attracting power of a particular atom. In an all-carbon π-electron system, α represents the energy of an electron in an isolated p-orbital before overlap. The second term, the *resonance integral*, is a measure of the stabilization gained by interaction of adjacent p-orbitals. It is denoted by the symbol β.

Calculation by the Hückel method of the energies of the six π-orbitals of benzene gives the values shown in Fig. 2.17(a). The two π-orbitals of ethylene are shown for comparison in Fig. 2.17(b). Six π-electrons occupying the three bonding orbitals of benzene will therefore have a total energy of $(6\alpha + 8\beta)$ whereas six π-electrons in three isolated bonding orbitals of ethylene will

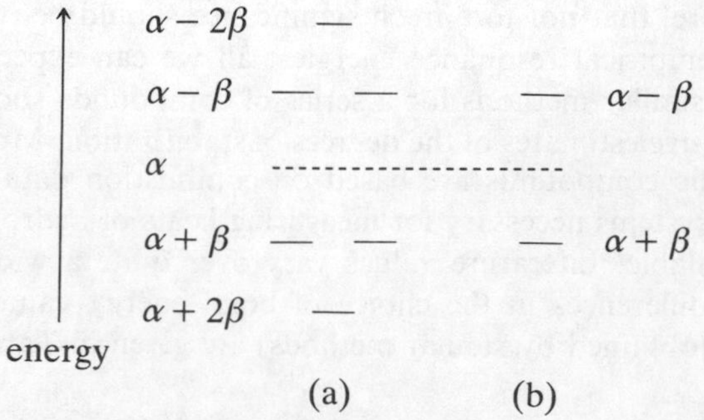

Fig. 2.17 Hückel energies of π-orbitals of benzene (a) and of ethylene (b).

have a total energy of $(6\alpha + 6\beta)$. The π-electron arrangement of benzene is therefore more stable by an amount 2β. This is called the *delocalization energy* of benzene. The delocalization energy is obviously the same for pyridine and for the other six-membered heteroaromatics if the effect of the replacement of carbon by nitrogen is ignored. In practice the effects of the introduction of heteroatoms can be accommodated by using adjustable parameters which allow for the uneven distribution of π-electron density.

This delocalization energy does not correspond to the empirical resonance energy because the latter is calculated relative to a model with alternating bond lengths whereas the former is based on a hypothetical localized model of identical geometry to that of the delocalized system. In order to relate the two we have to add to the empirical resonance energy the energy needed to compress the alternating bond structure to the nonalternating one. This *distortion energy* has been estimated at about $27\,\text{kcal}\,\text{mol}^{-1}$ ($113\,\text{kJ}\,\text{mol}^{-1}$) for benzene, so the value is significant in comparison with that of the empirical resonance energy. It is therefore more useful to recognize that delocalization energies represent a comparative measure of stabilization than to quantify them numerically.

2.2.6 Calculated resonance energies[13]

The problem with the measure of aromatic stabilization based on simple unconjugated π-electron systems as models is that 'delocalization energy' is not a unique property of cyclic systems. For example, butadiene can be shown to have a delocalization energy of 0.472β on the basis of the simple Hückel MO method, and all other acyclic conjugated systems also have some delocalization energy. In seeking a measure of aromatic character we need to estimate the *additional* contribution to the total delocalization energy which results from the compound having a cyclic structure. Dewar recognized this and proposed that the bond energies of reference structures in resonance energy calculations should be those of *nonaromatic* systems rather than those of *nonconjugated* systems.[14] It was shown that the π binding energy of linear polyenes is directly proportional to the length of the chain. Each additional carbon–carbon 'single' and 'double' bond in the polyene makes the same contribution to the total π energy as do those in butadiene or hexatriene. This does not, of course, imply that there is no conjugation, but it means that conjugation has the same energetic consequences per bond in noncyclic systems. It is therefore possible to calculate 'reference' π binding energies for any π system, cyclic or acyclic, by adding together the values appropriate to the types of bond present. This additivity principle applies to π-bonds to heteroatoms as well as to carbon–carbon bonds.

13 N. C. Baird, *J. Chem. Ed.*, 1971, **48**, 509.
14 M. J. S. Dewar, *The Molecular Orbital Theory of Organic Chemistry*, McGraw-Hill, New York, 1969, Chapter 5.

Cyclic systems that show *additional* π binding energy compared with the calculated reference value are termed 'aromatic'. This additional energy of stabilization has been called '*Dewar resonance energy*'[13] but the principle has been adopted by others to calculate resonance energies.[15] Cyclic systems which have a resonance energy close to zero [within about $2.5\,\text{kcal}\,\text{mol}^{-1}$ $(10\,\text{kJ}\,\text{mol}^{-1})$] are classed as 'nonaromatic'. A few cyclic systems are calculated to have a negative value (that is, they have less binding energy than the reference structure) and these are termed 'antiaromatic'. The interpretation of this last classification is discussed in Section 2.3.

Fortunately, resonance energies based on the Dewar model can be calculated by the simple Hückel MO method, despite the fact that the method ignores σ and π interactions. This is because the σ and π contributions to the binding energy are both directly proportional to the bond order of a given bond. The π resonance energies are therefore directly proportional to the total resonance energies.[16] In bonds to heteroatoms, values of the coulomb and resonance integrals have to be modified. Values have been obtained which give the best fit with experimental values of heats of atomization of known compounds, and these are then used to calculate the π-bond energies of various types of bond, in units of the resonance integral β. The total π-bond energy of a reference structure (i.e. a dominant valence structure) is calculated by adding up the contributions from the individual bonds, and this is then compared with the total π binding energy computed from a Hückel MO program.

In order to make a comparison of the aromaticities of different heterocycles, it is useful to calculate the resonance energies per π-electron (REPE) by dividing the resonance energy by the number of π-electrons in the molecule. For known systems the values correlate well with other criteria of aromaticity; some values for common heterocycles are shown in Table 2.4. The method can also be used to predict the degree of aromaticity of heterocycles which have not yet been synthesized. Table 2.4 also includes some calculated values of aromatic energies, which represent the difference in energy between an analogue with a localized bond and the delocalized structure.

2.2.7 General conclusions

The various criteria for determining aromatic character, including some that we have not discussed here, support the view that the azabenzenes shown in Fig. 2.3 are aromatic compounds having degrees of stabilization similar to those of benzene, or perhaps slightly lower. Benzo-fused heterocycles of this type show somewhat lower stabilization. The extent of delocalization in the five-membered heterocycles varies in relation to the heteroatoms present. Furan is a much more localized system than pyrrole or thiophene.

15 L. J. Schaad and B. A. Hess, *J. Chem. Ed.*, 1974, **51**, 640. For an alternative definition of resonance energies see J. Aihara, *J. Am. Chem. Soc.*, 1976, **98**, 2750.
16 L. J. Schaad and B. A. Hess, *J. Am. Chem. Soc.*, 1972, **94**, 3068.

Table 2.4 Resonance energies per π-electron (REPE) and aromatic energies of some heterocycles.[a]

Heterocycle	REPE (β)	Reference	Aromatic energy (kJ mol^{-1})[e]
Pyridine	0.058	b	107
Pyrimidine	0.049	b	104
Pyrazine	0.049	b	103
Quinoline	0.052	b	–
Isoquinoline	0.051	b	–
Pyrrole	0.039	c	94
Pyrazole	0.055	b	–
Imidazole	0.042	b	–
Thiophene	0.032	d	69
Furan	0.007	c	51
Indole	0.047	c	–
Benzofuran	0.036	c	–
Benzo[b]thiophene	0.044	d	–
Isoindole	0.029	c	–
Isobenzofuran	0.002	c	–
Benzo[c]thiophene	0.025	d	–

a Comparative values are: for benzene, REPE 0.065 and aromatic energy 118; for naphthalene, REPE 0.055. b B. A. Hess, L. J. Schaad, and C. W. Holyoke, *Tetrahedron*, 1975, **31**, 295. c B. A. Hess, L. J. Schaad, and C. W. Holyoke, *Tetrahedron*, 1972, **28**, 3657. d B. A. Hess and L. J. Schaad, *J. Am. Chem. Soc.*, 1973, **95**, 3907. e Values are from M. J. S. Dewar and A. J. Holder, *Heterocycles*, 1989, **28**, 1135.

The aromatic character of the five-membered ring of indole is rather less than that of pyrrole, and benzo fusion across the 3,4-bond of pyrrole or of furan reduces the aromatic character of the systems considerably, to the extent that isobenzofuran is best regarded as nonaromatic.

2.3 Aromatic character and other types of unsaturated heterocycles

2.3.1 Monocyclic systems that conform to Hückel's rule

In recent years many new heterocycles have been synthesized that formally obey the Hückel rule; that is, the structures can be considered to contain $(4n + 2)$ π-electrons in a monocyclic unsaturated system. The heteroatoms participate in these structures in two ways: the heteroatom (usually nitrogen) can be part of a formal double bond, like the nitrogen atom in pyridine, or it can donate an electron pair to the π system, like the nitrogen atom in pyrrole.

The heterocycles **6** are an extreme example of the second type, but there is no evidence that a planar cyclic structure (which would formally contain six

6   7

π-electrons) is favoured for such compounds. For example, ozone is noncyclic, and although a cyclic triaziridine (**6**, X = Y = NCHMe$_2$, Z = NCO$_2$Et) has been isolated, the nitrogen atoms are pyramidal and it rapidly reverts to an open-chain structure.[17] Of the four-membered ring systems **7**, only the dithiete (X = Y = S) seems to be stable, and there is no evidence that the ring system is aromatic.

Ten π-electron heterocycles of several different types have been synthesized. Monocyclic systems with *cis* double bonds can be constructed by incorporating one 'pyrrolic' nitrogen into a nine-membered ring, **8**,[18] or two into an eight-membered ring, **9**.[19] Despite the strain involved, both types of compound seem to achieve sufficient planarity to sustain a ring current, unless the substituents R are strongly electron withdrawing, in which case nonplanar structures are preferred. As with the five-membered hetero-aromatics, this can be ascribed to greater localization of the lone pairs on the heteroatoms when the groups R are electron withdrawing. A third type of structure is one containing two *transoid* double bonds linked by a methylene bridge, of which the parent system **10** and several derivatives have been made.[20] These can be regarded as aromatic on the basis of their bond lengths and NMR spectra.

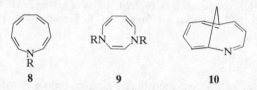

8   9   10

There are a few larger 'aza-annulenes' containing fourteen[21] or eighteen[22] π-electrons, which incorporate nitrogen atoms of the two types mentioned above. The evidence for the aromaticity of these ring systems is based mainly on the magnetic criterion. The NMR spectrum of aza[18]annulene shown in Fig. 2.18 illustrates the very large separation of the signals for the 'inner' and 'outer' hydrogens, which is typical of large annulenes with a diamagnetic ring current.

17 C. Leuenberger, L. Hoesch, and A. S. Dreiding, *J. Chem. Soc., Chem. Commun.*, 1980, 1197.
18 Review: A. G. Anastassiou, *Accounts Chem. Res.*, 1972, **5**, 281.
19 M. Breuninger, R. Schwesinger, B. Gallenkamp, K.-H. Müller, H. Fritz, D. Hunkler, and H. Prinzbach, *Chem. Ber.*, 1980, **113**, 3161.
20 M. Schäfer-Ridder, A. Wagner, M. Schwamborn, H. Schreiner, E. Devrout, and E. Vogel, *Angew. Chem. Int. Edn Engl.*, 1978, **17**, 853; R. Destro, M. Simonetta, and E. Vogel, *J. Am. Chem. Soc.*, 1981, **103**, 2863.
21 A. G. Anastassiou, R. L. Elliott, and E. Reichmanis, *J. Am. Chem. Soc.*, 1974, **96**, 7823; G. Schröder, G. Frank, H. Röttele, and J. F. M. Oth, *Angew. Chem. Int. Edn Engl.*, 1974, **13**, 205; H. Rötelle and G. Schröder, *Chem. Ber.*, 1982, **115**, 248.
22 W. Gilb and G. Schröder, *Chem. Ber.*, 1982, **115**, 240.

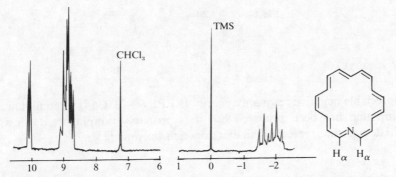

Fig. 2.18 Nuclear magnetic resonance spectrum of aza[18]annulene. The low field doublet is the signal for H_α; the high field multiplet is for the five 'inner' hydrogens. (Reproduced from ref. 22 by kind permission.)

2.3.2 *Other unsaturated heterocycles*

There are some fused heterocycles, other than simple benzo-fused compounds, which show evidence for cyclic delocalization. One such compound is indolizine, **5**, which is calculated to have appreciable resonance energy (REPE = 0.027β). The same is true of the stable compound **11** (REPE = 0.040β), which has a 10π-electron periphery and which shows evidence of a ring current in the NMR spectrum.[23] The 10π-electron periphery of the cyclopenta[*b*]pyridine, **12**, also sustains a ring current.[24]

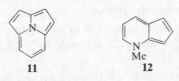

One interesting feature of the resonance energy calculations which was referred to in Section 2.2 is that some cyclic systems containing four or eight π-electrons are predicted to have negative resonance energies. Some 4π-electron heterocycles, **13** (X = S or O), have been generated in a solid matrix at low temperatures.[25] A few azacyclobutadienes have been isolated; the steric protection provided by the substituents in compound **14** make this derivative unusually stable.[26] 1*H*-Azepine, **15**, is predicted to have an

23 Review: W. Flitsch and U. Krämer, *Adv. Heterocycl. Chem.*, 1978, **22**, 321.
24 A. G. Anastassiou, E. Reichmanis, and S. J. Girgenti, *J. Am. Chem. Soc.*, 1977, **99**, 7392.
25 M. Torres, J. L. Bourdelande, A. Clement, and O. P. Strausz, *J. Am. Chem. Soc.*, 1983, **105**, 1698. Review: M. Torres, E. M. Lown, H. E. Gunning, and O. P. Strausz, *Pure Appl. Chem.*, 1980, **52**, 1623.
26 U. J. Vogelbacher, M. Ledermann, T. Schach, G. Michels, U. Hees, and M. Regitz, *Angew. Chem. Int. Edn Engl.*, 1988, **27**, 272.

appreciable negative resonance energy $(REPE = -0.036\beta)$ when planar. The compound has been prepared but it is probably nonplanar; it shows an NMR spectrum typical of an electron-rich polyene.[27]

2.4 Reactivity of heteroaromatic compounds

The concept of aromatic character was originally associated with the stability of benzene and its lack of reactivity relative to a typical polyene. There is a traditional view that in their reactions, aromatic compounds tend to 'revert to type'; that is, to give products of substitution rather than of addition or of ring opening. This view has some validity since the tendency to restore the ring system in the reaction product is related to the stability of that ring system. Reactivity cannot, however, be regarded as a criterion of aromaticity because it depends on the difference in energy between ground state and transition state. For example, the criteria discussed in Section 2.2 indicate that pyrrole is more aromatic than furan, yet pyrrole is much more reactive toward electrophiles than furan. This is because a nitrogen atom with a lone pair is much more polarizable than oxygen, so that electron release from the nitrogen atom is much easier. Silabenzene (Fig. 2.5) is an example of a heterocycle that has appreciable aromatic character yet which by any standards is highly reactive: it was originally detected in an argon matrix at 10 K. It is 'unreactive' only by comparison with compounds with isolated silicon–carbon π-bonds.

Heteroaromatic compounds differ widely in their reactivity toward reagents of different types and they also usually show selectivity for reaction at specific positions of the ring system. Obviously, it is useful to have guidelines to enable us to account for, and to predict, these important aspects of their chemistry. There are serious difficulties in attempting generalized explanations of the reactivity of heteroaromatic compounds in solution. Theories that ignore the role of solvent are of dubious validity: factors such as hydrogen bonding to solvents can clearly affect reactivity patterns, and reactivity in the gas phase sometimes differs significantly from that in solution: this is particularly the case with five membered heteroaromatic compounds.[28] Another common problem is that in substitution reactions, the site of attachment of the incoming group in the isolated product may not be the

27 E. Vogel, H.-J. Altenbach, J.-M. Drossard, H. Schmickler, and H. Stegelmeier, *Angew. Chem. Int. Edn Engl.*, 1980, **19**, 1016.
28 Review: M. Speranza, *Adv. Heterocycl. Chem.*, 1986, **40**, 25.

site of initial attack: this exemplified by the attack of some electrophiles on nitrogen heterocycles, where the initial attack on nitrogen is rapid but reversible and where the electrophile eventually binds to carbon.

With these reservations in mind, we can identify two main approaches to explaining relative reactivities of different heterocycles, and selectivity for attack at different positions within the same molecule.

1. One approach is based on some inherent property of the ground state heterocycle. For example, in considering ionic attack by a positively charged electrophile on the π-electron system of a heterocycle it may be appropriate to consider the overall distribution of π-electron density in the unperturbed heterocycle, the site of highest electron density being the initial point of attack. Reaction with a 'softer', more polarizable reagent may be better accommodated by considering the energies and coefficients of the so-called *frontier molecular orbitals* of the heterocycle.[29] These are the highest occupied molecular orbital (HOMO) and the lowest unoccupied orbital (LUMO). Attack by a polarizable electrophile involves mixing of the HOMO of the heterocycle with the LUMO of the electrophile. The activation energy for the reaction is then determined by the difference in energy between these two orbitals: the smaller the energy gap, the more efficient the mixing of the orbitals, and the lower the activation energy of the reaction. The energy of the π-HOMO of many heterocycles can be estimated experimentally from ionization potentials: some values are given in Table 2.5. For any particular electrophile we might expect reaction rates to be fastest for those electrophiles with the smallest ionization potentials; that is, the highest energy HOMOs.

Table 2.5 Ionization potentials associated with π-HOMO's (eV).

Benzene	9.24	*a*
Pyridine	9.73	*a*
Pyridazine	10.61	*a*
Pyrimidine	10.41	*a*
Pyrazine	10.18	*a*
1,3,5-Triazine	11.67	*a*
Pyrrole	8.2	*b*
Furan	8.9	*b*
Thiophene	8.9	*b*
Pyrazole	9.15	*c*
Imidazole	8.78	*c*

a F. Brogli, E. Heilbronner, and T. Kobayashi, *Helv. Chim. Acta*, 1972, **55**, 274, *b* J. A. Sell and F. Kupperman, *Chem. Phys. Lett.*, 1979, **61**, 355. *c* S. Cradock, R. H. Findlay, and M. H. Palmer, *Tetrahedron*, 1973, **29**, 2173.

29 Review: I. Fleming, *Frontier Orbitals and Organic Chemical Reactions*, Wiley-Interscience, London, 1976.

Also, attack is predicted to occur at a site which bears the largest orbital coefficient in the HOMO. For example, it can be seen from Fig. 2.7 that the HOMO of pyrrole has the largest orbital coefficient at the 2- and the 5-positions, so these should be the positions at which electrophilic attack occurs preferentially.

Conversely, nucleophilic attack is predicted to involve interaction of the LUMO of the heterocycle with a filled orbital on the nucleophile: the lower in energy the LUMO is, the faster the reaction with a given nucleophile. Again, attack is predicted to occur preferentially at the atom bearing the largest MO coefficient.

2. A second general approach is based on using a reaction intermediate as an appropriate model for the transition state. For example, in electrophilic substitution reactions the energies of the intermediate σ-complexes can be used to estimate the relative energies of the transition states. The assumption is that since the σ-complexes are likely to be closer in energy to the transition states than the starting materials, their relative energies are likely to reflect the order of activation energies for reactions of the same type.[30] The localization energies refer to the energy difference between the starting heterocycles and the transition state complexes: the relative energies of the σ-complexes should thus give the order of localization energies for a series of similar substitution reactions. For example, the three σ-complexes **16, 17,** and **18** are protonated forms of pyrrole which represent the intermediates formed in electrophilic attack on pyrrole at the 1-, 2-, and 3-positions (Fig. 2.19). Molecular orbital calculations of their energies indicate that the order of stability is **17 > 18 > 16**. This is also the order predicted qualitatively from simple resonance theory. As shown, more resonance forms can be written for the 2-protonated pyrrole **17** than for the 3-protonated species **18**, and the 1-protonated species **16** has no other resonance forms. This method is

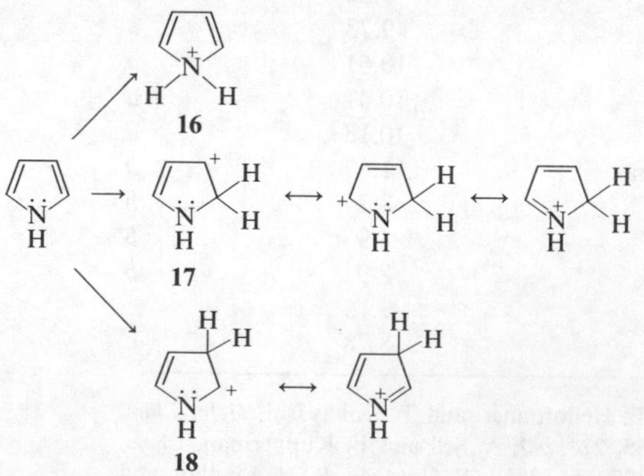

Fig. 2.19 Intermediates formed in the protonation of pyrrole.

30 For a review of quantitative aspects of reactivity in electrophilic substitution see A. R. Katritzky and R. Taylor, *Adv. Heterocycl. Chem.*, 1990, **47**.

not very good at predicting reactivity differences between heterocycles containing different heteroatoms. With a few exceptions, however, this valence bond approach provides a simple general method of predicting relative reactivities at different positions of a particular aromatic heterocycle, and it is used consistently in later chapters for this purpose.

2.5 Tautomerism of heteroaromatic compounds[31]

Many heteroaromatic compounds can exist in two or more tautomeric forms. When the alternative sites for protonation are both heteroatoms, equilibria are normally established in solution by rapid intermolecular proton transfer between a nitrogen atom within the ring and a nitrogen, oxygen, or sulphur substituent, as illustrated for 2- and 4-substituted pyridines in Fig. 2.20.

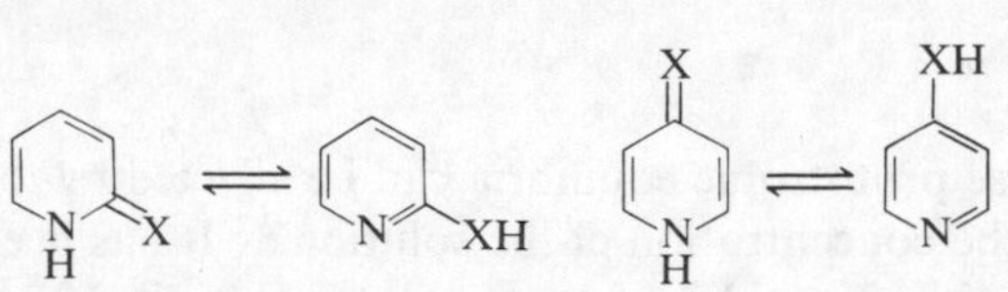

Fig. 2.20 Prototropic tautomerism in 2- and 4-substituted pyridines (X = NR, O, or S).

These equilibria are of particular importance in derivatives of pyrimidine and purine, since these heterocycles are incorporated into the structures of nucleic acids (Chapter 7). When one tautomer is predominant in solution, its structure can be assigned by comparison of its infrared, ultraviolet or NMR spectra with those of suitably alkylated derivatives. For example, the ultraviolet spectrum of 2-pyridone, **19**, strongly resembles that of 1-methyl-2-pyridone, **20**, in a variety of solvents, but is quite different from that of 2-methoxypyridine, **21**. It can thus be concluded that the equilibrium between 2-pyridone and 2-hydroxypyridine lies strongly (at least 9:1) in favour of the 2-pyridone tautomer in these solutions.

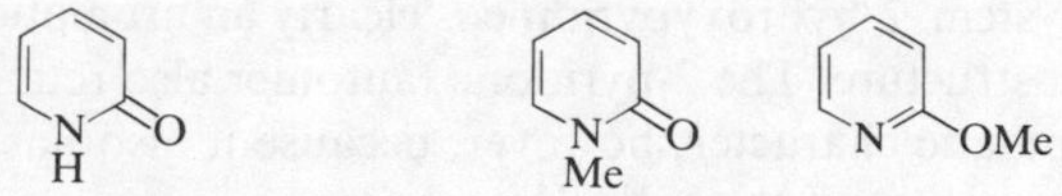

	19	**20**	**21**
λ_{max}(EtOH)	229 (log ε 3.85)	230 (3.78)	270 (3.55)
	300 (3.70)	305 (3.70)	

31 Review: J. Elguero, C. Marzin, A. R. Katritzky, and P. Linda, *Adv. Heterocycl. Chem.*, *Suppl. 1*, 1976.

The corresponding thione (X = S) also preferentially exists in solution in the left-hand tautomeric structure in Fig. 2.20. On the other hand, the preferred form of 2-aminopyridine is the right-hand structure (X = NH). These are also the predominant tautomers in solution for derivatives of other six-membered heteroaromatics. For example, pyrazine-2-thione, **22**, exists in solution mainly (about 97%) in this tautomeric form, as shown by comparison of its NMR spectra with those of the *N*-methyl compound **23** and the *S*-methyl compound **24**. Nitrogen NMR is a very useful tool for determining the positions of such equilibria.[32] The ^{15}N chemical shifts (relative to CD_3NO_2) are shown for *N*-1 in the three heterocycles in d_6-dimethyl sulphoxide.

The positions of these prototropic equilibria can be affected by the nature of the solvent and by the concentration of the solution.[33] It has been shown that for 2-hydroxypyridine–2-pyridone in very dilute solution (10^{-7} M) or in the gas phase, the hydroxypyridine form is the predominant tautomer. 2-Pyridone is associated (as a hydrogen-bonded dimer) in nonpolar solvents and it is this association, rather than the inherent stability of the monomer, which makes this tautomer the favoured one in nonpolar solvents at normal concentrations. With 4-pyridone it is impossible to prevent association in solution, even at very low concentrations, so the true equilibrium position for the monomeric species cannot be detected. Again, however, the hydroxypyridine tautomer appears to be the more stable one in the gas phase.[34] Polar solvents can affect the positions of the equilibria because the tautomers are solvated to different extents: the tautomers usually differ in polarity and in their ability to form hydrogen bonds to the solvent. Thus, it is only in the gas phase that realistic information about the differences in chemical binding energies of tautomers can be obtained.

The differences are due, in part, to differences in the degree of aromatic character of the tautomers. For example, in the 2-hydroxypyridine–2-pyridone system, 2-hydroxypyridine is clearly an aromatic compound with a 6π-electron structure. The 2-pyridone tautomer also retains a considerable degree of aromatic character, however, because it also has a cyclic array of six p-orbitals as part of the π-bonding system.

32 S. Tobias and H. Günther, *Tetrahedron Lett.*, 1982, **23**, 4785; see also B. C. Chen, W. von Philipsborn, and K. Nagarajan, *Helv. Chim. Acta*, 1983, **66**, 1537.
33 Review: P. Beak, *Accounts Chem. Res.*, 1977, **10**, 186.
34 P. Beak, J. B. Covington, S. G. Smith, J. M. White, and J. M. Zeigler, *J. Org. Chem.*, 1980, **45**, 1354.

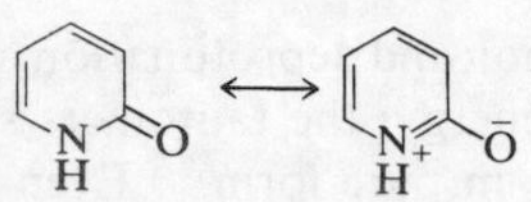

19

The difference in empirical resonance energy between the methylated analogues **20** and **21** has been estimated to be about $6.5 \, \text{kcal mol}^{-1}$ $(27 \, \text{kJ mol}^{-1})$ (about 20% of the empirical resonance energy of benzene)[35] so 1-methyl-2-pyridone has appreciable resonance energy. From the studies of the effects of the media on the positions of the equilibria it is clear that the relative aromaticity of the different tautomers has little relevance to the structure of the dominant tautomer in solution. This is determined much more by the concentration of the solution and by the nature of the solvent. The chemical reactions of the tautomeric mixture are, of course, not necessarily related to the structure of the predominant tautomer, particularly if proton transfers between the tautomers take place rapidly.

Several other types of prototropic tautomerism occur in heteroaromatic compounds; some examples are shown in Fig. 2.21. The 3-hydroxypyridine–pyridinium oxide equilibrium (a) is finely balanced in solution because no neutral 'pyridone' tautomer exists. The interconversion of fully aromatic structures by protonation and deprotonation on different ring nitrogen atoms, as for 1,2,4-triazole (b), is usually a very rapid low-energy process. In contrast,

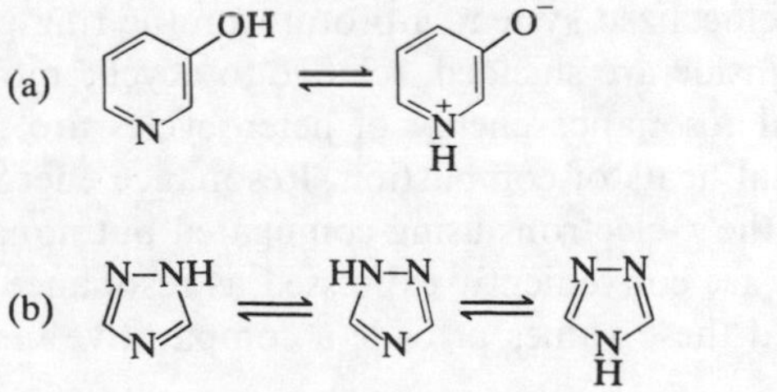

Fig. 2.21 Some other types of heteroaromatic tautomerism.

35 A. K. Burnham, J. Lee, T. G. Schmalz, P. Beak, and W. H. Flygare, *J. Am. Chem. Soc.*, 1977, **99**, 1836.

the interconversion of tautomers by protonation and deprotonation at carbon can be a process of appreciable activation energy; the tautomers shown in example (c) can both be isolated separately in pure form.[36] Even the enol tautomers of the substituted five-membered heterocycles shown in (d) can be generated and characterized in solution, although they are unstable. In the case of 3-hydroxythiophene (X = S) this is the predominant tautomer under equilibrium conditions, but the furan (X = O) rapidly tautomerizes to the more stable keto form. The keto forms of the corresponding pyrroles (X = NR) are also the more stable tautomers.[37]

Tautomers of heteroaromatic compounds will usually be drawn and named according to the structure of the predominant form in solution when they are discussed in subsequent chapters.

Summary

1. The important groups of aromatic heterocyclic compounds are (a) the analogues of benzene in which one or more of the CH groups of benzene is replaced by nitrogen; (b) ring systems containing five atoms and six π-electrons in a planar delocalized array, and (c) compounds in which ring systems of the first two types are fused to a benzene ring.
2. Useful criteria of aromaticity are (a) bond lengths, which tend not to alternate in aromatic rings, and (b) chemical shifts in ^{1}H NMR spectra. These reveal the existence of a diamagnetic ring current, which is a characteristic of a cyclic delocalized system, if protons on the ring periphery are deshielded and those inside are shielded, relative to acyclic models.
3. Values of the empirical resonance energy of heterocycles are normally obtained from experimental heats of combustion. Resonance energy values can also be calculated for the π-electrons using conjugated but nonaromatic systems as models; values are conveniently expressed as resonance energies per π-electron (REPE) and these values provide a comparative measure of aromatic character.
4. The Hückel rule, that cyclic π systems containing $(4n + 2)$ π-electrons are aromatic, also applies to heterocycles with 10, 14, and 18 electrons, on the basis of their diamagnetic ring currents. Compounds with four π-electrons are rare and highly unstable; potential eight-π-electron heterocycles avoid planar structures.
5. Six-membered heteroaromatic compounds bearing OH or SH groups in the 2- or 4-positions preferentially exist in solution in the tautomeric keto or thioketo forms. The positions of these equilibria are often quite different in the gas phase. 2-Pyridone has appreciable aromatic character. Compounds with amino groups in the 2- and 4-positions exist as such, rather than as the imino tautomers, in solution.

36 J. Harley-Mason and T. J. Leeney, *Proc. Chem. Soc.*, 1964, 368.
37 B. Capon and F.-C. Kwok, *J. Am. Chem. Soc.*, 1989, **111**, 5346.

Problems

1. Using Fig. 2.2 as a basis, sketch the three π-bonding molecular orbitals of pyrazine. Put them in order of increasing energy. How do you expect their energies to differ from the energies of the corresponding π-orbitals of pyridine?

2. Which of the following heterocycles can be classed as aromatic?

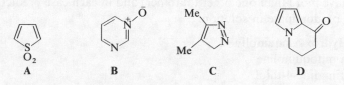

A   B   C   D

3. The methylene-bridged heterocycle **E** has been synthesized. Does the π-electron system obey the Hückel rule? What would you predict its properties to be?

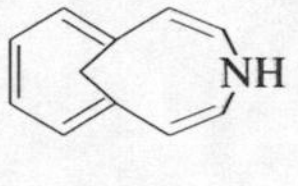

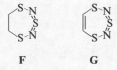

E

4. Whereas the heterocycle **F** contains a normal carbon–carbon single bond, its unsaturated analogue **G** has a carbon–carbon bond intermediate in length between a single and double bond; other bonds in the molecule are also not 'pure' double or single bonds. Suggest an explanation.

F   G

5. Compound **H** was readily oxidized, and, when oxidized and treated with perchloric acid, it gave a perchlorate salt $[C_{20}H_{12}O_4]^{++}2ClO_4^-$. This salt showed two signals in the ^1H NMR spectrum at $\delta 11.17$ and 12.13, with intensities in the ratio of 2:1. Draw a structure for the cation and account for its ready formation.

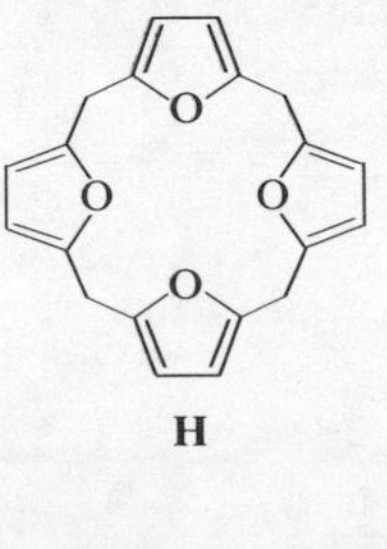

H

6. Draw another possible structure for planar aza[18]annulene and explain how this structure can be ruled out on the basis of the NMR spectrum shown in Fig. 2.18.

7. The ultraviolet spectrum of 4-hydroxypyridine–4-pyridone in cyclohexane shows only absorptions attributable to the 4-pyridone tautomer, whereas the spectrum of the 2,6-di-t-butyl derivative shows only absorptions due to the 4-hydroxypyridine tautomer. Suggest an explanation.

8. Draw reasonable alternative tautomeric structures for each of the following (several have more than one other tautomer) and in each case predict which form will predominate in solution.

(a) 1-Hydroxyisoquinoline
(b) 3-Aminoquinoline
(c) Pyrimidine-4-thiol
(d) 4-Hydroxycinnoline
(e) 2-Hydroxyfuran
(f) 2,4,6-Trihydroxypyrimidine

3 Nonaromatic heterocycles

3.1 Introduction

The fully unsaturated heterocycles discussed in Chapter 2 often have properties that are very different from those of their acyclic counterparts. In this chapter we shall consider saturated and partially saturated heterocycles, for which there is no possibility of cyclic delocalization. To what extent do the properties of these heterocycles differ from those of analogous open-chain compounds?

In attempting to answer this question we can recognize that the size of the ring system and the overall shape of the molecule are the most important factors that will distinguish the cyclic and acyclic compounds. The existence of a ring system imposes constraints on the molecule which may be absent in the acyclic model. The greater these constraints, the more likely it is that there will be differences in properties between the cyclic and acyclic systems. Flexible molecules preferentially adopt conformations in which the bonding interactions are maximized and repulsive nonbonding interactions are minimized. In these conformations there exist 'natural' bond angles and bond lengths which do not differ greatly from one compound to another, and preferred arrangements of substituents, as, for example, in the staggered arrangement of hydrogen atoms and of alkyl groups on adjacent carbon atoms. If the presence of a ring system compels the molecule to adopt a structure in which these preferred features cannot be attained, the molecule can be regarded as *strained*.[1] It will, of course, still preferentially adopt a shape that maximizes the attractive or bonding interactions and minimizes the repulsive interactions, but this may be of higher energy than that of the acyclic model.

The strain that exists in such molecules can be arbitrarily considered as being made up of different types; for example, bond angle strain, bond stretching or compression, bond torsion (twisting), and nonbonded interactions. The division is an arbitrary one because the components of strain are interdependent, and changes in any one component affect the others. In the following sections the restrictions on the mobility of saturated and partly

---

1 Review: A Greenberg and J. F. Liebman, *Strained Organic Molecules*, Academic Press, New York, 1978.

saturated heterocycles will be considered in terms of strain, and some of the physical and chemical consequences will be discussed.

3.2   Bond angle strain

Distortions of 'natural' bond angles are often found in cyclic systems. The energy V_α required to distort the bond angle away from the equilibrium position is called the *bond angle strain* or *Baeyer strain*. For small bond angle distortions it is proportional to the square of the angular displacement away from the equilibrium value. If α is the bond angle in the strained heterocycle and α_0 the equilibrium value:

$$V_\alpha \simeq 0.01(\alpha - \alpha_0)^2 \text{ kcal mol}^{-1} \text{ deg}^2 \, (0.04 \text{ kJ mol}^{-1} \text{ deg}^2)$$

Thus, a bond angle distortion of $10°$ requires only about 1 kcal mol^{-1} (4 kJ mol^{-1}) and a deformation of $20°$ requires about 4 kcal mol^{-1} (16 kJ mol^-). It is relatively easy for cyclic compounds to accommodate small distortions of bond angles.

3.2.1   *Angle strain and bonding in small-ring heterocycles*

The largest distortions from natural bond angles are found in three-membered rings. Saturated three-membered heterocycles have internuclear bond angles of about $60°$, which are considerably smaller than the normal values for an sp^3-hybridized centre (Table 3.1). In unsaturated three-membered hetero-cycles such as $2H$-azirines the angle distortion is even greater. The strain in the saturated three-membered rings, **1**, calculated on the basis of experimental and estimated enthalpies of formation, is shown in Table 3.1.

The effects of bond-angle distortion are minimized in these small ring compounds by changes in hybridization at atoms forming the rings. In the

Table 3.1 Dimensions of saturated three-membered rings, **1**, and their strain energies.

X in **1**	CXC (deg.)[a]	C—C (Å)[b]	C—X (Å)[b]	Strain[c] kcal mol^{-1} (kJ mol^{-1})
CH_2	60	1.510	1.510	27.5 (115)
NH	60	1.481	1.475	27.1 (113)
O	61	1.472	1.436	27.2 (114)
S	48.5	1.492	1.819	19.9 (83)

[a] Ref. 1, p. 281. [b] Average values for bond lengths in unstrained systems (sp^3—sp^3 bonds) are: C—C, 1.54; C—N, 1.47; C—O, 1.43, and C—S, 1.81 Å. [c] K. Pihlaja and E. Taskinen, in *Physical Methods in Heterocyclic Chemistry*, ed. A. R. Katritzky, Academic Press, New York, Vol. 6, 1974.

1

saturated three-membered rings, **1**, for example, the bonds forming the ring are not made up of sp³ hybrids, but have more p character. This allows effective overlap of orbitals directed outside the axes joining the nuclei of the ring atoms (Fig. 3.1). The result is that the *interorbital* angle distortion is much less than the *internuclear* angle distortion, because as the p character of the ring bonds increases, the 'natural' interorbital angle reduces from the sp³ value of about 110° toward the value of 90° for orthogonal p-orbitals. The ring atoms can be regarded as being linked by 'bent' or 'banana-shaped' bonds.

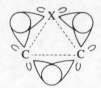

Fig. 3.1 Bonding overlap in three-membered rings, **1**.

This bonding system in three-membered heterocycles can be regarded as being constructed from the interaction of π and π* orbitals of an ethylene fragment with empty and filled orbitals on the heteroatom X, as shown in Fig. 3.2.[2] The first type of interaction (a) involves the transfer of electron density to the heteroatom, and is the dominant one for electronegative atoms or groups X; the second type (b) involves back-donation of electron density to the olefinic fragment. The electron distribution in the bonding systems of these rings is determined by the relative strengths of the two types of interaction.

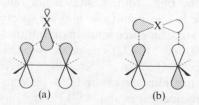

Fig. 3.2 Construction of bonding system of three-membered heterocycles (a) by interaction of ethylene π-orbital with unfilled orbital on X, (b) by interaction of π*-orbital with filled orbital on X.

2 M. J. S. Dewar and G. P. Ford, *J. Am. Chem. Soc.*, 1979, **101**, 783.

This change in ring bonding also affects other bonds formed by the ring atoms: those to the substituents. These have much more s character than normal (they are approximately sp² hybrids in saturated three-membered rings). Thus, the angles between the substituents are *greater* than the tetrahedral value.

A useful guide to the state of hybridization is provided by the ^{13}C—H coupling constants in NMR spectra. These coupling constants increase with increasing s character of the C—H bonds in a given series of compounds, and for three-membered rings the values are considerably higher than in open-chain analogues. Some values are shown in Fig. 3.3.

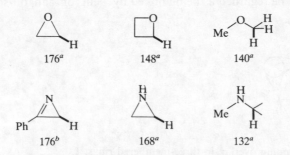

Fig. 3.3 ^{13}C—H coupling constants (Hz) for some small-ring heterocycles and open-chain analogues. [a] From C. Van Alsenoy, H. P. Figeys, and P. Geerlings, *Theor. Chim. Acta*, 1980, **55**, 87. [b] F. W. Fowler and A. Hassner, *J. Am. Chem. Soc.*, 1968, **90**, 2875.

3.2.2   Some consequences of bond angle strain in small rings

The most easily recognizable manifestation of bond angle strain in these small rings is their chemical reactivity, and particularly their ease of ring opening. This is the dominant feature of the chemistry of three-membered heterocycles, as will be shown in Chapter 9, but it is less important for four-membered rings.

The different hybridization of the ring atoms is shown not only in the ^{13}C—H coupling constants referred to above, but also in other spectroscopic properties. For example, the C=N stretching frequency in the infrared spectra of 2*H*-azirines is raised to about 1800 cm^{-1} (CCl$_4$), compared with the normal value in unstrained imines of about 1650 cm^{-1}. Similarly, the carbonyl stretching frequency of α-lactams (aziridinones) is at 1830–1859 cm^{-1}, and that of monocyclic β-lactams (azetidinones) is at 1745–1765 cm^{-1}, whereas unstrained amides absorb below 1680 cm^{-1}. The higher frequencies are associated with an increase in the bond orders because of rehybridization of the carbon atoms of the bonds.[3]

3 D. Dolphin and A. Wick, *Tabulation of Infrared Spectral Data*, Wiley-Interscience, New York, 1977.

The data of Table 3.1 show that the three-membered rings, and particularly oxirane and aziridine, have short carbon–carbon bonds. This, combined with the enhanced π-type characteristics of the ring bonds, permits the transmission of conjugation through the ring systems.[4] For example, the ultraviolet absorption spectra of the aziridines, **2**, provide evidence for the transmission of conjugation through the aziridine ring.[5]

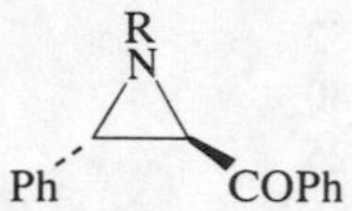

2

The increased s character of the lone pair on nitrogen in aziridines, compared with other amines, is revealed in several properties. Aziridine itself is less basic than aliphatic secondary amines ($pK_a = 7.98$), and the lone pair interacts less efficiently than in other amines with conjugative substituents, such as phenyl, attached to nitrogen.[6]

Another important property of aziridines is their greatly reduced rate of pyramidal inversion at nitrogen, compared with other nitrogen compounds (Fig. 3.4). Simple unstrained secondary amines have low barriers to nitrogen inversion [$\Delta G^{\ddagger} \simeq 6\,\text{kcal mol}^{-1}\ (25\,\text{kJ mol}^{-1})$] but the barrier is much higher in aziridine [$\Delta G^{\ddagger} \simeq 17\,\text{kcal mol}^{-1}\ (72\,\text{kJ mol}^{-1})$]. This is due to the increased angle strain in the planar transition state required for inversion. Whereas the bond angle in aziridine differs from the normal tetrahedral angle by about 50°, the angle in the planar transition state deviates from that in unstrained planar nitrogen by 60°; thus, there is increased angle strain in the transition state and an increased barrier to inversion.

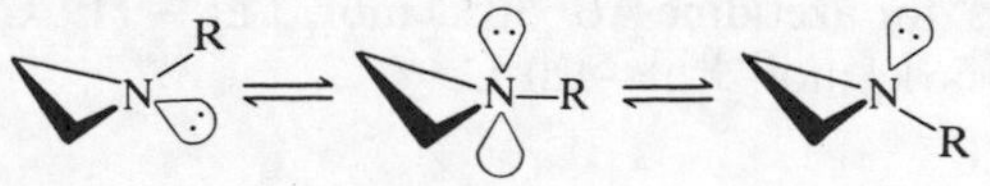

Fig. 3.4 Nitrogen inversion in aziridines.

The inversion barrier is further influenced by the nature of the substituent on nitrogen, as illustrated in Table 3.2. Conjugatively electron-withdrawing substituents, which stabilize the planar transition state, lower the energy barrier. There may also be a steric effect which destabilizes the pyramidal forms, as illustrated by the lower barrier for the t-butyl derivative.

4 A discussion of the transmission of conjugation, mainly by cyclopropane, is given by A. de Meijere, *Angew. Chem. Int. Edn Engl.*, 1979, **18**, 809.
5 N. H. Cromwell, R. E. Bambury, and J. L. Adelfang, *J. Am. Chem. Soc.*, 1960, **82**, 4241.
6 Review: O. C. Dermer and G. E. Ham, *Ethyleneimine and Other Aziridines*, Academic Press, New York, 1969.

Table 3.2 Influence of nitrogen substituents on barriers to nitrogen inversion in aziridines.

R	$\Delta G^{\ddagger}$ kcal mol^{-1}	(kJ mol^{-1})	Temperature (°C)	Refs.
H	17.3	(72)	68	a
Et	19.4	(81)	108	b
CMe$_3$	17.0	(71)	52	b
Ph	11.7	(49)	−40	b
SO$_2$Me	12.8	(53.5)	−25	b
CONMe$_2$	9.9	(41)	−86	b
CO$_2$Me	7.1	(30)	−138	b
NH$_2$	>22	(>90)	150	c
Cl^d	>21	(>90)	120	d
OMe^e	>22	(>90)	130	e

a R. E. Carter and T. Drakenberg, *Chem. Commun.*, 1972, 582. *b* From J. B. Lambert, *Top. Stereochem.*, 1971, **6**, 52. *c* S. J. Brois, *Tetrahedron Lett.*, 1968, 5997. *d* Determined for 1-chloro-2-methylaziridine: S. J. Brois, *J. Am. Chem. Soc.*, 1968, **90**, 506. *e* Determined for 1-methoxy-2,2,3,3-tetramethylaziridine: S. J. Brois, *J. Am. Chem. Soc.*, 1970, **92**, 1079.

Inductively electron-withdrawing groups with an adjacent lone pair (NH$_2$, Cl, and OMe) raise the barrier considerably. Another factor, apart from angle strain, is obviously involved: the raised barrier is probably due in part to unfavourable lone pair–lone pair interactions in the planar transition state (Section 3.3). The barrier can be large enough for separate invertomers to be isolated. For example, the invertomers **3** and **4** have been separated by GLC and their individual NMR spectra recorded.[7]

The effect of angle strain on the barrier to inversion in azetidines is much smaller than in aziridines: for azetidine $\Delta G^{\ddagger}$ 30 kJ mol^{-1} at −119°C, and for 1-chloroazetidine $\Delta G^{\ddagger} = 56$ kJ mol^{-1} at −20°C.

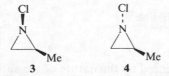

3.2.3 *Angle strain in larger rings*

In small rings the bond angles are restricted to values smaller than the natural angles. There are also a few heterocycles with bond angle strain of the opposite type; that is, with larger than normal bond angles. An example is

7 Review: W. B. Jennings and D. R. Boyd, in *The Stereodynamics of Organonitrogen Compounds*, eds J. B. Lambert and Y. Takeuchi, VCH, New York, 1992, p. 119.

the tertiary amine 1-azabicyclo[3.3.3]undecane, **5**, which has a near-planar arrangement of groups about nitrogen.[8] The constraining effect of the cage structure is even more pronounced in the larger analogue, **6**.[9] This compound preferentially adopts a conformation in which the lone pair is essentially directed towards the inside of the cage; as a result, it is difficult to protonate the nitrogen on the outside face, and it is a very weak base ($pK_a = 0.6$). The 'penalty' for pyramidalization of the nitrogen on the outside face is calculated to be $17 \, \text{kcal mol}^{-1}$ ($71 \, \text{kJ mol}^{-1}$).

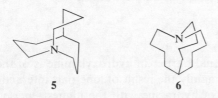

    5   6

3.3 Torsional energy barriers

Bond torsion refers to the rotation about the axis linking two nuclei. The *torsion angle* or *dihedral angle* is the measure of the rotation of the bond from the position in which substituents on the nuclei are eclipsed. In flexible systems the energy of the molecule varies as the bond is rotated: for ethane, the difference between the highest energy (eclipsed) and lowest energy (staggered) conformations is about $2.9 \, \text{kcal mol}^{-1}$ ($12 \, \text{kJ mol}^{-1}$). The barrier between the two chair forms of cyclohexane is about $10.5 \, \text{kcal mol}^{-1}$ ($44 \, \text{kJ mol}^{-1}$). The source of this torsional energy barrier is complex and can include contributions from the interaction of nuclei and of lone pairs.

3.3.1 Single bonds

Saturated heterocycles contain carbon–heteroatom bonds and, in some cases, heteroatom–heteroatom bonds as part of the ring system, for which the barriers to rotation are different from those about carbon–carbon bonds.[10] Single bonds between carbon and heteroatoms generally have a lower barrier to rotation than for carbon–carbon bonds, as illustrated in Table 3.3.

On the other hand, heteroatom–heteroatom bonds have a much higher barrier to rotation than carbon–carbon bonds. For example, hydrazine is calculated to have a rotational energy barrier of $11.9 \, \text{kcal mol}^{-1}$

8 J. C. Coll, D. R. Crist, M. del C. G. Barrio, and N. J. Leonard, *J. Am. Chem. Soc.*, 1972, **94**, 7092.

9 Review: R. W. Alder, *Chem. Rev.*, 1989, **89**, 1215.

10 Review: F. G. Riddell, *The Conformational Analysis of Heterocyclic Compounds*, Academic Press, London, 1980.

Table 3.3 Rotational energy barriers for single bonds.

Compound	Rotational barrier[a] kcal mol^{-1}	(kJ mol^{-1})
CH_3—CH_3	2.9	(12.2)
CH_3—NH_2	2.0	(8.3)
CH_3—OH	1.1	(4.6)
CH_3—SH	1.3	(5.3)

[a] From ref. 10, p. 15.

(49.7 kJ mol^{-1}), and the rotational barrier in hydroxylamine is of the same order. The increased barrier is mainly the result of lone-pair interactions on the heteroatoms. The rotamer of hydrazine with the highest energy is the one with the lone pairs eclipsed, and the minima on the rotational energy profile occur with the lone pairs at 90°. There is a smaller energy barrier [of 3.7 kcal mol^{-1} (15.5 kJ mol^{-1})] to rotation to a conformer with the lone pairs *trans* to each other. The rotational energy profile for hydrazine is illustrated in Fig. 3.5.

The consequences of these differing torsional energy barriers are mainly seen in heterocycles containing heteroatom–heteroatom bonds. These compounds show a preference for conformations in which the lone-pair orbitals are, as nearly as possible, perpendicular. This is sometimes referred to as *the gauche effect*. Compounds that are restricted to conformations in which the lone pairs are eclipsed, or nearly so, are destabilized.

An example of the latter is provided by the properties of diaziridinones. Several monocyclic diaziridinones have been made which are thermally quite stable. They adopt conformations, **7**, in which the substituents on nitrogen,

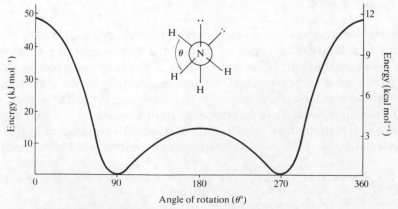

Fig. 3.5 Rotational energy profile for hydrazine (based on molecular orbital calculations: P. B. Ryan and H. D. Todd, *J. Chem. Phys.*, 1977, **67**, 4787).

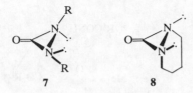

7   **8**

and the lone pairs, are *trans*. The bicyclic diaziridinone, **8**, is different in that it is forced to adopt a structure with the lone pairs eclipsed. Its properties are strikingly different from those of the monocyclic compounds: it readily loses carbon monoxide above room temperature, and it is easily cleaved by reaction with nucleophiles.[11]

3.3.2 Double bonds and partial double bonds

π-Bonds have a very high barrier to rotation, and are normally regarded as rigid. Some cyclic compounds incorporate π-bonds, or, more commonly, systems of electrons which involve π-type overlap, in which the structures of the molecules compel the π-bonding system to be twisted. In carbocyclic systems the simplest examples are those containing a *trans*-fused double bond in a ring of less than eight atoms, and particularly those in which the double bond is at a bridgehead position. Many years ago Bredt recognized the strain involved in constructing such bridgehead olefins and formulated the so-called *Bredt rule* by which the formation of strained bridgehead olefins is prohibited. It was recognized later that the criterion of strain in such compounds is the size of the smallest ring containing a *trans* olefinic linkage, and if this is large enough, it is feasible for bridgehead olefins to exist.[12]

A few heterocycles are known with a C=N π-bond at the bridgehead, in which the groups attached to the π-bonds cannot lie coplanar. As a consequence, the compounds are unusually reactive. An example is the imine **9** which has only a short lifetime in solution and which reacts readily with methanol to give the adduct **10**.[13]

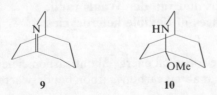

9   **10**

11 C. A. Renner and F. D. Greene, *J. Org. Chem.*, 1976, **41**, 2813.
12 Review: P. M. Warner, *Chem. Rev.*, 1989, **89**, 1067.
13 K. B. Becker and C. A. Gabutti, *Tetrahedron Lett.*, 1982, **23**, 1883; T. Sasaki, S. Eguchi, T. Okano, and Y. Wakata, *J. Org. Chem.*, 1983, **48**, 4067.

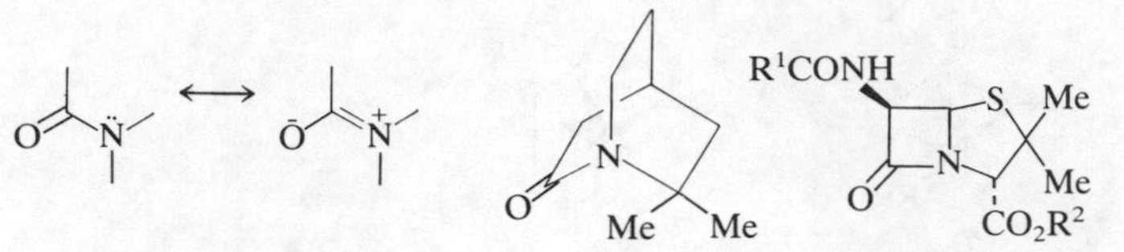

Fig. 3.6 (a) Delocalization in planar amides; (b) examples of lactams in which delocalization is restricted.

Examples of torsional strain in systems of partial double bonds are much more common. The amide linkage in unstrained systems is a delocalized planar system (Fig. 3.6) with an appreciable stabilization energy of about $18\,\text{kcal mol}^{-1}$ $(75\,\text{kJ mol}^{-1})$. The extent of the delocalization is shown by the C—N bond lengths of about $1.32\,\text{Å}$, intermediate between the values for single and double bonds, and by the low carbonyl stretching frequencies $(1680–1600\,\text{cm}^{-1})$ in the infrared spectra. The delocalized system is disrupted in some types of lactam, particularly those in which the nitrogen atom is at a bridgehead position. For example, in the lactam **11** the lone pair is almost orthogonal to the carbonyl π-bond and cannot interact effectively with it. As a result the compound is unusually basic ($pK_a = 5.33$, compared with $pK_a = 0$ for similar unstrained amides) and the carbonyl stretching frequency is appreciably higher than normal (ν_{max} $1762\,\text{cm}^{-1}$).[14] In the penicillin ring system, **12**, the bridgehead nitrogen is not planar, since sp^2-hybridized nitrogen would impose greatly increased bond angle strain on the system. The carbonyl stretching frequency in penicillins $(1780–1770\,\text{cm}^{-1})$ is higher than for other azetidinones $(1765–1745\,\text{cm}^{-1})$ and the carbonyl group is readily attacked by nucleophiles. It has been suggested that the increased electrophilicity of the carbonyl group may be associated with the antibiotic activity of penicillins, but the bicyclic structure is not essential for biological activity and the evidence is not conclusive (Section 9.8).[15]

3.4   Influence of bond lengths and van der Waals radii: conformational preferences of flexible heterocycles[10,16]

In Section 3.3 we noted that the torsional energy about carbon–heteroatom bonds is generally lower than that about carbon–carbon bonds, whereas that

14 H. Pracejus, M. Kehlen, H. Kehlen, and H. Matschiner, *Tetrahedron*, 1965, **21**, 2257.
15 The validity of the concept of resonance stabilization of amides has recently been questioned, and alternative explanations have been put forward for the increased reactivity of nonplanar lactams. See A. J. Bennet, Q.-P. Wang, H. Slebocka-Tilk, V. Somayaji, R. S. Brown, and B. Santarsiero, *J. Am. Chem. Soc.*, 1990, **112**, 6383.
16 Review: W. L. F. Armarego, *Stereochemistry of Heterocyclic Compounds*, Wiley-Interscience, New York, 1977.

about heteroatom–heteroatom bonds is higher. These differences are potentially important in determining the preferred conformations of flexible heterocycles, but there are some other factors which can also affect these conformations. These are (a) differences in the lengths of bonds to heteroatoms compared with carbon–carbon bonds, (b) differences in van der Waals radii of heteroatoms, and (c) the presence of lone pairs on the heteroatoms.

Carbon–oxygen single bonds (average length 1.43 Å) and carbon–nitrogen bonds (1.47 Å) are shorter than carbon–carbon bonds (1.51 Å). The van der Waals radii also increase in the same order, $O < NH < CH_2$. Thus, other nonbonded atoms are able to approach closer to oxygen than to an NH group, and closer to an NH than to a CH_2 group. The combined effect of these differences is usually quantitative rather than qualitative; that is, the preferred conformations of saturated oxygen and nitrogen heterocycles are normally qualitatively similar to those of the corresponding carbocyclic compounds. The differences are found mainly in bond lengths, bond angles, and inversion barriers.

3.4.1 Saturated six-membered heterocycles

The chair conformations adopted by cyclohexane and its derivatives are also the preferred ones for six-membered heterocycles. The conformational energy barrier to ring inversion in tetrahydropyran [9.9 kcal mol^{-1} (41.4 kJ mol^{-1}) at $-65°C$] is very similar to that in cyclohexane, as are those in other simple six-membered heterocycles (Fig. 3.7).

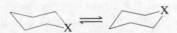

Fig. 3.7 Ring inversion in six-membered heterocycles. Inversion barriers ($\Delta G^{\ddagger}$) are: X = O, 9.9 kcal mol^{-1} (41.4 kJ mol^{-1}) at $-65°C$; X = NH, 10.4 kcal mol^{-1} (43.6 kJ mol^{-1}) at $-62.5°C$; X = S, 9.0 kcal mol^{-1} (37.7 kJ mol^{-1}) at $-93°C$ (J. B. Lambert and S. I. Featherman, *Chem. Rev.*, 1975, **75**, 611).

The effects of changes in bond lengths and nonbonded interactions are most easily discernible in substituted oxygen heterocycles. Consider, for example, 2-methyl-1,3-dioxane, **13**.[10,17] The conformation **13a** with an equatorial methyl group is even more favoured than in methylcyclohexane

17 Reviews: E. L. Eliel, *Accounts Chem. Res.*, 1970, **3**, 1; *Angew. Chem. Int. Edn Engl.*, 1972, **11**, 739.

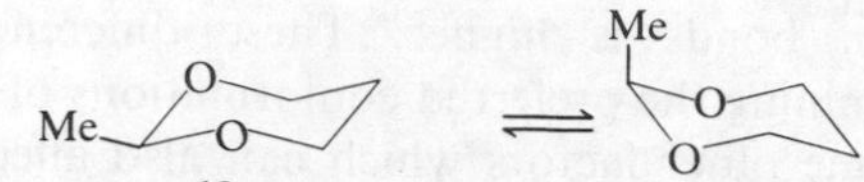

[$\Delta G = 4.0 \, \text{kcal mol}^{-1}$ (16.6 kJ mol^{-1}) at 25°C in ether for **13a** over **13b**] because the shorter C—O bonds cause larger 1,3-diaxial interactions in **13b** than in methylcyclohexane. On the other hand, 5-methyl-1,3-dioxane, **14**, shows only a slight preference for the conformation **14a** with an equatorial methyl group [$\Delta G = 0.8 \, \text{kcal mol}^{-1}$ (3.3 kJ mol^{-1}) at 25°C]. This is because in the axial conformer **14b** the interactions of the methyl group with the lone pairs on oxygen are much smaller than the 1,3-diaxial interactions with CH bonds which occur in axial methylcyclohexane. It is even possible for bulky alkyl groups such as t-butyl to occupy the 5-position as axial substituents.[17]

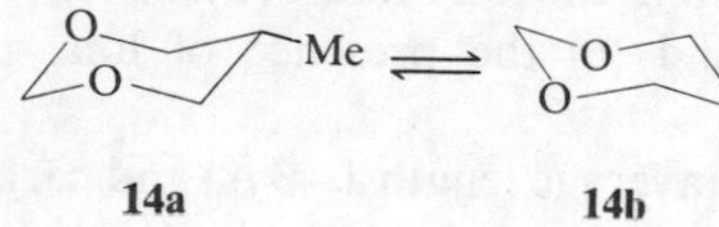

Another important difference is that electronegative 2-substituents on tetrahydropyrans often show a preference for axial positions: this preference, which is called *the anomeric effect*, is discussed in Section 3.5.

Piperidine and other nitrogen-containing heterocycles undergo a more complicated set of conformational changes because two different types of change occur: tetrahedral inversion of the nitrogen substituents, and ring inversion. These conformational changes are shown for piperidine in Fig. 3.8. Tetrahedral inversion at nitrogen is the lower energy process. The invertomer in which the NH group is equatorial is more stable both in the gas phase and in solution. Alkyl groups attached to nitrogen also preferentially occupy equatorial positions: for *N*-methylpiperidine the energy difference between conformers having axial and equatorial methyl is 2.7 kcal mol^{-1} (11.3 kJ mol^{-1}).

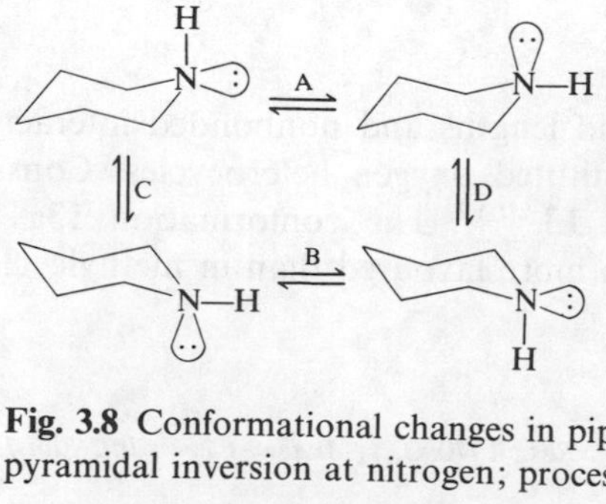

Fig. 3.8 Conformational changes in piperidine. Processes A and B involve pyramidal inversion at nitrogen; processes C and D are ring inversions.

The strong preference for conformations with an axial lone pair and an equatorial alkyl substituent at nitrogen is also seen in the bicyclic system **15**, for which the *transoid* conformation **15a** is more stable [by about 4.5 kcal mol^{-1} (19 kJ mol^{-1})] than the *cisoid* conformation **15b**.[18]

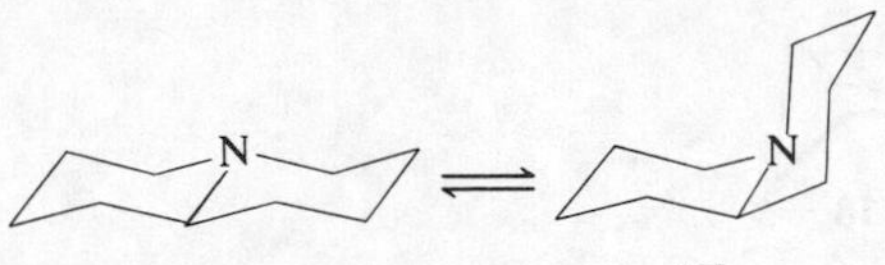

15a	**15b**

Barriers to nitrogen inversion are considerably raised by attached heteroatoms inside or outside the ring.[19] For example, the barrier to inversion (eq, eq to eq, ax) for the hexahydropyridazine **16** is 11.6 kcal mol^{-1} (48.5 kJ mol^{-1}).

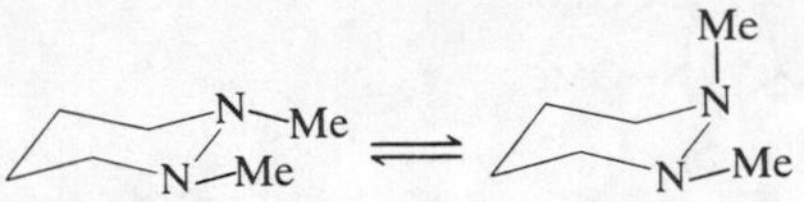

16a	**16b**

3.4.2 Four- and five-membered heterocycles

Cyclobutane is a puckered molecule with a barrier to ring puckering of 1.5 kcal mol^{-1} (6.2 kJ mol^{-1}). The barrier is slightly lower in azetidine [1.3 kcal mol^{-1} (5.3 kJ mol^{-1})] and the two conformers are unequal in energy, that having the NH bond equatorial being the more stable (Fig. 3.9).[20]

In oxetane, the energy barrier to puckering is lower than the ground vibrational energy level, so the molecule is essentially planar and freely vibrating. This is a consequence of the lower torsional energy barrier about the C—O bonds.

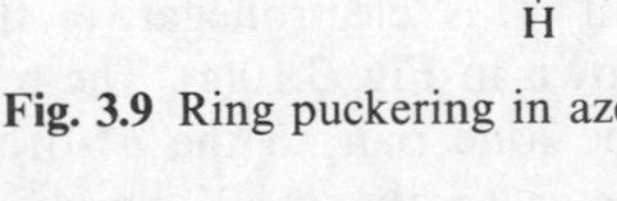

Fig. 3.9 Ring puckering in azetidine.

18 C. D. Johnson, R. A. Y. Jones, A. R. Katritzky, C. R. Palmer, K. Schofield, and R. J. Wells, *J. Chem. Soc.*, 1965, 6797.
19 Reviews: S. F. Nelson, *Accounts Chem. Res.*, 1978, **11**, 14; A. R. Katritzky, R. C. Patel, and F. G. Riddell, *Angew. Chem. Int. Edn Engl.*, 1981, **20**, 521; T. A. Crabb and A. R. Katritzky, *Adv. Heterocycl. Chem.*, 1984, **36**, 1.
20 L. A. Carreira and R. C. Lord, *J. Chem. Phys.*, 1969, **51**, 2735; R. Dutler, A. Rauk, and T. S. Sorensen, *J. Am. Chem. Soc.*, 1987, **109**, 3224.

Simple five-membered ring heterocycles can be represented as a series of freely interconverting nonplanar structures, such as the envelope form **17** and the half-chair form **18**, in which the eclipsing interactions of adjacent C—C bonds are relieved.

17 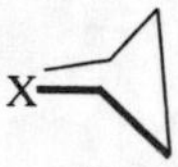   **18**

As with other heterocycles, torsional barriers are higher in bonds linked to two heteroatoms, and rates of inversion at tetrahedral nitrogen are reduced. An extreme example is provided by the *cis* and *trans* invertomers **19** and **20**, which are stable enough to be isolated without interconversion.[21]

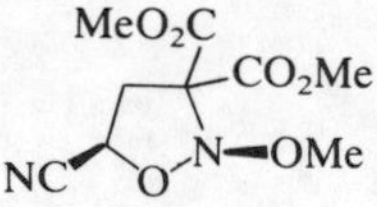

19 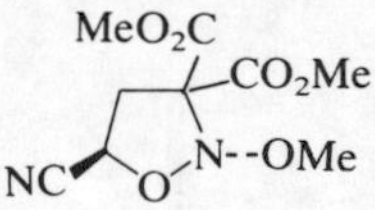   **20**

3.5   Other types of interaction in saturated heterocycles

3.5.1   'Through-bond' orbital interactions: the anomeric effect[22]

The presence of lone pairs in heterocyclic compounds can have subtle effects on the bonding system in other parts of the molecule. A suitably oriented lone-pair orbital can 'mix in' with bonding or antibonding orbitals which are close in energy, and can affect the strengths of those bonds. The most common interaction of this type is the one between a lone pair and an antiperiplanar σ-bond on an adjacent atom, as illustrated in Fig. 3.10(a). Because of the alignment of the lone pair and the σ-bond, a strong interaction is possible. This can be represented by the resonance structures shown in Fig. 3.10(b), or better, by an orbital interaction diagram as shown in Fig. 3.10(c). The lone pair can interact with both the σ-bonding and the σ*-antibonding orbital of C—Y, but if Y is electronegative, the latter interaction is the more important, as shown in Fig. 3.10(c). The results are the transfer of electron density from the lone pair to the σ*-orbital, the introduction of some π-bonding character to the C—X bond, which is strengthened, and the weakening of the C—Y bond.

21 K. Müller and A. Eschenmoser, *Helv. Chim. Acta*, 1969, **52**, 1823.
22 Reviews: W. A. Szarek and D. Horton (eds.), *The Anomeric Effect: Origin and Consequences*, ACS Symposium Series no. 87, Am. Chem. Soc., Washington D.C., 1979; A. J. Kirby, *The Anomeric Effect and Related Stereoelectronic Effects at Oxygen*, Springer-Verlag, Heidelberg, 1983; see also ref. 17.

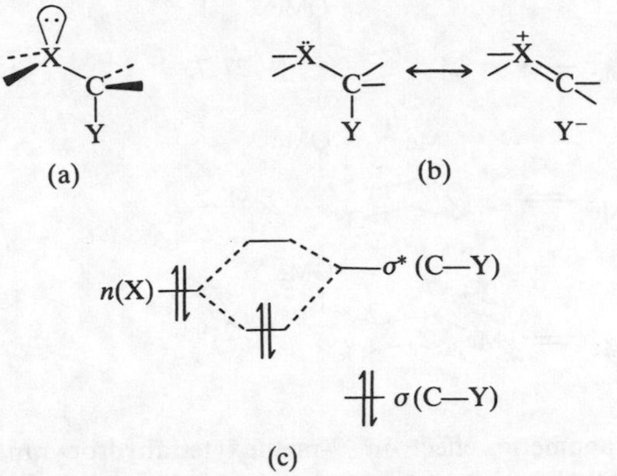

Fig. 3.10 Antiperiplanar interaction of lone pair and σ-bond.

The existence of a significant interaction of this type should be discernible in several ways. If the interaction is overall a stabilizing one (as it is in Fig. 3.10), flexible molecules should preferentially adopt conformations that maximize it. There should also be evidence in terms of structure (for example, bond lengths) and reactivity for the bonding changes that result.

An important feature of the stereochemistry of tetrahydropyrans and other reduced oxygen and sulphur heterocycles is that electronegative groups (F, Cl, Br, OR) at C-2 tend to preferentially occupy an axial position even though this is sterically more hindered. The conformational preference was first recognized in carbohydrate chemistry, many sugars having pyranose ring structures with oxygen substituents at the 2-position. It became known as *the anomeric effect* because it was found in anomeric equilibria between α- and β-glycosides. It is the free energy difference between the axial and the equatorial conformer, after allowing for the usual equatorial preference (A-value) of the substituent. The values are about $2.6–3.1 \, kcal \, mol^{-1}$ $(11–13 \, kJ \, mol^{-1})$ for chlorine and bromine, and about $1.0–1.7 \, kcal \, mol^{-1}$ $(4–7 \, kJ \, mol^{-1})$ for oxygen substituents. The values are therefore not large, and the equilibrium position is influenced by the presence of other ring substituents (Fig. 3.11) and by solvent effects. The proportion of axial conformer in solution decreases as the dielectric constant of the solvent increases, because the equatorial conformers are more polar. Strongly hydrogen-bonding solvents also preferentially stabilize conformers with equatorial alkoxy groups.

The anomeric effect is much stronger with elements in the first row of the Periodic Table (F, OR, NR_2) than with elements in the second and later rows. The only significant anomeric effects of the heavier elements are found when the halogens (Cl and Br) act as electron acceptors.[23]

The X-ray crystal structures of several heterocycles of these types bearing

23 P. von R. Schleyer, E. D. Jemmis, and G. W. Spitznagel, *J. Am. Chem. Soc.*, 1985, **107**, 6393.

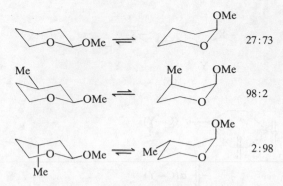

Fig. 3.11 The anomeric effect in 2-methoxytetrahydropyran and the influence of a methyl substituent on the position of equilibrium (M. Gelin, Y. Bahurel, and G. Descotes, *Bull. Soc. Chim. Fr.*, 1970, 3723).

2-halo substituents have been determined and those with axial halogen groups have bonds that are shortened or lengthened in the appropriate direction. For example, in *cis*-2,3-dichloro-1,4-dioxane, which contains both an axial and an equatorial chloro substituent, the axial C—Cl bond is abnormally long and the adjacent C—O bond is shorter than normal. The bonds to the carbon atom bearing the equatorial chlorine are normal in length (Fig. 3.12).

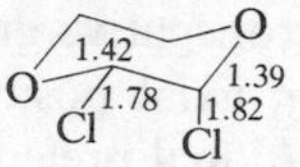

$$\begin{array}{c}
\text{O} \quad\overset{\displaystyle 1.42}{\diagup}\quad \text{O} \\
\overset{\displaystyle 1.78}{\underset{\displaystyle \text{Cl}}{}} \quad \overset{\displaystyle 1.39}{\underset{\displaystyle 1.82}{}} \\
\text{Cl}
\end{array}$$

Fig. 3.12 Bond lengths (Å) in *cis*-2,3-dichloro-1,4-dioxane (C. Altona and C. Romers, *Acta Cryst.*, 1963, **16**, 1225).

With alkoxy-substituted tetrahydropyrans the bond-length changes are less distinct. This is because the substituent can rotate so that a lone pair on the exocyclic oxygen can be antiperiplanar to the ring C—O bond and thus exert an anomeric effect in the opposite direction. This can occur with both axial and equatorial alkoxy substituents. Since the lone pair on the ring oxygen can only exert an influence on an axial 2-substituent, the overall anomeric effect is still greater for the axially substituted conformers, however. The three possible modes of interaction are shown in Fig. 3.13.

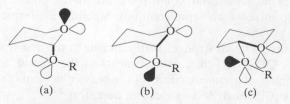

Fig. 3.13 Antiperiplanar interactions of lone pairs and C—O bonds. Conformers (a) and (b) are equivalent and have axial OR; conformer (c) has equatorial OR.

There are significant effects on chemical reactivity, the bonds antiperiplanar to the lone pairs being weakened by the interaction. For example, the cyclic acetal **21**, in which the 2-aryloxy substituent is equatorial, is much more resistant to hydrolysis than its isomer **22** with an axial aryloxy group. Isomer **21** is unaffected by 0.1 M HCl after several weeks at 39°C whereas isomer **22** is rapidly hydrolysed in the same conditions (half-life 7.7 min at 39°C).

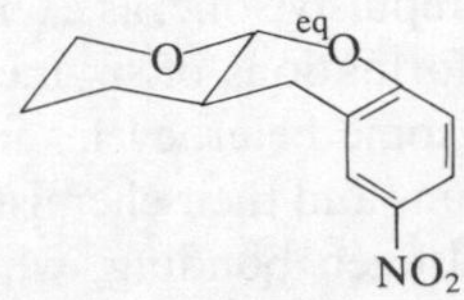

21 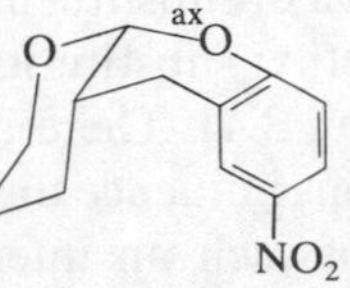   **22**

The bond-weakening effect of antiperiplanar lone pairs is also seen in other types of ring system. A striking example is provided by the tricyclic orthoamide **23**.[24] The conformation of this compound is such that all three nitrogen lone pairs are antiperiplanar to the central C—H bond. The central hydrogen atom appears at abnormally high field in the ^{1}H NMR spectrum ($\delta 2.33$ in CDCl$_3$), which indicates the high electron density at hydrogen. The C—H bond is so weakened (by electron donation from the lone pairs into the σ^*-orbital) that it can be broken by reaction with a proton to give molecular hydrogen.

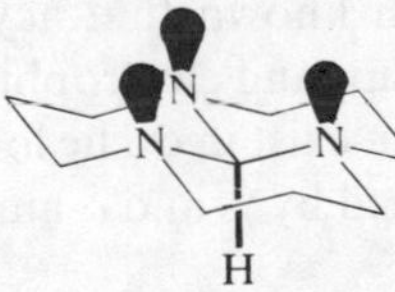

23

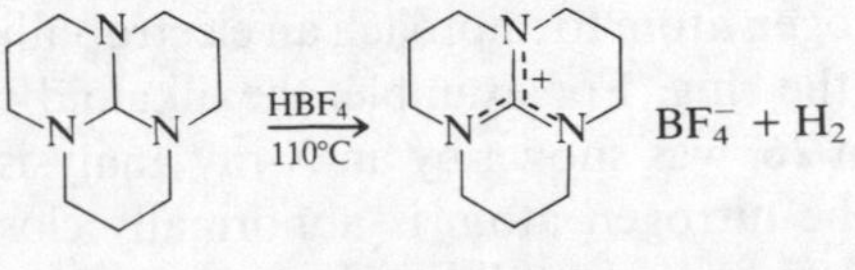

This bond-weakening effect can be transmitted further through a σ-bond framework if the steric arrangement is correct. An example is provided by the rapid solvolysis of 4-halo-1-azabicyclo[2.2.2]octanes, **24**, in which both the antiperiplanar C—C σ-bond and the C—X σ-bond are broken by

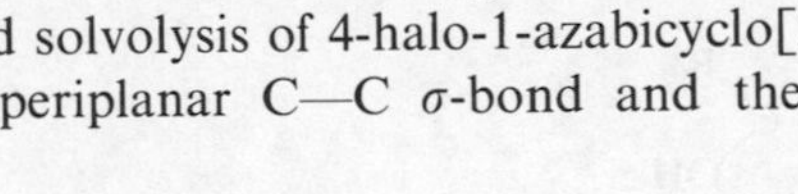

24

24 T. J. Atkins, *J. Am. Chem. Soc.*, 1980, **102**, 6364; J. M. Erhardt, E. R. Grover, and J. D. Wuest, *J. Am. Chem. Soc.*, 1980, **102**, 6365.

interaction with the nitrogen lone pair. This is one of a group of similar fragmentations in which the antiperiplanar arrangement of lone pairs and σ-bonds is a characteristic feature.[25]

3.5.2   *Attractive 'through-space' interactions*

The steric interaction of substituents is usually a repulsive one, as exemplified by the effects of methyl substitution on the conformations of six-membered heterocycles (Section 3.4). There are, however, some heterocycles in which attractive interactions influence their conformations and their chemistry. The simplest example of such an interaction is hydrogen bonding, which can occur, for example, between a ring heteroatom and a hydroxyl substituent. Thus, 3-hydroxypiperidines, **25**, can exist partly in the conformations **25a** with the hydroxyl group axial.

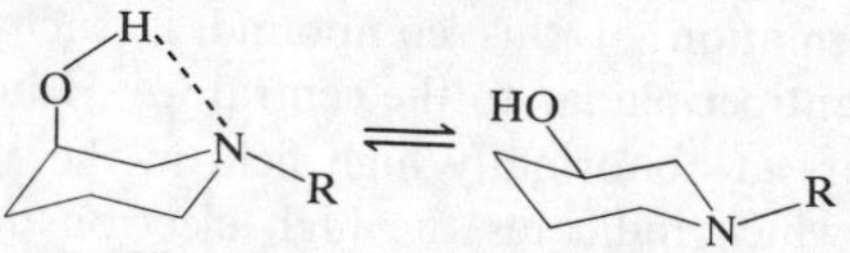

Another type of attractive interaction exists in compounds which have nucleophilic heteroatoms and electrophilic carbon atoms in appropriate positions in medium-sized rings. It is well known that acyclic compounds having nucleophilic amino or hydroxy groups and electrophilic carbon atoms separated by a three- or four-atom chain can exist in cyclic forms. An example of this 'ring-chain tautomerism'[26] is provided by γ- and δ-aminoketones (Fig. 3.14).

Several medium-ring aminoketones preferentially adopt conformations that allow the nucleophilic nitrogen atom to approach an electrophilic carbon atom on the opposite side of the ring. For example, the alkaloid clivorine, which contains the ring system **26**, was shown by an X-ray analysis to have a crystal structure in which the nitrogen atom is abnormally close to the carbonyl group (separation 1.99 Å) and is directed towards the carbonyl

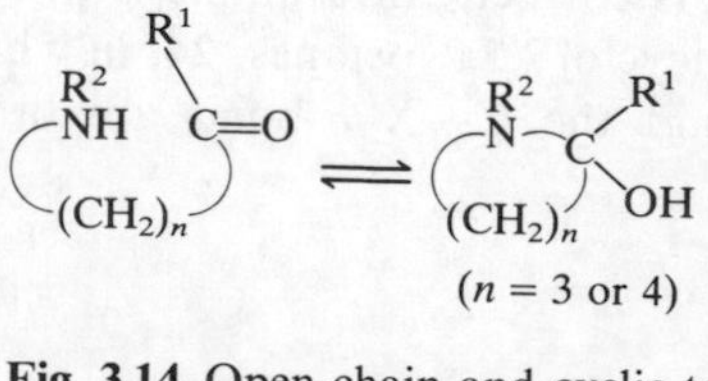

$$(n = 3 \text{ or } 4)$$

Fig. 3.14 Open-chain and cyclic tautomers of aminoketones.

25 Review: C. A. Grob and P. W. Schiess, *Angew. Chem. Int. Edn Engl.*, 1967, **6**, 1.
26 Reviews: P. R. Jones, *Chem. Rev.*, 1963, **63**, 461; R. Valters, *Russ. Chem. Rev.*, 1974, **43**, 665.

group at the angle required for incipient formation of a σ-bond.[27] The carbonyl bond is unusually long and has a very low stretching frequency in the infrared (v_{max} 1603 cm^{-1}), which indicates its low bond order. The attractive 'through-space' interaction of the nitrogen lone pair and the carbonyl group overrides any unfavourable steric interactions in other parts of the ring skeleton. The chemical consequences are apparent in simpler compounds of this type: for example, the aminoketone **27** is weakly basic, and is protonated on oxygen rather than on nitrogen to give the bicyclic ammonium salt **28**.[28] Similar interactions can occur in other medium-ring heterocycles which are sufficiently flexible to allow a transannular interaction between a nucleophilic and an electrophilic centre.[29]

Summary

1. Bond angle strain is most important in three-membered rings. In these rings, hybridization at the ring atoms is altered from the normal sp^3 type to minimize orbital distortion. Ring bonds have more p character, and bonds to substituents have more s character. The consequences of bond angle strain are increased reactivity towards ring opening, and changes in spectroscopic properties. Barriers to inversion in aziridines are much larger than in simple amines.
2. Torsional energy barriers about C—O and C—N bonds are somewhat lower than about C—C, but torsional barriers about N—O and N—N bonds are much higher, since conformations are preferred in which lone pairs on adjacent heteroatoms are orthogonal (the gauche effect). Conformations of saturated heterocycles are qualitatively similar to those of analogous carbocycles but with modifications resulting from different torsional barriers. Nitrogen-containing rings exist preferentially in conformations with nitrogen substituents equatorial.
3. Lone pairs on heteroatoms can interact with antiperiplanar σ-bonds. When the antiperiplanar substituents are electronegative, the interaction is a stabilizing one and the groups preferentially adopt axial positions when adjacent to ring oxygen (the anomeric effect). The σ-bonds are weakened by the interaction and this can lead to increased reactivity of the substituents.

27 K. B. Birnbaum, *Acta Cryst.* (B), 1972, 2825; G. I. Birnbaum, *J. Am. Chem. Soc.*, 1974, **96**, 6165.
28 N. J. Leonard and J. A. Klainer, *J. Org. Chem.*, 1968, **33**, 4269.
29 Review: N. J. Leonard, *Rec. Chem. Progr.*, 1956, **17**, 243.

4. Attractive interactions between groups can modify conformations of flexible heterocycles. Hydrogen bonding to hydroxyl groups and trans-annular interactions of nucleophilic and electrophilic atoms in medium-sized rings are examples of such interactions.

Problems

1. Calculate the angular deviation from unstrained values in the bond angles marked below. Assume 'natural' values of 109° (sp^3) and 120° (sp^2) and assume that the heterocycles have regular (triangular, square, or pentagonal) geometry.

2. The oxaziridine **A** (R=CMe$_3$) has been isolated in optically pure form but racemizes when heated in solution at 60–100°C. The corresponding oxaziridine with R = Me racemizes 8000 times more slowly at 100°C. Discuss the ability of these compounds to exist in enantiomeric forms and suggest a reason for the higher rate of racemization of the *t*-butyl derivative.

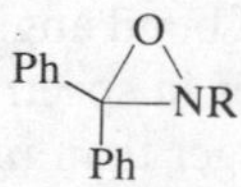

Ph
   \
    —O
Ph—⟨ |
    —NR

A

3. The methylene signal of the diazetidinone **B** appears in the ^{1}H NMR spectrum as a sharp singlet (δ 4.84) at +35°C but as two AB doublets (δ 4.53 and 5.25, *J* 13 Hz) at −50°C. Explain.

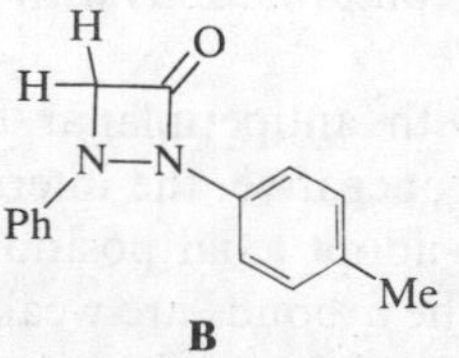

H   O
 \   //
H—|‒‒
   N—N
  /   \
Ph

Me

B

4. 1-Methylpiperidine exists preferentially in a chair conformation with the methyl group equatorial. 1,2-Dimethylhexahydropyridazine, **C**, contains a significant proportion of the conformer shown, with one methyl group axial.

Suggest an explanation.

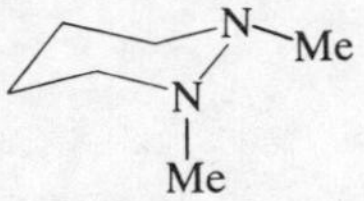

C

5. 1-Azidoadamantane, **D**, was photolysed in cyclohexane and gave, in high yield, a compound $C_{20}H_{30}N_2$. Suggest a structure for the product and a mechanism for its formation.

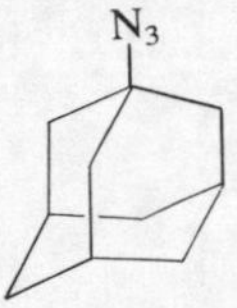

D

6. Predict the most stable conformation (in a nonpolar solvent) for each of the structures shown.

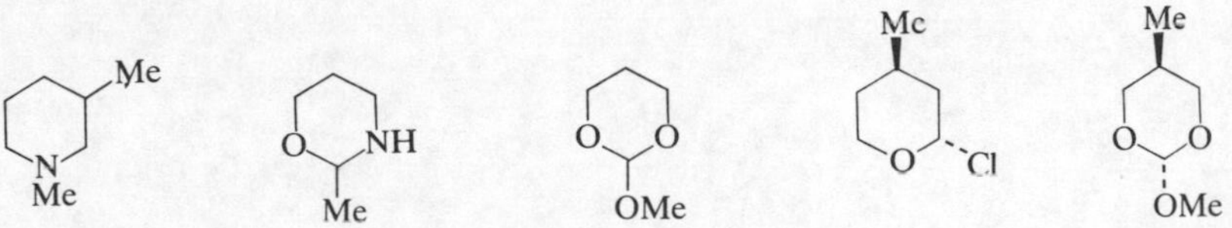

7. Suggest explanations for each of the following.

(a) The diamine **E** titrates as a monoamine, and is strongly basic (the pK_a of its salt is 11.88).

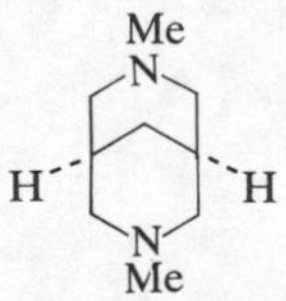

E

(b) Reaction of the *N*-oxide **F** with benzoyl chloride in chloroform gave a precipitate which, on treatment with 1 M HCl, gave the products shown.

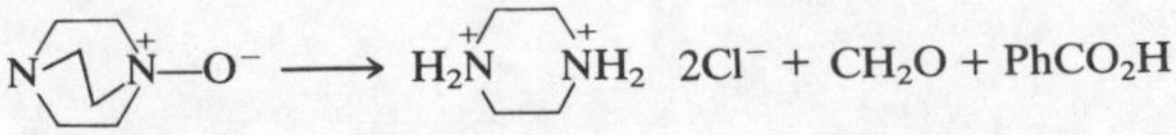

F

(c) The salt **G** rapidly acetylates piperidine in dichloromethane at $-20°C$.

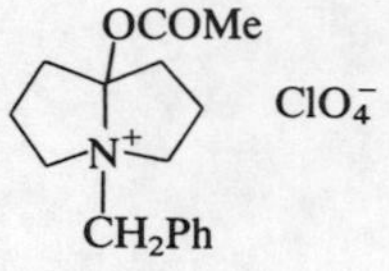

OCOMe

ClO_4^-

N$^+$

CH$_2$Ph

G

8. The pentacyclic amine **H** is one of the strongest organic bases known (its pK_a in acetonitrile is 29.22). At which nitrogen will it protonate? What factors might contribute to its very high basicity?

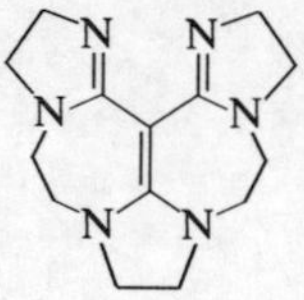

H

4 Ring synthesis

4.1 Introduction

In this chapter we shall consider the general methods which are available for the construction of heterocyclic compounds from open-chain precursors. For many of the common ring systems there is a wide range of practicable synthetic routes available. These can vary in complexity from one-step syntheses using a single reaction component, to multicomponent procedures with a large number of steps. Some useful heterocyclic syntheses involve the conversion of one ring system into another, by heat, by irradiation, or by the use of additional reagents. The method of choice in a particular case usually depends upon the pattern of substituents required in the product.

This chapter is primarily concerned with the *ring-forming reactions* used in these diverse processes. The types of ring-forming reaction available are not usually limited to a particular heterocycle but apply to a range of structures. The precursor components are varied to incorporate the appropriate heteroatoms and substituents, but the nature of the ring-closure reaction is usually more dependent upon the size of the ring being formed and its degree of unsaturation than upon the types of heteroatom present. It is therefore appropriate to treat these ring-forming reactions as general methods.

The types of ring-forming reaction available can be divided into two broad groups. Reactions in which a single ring bond is formed in the ring-closure process are called *cyclization reactions*, whereas those in which two ring bonds are formed, and in which no small molecules are eliminated in the process, are called *cycloaddition reactions*. We shall make use of this classification even though it is an oversimplified one. In practice there is not always a clear-cut distinction between the two groups, and in many heterocyclic syntheses the timing of the steps is uncertain. It is also important to appreciate that the ring-closure step in a multistep reaction sequence may not be the most difficult one; indeed, many such ring closures are fast relative to the other steps in the sequence and so the 'cyclization precursor' may not be detectable.

In designing a synthesis of a particular heterocycle, we need to consider (a) which bonds are most conveniently made in the ring-forming steps, (b) what degree of unsaturation is required in the heterocycle, and whether or not the oxidation state can easily be altered after the formation of the ring, and (c) what functional groups are required, and whether they are best introduced before, during, or after the ring-closure step. The remainder of this chapter is concerned mainly with the first point: we shall consider the major ring-forming reactions and some of their limitations. As for the second point, we can recognize that further unsaturation can often be introduced after formation of the ring, especially when the target compound is aromatic in character, and that double bonds can often be reduced without causing ring cleavage. It is thus not always necessary to arrive at the correct oxidation state in the ring-forming step. In order to deal with point (c) we need to know more about the properties of the particular heterocycle, but some useful generalizations will be illustrated in this chapter. For example, a carbonyl group in a ring system is often conveniently introduced in a ring-closure step involving intramolecular nucleophilic attack on an ester or a related acid derivative. C-Amino groups can often be introduced by a cyclization reaction in which a nucleophile adds to a nitrile group (Section 4.2.5).

The possible approaches to a particular target molecule may be analysed by the retrosynthetic method, in which reasonable bond construction processes are devised in reverse order from the target system. As an example, consider the construction of a thiazole ring by cyclization. Three of the possible ring-closure reactions are analysed retrosynthetically in Fig. 4.1(a–c). By taking the analysis one step further we can identify reasonable precursor molecules from which the ring could be constructed. In fact, there are examples of the use of all three of these reaction sequences for the synthesis of thiazoles. The intermediates in such multistep reactions may not be isolated, and indeed the sequence in which the reactions take place may be variable. For example, it is possible to prepare 2,5-dihydrothiazoles in good yield in a one-pot process by mixing ketones with elementary sulphur and ammonia at room temperature. We can formally analyze this process retrosynthetically as in Fig. 4.1(d), although in practice the bonds can be made with equal efficiency in a different order (for example, the C—S bonds can be made after the C=N bond).

4.2   Cyclization reactions

4.2.1   Reaction types

Cyclization reactions can involve any intramolecular versions of the common σ-bond-forming processes. By far the most common are those in which a nucleophilic atom interacts with an electrophile. The predominant reaction types are nucleophilic displacement at a saturated carbon atom, nucleophilic addition to unsaturated carbon, and nucleophilic addition–elimination.

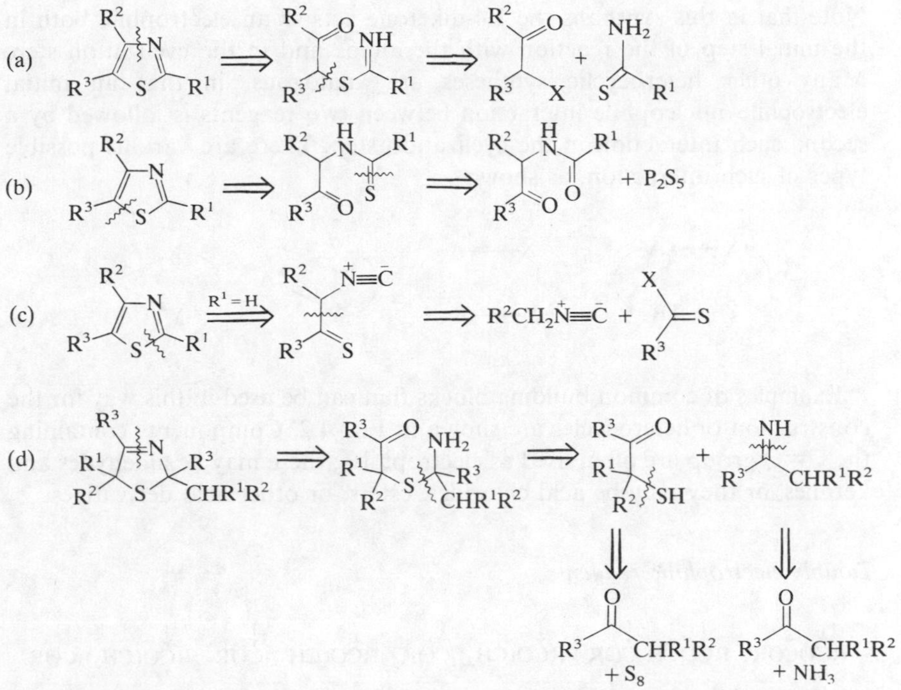

Fig. 4.1 Retrosynthetic analysis of thiazole ring synthesis. For route (d) see F. Asonger and H. Offermanns, *Angew. Chem. Int. Edn Engl.*, 1967, **6**, 907.

Heterocyclic rings can also be constructed by intramolecular radical, carbene, and nitrene reactions, and by electrocyclic ring closure of conjugated π-electron systems. All these processes are illustrated in later sections.

Although cyclization reactions involve the formation of one bond in the ring-forming step, in practice the cyclization intermediate is usually constructed *in situ* from two or more simpler reagents. This was illustrated for thiazole synthesis in Fig. 4.1. Another example is provided by a synthesis which is typical of many other routes to heterocycles: the Paal–Knorr synthesis of pyrroles from 1,4-dicarbonyl compounds and primary amines. The cyclization step involves nucleophilic attack on a carbonyl group by the imino nitrogen of an intermediate **1**. This intermediate has been detected and characterized by ^{13}C NMR spectroscopy.[1]

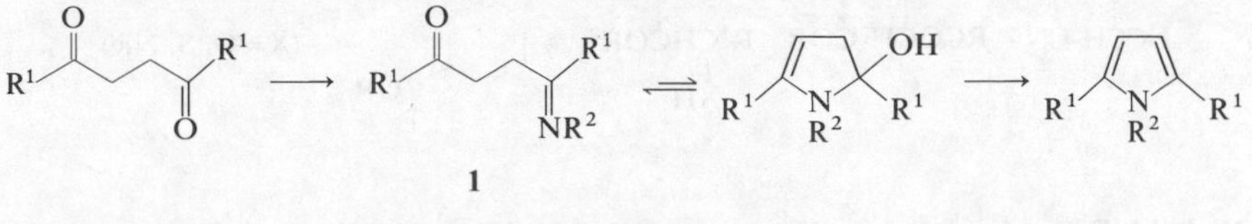

1

1 Review: A. R. Katritzky, D. L. Ostercamp, and T. I. Yousaf, *Tetrahedron*, 1987, **43**, 5171.

Note that in this synthesis the 1,4-diketone acts as an electrophile both in the initial step of the reaction with the amine and in the cyclization step. Many other heterocyclic syntheses are analogous, in that an initial electrophile–nucleophile interaction between two reagents is followed by a second such interaction in the cyclization step. There are various possible types of such interaction, as shown.

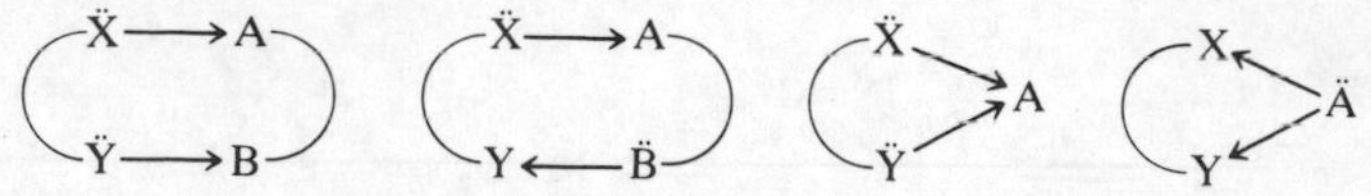

$$\ddot{X} \longrightarrow A \qquad \ddot{X} \longrightarrow A \qquad \ddot{X} \searrow A \qquad X \searrow \ddot{A}$$
$$\ddot{Y} \longrightarrow B \qquad Y \longleftarrow \ddot{B} \qquad \ddot{Y} \qquad Y$$

Examples of common building blocks that can be used in this way for the construction of heterocycles are shown in Fig. 4.2. Components containing the C=O group are often used as electrophiles; these may be aldehydes and ketones, or they may be acid chlorides, esters, or other acid derivatives.

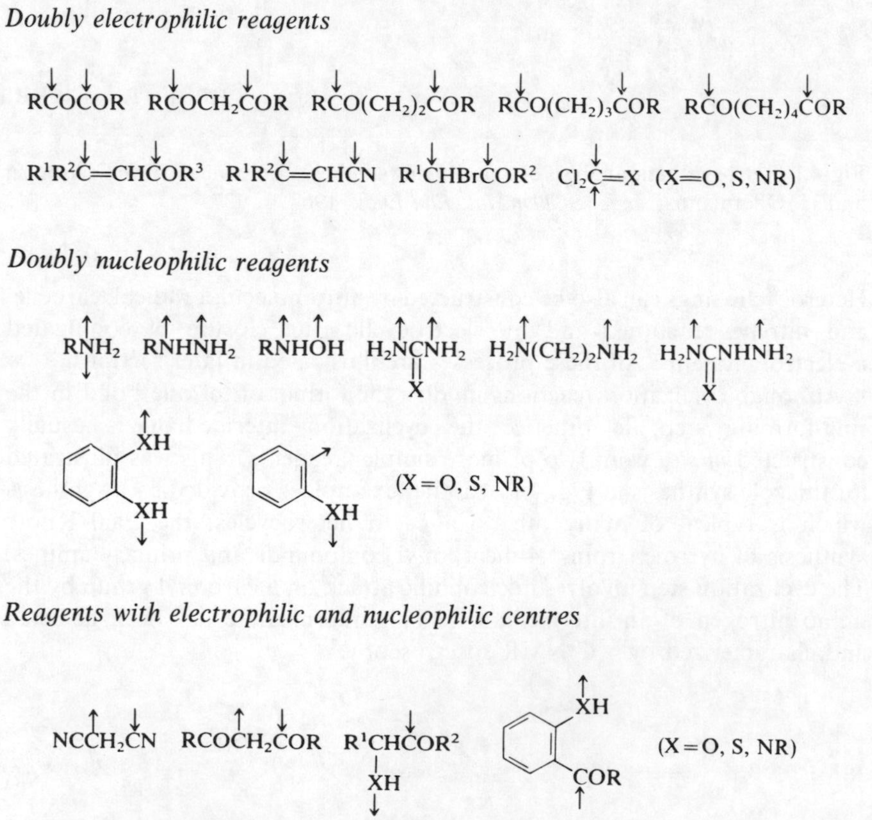

Doubly electrophilic reagents

$$RCOCOR \quad RCOCH_2COR \quad RCO(CH_2)_2COR \quad RCO(CH_2)_3COR \quad RCO(CH_2)_4COR$$

$$R^1R^2C{=}CHCOR^3 \quad R^1R^2C{=}CHCN \quad R^1CHBrCOR^2 \quad Cl_2C{=}X \ (X{=}O, S, NR)$$

Doubly nucleophilic reagents

$$RNH_2 \quad RNHNH_2 \quad RNHOH \quad H_2N\overset{||}{\underset{X}{C}}NH_2 \quad H_2N(CH_2)_2NH_2 \quad H_2N\overset{||}{\underset{X}{C}}NHNH_2$$

$(X=O, S, NR)$

Reagents with electrophilic and nucleophilic centres

$$NCCH_2CN \quad RCOCH_2COR \quad R^1CHCOR^2 \quad$$

$(X=O, S, NR)$

Fig. 4.2 Examples of the different types of building blocks used in heterocyclic synthesis.

The synthesis of benzo-fused heterocycles deserves special comment because most important preparations involve annelation of a heterocycle to a benzene ring. There are two general approaches to their synthesis, one based on an *ortho*-disubstituted benzene and the other based on a monosubstituted benzene in which a free *ortho* position acts as a nucleophilic centre (i.e. it is subject to electrophilic attack). For example, the quinoline ring system can be prepared from 2-aminobenzaldehyde and a two-carbon reagent with adjacent electrophilic and nucleophilic centres [Fig. 4.3(a)]. Alternatively it can be prepared from aniline and a three-carbon, doubly electrophilic reagent such as an αβ-unsaturated ketone [Fig. 4.3(b)]. The second approach has the disadvantage that in the synthesis of more complex derivatives starting with an aniline in which the two free *ortho*-positions are not equivalent, there is the possibility of obtaining isomeric products [Fig. 4.3(c)].

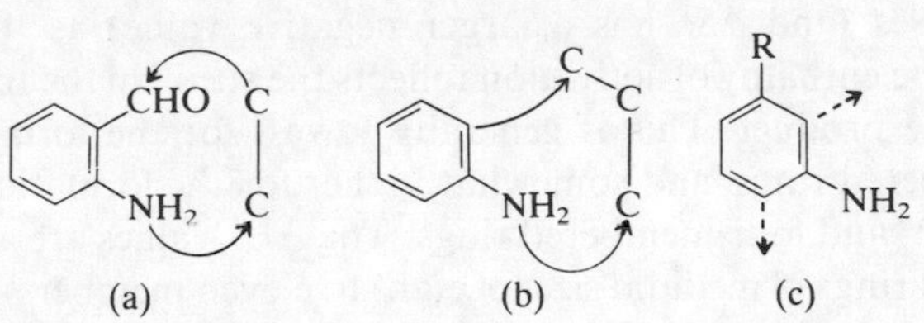

Fig. 4.3 Approaches to the synthesis of the quinoline ring. Possibilities of isomerism using the second approach are illustrated in (c).

A system of nomenclature for describing possible types of cyclization process is illustrated in Fig. 4.4. This is based on the state of hybridization of the atom attacked by the nucleophile and on whether the shift of electrons away from that atom in the cyclization reaction is within (*endo-*) or outside (*exo-*) the ring being formed.

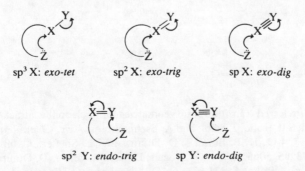

Fig. 4.4 Types of nucleophile–electrophile cyclization.

Intramolecular displacement at a saturated carbon atom is an example of an *exo-tet* process, and nucleophilic addition and addition–elimination reactions of carbonyl compounds are *exo-trig* processes. These are the most common types of cyclization and are considered further below. Examples of each of the other types of cyclization are also given.

In order to decide whether a particular ring system can be prepared efficiently by one of these cyclization processes, it is necessary to take into account the size of the ring being formed and the nature of the transition state leading to its formation. The free energy of activation $\Delta G^{\ddagger}$ for the cyclization process is made up of an enthalpy term ($\Delta H^{\ddagger}$) and an entropy term ($T\Delta S^{\ddagger}$) (T being the absolute temperature).

$$\Delta G^{\ddagger} = \Delta H^{\ddagger} - T\Delta S^{\ddagger}$$

The entropy of activation for the intramolecular process is related to the probability of the two ends of the chain approaching each other. This probability decreases (and $\Delta S^{\ddagger}$ has a larger negative value) as the chain length increases. The enthalpy of activation reflects the strain in the transition state leading to the product. This is generally lowest for the formation of five- and six-membered rings, and somewhat higher for the formation of the more strained three- and four-membered rings. The $\Delta H^{\ddagger}$ values are also high for the formation of rings of medium size (of eight to eleven members) because of the nonbonded interactions in these rings. The free energies of activation for medium-sized rings are also high, and it is therefore difficult to form these rings efficiently by cyclization processes.

Another significant factor in determining the feasibility of a particular cyclization process is the geometry of approach of the nucleophile to the electrophile in the transition state. It is well known, for example, that bimolecular nucleophilic displacement at saturated carbon requires a transition state in which the nucleophile attacks at the side opposite to that occupied by the leaving group. The attack of a nucleophile on a carbonyl group takes place preferentially from above or below the plane of the molecule, at an angle with the C=O bond close to the tetrahedral value. The preferred directions of nucleophilic attack on these and other electrophilic centres are shown in Fig. 4.5. It is likely that reasonable deviations from

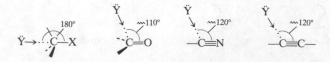

Fig. 4.5 Preferred transition state geometries for nucleophilic attack. C—X: L. Tenud, S. Farooq, J. Seibl, and A. Eschenmoser, *Helv. Chim. Acta*, 1970, **53**, 2059. C=O: H. B. Bürgi, J. D. Dunitz, and E. Shefter, *J. Am. Chem. Soc.*, 1973, **95**, 5065. C≡N: G. Procter, D. Britton, and J. D. Dunitz, *Helv. Chim. Acta*, 1981, **64**, 471. C≡C: O. Eisenstein, G. Procter, and J. D. Dunitz, *Helv. Chim. Acta*, 1978, **61**, 2538.

these preferred geometries of approach can be accommodated,[2] but the constraints may make some of the reaction types shown in Fig. 4.4 difficult to achieve in the formation of five-membered or smaller rings. For example, it has been suggested that the *endo-trig* process may be difficult in the formation of rings of less than six atoms. Certainly such reactions are not very common. The *endo-dig* process, which appears at first sight to be even more difficult to achieve, does occur in the formation of five-membered rings. This may be because the triple-bonded group has a π-bond in the same plane as the nucleophile, whereas in the *endo-trig* reaction the nucleophile has to approach the π-bond from above or below the molecular plane.[2]

4.2.2 Displacement at saturated carbon

Intramolecular S_N2 reactions are widely used for the construction of saturated heterocyclic compounds. Five-membered rings are generally the easiest to form, as this ring size represents the best balance between enthalpy and entropy terms: the rings are strain-free and the transition states are accessible. The influence of chain length on the rates of two types of S_N cyclization is shown in Table 4.1.

It can be seen that in the bromoalkylamine cyclizations the formation of the five- and six-membered rings is especially favourable. The formation of aziridine ($n=3$) is also a relatively favoured process. This is despite the fact that the product is strained; the small degree of ordering required in the transition state facilitates the reaction. (The cyclization of bromoacetate anions to give an α-lactone is relatively slower, as shown in Table 4.1, possibly because of the additional strain involved in incorporating an sp^2-hybridized atom into a three-membered ring.)

These reactions have been shown in many diverse cases to involve displacement with inversion at carbon. An example, shown in Fig. 4.6, is the stereoselective formation of *cis-* and *trans-*2,3-diphenylaziridines from the *threo-* and *erythro-*chloroamines.

In cases where the rates of ring closure are low, they may not compete effectively with intermolecular reactions and the yields of cyclic products may therefore be poor. An example is provided by the reaction of 3-chloropropanol with sodium hydroxide in aqueous methanol (Fig. 4.7). The reaction gave oxetane in 14% yield and open-chain solvolysis products in 78% yield.

The efficiency of formation of three- and four-membered rings is improved if the carbon atom bearing the nucleophile is heavily substituted. This is illustrated in Table 4.2 for the formation of oxiranes from chlorohydrins. The rate may increase with substitution because, as was discussed in Chapter 3, the angle between the bonds to external substituents on small rings is

2 Review: F. M. Menger, *Tetrahedron*, 1983, **39**, 1013.

Table 4.1 Influence of chain length on rates of intramolecular nucleophilic displacement.

(a) Cyclization of bromoalkylamines
$Br(CH_2)_{n-1}NH_2$ (water, 25°C).[a]

Ring size (n)	Relative rate
3	70
4	1
5	6×10^4
6	10^3
7	2

(b) Cyclization of ω-bromocarboxylate anions
$Br(CH_2)_{n-2}CO_2^-$ (99% aq. Me_2SO, 50°C).[b]

Ring size (n)	Relative rate
3	21.7
4	2.4×10^4
5	2.8×10^6
6	2.6×10^4
7	97.3
8	1.00
9	1.12
10	3.35
11	8.51
15	45.1
23	60.4

a Data adapted from ref. 3. *b* C. Galli, G. Illuminati,
L. Mandolini, and P. Tamborra, *J. Am. Chem. Soc.*,
1977, **99**, 2591.

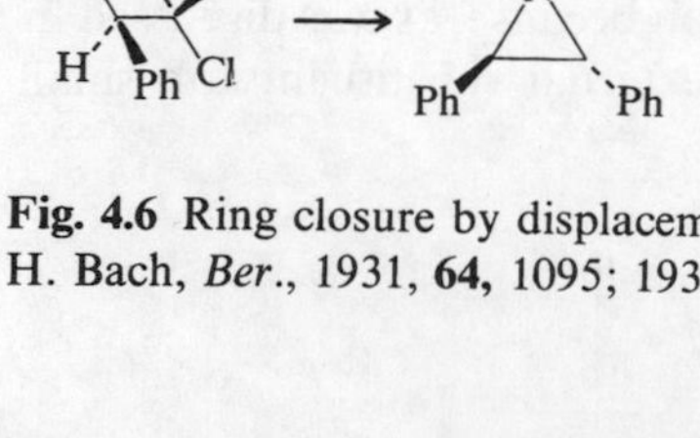

Fig. 4.6 Ring closure by displacement with inversion (A. Weissberger and
H. Bach, *Ber.*, 1931, **64**, 1095; 1932, **65**, 631).

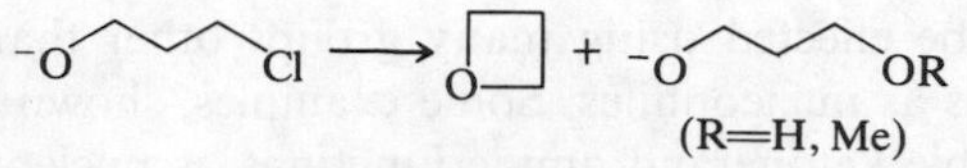

$$\text{(R=H, Me)}$$

Fig. 4.7 Competing ring closure and solvolysis (W. H. Richardson, C. M. Golino, R. H. Wachs, and M. B. Yelvington, *J. Org. Chem.*, 1971, **36**, 943).

Table 4.2 Relative rates of ring closure of chlorohydrins (aq. NaOH, 18°C).[a]

Substrate	Relative rate
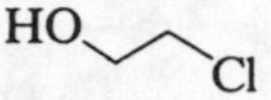	1
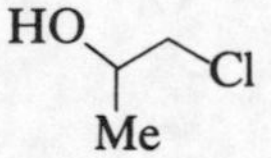	325
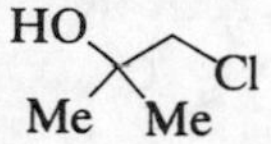	39 000

a Data adapted from ref. 3.

greater than the tetrahedral value, and there is some relief of steric crowding in forming the product.[3]

As the examples in Section 4.2.1 showed, the synthetic scope of cyclization reactions is greatly increased by methods for the *in situ* formation of suitable precursors. Two important methods for the synthesis of oxiranes from carbonyl compounds further illustrate the point. The reaction of carbonyl compounds with anions of α-haloketones or α-haloesters[4] (the Darzens reaction) or with sulphonium ylides[5] leads to the formation of intermediates which can undergo cyclization by internal nucleophilic displacement (Fig. 4.8).

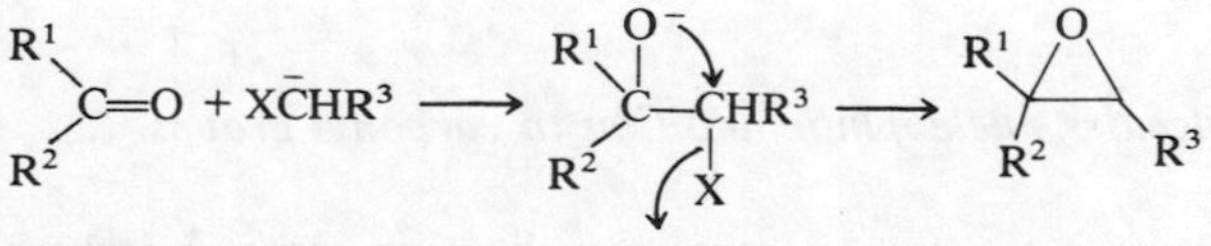

Fig. 4.8 Formation of oxiranes from carbonyl compounds by the Darzens reaction (X = Cl, Br) and from sulphur ylides (X = $\overset{+}{S}R_2^4$).

3 Review: A. J. Kirby, *Adv. Phys. Org. Chem.*, 1980, **17**, 183.
4 Review: M. Ballester, *Chem. Rev.*, 1955, **55**, 283.
5 Review: B. M. Trost and L. S. Melvin, *Sulfur Ylides*, Academic Press, New York, 1975.

The ring closure can be effected using many groups other than simple amino or hydroxy groups as nucleophiles. Some examples, shown in Table 4.3, make use of enolizable ketone and amide functions as nucleophiles in the presence of bases.

Table 4.3 Examples of cyclization by nucleophilic displacement at saturated carbon.[a]

Reagents	Likely cyclization intermediates	Products	Refs. and notes
(i) RCONH—CH₂Cl, —NHOCH₂Ph, NaH	RCONH, Cl, N⁻, OCH₂Ph	RCONH, NOCH₂Ph	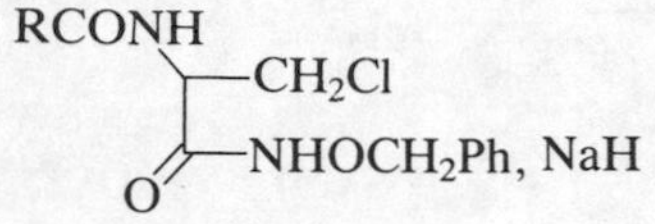 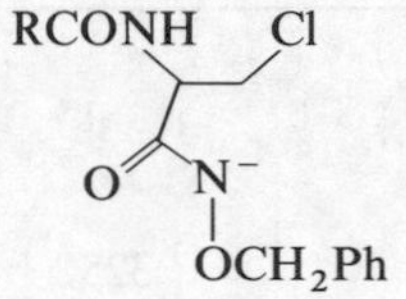 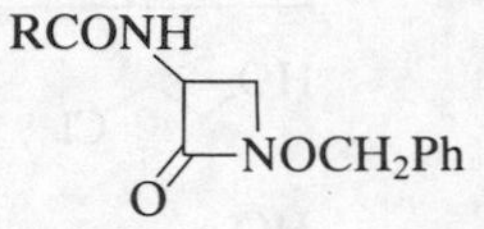 b
(ii) R¹ OH, R², H, HN, O, H, R³ , SOCl₂, Et₃N	R¹ OSOCl, R², H, HN, O, H, R³	R¹ H, R², H, N, O, R³	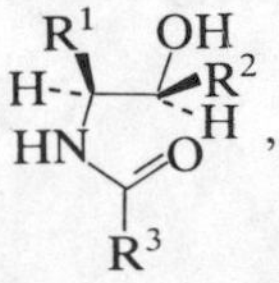 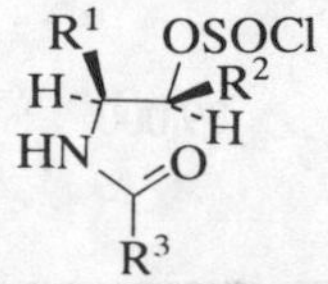 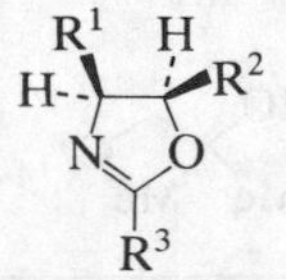 c
(iii) R¹COCH₂Cl, R²COCH₂CO₂Et, pyridine	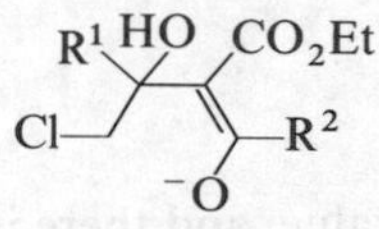 R¹ HO CO₂Et, Cl, R², ⁻O	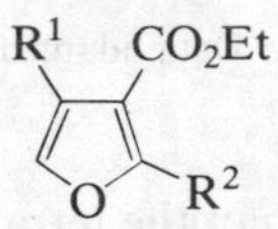 R¹ CO₂Et, O, R²	d
(iv) Me₂S⁺CH=C=CH₂Br⁻, MeCOCH₂COMe, ⁻OEt	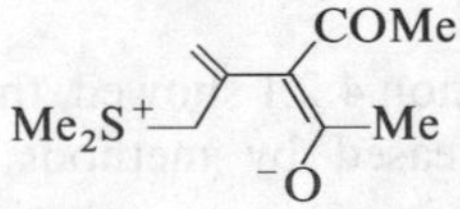 COMe, Me₂S⁺, Me, ⁻O	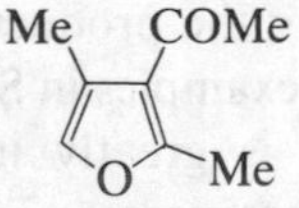 Me COMe, O, Me	e

a All are *exo-tet* cyclizations. *b* M. J. Miller, P. G. Mattingly, M. A. Morrison, and J. F. Kerwin, *J. Am. Chem. Soc.*, 1980, **102**, 7026. *c* J. Sicher and M. Pánková, *Coll. Czech. Chem. Commun.*, 1955, **20**, 1409. *d* The Feist–Benary furan synthesis; F. M. Dean, *Adv. Heterocycl. Chem.*, 1982, **30**, 167. *e* J. W. Batty, P. D. Howes, and C. J. M. Stirling, *J. Chem. Soc. Perkin Trans. 1*, 1973, 65.

4.2.3 *Intramolecular nucleophilic addition to carbonyl groups*

This type of process is the most common cyclization reaction in heterocyclic synthesis. Internal nucleophilic attack at the carbonyl group of esters, acid chlorides, etc. is followed by displacement of a leaving group, and the carbonyl function is retained in the cyclic product. Attack by a nucleophile on an aldehydic or ketonic carbonyl group is often followed by dehydration of the intermediate, especially when it leads to the formation of a heteroaromatic ring system. Such cyclizations may be acid-catalysed when the nucleophile is a weak one, and the attack is then probably on the protonated carbonyl function.

Three types of intramolecular cyclization onto aldehydic and ketonic carbonyl groups are illustrated in Table 4.4. Aldol-type cyclization involving attack by nucleophilic carbon (a) provides routes to a variety of hetero-aromatic compounds. Cyclization through nucleophilic heteroatoms is exemplified in (b). Routes to benzo-fused heterocycles in which the *ortho* carbon atom of a substituted benzene acts as the nucleophile are illustrated in (c).

Example (i) in Table 4.4 is typical of many cyclizations of this type in requiring only a weak base, such as potassium carbonate. The carbanion generated by deprotonation of the methylene group undergoes a very favourable intramolecular aldol reaction, which leads to the formation of a benzofuran by dehydration. In the Knorr pyrrole synthesis [example (ii)] the cyclization intermediate is the enamine shown; this is converted into the pyrrole by intramolecular attack on the carbonyl group of the acetyl function followed by dehydration. Practical aspects of the Knorr synthesis, which is an important method of preparation of pyrroles, are discussed in Chapter 6.

The method of preparation of 2-substituted indoles shown in (iii) is one of many examples of heterocyclic synthesis which involve intramolecular attack of an amino function on a carbonyl group. In this type of synthesis the amine is rarely isolated but instead is generated *in situ* by reduction. Intermediates bearing other suitably placed nucleophilic groups, such as −OH and −SH groups, also cyclize readily. The isoxazole synthesis (iv) is an example: reaction of a β-dicarbonyl compound with hydroxylamine gives a detectable mono-oxime, but this rapidly cyclizes, the isoxazole being formed by elimination of water from the cyclization product.

A large number of syntheses of benzo-fused heterocyclic compounds involve cyclization of a monosubstituted benzene. A free *ortho* position of the benzene ring acts as the nucleophilic centre and in many cases a carbonyl group on the side chain acts as the electrophilic centre. Such reactions usually require a catalyst (a proton acid or a Lewis acid) to activate the carbonyl group. In the Combes quinoline synthesis [example (v)] an intermediate suitable for cyclization is generated *in situ* from aniline and a diketone, but in the benzofuran synthesis shown in (vi) the intermediate aryloxyketone is normally isolated. The cyclization steps are followed by rapid dehydration to produce the aromatic heterocycles.

4.2.4 *Intramolecular addition of nucleophiles to other double bonds*

Cyclization by nucleophilic addition to double bonds other than carbonyl groups is illustrated in Table 4.5. Activated C=S and C=N bonds can act as the electrophiles, as in examples (i) and (ii). In example (iii) the electrophile is an activated C=C bond to which an internal conjugate addition reaction can take place. It is worth noting that in this example, the kinetically favoured

Table 4.4 Examples of heterocyclic ring synthesis involving cyclization by nucleophilic attack on a carbonyl group.[a]

Reagents	Cyclization intermediates	Products	Refs. and notes

(a) *Aldol-type cyclization*

(i) COR^1 / OH benzene, $BrCH_2COR^2$, base → intermediate: 2-(R^1CO)phenyl-O-CH_2COR^2 → product: benzofuran with R^1 and COR^2 ... b

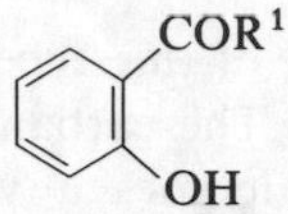

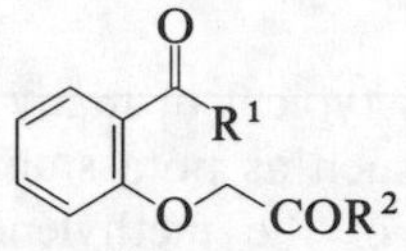

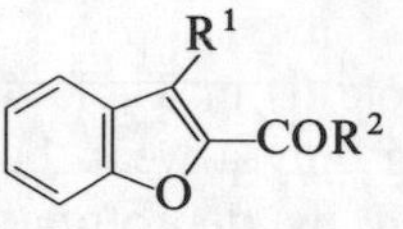

(ii) $MeCOCH(NH_2)CO_2R$, $MeCOCH_2CO_2R$ → intermediate RO_2C—$N(H)$ enamine with Me, CO_2R → product pyrrole RO_2C, $N(H)$, Me, Me, CO_2R ... c

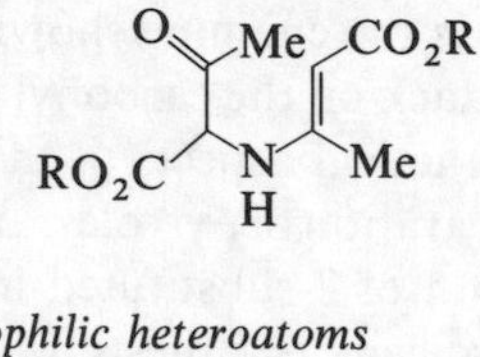

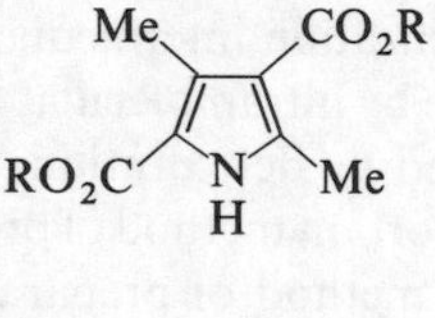

(b) *Cyclization through nucleophilic heteroatoms*

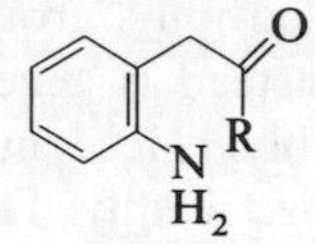

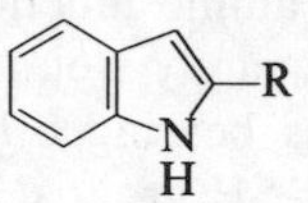

(iii) CH_2COR / NO_2, H_2/Ni → intermediate o-amino phenyl CH_2COR (NH_2) → product indole 2-R, $N(H)$ ... d

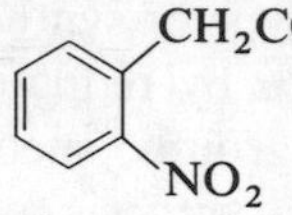

(iv) $RCOCH_2COR$, H_2NOH → intermediate oxime R—CO—CH=... N—OH → product isoxazole 3-R, 5-R ... e

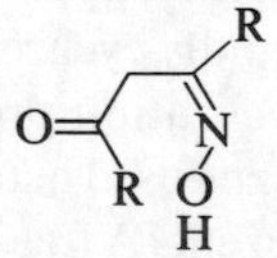

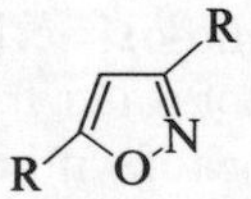

(c) *Cyclization onto an* ortho *position of a benzene ring*

(v) $RCOCH_2COR$, $PhNH_2$, H^+ → intermediate → product quinoline with R at 4, 2-R ... f

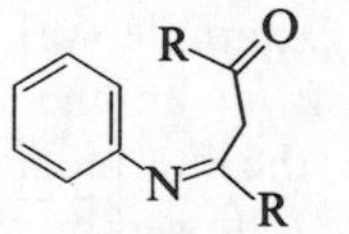

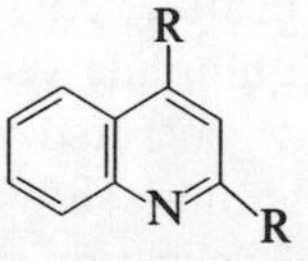

(vi) $RCOCH(Cl)R$, $PhOH$, base; then $ZnCl_2$ → intermediate → product benzofuran 3-R, 2-R ... b

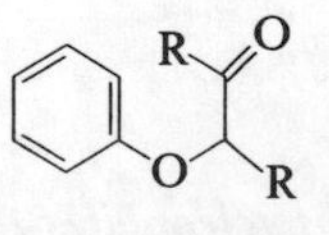

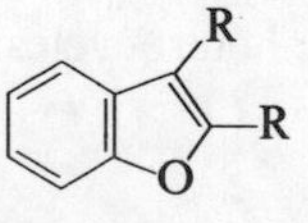

---

[a] All are *exo-trig* cyclizations. [b] A. Mustafa, *Benzofurans*, Wiley-Interscience, New York, 1974. [c] R. J. Sundberg, in *Comprehensive Heterocyclic Chemistry*, Vol. 4, Chapter 3.06. The α-aminoketone is normally generated *in situ* from the corresponding α-oximinoketone. [d] R. K. Brown, in *Indoles*, ed. W. J. Houlihan, Wiley-Interscience, New York, 1972. [e] S. A. Lang and Y. I. Lin, in *Comprehensive Heterocyclic Chemistry*, Vol. 6, Chapter 4.16. [f] G. Jones, *Quinolines*, Vol. 1, Wiley-Interscience, New York, 1977.

Table 4.5 Examples of other types of *exo*-cyclization onto trigonal centres.

Reagents	Likely cyclization intermediates	Products	Refs. and notes
(i) $R_2C(OH)C(OH)R_2$, $Cl_2C{=}S$	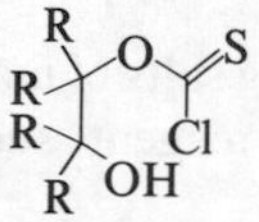	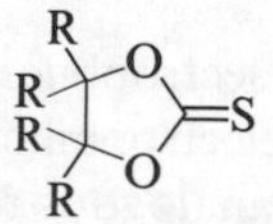	*a*
(ii) $PhNHCOCH_2R$, $POCl_3$, Me_2NCHO	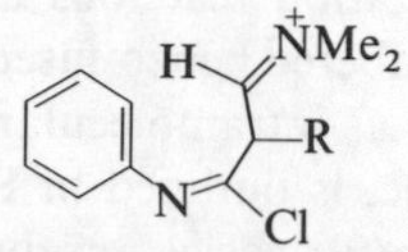	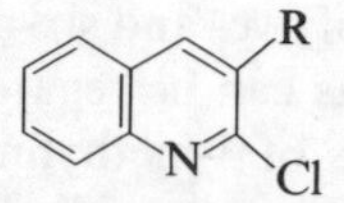	*b*
(iii)			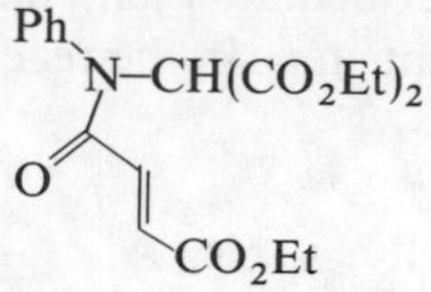 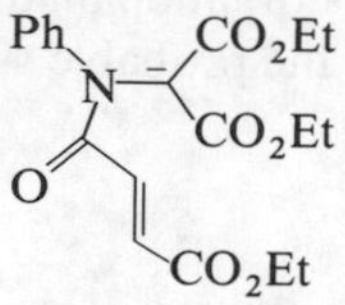 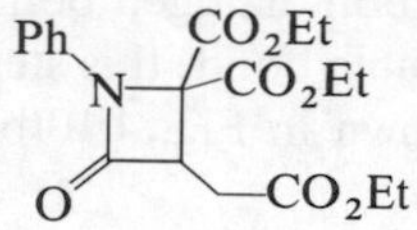 *c*
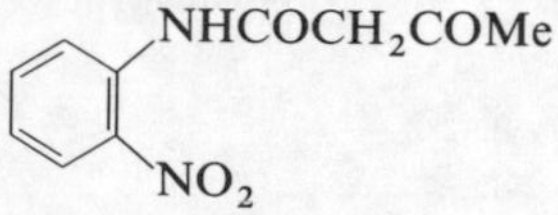 (iv) NaOH aq.	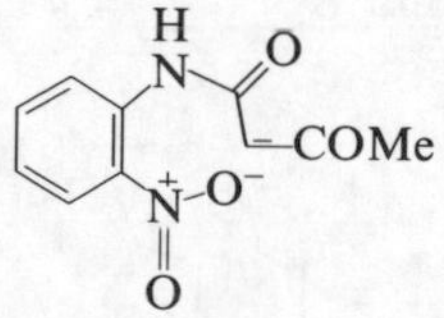	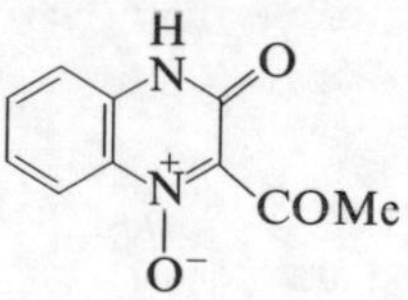	*d*

a S. Sharma, *Synthesis*, 1978, 803. *b* O. Meth-Cohn and B. Tarnowski, *Adv. Heterocycl. Chem.*, 1982, **31**, 207. *c* A. K. Bose, M. S. Manhas, and R. B. Ramer, *Tetrahedron*, 1965, **21**, 449. *d* P. N. Preston and G. Tennant, *Chem. Rev.*, 1972, **72**, 627.

4-*exo-trig* reaction is taking place, rather than the feasible alternative, a 5-*endo-trig* addition.

The great majority of cyclizations take place by reaction at an electrophilic carbon centre, but there are a few heterocyclic syntheses which involve cyclization onto electrophilic nitrogen. One such reaction, in which a nitro group acts as the electrophile, is shown in example (iv).

In addition to these *exo-trig* cyclizations there are several related reactions in which a ring system is formed by intramolecular addition to a simple carbon–carbon double bond. Most of these reactions are initiated by attack on the double bond by an external electrophile. The cation intermediate is then captured by an internal nucleophile. An example of such a reaction is shown in Fig. 4.9.[6] In this example the reaction is initiated by bromine.

6 E. Vedejs and G. A. Krafft, *Tetrahedron*, 1982, **38**, 2857.

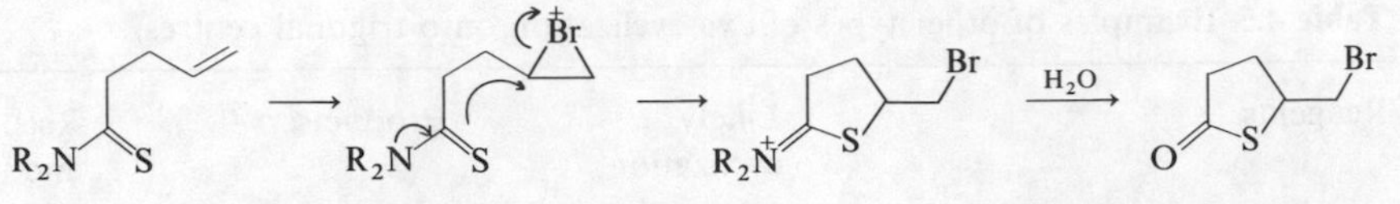

Fig. 4.9 Cyclization onto a carbon–carbon double bond promoted by an electrophile.

Many other electrophiles, including salts of mercury(II), nickel(II), and other metals[7] and electrophilic selenium reagents such as benzeneselenyl chloride,[8] bring about analogous reactions.

Palladium-catalysed cyclization reactions are particularly useful for the synthesis of five- and six-membered benzo-fused heterocycles.[9] Many of these cyclizations can be regarded as intramolecular Heck reactions: an example, a synthesis of 3-methylindole, is outlined in Fig. 4.10(a).[10] The catalyst in such reactions is a palladium(0) species, produced *in situ*, which inserts into the carbon–halogen bond. This organopalladium intermediate then adds to the double bond: this step and the probable subsequent steps in the reaction are shown in Fig. 4.10(b).

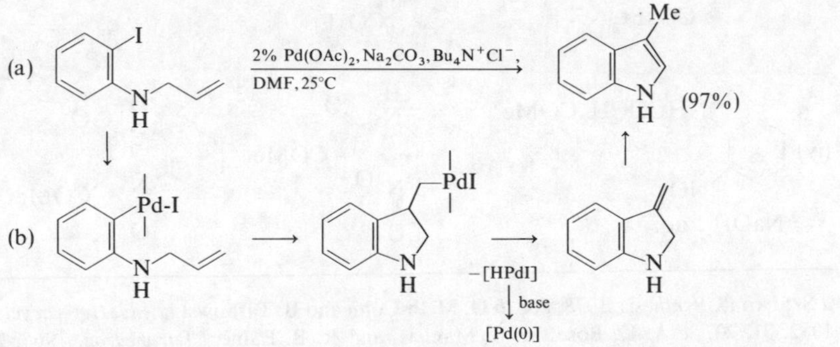

Fig. 4.10 (a) Palladium-catalysed cyclization of *N*-(2-iodophenyl)allylamine; (b) probable reaction sequence.

The majority of cyclization reactions onto trigonal centres are *exo* processes. There is one important group of reactions, useful for making five- and six-membered nitrogen heterocycles, which are *endo* cyclizations. These reactions involve the generation and capture by internal nucleophiles of iminium ions. Simple iminium ions require good nucleophiles to intercept them efficiently. A well-established example, shown in Fig. 4.11(a), is the

7 M. B. Gasc, A. Lattes, and J. J. Perie, *Tetrahedron*, 1983, **39**, 703.

8 C. Paulmier, *Selenium Reagents and Intermediates in Organic Synthesis*, Pergamon Press, Oxford, 1986, Chapter VIII.

9 Review: T. Sakamoto, Y. Kondo, and H. Yamanaka, *Heterocycles*, 1988, **27**, 2225.

10 L. S. Hegedus, G. F. Allen, J. J. Bozell, and E. L. Waterman, *J. Am. Chem. Soc.*, 1978, **100**, 5800.

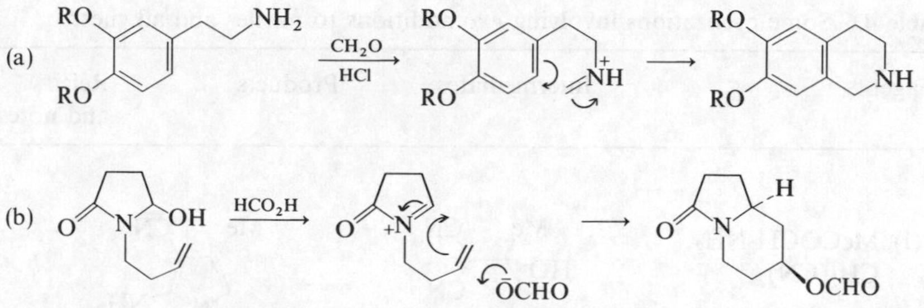

(a)

(b)

Fig. 4.11 (a) Pictet–Spengler isoquinoline synthesis; (b) an example of *N*-acyliminium ion cyclization.

Pictet–Spengler synthesis of isoquinolines. Here the nucleophile is an electron-rich benzene ring. *N*-Acyliminium ions are much more powerful electrophiles which can be generated in several different ways;[11] one common method, illustrated in Fig. 4.11(b), is the reduction of an imide followed by acid-catalysed dehydration. In the presence of a suitable internal nucleophile (here a carbon–carbon double bond) the ion is captured to produce a new ring system.

4.2.5 *Cyclization onto triple bonds*

Nucleophilic addition to cyano groups provides an important method of synthesis of *C*-amino-substituted heterocycles, as exemplified by reactions (i) to (iv) in Table 4.6.

In these reactions the initial product of cyclization is an imine, as shown in Fig. 4.12. Proton shifts then take place to convert this initial product into a more stable, aromatic, *C*-amino compound. If such proton shifts cannot occur the imino group is often hydrolysed to a carbonyl group during workup. The *exo* addition to carbon–carbon triple bonds is not so common, but it has been used to synthesize some five- and six-membered heterocycles, as shown in examples (v) and (vi). The reactive intermediate benzyne (1,2-didehydrobenzene) can be regarded as a cyclic acetylene, and intramolecular nucleophilic additions to arynes are useful for the synthesis of some benzo-fused heterocycles. An example of this type of cyclization is shown in (vii).

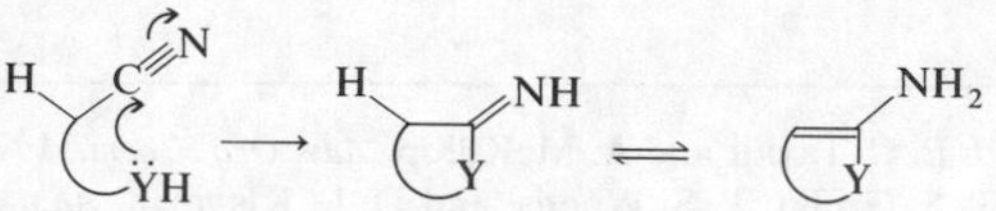

Fig. 4.12 *C*-Amino compounds by cyclization of nitriles.

11 Reviews: W. N. Speckamp and H. Hiemstra, *Tetrahedron*, 1985, **41**, 4367; T. A. Blumenkopf and L. E. Overman, *Chem. Rev.*, 1986, **86**, 857.

Table 4.6 Some cyclizations involving *exo* additions to nitriles and alkynes.[a]

Reagents	Intermediates	Products	Refs. and notes
(i) $MeCOCH_2NH_2$, $CH_2(CN)_2$	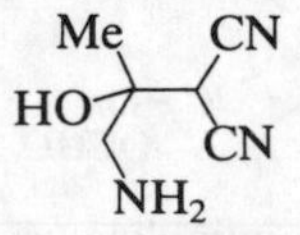 Me, CN; HO, CN; NH_2	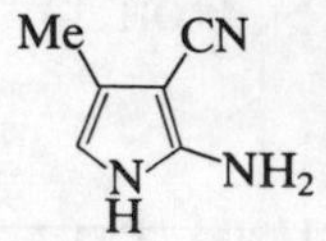 Me, CN; pyrrole (N—H) ring with NH_2	b
(ii) $(H_2N)_2CS$, $PhCHBrCN$	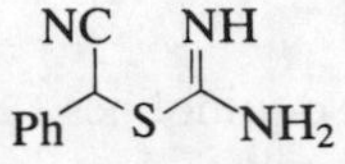 NC, NH; Ph—S, NH_2	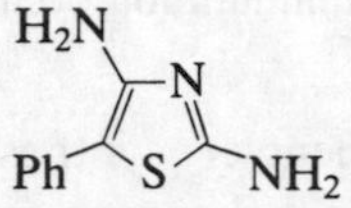 H_2N; thiazole ring; Ph—S, NH_2	c
(iii) $EtO_2CC{=}NH.OEt$, H_2NCH_2CN	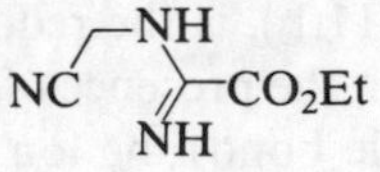 —NH; NC, CO_2Et; NH	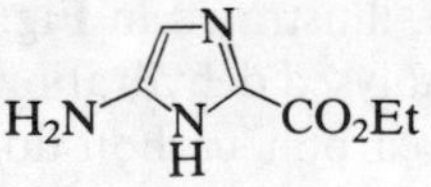 imidazole ring; H_2N, CO_2Et (N—H)	d
(iv) $H_2NCR{=}CRCN$, $H_2NCH{=}NH$	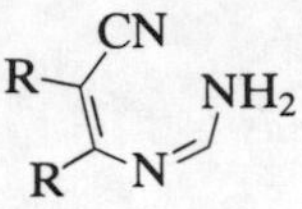 CN; R, NH_2; R, N	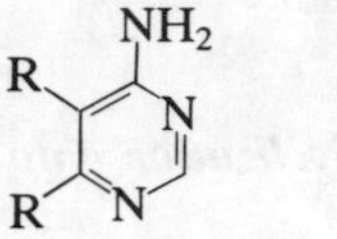 NH_2; R, pyrimidine ring, R	b
(v) $HC{\equiv}C(CH_2)_3OH$, $NaNH_2$	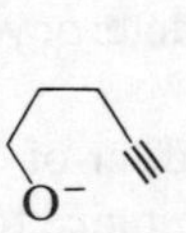 O^- with alkyne	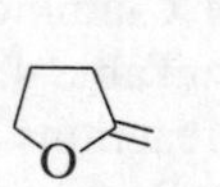 tetrahydrofuran ring with exocyclic $=CH_2$	e
(vi) $HC{\equiv}CCMeOH(CH_2)_3NEt_2$, base	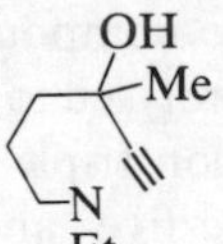 OH, Me; alkyne; N, Et_2	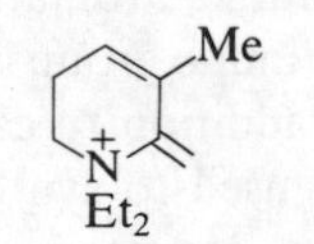 Me; ring N^+Et_2 with exocyclic methylene	f
(vii) 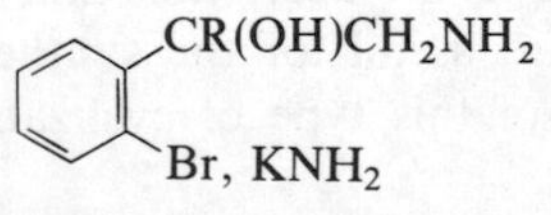 $CR(OH)CH_2NH_2$ benzene ring, Br, KNH_2	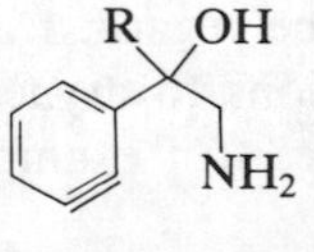 R, OH; benzene; NH_2	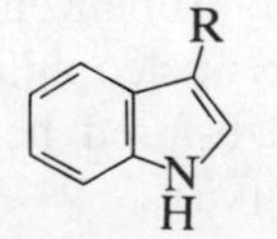 R; indole ring (N—H)	g

---

[a] All are *exo-dig* cyclizations. b E. C. Taylor and A. McKillop, *Adv. Org. Chem.*, 1970, **7**. c The Hantzsch thiazole synthesis; T. S. Griffin, T. S. Woods, and D. L. Klayman, *Adv. Heterocycl. Chem.*, 1975, **18**, 100. d A. McKillop, A. Henderson, P. S. Ray, C. Avendano, and E. G. Molinero, *Tetrahedron Lett.*, 1982, **23**, 3357. e R. Fuks and H. G. Viehe, in *Chemistry of Acetylenes*, ed. H. G. Viehe, Marcel Dekker, New York, 1969, p. 425. f M. Miocque, M. Duchon-d'Engenières, and J. Sauzières, *Bull. Soc. Chim. Fr.*, 1975, *1777*. g I. Fleming and M. Woolias, *J. Chem. Soc., Perkin Trans. 1*, 1979, 827.

There are also a significant number of examples of heterocyclic syntheses which involve *endo* cyclization onto a triple bond. Although such reactions appear to be sterically unfavourable because of the linear nature of the triple bond, it is easy to distort the triple bond to achieve the required transition-state geometry. Some examples of cyclization which can be rationalized as *endo* additions to carbon–carbon triple bonds are shown in Table 4.7.

Table 4.7 Examples of ring formation by *endo* attack on carbon–carbon triple bonds.[a]

Reagents	Intermediates	Products	Refs.
(i) $R^1COC{\equiv}CR^2$, NH_2NH_2	R^1 group with N, ${\equiv}{-}R^2$, NH_2	R^2, R^1 pyrazole with $N{-}N{-}H$	b
(ii) $RC{\equiv}CC{\equiv}CR$, H_2S, $Ba(OH)_2$	R, ${\equiv}{-}R$, S^-	R thiophene R (with S)	c
(iii) $R^1C{\equiv}CCO_2R^2$, H_2NOH	$R^1{-}{\equiv}{-}$, O, NH, HO	R^1, O, $O{-}NH$ ring	d

a All are 5-*endo-dig* cyclizations. *b* A. N. Kost and I. I. Grandberg, *Adv. Heterocycl. Chem.*, 1966, **6**, 347. *c* A. Ulman and J. Manassen, *J. Am. Chem. Soc.*, 1975, **97**, 6540. *d* I. Iwai and N. Nakamura, *Chem. Pharm. Bull.*, 1966, **14**, 1277.

Isonitriles also undergo *endo* cyclization reactions readily, and these reactions provide useful methods for the preparation of several five-membered heterocycles containing nitrogen. The isonitrile cyclizations often give heterocycles with substitution patterns which are not easily available by other methods of ring synthesis. The most common reaction sequence using isonitriles is shown in Fig. 4.13. A simple isonitrile XCH_2NC is deprotonated by a base, and the anion is then made to react with an unsaturated electrophile. The intermediate so generated can cyclize in a 5-*endo-dig* process to give the heterocycle, which is unsubstituted at the 2-position.

Examples of these and a related isonitrile cyclization are given in Table 4.8. Tosylmethyl isocyanide ('TOSMIC') has found the widest use because of the mild conditions required for its reaction and because the tosyl substituent is often lost in an aromatization step, after cyclization.

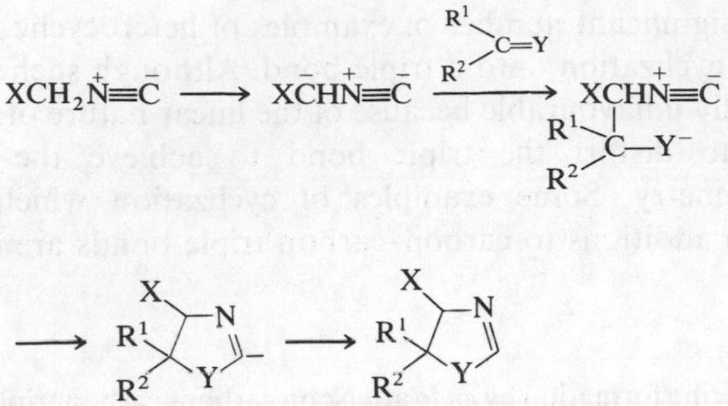

Fig. 4.13 Construction of five-membered heterocycles through isonitriles.

4.2.6   *Radical cyclization*[12]

The intramolecular addition of a radical to a π bond leads to the formation of a new ring system. Most of the ring systems produced by radical cyclization are five- or six-membered, and either partly or fully saturated. The method usually leads to the formation of heterocycles by a process in which a carbon-centred radical becomes bonded to the carbon atom of a π bond. This π bond may be a carbon–carbon double or triple bond, or it may be part of an aromatic ring; there are also a few examples of cyclization onto π bonds containing heteroatoms. A heterocycle is formed if there is a heteroatom present in the linking chain. Less commonly one of the atoms forming the new bond, usually at the radical centre, is a heteroatom. Unless the radicals are highly stabilized the intramolecular addition step is irreversible. Such reactions are thus kinetically controlled. Five- and six-membered rings are most commonly formed by preferential *exo* cyclization.

The final products isolated from these cyclizations depend on the method used to generate the radicals. One of the most common methods of carrying out these reactions is illustrated by the example shown in Fig. 4.14. The reaction is a reductive cyclization brought about by tributyltin hydride. A radical initiator, here azobis(isobutyronitrile), decomposes to produce radicals (step 1) which abstract a hydrogen atom from tributyltin hydride, breaking the weak tin–hydrogen bond (step 2). The tributyltin radical so formed abstracts bromine from the substrate (step 3). The carbon radical then cyclizes to produce a new alkyl radical (step 4) which abstracts hydrogen from tributyltin hydride (step 5). Steps 3–5 are then continued, as a radical chain reaction.

Two more examples of cyclization using tributyltin hydride are shown in Table 4.9. In example (i) an iminyl radical is generated by the cleavage of an N—S bond. Example (ii) illustrates the great power of this method in

12 Reviews: D. P. Curran, *Synthesis*, 1988, 417, 489; B. Giese, *Radicals in Organic Synthesis*, Pergamon Press, Oxford, 1986, Chapter 4.

Table 4.8 Cyclizations of isonitriles.[a]

Reagents	Intermediates	Products	Refs. and notes
(i) TsCH₂NC,K₂CO₃, RCHO	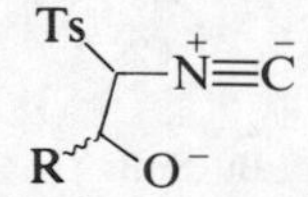	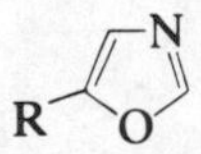	b
(ii) TsCH₂NC, K₂CO₃, R¹CH=NR²	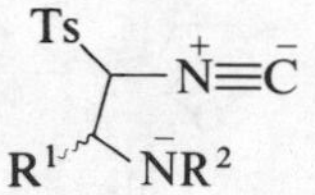	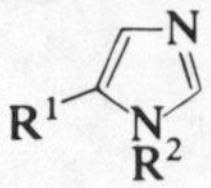	b
(iii) TsCH₂NC, NaH, R¹CH=CHCOR²	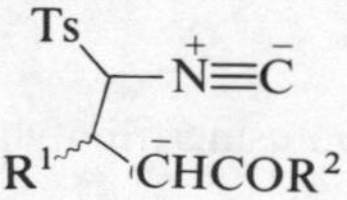	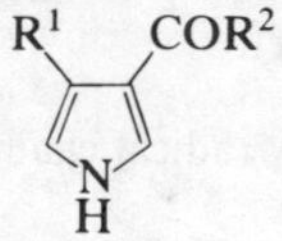	b
(iv) TsCH₂NC, R₄NOH, CS₂	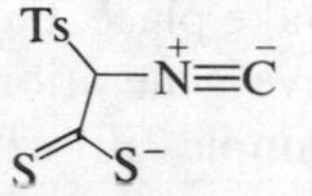	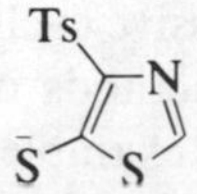	b
(v) R¹O₂CCH₂NC, NaH, R²COCl	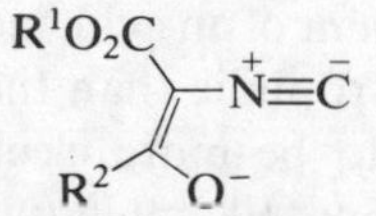	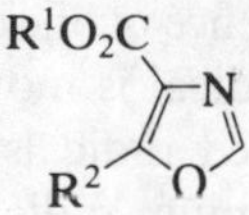	c, d
(vi) MeNC, BuLi, PhCN	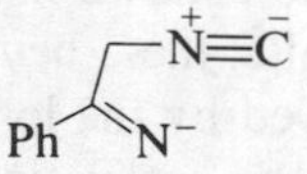	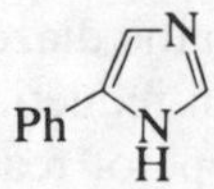	c
(vii) PhCH₂NC, BuLi, PhNCS	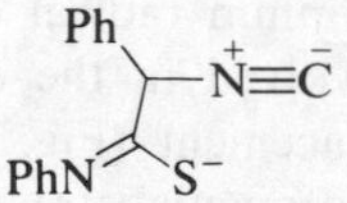	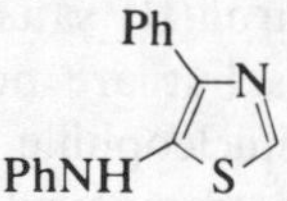	c
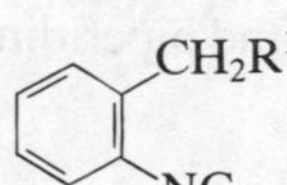 (viii) ...CH₂R¹ ...NC, LiNR²₂	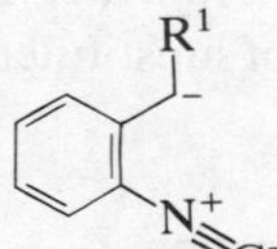	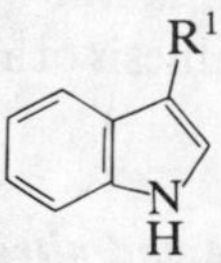	d, e

a All can be regarded as 5-*endo-dig* cyclizations. *b* A. M. van Leusen, *Lect. Heterocycl. Chem.*, 1980, **5**, S111; H. M. Walborsky and M. P. Periasamy, in *The Chemistry of Functional Groups*, suppl. C, eds S. Patai and Z. Rappoport, Wiley-Interscience, 1983, p. 835. *c* D. Hoppe, *Angew. Chem. Int. Edn Engl.*, 1974, **13**, 789; U. Schöllkopf, *Angew. Chem. Int. Edn Engl.*, 1977, **16**, 339. *d* This process can alternatively be regarded as 1,5-anionic electrocyclization; see Section 4.2.8. *e* Y. Ito, K. Kobayashi, and T. Saegusa, *J. Am. Chem. Soc.*, 1977, **99**, 3532.

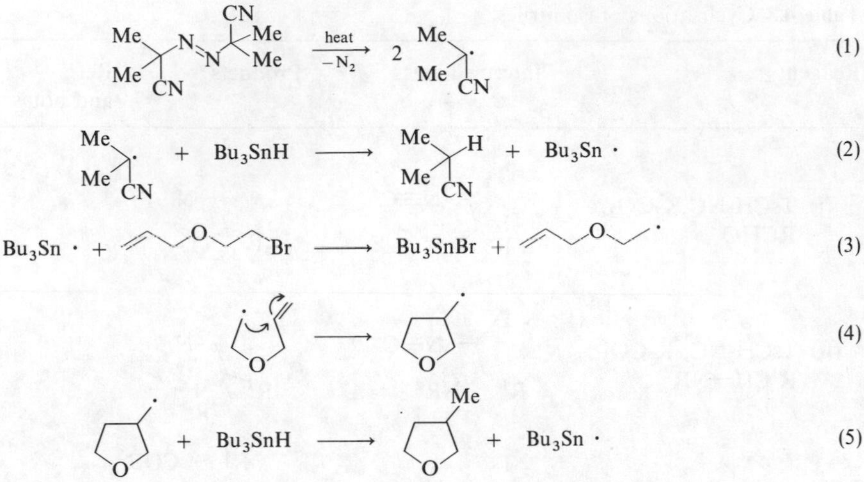

(1)

(2)

(3)

(4)

(5)

Fig. 4.14 A radical chain cyclization using tributyltin hydride.

that two successive cyclizations take place, the product being formed in high yield. Other methods of reductive generation of radicals are illustrated in the remaining examples. In example (iii) samarium iodide is acting as a one-electron reducing agent. This cyclization gives better yields if carried out in the presence of one equivalent of an acid, indicating that the protonated aminoalkyl radical is more electrophilic than the neutral species. Similarly, nitrogen-centred radicals tend to be more electrophilic when protonated; that is, as aminium radical cations. Example (iv) shows the cyclization of a radical of this type. The cyclization shown in example (v) is typical of many based on aromatic diazonium salts, these being converted into aryl radicals by one-electron reduction followed by the loss of nitrogen.

A related type of reaction is the cyclization of saturated N-chloroalkyl-amines, which is carried out in strongly acidic media by photolysis or by the action of iron(II) salts. Aminium radical cations are intermediates in these reactions but are not involved in the cyclization step, which is a conventional nucleophilic displacement (Fig. 4.15). The key step in the reaction is an intramolecular hydrogen abstraction by the aminium radical. The cyclization is known as the *Hofmann–Löffler reaction*[13] and has been used for the synthesis of a variety of substituted pyrrolidines and piperidines.

4.2.7   *Carbene and nitrene cyclization*

Monovalent nitrogen intermediates (nitrenes) and divalent carbon inter-mediates (carbenes) are highly reactive species which can undergo addition

13 Reviews: P. Kovacic, M. K. Lowery, and K. W. Field, *Chem. Rev.*, 1970, **70**, 639; M. E. Wolff, *Chem. Rev.*, 1963, **63**, 55.

Table 4.9 Examples of radical cyclization.

	Reagents	Intermediates	Products	Refs.
(i)	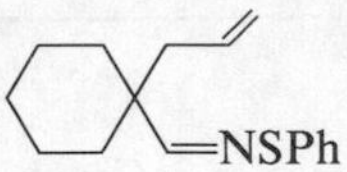 , Bu_3SnH	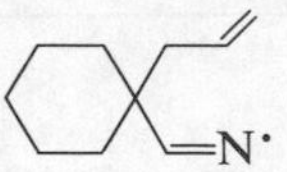	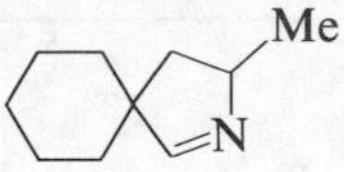	*a*
(ii)	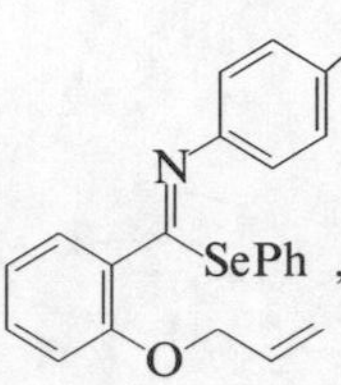 , Bu_3SnH	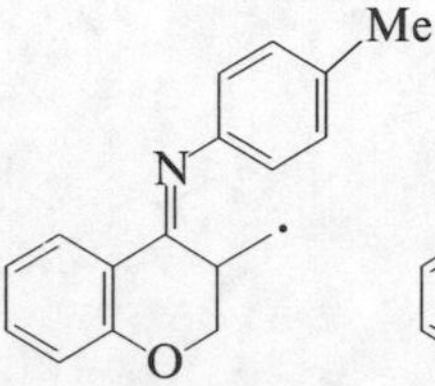	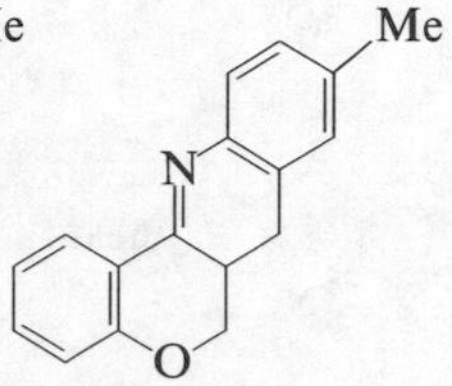	*b*
(iii)	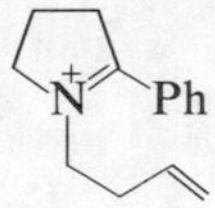 ClO_4^- , SmI_2, H^+	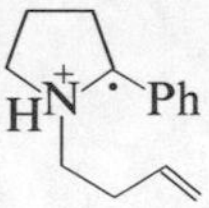	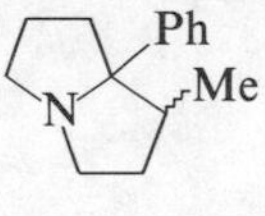	*c*
(iv)	$Ph(CH_2)_3NMeCl$, H_2SO_4, $MeCO_2H$, Fe^{2+}	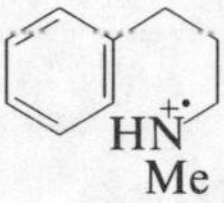	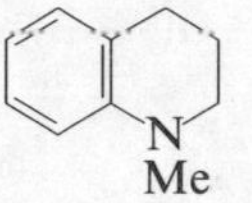	*d*
(v)	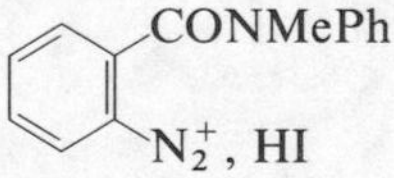 , HI	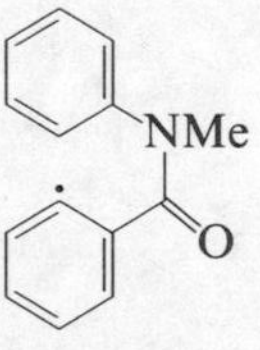	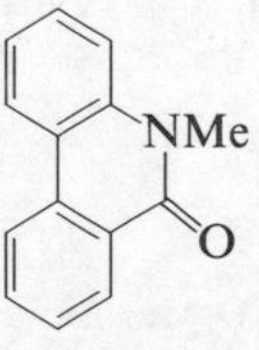	*e*

a J. Boivin, E. Fouquet, and S. Z. Zard, *Tetrahedron Lett.*, 1990, **31,** 85. *b* M. D. Bachi and D. Denemark, *J. Am. Chem. Soc.*, 1989, **111,** 1886. *c* S. F. Martin, C-P. Yang, W. L. Laswell, and H. Rüeger, *Tetrahedron Lett.*, 1988, **29,** 6685. *d* Review: L. Stella, *Angew. Chem. Int. Edn Engl.*, 1983, **22,** 337. *e* D. H. Hey, G. H. Jones, and M. J. Perkins, *J. Chem. Soc., Perkin Trans. 1*, 1972, 113.

$$Me(CH_2)_3\overset{+}{N}HRCl \longrightarrow Me(CH_2)_3\overset{+\cdot}{N}HR + Cl^\cdot$$
$$Me(CH_2)_3\overset{+\cdot}{N}HR \longrightarrow \dot{C}H_2(CH_2)_3\overset{+}{N}H_2R$$
$$\dot{C}H_2(CH_2)_3\overset{+}{N}H_2R + Me(CH_2)_3\overset{+}{N}HRCl \longrightarrow Cl(CH_2)_4\overset{+}{N}H_2R + Me(CH_2)_3\overset{+\cdot}{N}HR$$
$$Cl(CH_2)_4\overset{+}{N}H_2R \xrightarrow{\text{base}}$$
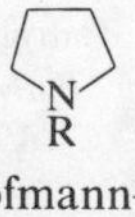

Fig 4.15 Formation of pyrrolidines by the Hofmann–Löffler reaction.

Table 4.10 Examples of the formation of heterocycles by carbene and nitrene cyclization.

Reagents	Intermediates	Products	Refs.
(i)			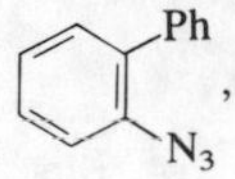 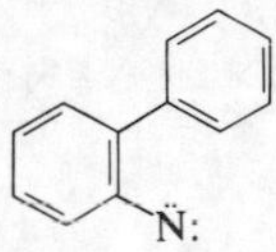 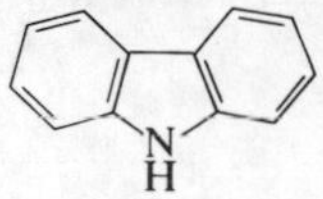 a
(ii)			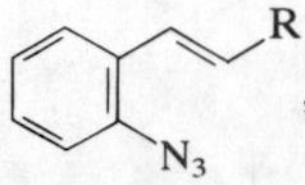 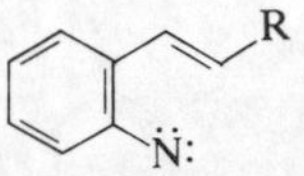 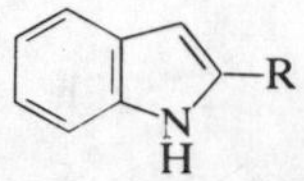 a, b
(iii)			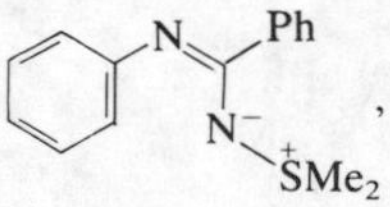 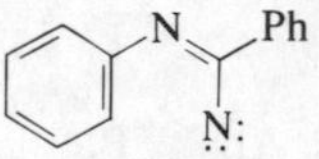 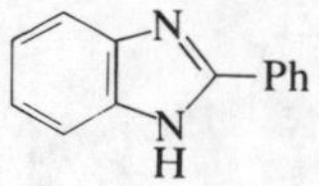 c
(iv)			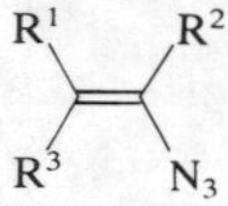 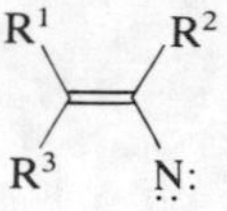 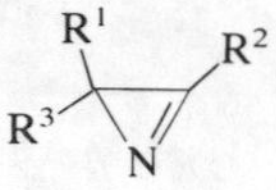 a, d
(v)			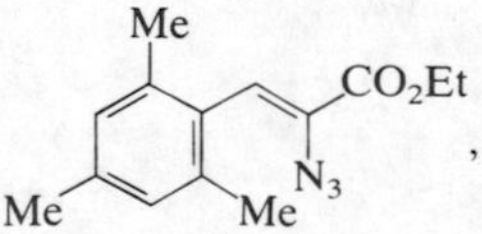 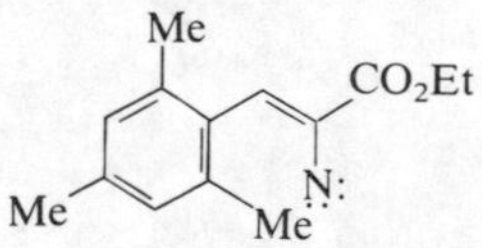 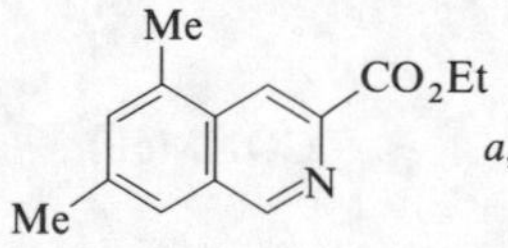 a, e
(vi)			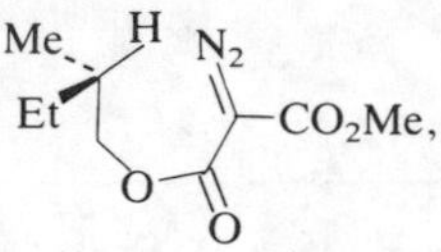 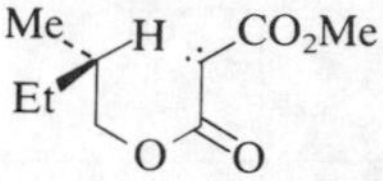 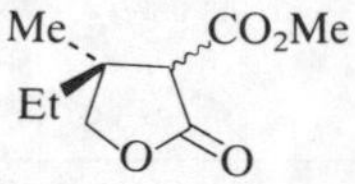 f
(vii)			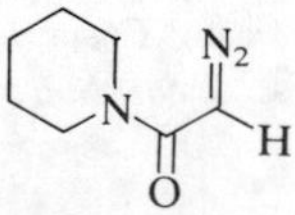 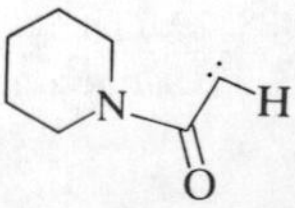 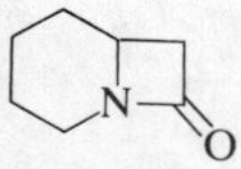 g

a E. F. V. Scriven and K. Turnbull, *Chem. Rev.*, 1988, **88**, 297. b R. J. Sundberg, *The Chemistry of Indoles*, Academic Press, New York, 1970, Chapter 3. c T. L. Gilchrist, C. J. Moody, and C. W. Rees, *J. Chem. Soc., Perkin Trans. 1*, 1975, 1964. d V. Nair, in *Small Ring Heterocycles*, Part 1, ed. A. Hassner, Wiley-Interscience, New York, 1983, p. 215. e T. L. Gilchrist, C. W. Rees, and J. A. R. Rodrigues, *J. Chem. Soc., Chem. Commun*, 1979, 627. f H. Ledon, G. Instrumelle, and S. Julia, *Bull. Soc. Chim. Fr.*, 1973, 2071. g D. M. Brunwin, G. Lowe, and J. Parker, *J. Chem. Soc. (C)*, 1971, 3756; see also R. R. Rando, *J. Am. Chem. Soc.*, 1972, **94**, 1629.

reactions with multiple bonds and can insert into unactivated carbon–hydrogen bonds. Some examples of intramolecular versions of these reactions, leading to heterocycles, are shown in Table 4.10. The thermal or photochemical decomposition of 2-azidobiphenyl [reaction (i)] is an important route to carbazole. It has been shown to go by way of the singlet (spin-paired) nitrene, which cyclizes onto the *ortho* position of the attached phenyl substituent to give an intermediate, **2**. This can then aromatize to give carbazole by a hydrogen shift to nitrogen. It is reasonable to assume that similar modes of cyclization are involved in related processes such as those in examples (ii) and (iii).

2

The photodecomposition of vinyl azides to give azirines [example (iv)] can be regarded as an intramolecular nitrene addition to a double bond. The ability of singlet carbenes and nitrenes to insert into unactivated CH bonds is a valuable characteristic of the intermediates: three such processes, which lead to the formation of heterocycles, are shown in examples (v) to (vii).

4.2.8 *Electrocyclic reactions*[14]

The cyclization reactions that we have considered so far are all intramolecular versions of well-known σ-bond-forming processes. Electrocyclic reactions are different, in that they have no direct intermolecular counterpart. The open-chain reagent used in an electrocyclic ring closure must be a fully conjugated π-electron system. Electrocyclic ring closure is the reaction in which a σ-bond is formed at the termini of the π system. The reactions are normally brought about by the input of energy in the form of heat or light, and without any additional reagents. An equilibrium is set up between the acyclic and cyclic isomers. In many cases the acyclic isomer predominates, so that the electrocyclic reaction may be a ring opening rather than a ring formation.

The four most important types of electrocyclic reaction found in heterocyclic chemistry are illustrated schematically in Fig. 4.16. Reactions (a) and (b) involve the use of open-chain reagents containing four π-electrons, in a 1,3-dipolar species (a) or in a heterodiene (b). Reactions (c) and (d) are the six-π-electron analogues of (a) and (b). The open-chain species can thus be precursors of saturated or partially saturated heterocycles containing from three to six atoms. Higher-order electrocyclic reactions of systems with more than six π-electrons are also feasible, but they are not so commonly encountered.

14 Review: E. N. Marvell, *Thermal Electrocyclic Reactions*, Academic Press, New York, 1980.

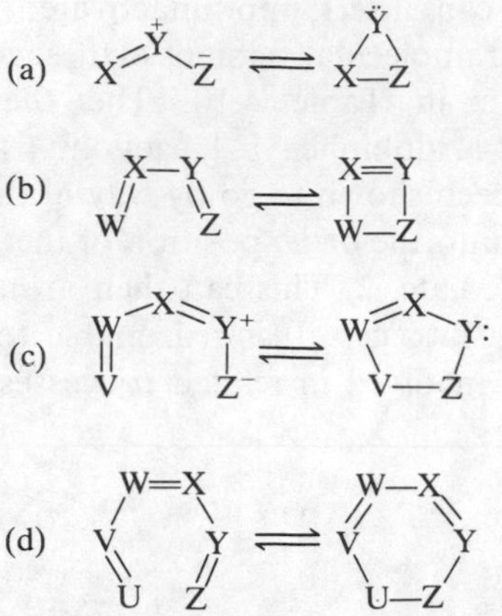

(a) $X\overset{\overset{+}{Y}}{=}Z \rightleftharpoons X\overset{\ddot{Y}}{-}Z$

(b) $\overset{X-Y}{\underset{W\quad Z}{\diagup\quad\diagdown}} \rightleftharpoons \overset{X=Y}{\underset{W-Z}{|\quad|}}$

(c) $\overset{W}{\underset{V}{\parallel}}\overset{X}{\underset{Z^-}{=}}Y^+ \rightleftharpoons \overset{W}{=}\overset{X}{\underset{V-Z}{\diagdown}}Y:$

(d) $\overset{W=X}{\underset{U\quad Z}{\diagup\qquad\diagdown}}\overset{}{\underset{}{Y}} \rightleftharpoons \overset{W-X}{\underset{U-Z}{\diagup\qquad\diagdown}}\overset{}{\underset{}{Y}}$

Fig. 4.16 Electrocyclic reactions involving open-chain isomers containing four or six π-electrons.

The rationalization of the stereochemistry of electrocyclic ring closures and ring openings was the first achievement of the principle of orbital symmetry conservation developed by Woodward and Hoffmann. Two types of ring closure can be recognized, one in which the p-orbitals of the π-electron system rotate in the same direction to form the new σ-bond, and the other in which they rotate in opposite directions. The first is called *conrotatory* ring closure [Fig. 4.17(a)] and the second, *disrotatory* ring closure [Fig. 4.17(b)]. The Woodward–Hoffmann rules state which type of ring closure is preferred, and they predict the stereochemistry of the product. These rules are determined by the numbers of π-electrons in the open-chain species and by whether the reaction takes place in the ground state (thermal reactions) or in the first excited state of the polyene (photochemical reactions). The rules are shown under Fig. 4.17.

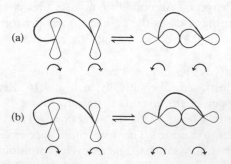

Fig. 4.17 Modes of electrocyclic ring closure. The conrotatory mode (a) is favoured for thermal reactions of 4π-electron systems and for photochemical reactions of 6π-electron systems. The disrotatory mode (b) is favoured for photochemical reactions of 4π-electron systems and for thermal reactions of 6π-electron systems.

In principle, any electrocyclic reaction can be brought about either thermally or photochemically and only the stereochemistry of the process will be different. In practice, the position of equilibrium is also a crucial factor since it determines whether or not the cyclic isomer can be isolated in good yield from the reaction, and this can be very different in thermal and photochemical processes. The stereochemical distinction between conrotatory and disrotatory processes is lost when one of the terminal atoms of the open π system is a heteroatom, so in many electrocyclic reactions leading to the preparation of heterocycles it is not possible to verify the predictions of the Woodward–Hoffmann rules.

The four-electron electrocyclic reactions (a) and (b) of Fig. 4.16 undoubtedly take place with a variety of substrates containing heteroatoms. With three-membered heterocycles of types **3** and **4**, ring opening is the more common process and can be brought about by the action of heat or light; ring closure is restricted to fewer examples and is usually a light-induced process.

$$\overset{\overset{\cdot\cdot}{Y}}{\underset{X-Z}{\triangle}} \rightleftharpoons X{\equiv}\overset{+}{Y}{-}\bar{Z}$$

3

X	Y	Z	
C	N	C	2*H*-azirines → nitrile
N	N	C	3*H*-diazirines ⇌ diazo compounds

$$\overset{\overset{\cdot\cdot}{Y}}{\underset{X-Z}{\triangle}} \rightleftharpoons \overset{\overset{+}{Y}}{X\diagdown Z}$$

4

X	Y	Z	
C	N	C	aziridines ⇌ azomethine ylides
C	O	C	oxiranes ⇌ carbonyl ylides
C	S	C	thiiranes ← thiocarbonyl ylides
C	N	N	diaziridines ⇌ azomethine imides
C	N	O	oxaziridines ⇌ nitrones

Three examples of these reactions are shown in Fig. 4.18. The ring opening of 3-aryl-2*H*-azirines to nitrile ylides on irradiation [example (a)] is an important method of generation of these 1,3-dipoles.[15] Some arylnitrones undergo ring closure to oxaziridines on irradiation [example (b)][16] and thiocarbonyl ylides, derived from the pyrolysis of thiadiazolines, are cyclized to thiiranes [example (c)].[17] Further examples of these reactions are given in Chapter 9.

15 Review: A. Padwa and P. H. J. Carlsen, in *Reactive Intermediates*, Vol. 2, ed. R. A. Abramovitch, Plenum, New York, 1982, p. 55.
16 Review: E. Schmitz, *Adv. Heterocycl. Chem.*, 1979, **24**, 63.
17 R. M. Kellog, M. Noteboom, and J. K. Kaiser, *J. Org. Chem.*, 1975, **40**, 2573.

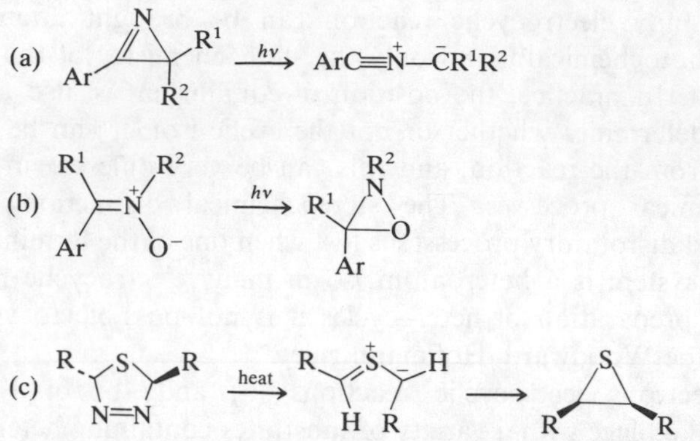

Fig. 4.18 Ring-opening and cyclization reactions involving three-membered heterocycles.

Six-electron cyclizations of type (c) in Fig. 4.16 are much more common and have been given the general description of 1,5-*dipolar cyclizations*.[18] Again, both ring-opening and ring-closure processes can be observed. Unstable 1,5-dipoles, generated *in situ*, cyclize thermally to the five-membered heterocycles. The cyclic isomers can also be removed from the equilibria by irreversible tautomerization to a more stable (often aromatic) structure. Examples of 1,5-dipolar cyclization are shown in Table 4.11. The first example shows the cyclization of an unstable 1,5-dipole. In examples (ii) and (iii) the primary cyclization products tautomerize to aromatic systems and so displace the equilibria in favour of the cyclic forms. Aromatic heterocycles are formed directly in cyclizations of the types shown in examples (iv) and (v).

The electrocyclic ring closure of heterotrienes [reaction (d) in Fig. 4.16] can be used as a method of preparing some six-membered heterocycles, particularly when the initial cyclization product can aromatize. As with the other electrocyclic processes, the reverse (ring-opening) reactions also occur and are sometimes more useful from a preparative point of view. Examples are the reversible ring opening of 2*H*-pyrans (**5**, X=O) and of 1,2-dihydropyridines (**5**, X=NR) in which the cyclic tautomers can be used to prepare the open-chain tautomers.[14] The preparation and ring opening of 1,2-dihydropyridines are discussed further in Chapter 5.

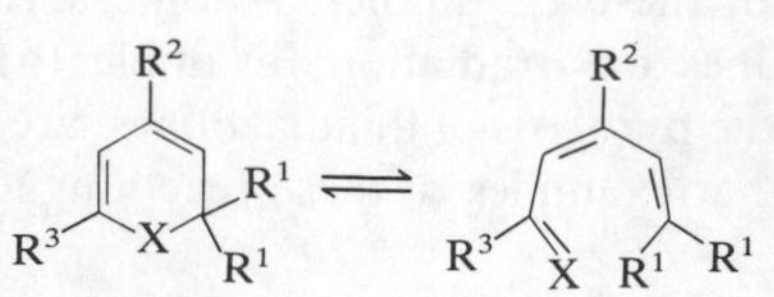

5

18 Reviews: E. C. Taylor and I. J. Turchi, *Chem. Rev.*, 1979, **79**, 181; R. Huisgen, *Angew. Chem. Int. Edn Engl.*, 1980, **19**, 947.

Table 4.11 1,5-Dipolar cyclizations.[a]

Precursors	1,5-Dipolar intermediates	Cyclization products	Final products (if different)
(i) , heat			—
(ii) b			
(iii) hv			
(iv) hv			—
(v) , HONO			—

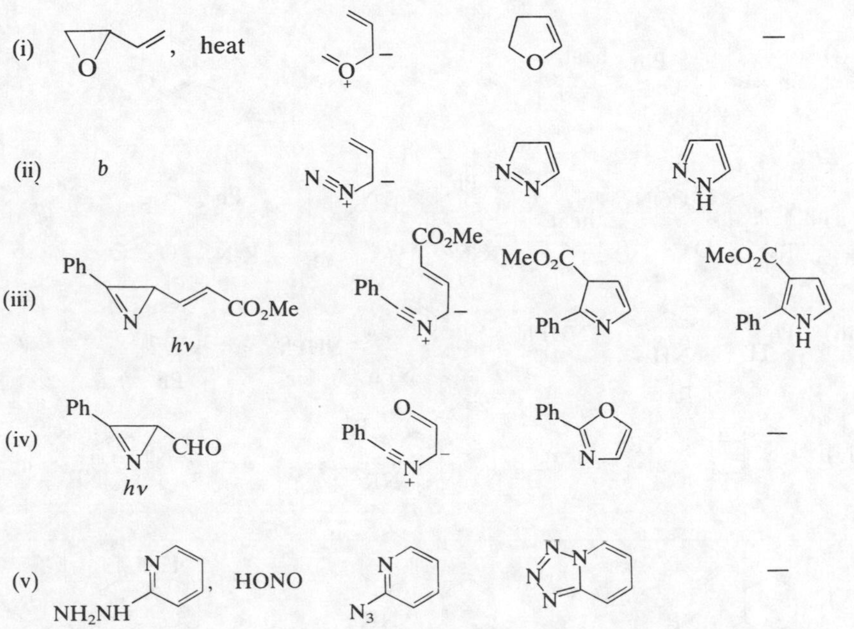

a All cyclization steps are thermal processes. See ref. 18 for primary literature references.
b Vinyldiazomethane is isolable but cyclizes above 25°C.

Some examples of the preparation of six-membered heterocycles by electrocyclic ring-closure reactions are given in Table 4.12. The isoquinoline synthesis shown in example (iv) involves two electrocyclic reactions: the opening of the benzocyclobutane to an *ortho*-xylylene intermediate, and the 6π-electron recyclization of the intermediate which is highly favourable because of the rearomatization of the benzenoid part of the molecule.

A different type of electrocyclization of heterotrienes, leading to the formation of five-membered rings, sometimes takes precedence over the usual type of ring closure. The general form of this reaction is shown in Fig. 4.19, together with two examples.[19] The reaction appears to be favoured by the presence of electronegative heteroatoms at the terminal positions of the heterotriene.

19 Review: M. V. George, A. Mitra, and K. B. Sukumaran, *Angew. Chem. Int. Edn Engl.*, 1980, **19**, 973.

Table 4.12 Formation of six-membered heterocycles by electrocyclic ring closure.

Precursors	Intermediates	Products	Refs. and notes
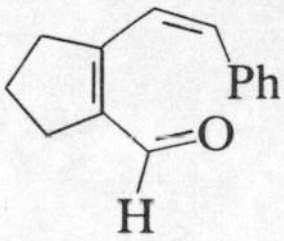 (i) Ph, heat	—	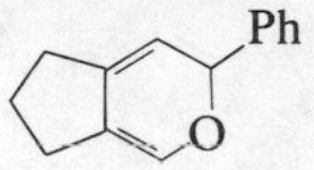	a
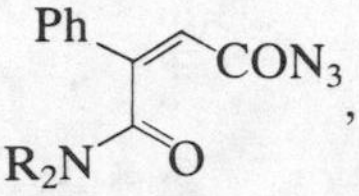 (ii) , heat	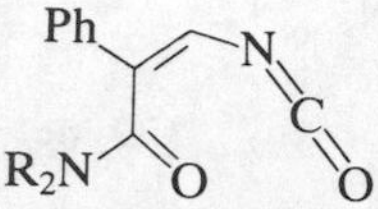	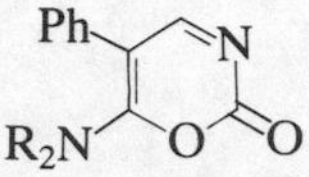	a
(iii)			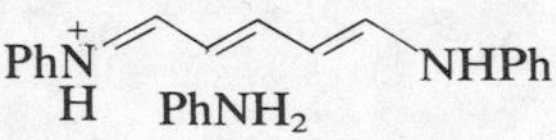 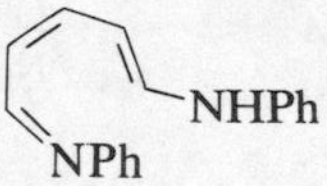  a
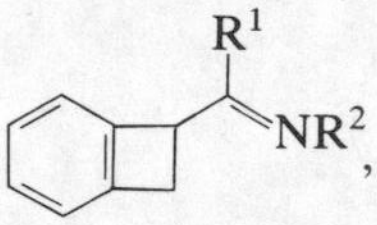 (iv) , heat	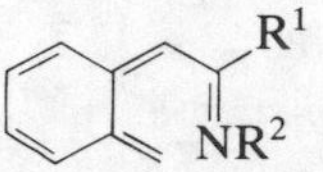	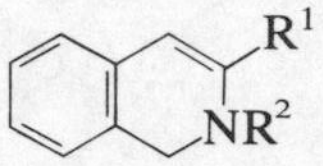	a
(v) , $h\nu$			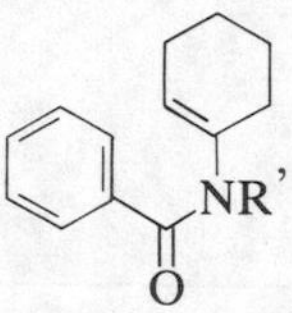 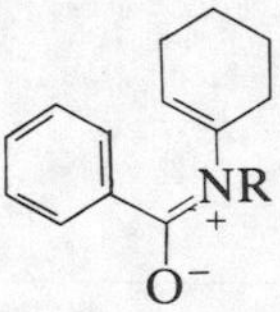 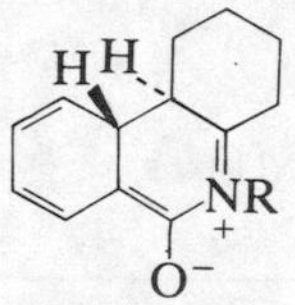 b

a See ref. 14 for primary literature references. *b* The photocyclization product shown is transformed into isolable products by rearrangement or by *in situ* reduction: I. Ninomiya, T. Kiguchi, S. Yamauchi, and T. Naito, *J. Chem. Soc., Perkin Trans. 1*, 1980, 197.

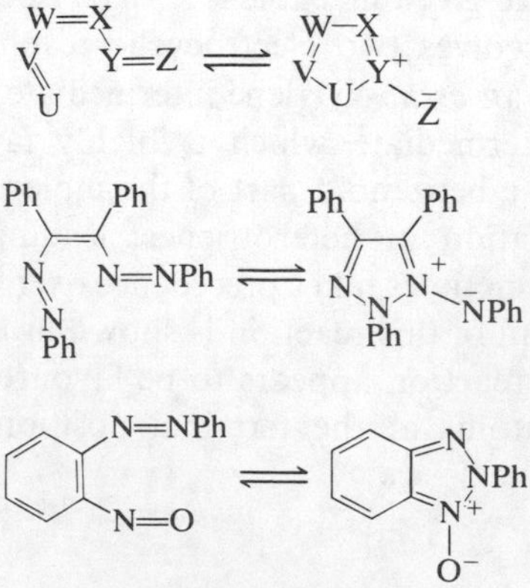

Fig. 4.19 Alternative mode of cyclization of heterotrienes.

4.3 Cycloaddition reactions

4.3.1 Reaction types

Most of the classical methods of heterocyclic synthesis are cyclization reactions. Methods of synthesis based on cycloaddition reactions have become increasingly important in recent years, however. Cycloaddition reactions can provide routes to heterocycles with well-defined substitution patterns, in many cases with great stereochemical control. The growth in the use of these reactions in heterocyclic synthesis has been stimulated by new theories of their mechanisms. The Woodward–Hoffmann theories of orbital symmetry conservation provided a basis for understanding the mechanisms of the various classes of cycloaddition, and the application of frontier orbital theory provided an explanation for the effects of substituents on the rates and selectivities of the reactions.[20]

Cycloaddition reactions provide useful synthetic routes to a wide range of heterocycles, especially those containing four, five, or six atoms in the ring. The most important types of reaction are (a) 1,3-dipolar cycloadditions, (b) hetero-Diels–Alder reactions, (c) [2 + 2] cycloadditions, and (d) cheletropic reactions. These are shown in outline in Fig. 4.20; reactions (d) and (e) are both cheletropic processes, involving four- and six-electron transition states, respectively.

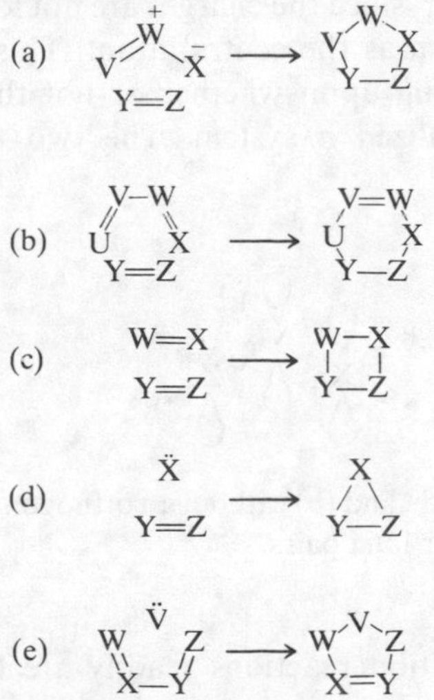

Fig. 4.20 The major types of cycloaddition process used in heterocyclic synthesis.

20. I. Fleming, *Frontier Orbitals and Organic Chemical reactions*, Wiley-Interscience, London, 1976.

1,3-Dipolar cycloaddition provides an excellent method of constructing five-membered rings because a wide variety of 1,3-dipoles are available which undergo addition to carbon–carbon multiple bonds and to multiple bonds containing heteroatoms. The Diels–Alder reaction is best known as a method of forming cyclohexane derivatives, but heteroatoms can be incorporated successfully into the skeleton of the diene or of the dienophile to provide important routes to several six-membered heterocycles. The third group, [2 + 2] cycloaddition, is a common method of synthesis of four-membered heterocycles when at least one reaction partner is a cumulene (for example, a ketene or an isocyanate). In cheletropic reactions one of the partners reacts through an atom which can both donate and accept a pair of electrons to form two new σ-bonds. This class is not extensive but it includes such reactions as nitrene additions to olefins and the addition of sulphur dioxide to dienes.

4.3.2   1,3-Dipolar cycloaddition

A 1,3-dipole is a three-atom π-electron system with four π-electrons delocalized over the three atoms. The name '1,3-dipole' is derived from the fact that it is impossible to write electron-paired resonance structures for these species without incorporating charges; this does not, however, imply that the compounds are particularly polar, since the charges are not localized. 1,3-Dipolar species contain a heteroatom as the central atom. This can be formally sp- or sp²-hybridized, depending upon whether or not there is a double bond orthogonal to the delocalized π system. The two types of 1,3-dipole are shown in Fig. 4.21.

Fig. 4.21 Types of 1,3-dipole: (a) with, and (b) without an orthogonal double bond. R^1–R^5 can be substituents or lone pairs.

1,3-Dipoles which undergo cycloaddition reactions readily are listed in Table 4.13. The first six species listed are dipoles of type (a), which formally have a central sp-hybridized atom, but the species are easily bent to permit cycloaddition reactions at the termini.[21] In addition to those shown in the table, there are several other 1,3-dipoles which can participate in cyclo-addition reactions. Ozone is a good partner in such reactions but the cycloadducts (primary ozonides) are very unstable.

21 P. Caramella and K. N. Houk, *J. Am. Chem. Soc.*, 1976, **98**, 6397.

Table 4.13 1,3-Dipoles useful in cycloaddition reactions.[a]

$X\equiv\overset{+}{Y}-\overset{-}{Z}$		$\overset{\overset{\displaystyle\mid}{}}{\underset{X\diagdown\quad\diagup Z^-}{Y^+}}$	
$N\equiv\overset{+}{N}-\overset{-}{N}-$	azides	nitrones	
$N\equiv\overset{+}{N}-\overset{-}{C}\diagup$	diazo compounds	azomethine imides	
$-C\equiv\overset{+}{N}-\overset{-}{O}$	nitrile oxides	azomethine ylides	
$-C\equiv\overset{+}{N}-\overset{-}{N}-$	nitrile imides	carbonyl ylides	
$-C\equiv\overset{+}{N}-\overset{-}{S}$	nitrile sulphides	thiocarbonyl ylides	
$-C\equiv\overset{+}{N}-\overset{-}{C}\diagup$	nitrile ylides		

a Comprehensive review: *1,3-Dipolar Cycloaddition Chemistry*, Vols. 1 and 2, ed. A. Padwa, Wiley-Interscience, New York, 1984. See also R. Huisgen, C. Fulka, I. Kalwinsch, X. Li, G. Mloston, J. R. Moran, and A. Proebstl, *Bull. Soc. Chim. Belg.*, 1984, **93**, 511 (thiocarbonyl ylides); K. G. B. Torssell, *Nitrile Oxides, Nitrones, and Nitronates in Organic Synthesis*, VCH, Weinheim, 1988; P. N. Confalone and E. M. Huie, *Org. React.*, 1988, **36**, 1 (nitrones); E. Vedejs, in *Advances in Cycloaddition*, Vol. 1, ed. D. P. Curran, JAI Press, Greenwich, Conn., 1988, p. 33 (azomethine ylides); O. Tsuge and S. Kanemasa, *Adv. Heterocycl. Chem.*, 1989, **45**, 232 (azomethine ylides); R. M. Paton, *Chem. Soc. Rev.*, 1989, **18**, 33 (nitrile sulphides); ref. *a* in Table 4.10 (azides); A. Padwa, *Acc. Chem. Res.*, 1991, **24**, 22 (carbonyl ylides).

Compounds which can react with these species in cycloaddition reactions are commonly called dipolarophiles. These contain unsaturated functional groups such as $C\equiv C$, $C=C$, $C\equiv N$, $C=N$, $C=O$, and $C=S$. In considering the viability of 1,3-dipolar cycloaddition as a route to a particular heterocycle, it is desirable to be able to estimate (a) the *reactivity* of the components under a given set of conditions and (b) the *selectivity* of the reaction in giving a single isomer where more than one might be formed.

Reactivity

The currently most widely accepted view of the mechanisms of 1,3-dipolar cycloadditions is that they are concerted processes (that is, with no distinct reaction intermediates) that proceed through unsymmetrical transition states in which the formation of one of the new σ-bonds is more advanced than the

6

other. The reactions can be represented as going through a transition state, **6**, in which the 4π-electron system of the dipole interacts with the 2π-electron system of the dipolarophile. This is a 'thermally allowed' process on the basis of the Woodward–Hoffmann rules. The reaction rates are only slightly influenced by the polarity of the solvent, showing that there is little change in polarity between the reactants and the transition state. The lack of solvent effects and the instability of many 1,3-dipolar species impose practical limitations on the choice of conditions to bring about 1,3-dipolar additions.

The reactivity of 1,3-dipoles towards different dipolarophiles often varies considerably. Frontier orbital theory provides a means of accounting for these variations and is also useful for predicting the reactivity of new dipolarophiles.[20] In outline the theory proposes that reaction through a transition state such as **6** is favoured if there is a favourable interaction between a filled π-orbital of one reactant and an empty π^*-orbital of the other. The orbitals must be of the correct phase to interact, the interaction must be sterically feasible, and the interaction will be stronger, the closer in energy the orbitals are. These interactions are therefore dominated by the highest occupied π-orbitals (HOMO) and the lowest unoccupied π^*-orbitals (LUMO) of the two reactants. The relative energies of these so-called *frontier orbitals* can vary, as shown in Fig. 4.22.

Reactions are therefore favoured if one component is strongly 'nucleophilic' and the other strongly 'electrophilic'. The orbital energies are determined both by the nature of the skeletal atoms in the two components and by their substituents. Estimated HOMO and LUMO energies for different types of carbon dipolarophiles are given in Table 4.14.

The more electrophilic dipolarophiles have the lower energy LUMO values whereas the more nucleophilic species have higher energy HOMO values. Thus, for a given 1,3-dipole, the dominant interaction is more likely to be of type (i) (Fig. 4.22) for dipolarophiles with conjugative electron-withdrawing groups, but of type (ii) for enol ethers, enamines, and other electron-rich dipolarophiles.

It is found in practice that the reactivity of 1,3-dipoles towards these various types of dipolarophile varies widely. For example, ozone reacts preferentially with electron-rich dipolarophiles but the rates of reaction of diazoalkanes are greater for electron-deficient dipolarophiles. Azides, nitrile

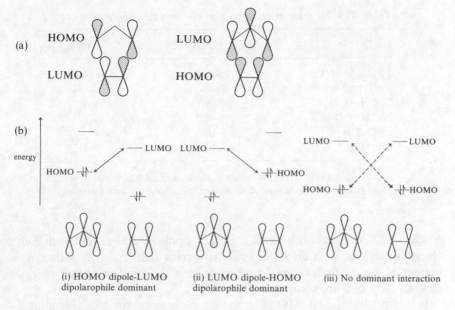

Fig. 4.22 (a) Frontier orbital combinations in 1,3-dipolar cycloaddition. (b) Types of frontier orbital interaction.

Table 4.14 Estimated frontier orbital energies of carbon dipolarophiles.[a]

Dipolarophile	HOMO (eV)	LUMO (eV)
(i) ÖR, NR$_2$	−8	+3
(ii) Alk	−9	+2
(iii) ≡—Alk	−10	+2
(iv) , Ph	−9	+1
(v) ethylene	−10.5	+1.5
(vi) COR, CO$_2$R	−11	0
(vii) ≡—CO$_2$R	−11	0

[a] For more extensive compilations of values see ref. 20 and ref. *a* in Table 4.13.

Table 4.15 Frontier orbital energies of some 1,3-dipoles.[a]

1,3-Dipole	HOMO (eV)	LUMO (eV)
$PhC\equiv\overset{+}{N}-\overset{-}{N}Ph$	−7.5	−0.5
$PhCH=\overset{+}{N}Me-\overset{-}{O}$	−8	−0.4
$N\equiv\overset{+}{N}-\overset{-}{C}H_2$	−9	+1.8
$N\equiv\overset{+}{N}-\overset{-}{N}Ph$	−9.5	−0.2
$PhC\equiv\overset{+}{N}-\overset{-}{O}$	−10	−1
$O=\overset{+}{O}-\overset{-}{O}$	−13.5	−2.2

a These values refer to the π-orbitals of the delocalized 1,3-dipolar systems. For more extensive compilations of values see ref. 20 and ref. *a* in Table 4.13.

oxides, and several other types of 1,3-dipole react faster both with electron-rich and with electron-deficient species than they do with simple olefins. These differences can be understood if the frontier orbital energies of the 1,3-dipoles are compared (Table 4.15).

It is clear that the HOMO of ozone is so low in energy that it is unlikely to interact strongly with the LUMO of any of the dipolarophiles in Table 4.14. The reactions of ozone will therefore be dominated by the interaction of type (ii) in Fig. 4.22 and the electron-rich dipolarophiles will react most rapidly (since the energy gap is smallest, and the interaction strongest). In contrast, diazomethane has a high-energy LUMO and so its reactions with dipolarophiles are dominated by interactions of the HOMO with the LUMO of the dipolarophile [type (i) in Fig. 4.22]. Dipolarophiles with low-energy LUMO values thus react most rapidly. With a dipole such as azidobenzene, interactions either of type (i) or of type (ii) may be dominant, depending upon the nature of the dipolarophile. Addition to ethylene is slow because neither interaction is strong [type (iii)]. The reactivities of the 1,3-dipoles are, of course, modified by functional groups. For example, the HOMO and LUMO energies of diazoketones and diazoesters are lower than those of simple diazoalkanes, and their reactivities are modified accordingly.

Selectivity

Many 1,3-dipolar cycloadditions can, in principle, give mixtures of isomers. Fortunately, most such reactions show considerable selectivity in that they give predominantly or exclusively one of the possible products. The formation of possible *regioisomers* is shown in Fig. 4.23(a) and possible *stereoisomers* are shown in (b) and (c). The first of the two products shown in each of (b) and (c) have retained the stereochemistry of the original reagents.

Electronic effects are of major importance in determining the regioselectivity of the cycloaddition, although steric effects also have some influence. As indicated above, the interaction of the dipole with the dipolarophile is usually dominated by one of the two possible frontier orbital combinations. Consider,

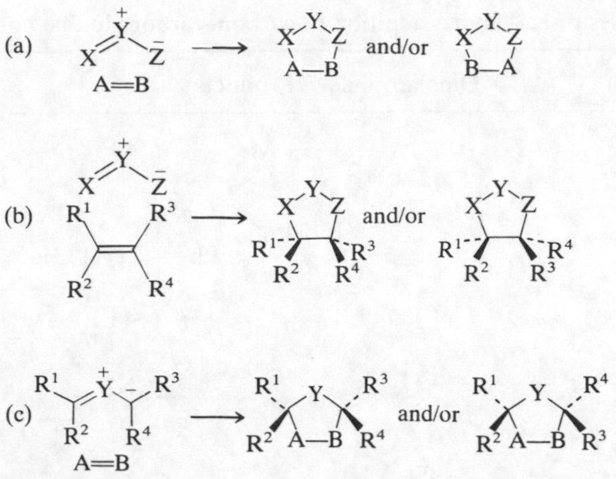

(a)

(b)

(c)

Fig. 4.23 Possibilities for isomer formation in 1,3-dipolar cycloaddition.

as an example, the addition of diazomethane to methyl methacrylate, which is dominated by the interaction of the HOMO of the dipole with the LUMO of the dipolarophile. Two regioisomers **7** and **8** could be formed, but only the first is isolated. This can be explained by the fact that in the dominant HOMO–LUMO combination shown in Fig. 4.24 the interaction leading to product **7** is much stronger than that leading to its isomer **8**, because the coefficients on the termini of the reaction partners are not equal.[20]

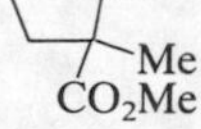

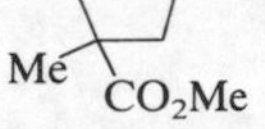

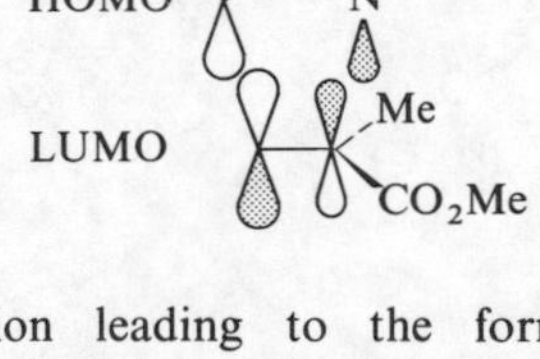

Fig. 4.24 HOMO–LUMO interaction leading to the formation of regioisomer **7**. The more favourable direction of combination is that in which the two terminal atoms with the larger orbital coefficients interact (see ref. 20).

The frontier orbital coefficients have been calculated for a large number of dipoles and dipolarophiles, and compilations of the data are available.[20] Examples showing the regioselectivities observed with different types of dipolarophile are given in Tables 4.16 to 4.18; these can usually be explained by the use of frontier orbital energies and coefficients, as indicated in the example above.

The stereoselectivity of the addition to dipolarophiles [Fig. 4.23(b)] is complete; only adducts having the stereochemistry of the original

Table 4.16 Examples of 1,3-dipolar addition to carbon–carbon double bonds.

Dipole (conditions)	Dipolarophile	Products	Refs.
(i) Ph$-\bar{N}-\overset{+}{N}\equiv$N (80°C, 7 h)	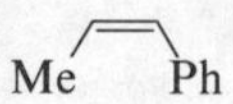 Me⎯Ph	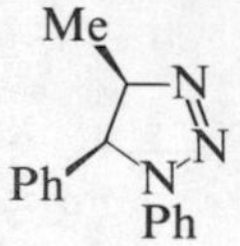 (22%)	a
(ii) Me$_2\bar{C}-\overset{+}{N}\equiv$N (0°C, 12 h)	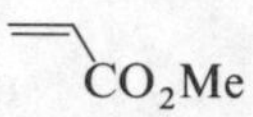 CO$_2$Me	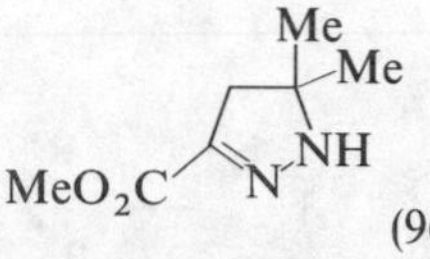 (96%)	b
(iii) Ph$-\overset{O}{\underset{O}{\overset{\|}{\underset{\|}{S}}}}-C\equiv\overset{+}{N}-\bar{O}$ (PhSO$_2$⎯=NOH, Br Na$_2$CO$_3$, r.t.)		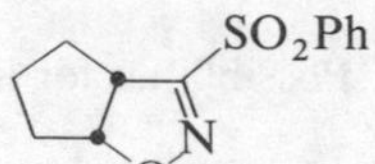 SO$_2$Ph (81%)	c
(iv) 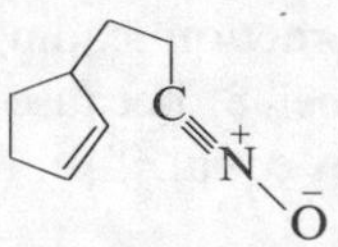 $\overset{+}{\underset{\bar{O}}{N}}$ (RCH$_2$NO$_2$, PhNCO, 80°C)	internal	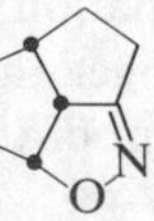 (91%)	d
(v) 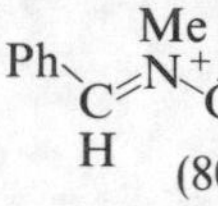 $\underset{H}{\overset{Me}{\underset{Ph}{\diagdown}C=\overset{+}{N}-\bar{O}}}$ (80°C)	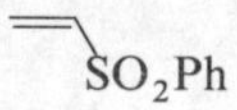 SO$_2$Ph	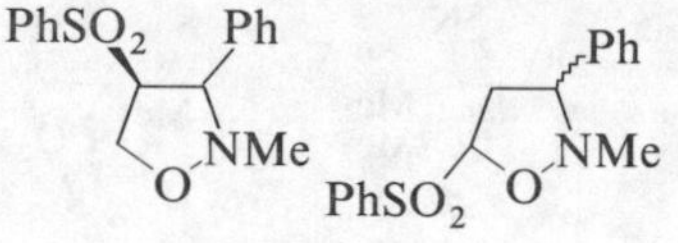 PhSO$_2$ Ph ...NMe / ...Ph PhSO$_2$ ...NMe (68:32)	e
(vi) 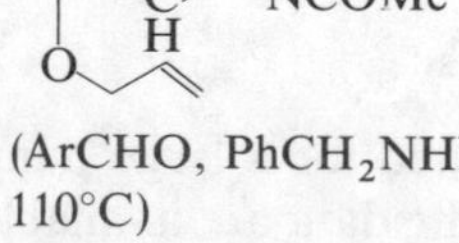 $\underset{H}{\overset{CH_2Ph}{C=\overset{+}{N}-\bar{N}COMe}}$ (ArCHO, PhCH$_2$NHNHCOMe, 110°C)	internal	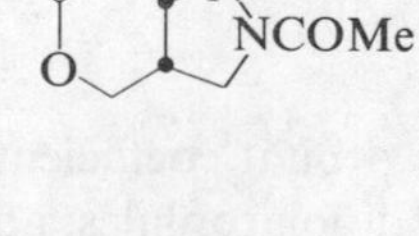 CH$_2$Ph NCOMe (91%)	f
(vii) 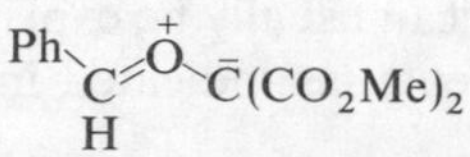 $\underset{H}{\overset{Ph}{C=\overset{+}{O}-\bar{C}(CO_2Me)_2}}$ (PhCHO, N$_2$C(CO$_2$Me)$_2$, Cu, 125°C)	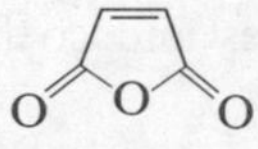	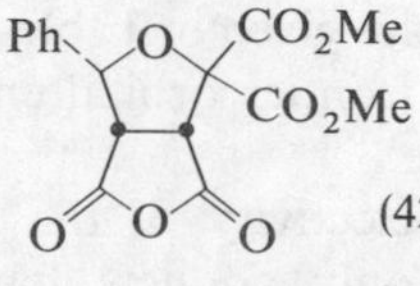 Ph CO$_2$Me / CO$_2$Me (42%)	g

dipolarophile are isolated.[22] This is strong evidence in favour of a concerted mechanism for the reactions. The additions are also stereoselective with respect to the dipole (Fig. 4.23(c)], although some dipoles such as azomethine ylides and carbonyl ylides can undergo *cis–trans* isomerization in the reaction conditions required to bring about cycloaddition. The stereoselectivity of 1,3-dipolar addition is an important advantage of the reaction as a synthetic method.

Scope of the reaction

Azides. Many azides are isolable compounds. Azides with a wide range of substituents can be prepared, the two most commonly used preparative methods being the nucleophilic displacement by azide ions of halides, etc., and the reaction of monosubstituted hydrazines with nitrous acid. Vinyl and acyl azides and some alkyl azides decompose easily when heated and this limits their usefulness as 1,3-dipoles. Nevertheless, 1,3-dipolar cycloaddition reactions of azides have been widely used in heterocyclic synthesis. Cycloaddition to acetylenes or to electron-rich alkenes bearing a leaving group (such as enamines) provides the most general route to 1,2,3-triazoles. Dihydrotriazoles can similarly be prepared by cycloaddition of azides to alkenes, but these primary products often react further, by loss of nitrogen, under the reaction conditions. Examples (i) in Tables 4.16 and 4.17 illustrate the intermolecular reactions. Example (ii) in Table 4.17 is an intramolecular cycloaddition to a triple bond; there are many similar examples of intramolecular addition to alkenes and alkynes, and a few examples of intramolecular addition to cyano groups.

Diazo compounds. Diazomethane and simple diazoalkanes are isolable but thermally unstable compounds which participate in cycloaddition reactions with a wide variety of multiple bonds. Many functionalized diazo compounds are also readily available; for example, from activated methylene compounds by diazo transfer reactions. The initial products of 1,3-dipolar addition of diazo compounds to multiple bonds are not aromatic and they incorporate an azo group in the new ring system. Many of these products are thermally unstable and are not isolated. They react further either by loss of nitrogen or by tautomerization. Examples (ii) in Table 4.16 and (iii) in Table 4.17 show reactions in which the isolated products result from proton shifts in the initial 1,3-dipolar adducts.

---

22 This is the norm, but there is an exception to every rule! See R. Huisgen, G. Mloston, and E. Langhals, *J. Am. Chem. Soc.*, 1986, **108**, 6401.

Footnotes to Table 4.16
a P. Scheiner, *J. Am. Chem. Soc.*, 1968, **90**, 988. b S. D. Andrews, A. C. Day, and A. N. McDonald, *J. Chem. Soc.* (*C*), 1969, 787. c P. A. Wade and H. R. Hinney, *J. Am. Chem. Soc.*, 1979, **101**, 1319. d R. H. Wollenberg and J. E. Goldstein, *Synthesis*, 1980, 757. e J. Sims and K. N. Houk, *J. Am. Chem. Soc.*, 1973, **95**, 5798. f W. Oppolzer, *Angew. Chem. Int. Edn Engl.*, 1977, **16**, 10. g R. Huisgen and P. de March, *J. Am. Chem. Soc.*, 1982, **104**, 4953.

Table 4.17 Examples of 1,3-dipolar addition to carbon–carbon triple bonds.

Dipole (conditions)	Dipolarophile	Products	Refs.
(i) $Ph-\bar{N}-\overset{+}{N}\equiv N$ (110°C, 18 h)	$PhC\equiv CH$	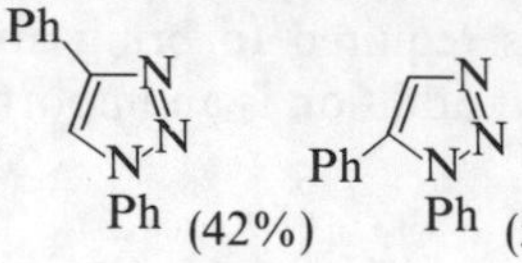 (42%) (52%)	a
(ii) 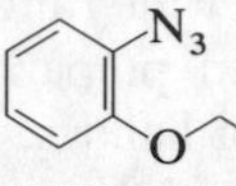 (140°C, 2 h)	internal	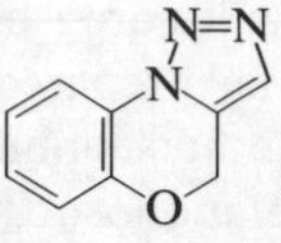 (45%)	b
(iii) $PhCO\bar{C}H-\overset{+}{N}\equiv N$ (40°C)	 from 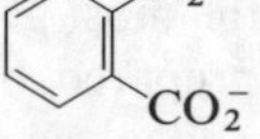	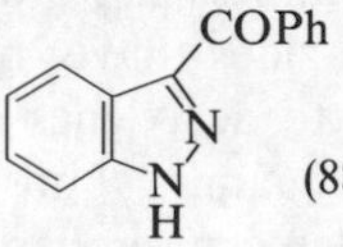 (88%)	c
(iv) $Ph-C\overset{+}{\equiv}N-\bar{N}Ph$ 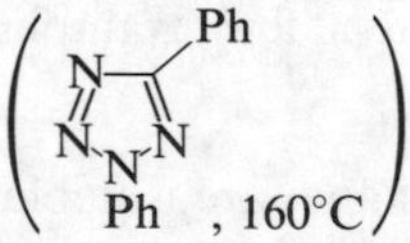 (, 160°C)	$PhC\equiv CCO_2Et$	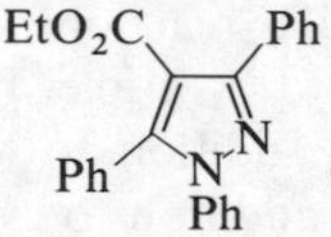 (84%)	d
(v) $Ph-C\overset{+}{\equiv}N-\bar{C}Me_2$ 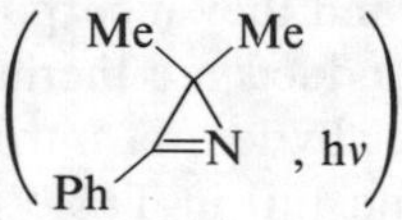 (, hν)	$MeO_2CC\equiv CCO_2Me$	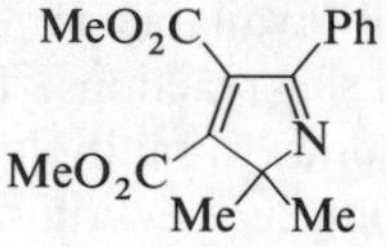 (69%)	e
(vi) 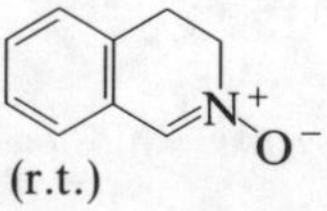 (r.t.)	$HC\equiv CCO_2Me$	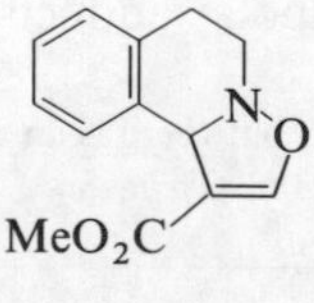 (83%)	f

a W. Kirmse and L. Horner, *Liebigs Ann. Chem.*, 1958, **614**, 1. b R. Fusco, L. Garanti, and G. Zecchi, *J. Org. Chem.*, 1975, **40**, 1906. c W. Ried and M. Schön, *Liebigs Ann. Chem.*, 1965, **689**, 141. d R. Huisgen, M. Seidel, G. Wallbillich, and H. Knupfer, *Tetrahedron*, 1962, **17**, 3. e N. Gakis, M. Märky, H.-J. Hansen, H. Heimgartner, H. Schmid, and W. Oberhänsli, *Helv. Chim. Acta*, 1976, **59**, 2149. f H. Seidl, R. Huisgen, and R. Knorr, *Chem. Ber.*, 1969, **102**, 904.

Nitrile oxides. A few nitrile oxides with bulky aryl substituents are isolable but most are generated and trapped *in situ*. The two routes to nitrile oxides which are most comonly used are shown in Fig. 4.25. Aldoximes are the starting materials in method (a). They are chlorinated at the α-position and the α-chloro-oximes then give the nitrile oxides on reaction with a base (usually triethylamine). Some nitrile oxides have also been generated directly from aldoximes by oxidation. Method (b) has nitroalkanes as the starting materials. These are dehydrated by reaction with phenyl isocyanate. Nitrile oxides dimerize easily and it is usually beneficial to generate the nitrile oxide slowly, in the presence of the trapping agent, so that its concentration remains low.

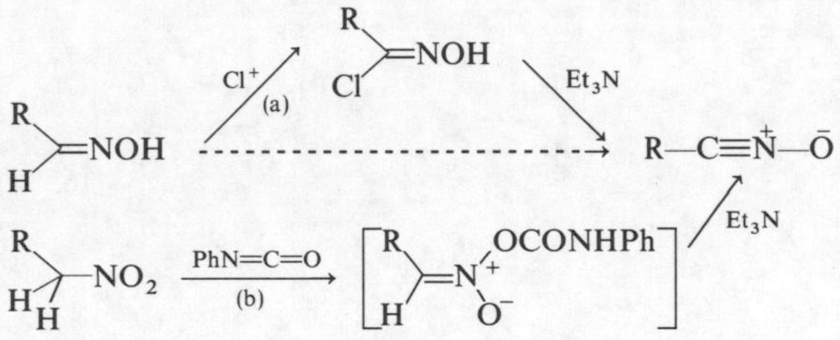

Fig. 4.25 Routes to nitrile oxides.

Aryl substituted nitrile oxides are relatively long lived in solution and their cycloaddition reactions have been extensively studied. They undergo cycloaddition to alkynes, alkenes, and heterodipolarophiles such as thio-carbonyl and activated azo compounds; they also add to carbonyl groups in the presence of a Lewis acid. Nitrile oxides with a wide range of other substituents (including H, alkyl, Br, CO_2R, and SO_2Ph) have also been shown to participate in 1,3-dipolar cycloaddition reactions. Additions to alkynes and alkenes have proved valuable as a means of preparing isoxazoles and dihydroisoxazoles, which are then used as intermediates in target synthesis because the N—O bond is easily cleaved. Intramolecular 1,3-dipolar additions of nitrile oxides are particularly important for this purpose. Examples (iii) and (iv) in Table 4.16 illustrate intermolecular and intra-molecular cycloaddition to alkenes. Two examples of addition to hetero-dienophiles are shown in Table 4.18 [entries (i) and (ii)].

Nitrile imides. These 1,3-dipoles are transient intermediates. The most important methods of generation are shown in Fig. 4.26. Method (a) is analogous to that used for the generation of nitrile oxides from oximes. Method (b) has as the starting material a heterocyclic compound (here a 2*H*-tetrazole) which decomposes thermally or under the influence of ultraviolet light with the loss of a small stable molecule. In the example shown nitrogen is lost, but there are other heterocyclic precursors which lose carbon dioxide or sulphur dioxide. The tetrazole route is illustrated in Table 4.17 (iv).

Table 4.18 Examples of 1,3-dipolar addition to multiple bonds containing hetero-atoms.

Dipole (conditions)	Dipolarophile	Products	Refs. and notes
(i) 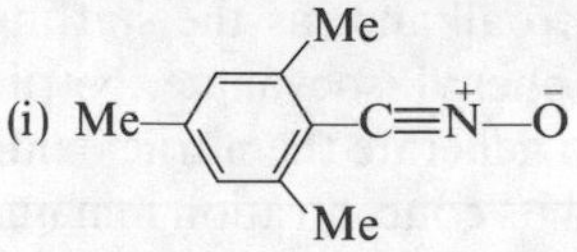 Me—[2,6-Me$_2$C$_6$H$_3$]—C≡$\overset{+}{N}$—$\overset{-}{O}$ (50°C, 10 days)	PhC≡N	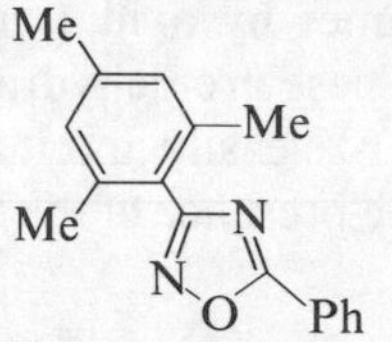 [dihydro-1,2,4-oxadiazole ring, mesityl and Ph substituents] (95%)	a
(ii) Ph—C≡$\overset{+}{N}$—$\overset{-}{O}$ 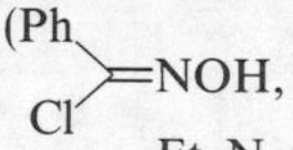 (Ph—C(Cl)=NOH, Et$_3$N, r.t.)	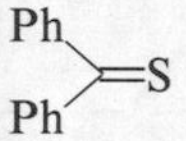 Ph$_2$C=S	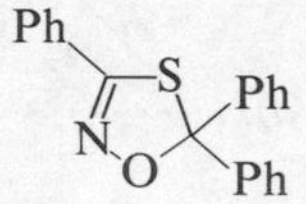 [1,4,2-oxathiazoline ring, Ph substituents] (65%)	b
(iii) Ph—C≡$\overset{+}{N}$—S 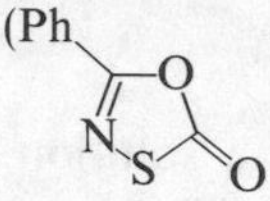 (Ph, N,S,O ring with O, 140°C, 26 h)	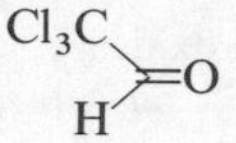 Cl$_3$C—CHO	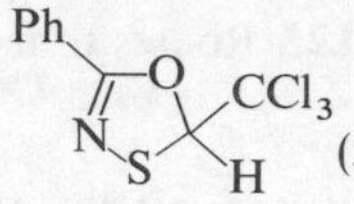 [Ph, O, N, S ring with CCl$_3$, H] (59%)	c
(iv) Ph—C≡$\overset{+}{N}$—$\overset{-}{C}$HC$_6$H$_4$.NO$_2$-4 (Ph—C(Cl)=NCH$_2$C$_6$H$_4$.NO$_2$-4, 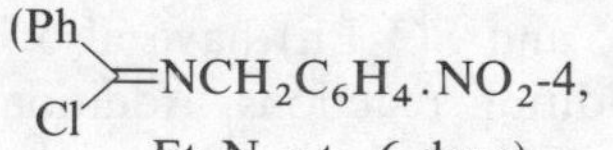 Et$_3$N, r.t., 6 days)	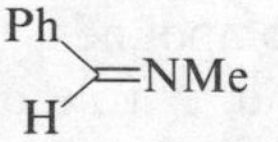 Ph—CH=NMe	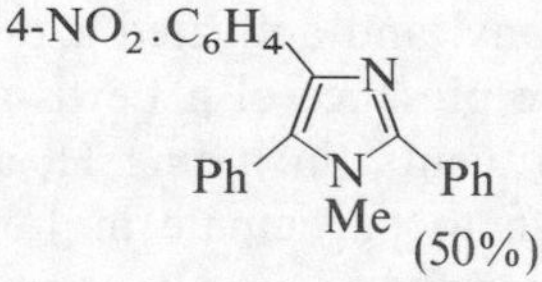 [imidazole ring: 4-NO$_2$.C$_6$H$_4$, Ph, Me, Ph] (50%)	d
(v) 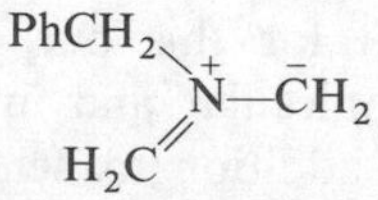 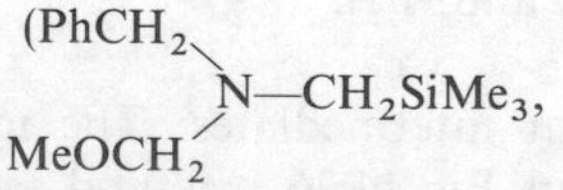 PhCH$_2$—$\overset{+}{N}$(=CH$_2$)(CH$_2^-$) (PhCH$_2$(MeOCH$_2$)N—CH$_2$SiMe$_3$, LiF, ultrasound, 35°C, 5 h)	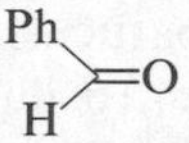 Ph—CHO	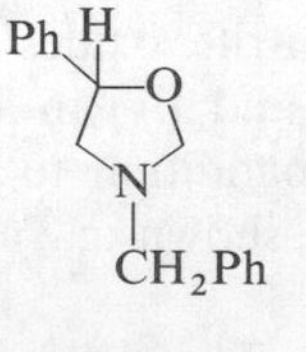 [oxazolidine ring: Ph, H, O, N—CH$_2$Ph] (80%)	e

a A. Dondoni and G. Barbaro, *Gazz. Chim. Ital.*, 1975, **105,** 701. *b* R. Huisgen and W. Mack, *Chem. Ber.*, 1972, **105,** 2815. *c* A. M. Damas, R. O. Gould, M. M. Harding, R. M. Paton, J. F. Ross, and J. Crosby, *J. Chem. Soc.*, *Perkin Trans. 1*, 1981, 2991. *d* The primary product is the 4,5-dihydroimidazole, but this is oxidized before isolation: K. Bunge, R. Huisgen, R. Raab, and H. J. Sturm, *Chem. Ber.*, 1972, **105,** 1307. *e* A. Padwa and W. Dent, *J. Org. Chem.*, 1987, **52,** 235.

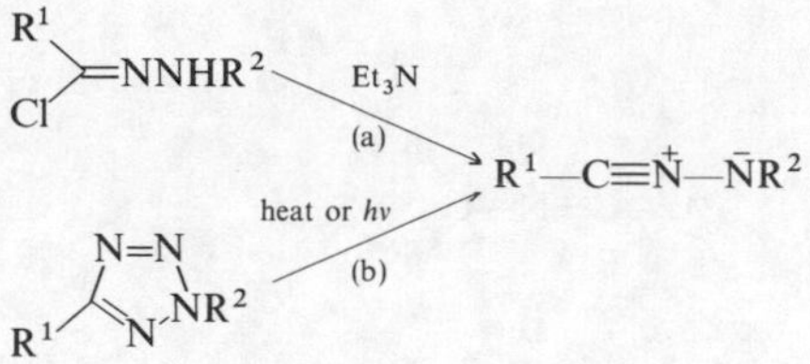

Fig. 4.26 Routes to nitrile imides.

Aryl substituted nitrile imides are the best known and the mechanism of their cycloaddition to alkenes and alkynes has been thoroughly studied. The reactions are used for the preparation of pyrazoles but, unlike the isoxazoles derived from nitrile oxides, these have not been used much as intermediates in target synthesis because of the relative strength of the N—N bond. Nitrile imides also add to heterodienophiles, including activated nitriles and carbonyl compounds.

Nitrile sulphides. These 1,3-dipoles are generated and trapped *in situ* by heating 1,3,4-oxathiazol-2-ones in solution at 110–160°C (Fig. 4.27). Cycloaddition reactions take place with several types of electron-deficient dipolarophiles, such as acetylenic and vinylic esters, activated nitriles, and activated carbonyl compounds. The reactions provide routes to five-membered heterocycles containing the C=N—S structural unit. An example of the reaction, the addition to the carbonyl group of trichloroacetaldehyde, is shown in Table 4.18 (iii).

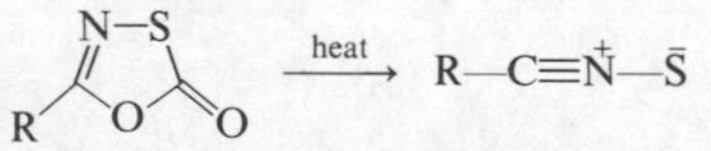

Fig. 4.27 Nitrile sulphides from 1,3,4-oxathiazol-2-ones.

Nitrile ylides. Nitrile ylides are transient intermediates in solution. Two methods of generating these species are illustrated in Fig. 4.28. Method (a), which is applicable mainly to the generation of aromatic nitrile ylides, is similar to those used for nitrile oxides and nitrile imides. Method (b), the photochemical ring-opening of a 2*H*-azirine, is also a good general procedure for aryl-substituted nitrile ylides. Nitrile ylides without stabilizing substituents are uncommon, but two routes to such intermediates are illustrated in Fig. 4.29.[23]

Nitrile ylides undergo cycloaddition to several types of electrophilic double and triple bonds. The synthesis of a 2*H*-pyrrole [Table 4.17 (v)] and of an imidazole [Table 4.18 (iv)] are examples of nitrile ylide cycloaddition.

23 A. Padwa, J. R. Gasdaska, M. Tomas, N. J. Turro, Y. Cha, and I. R. Gould, *J. Am. Chem. Soc.*, 1986, **108**, 6739.

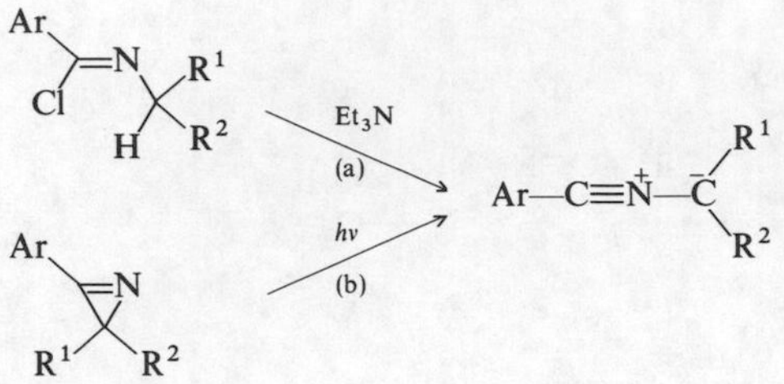

Fig. 4.28 Routes to aromatic nitrile ylides.

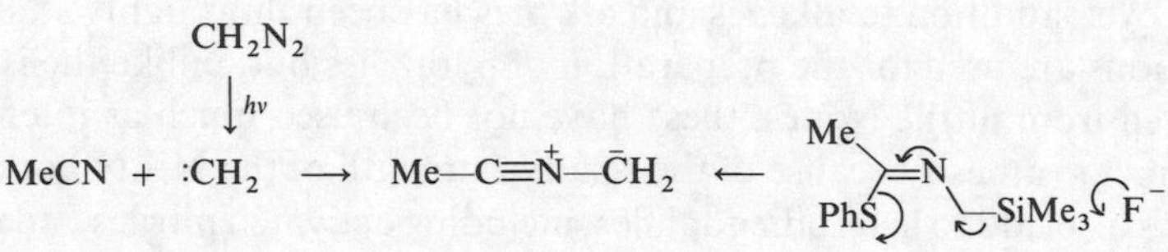

Fig. 4.29 Generation of acetonitrile methylide.

Nitrones. Many nitrones are isolable and stable compounds, especially those with *C*-aryl substituents. Those which lack stabilizing substituents tend to dimerize or trimerize easily and are therefore best generated *in situ*. The two most general methods of forming nitrones are shown in Fig. 4.30. Oxidation of an *NN*-disubstituted hydroxylamine [method (a)] is particularly useful for the formation of cyclic nitrones. Method (b) is the condensation of *N*-alkylhydroxylamines with aldehydes or ketones.

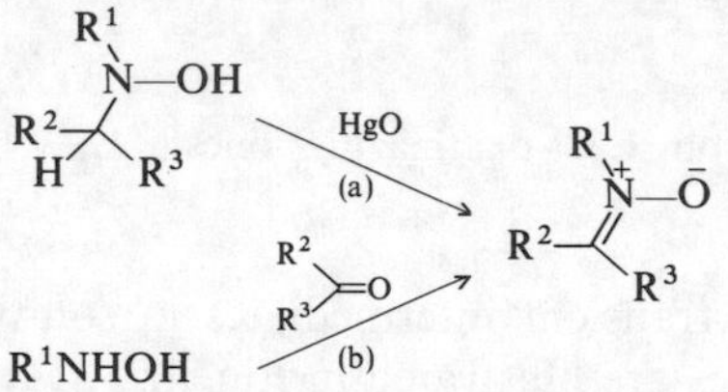

Fig. 4.30 Routes to nitrones from hydroxylamines.

There are many examples of cycloaddition of nitrones to carbon–carbon double and triple bonds, especially to those bearing electron-withdrawing groups. As with the nitrile oxide adducts, the N–O bond in the products is easily cleaved: for this reason, the cycloaddition has often been used as a step in a target synthesis. Two examples of the intermolecular reaction are shown in Table 4.16 (v) and in Table 4.17 (vi). The intramolecular nitrone addition to alkenes and the alkynes is also a versatile reaction since it proceeds well even with unactivated dipolarophiles.

Azomethine imides. Acyclic azomethine imides can be generated by the reaction of *N*-acyl-*N'*-alkylhydrazines with aldehydes (Fig. 4.31). They can be intercepted by alkenes, especially those with electron-withdrawing groups, and by other electrophilic dipolarophiles. It is also possible to achieve intramolecular cycloaddition to unactivated double bonds. Table 4.16 (vi) shows an example of this intramolecular process. *N*-Imides of aromatic heterocycles such as pyridine can also be classed as azomethine imides.[24] These imides are generated by deprotonation of the heterocyclic quaternary *N*-amino compounds (Fig. 4.32). Some of them are isolable. They participate in cycloaddition reactions with electrophilic alkenes and with other electron-deficient dipolarophiles.

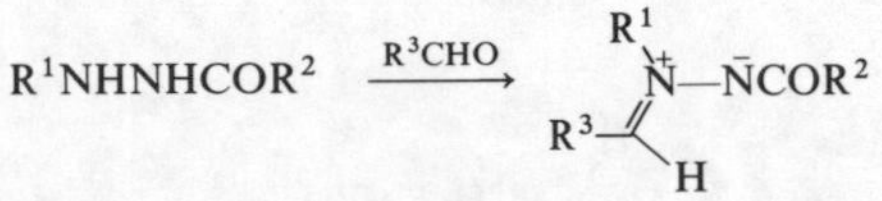

$$R^1NHNHCOR^2 \xrightarrow{R^3CHO} \underset{\underset{H}{\overset{R^3}{\diagdown}}}{\overset{R^1}{\underset{|}{N}}}\!\!\!^+\!\!-\bar{N}COR^2$$

Fig. 4.31 Azomethine imides from acylhydrazines.

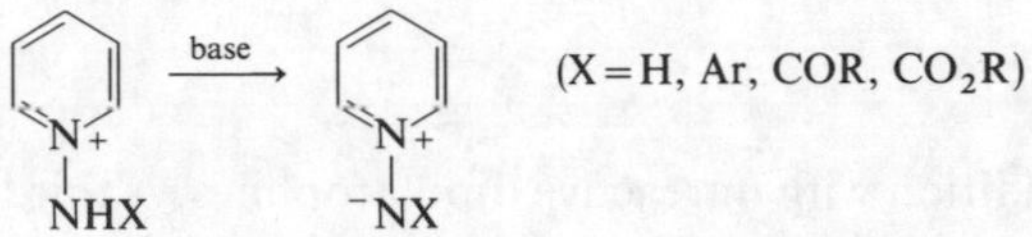

$(X = H, Ar, COR, CO_2R)$

Fig. 4.32 Pyridinium *N*-imides, examples of heteroaromatic *N*-imides.

Azomethine ylides. With the exception of a few species with stabilizing substituents, azomethine ylides are not isolable. The intermediates are valuable partners in 1,3-dipolar cycloaddition reactions because they provide access to pyrroles and reduced pyrroles bearing a wide range of substituents. There are several good methods for generating azomethine ylides, some of which are outlined in Fig. 4.33. One of the most versatile routes (a) starts from imines, which are alkylated with trimethylsilylmethyl triflate. The resulting salts are desilylated with fluoride ions to produce azomethine ylides which are unsubstituted at one of the terminal carbon atoms. A second method, which leads to the formation of azomethine ylides bearing stabilizing substituents, starts from aziridines [method (b)]. Aziridines bearing electron withdrawing groups on carbon undergo electrocyclic ring opening on heating or on ultraviolet irradiation. The thermal ring opening is stereoselective and conrotatory; the photolysis is stereoselective and disrotatory. The resulting azomethine ylides can, however, undergo *cis–trans* isomerization before their capture in cycloaddition reactions and this complicates the stereochemical

24 Review: Y. Tamura and M. Ikeda, *Adv. Heterocycl. Chem.*, 1981, **29**, 71.

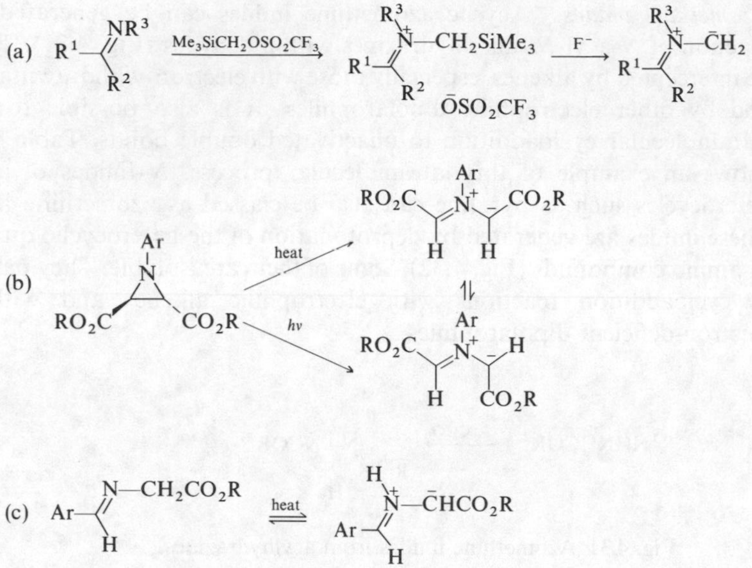

Fig. 4.33 Routes to azomethine ylides.

outcome of the cycloaddition with unreactive dipolarophiles. A third method (c), which leads to azomethine ylides unsubstituted on nitrogen, is the tautomerization of imines of α-aminoesters and related compounds. Such imines, when heated in the presence of suitable dipolarophiles, give cycloadducts. These are formed by the trapping of the 1,3-dipolar tautomer of the imine, which is reversibly generated by proton transfer from carbon to nitrogen.

Most azomethine ylides can be intercepted in cycloaddition reactions by highly electrophilic dipolarophiles such as N-phenylmaleimide and dimethyl acetylenedicarboxylate. Cycloadditions to less activated dipolarophiles, such as acrylate esters, are also common. An example of the cycloaddition of a nonstabilized azomethine ylide to the C=O bond of benzaldehyde is given in Table 4.18 (v): this 1,3-dipole also adds to thiobenzophenone and to electrophilic alkenes. There are also several examples of the 1,3-dipolar cycloaddition of azomethine ylides to unactivated dipolarophiles, including some useful intramolecular additions.

Carbonyl ylides. Two routes to these transient 1,3-dipoles are shown in Fig. 4.34. Method (a) is the thermolysis or photolysis of oxiranes bearing electron-withdrawing substituents; it is rather limited in scope. Method (b) is the reaction of carbenoids, produced by the metal catalysed decomposition of diazo compounds, with carbonyl compounds. This method is particularly useful when the carbonyl group and the diazo function are within

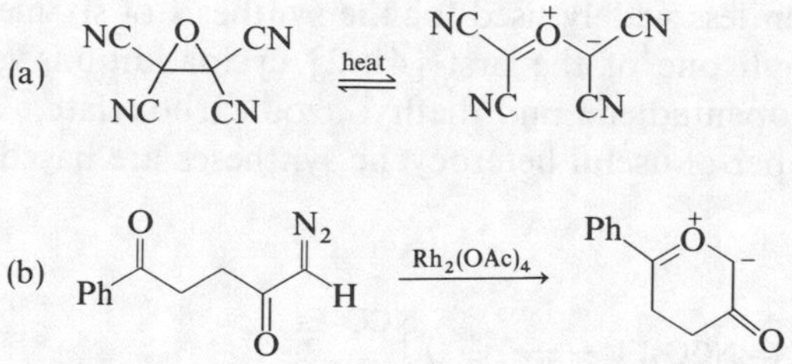

(a)

(b)

Fig. 4.34 Examples of methods of generation of carbonyl ylides.

the same molecule, as in the example shown.[25] The 1,3-dipoles can be intercepted intermolecularly by electrophilic alkenes [Table 4.16 (vii)] and by carbonyl compounds. There are also examples of intramolecular cycloaddition to unactivated alkenes.

Thiocarbonyl ylides. An example of the generation of a 1,3-dipole of this type is provided by the reaction of thiobenzophenone with diazomethane. Reaction at $-78°C$ gives a cycloadduct (Fig. 4.35) which is unstable above $-40°C$. It gives nitrogen and the 1,3-dipole, which was intercepted with maleic anhydride. A few other thiadiazolines are known to decompose in the same way.

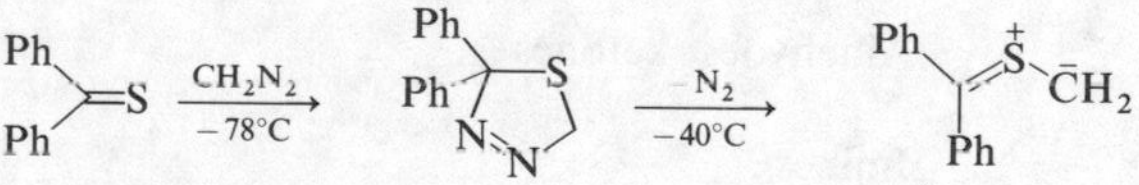

Fig. 4.35 Generation of a thiocarbonyl ylide.

4.3.3 Hetero–Diels–Alder reactions

The cycloaddition of conjugated dienes to activated olefins and acetylenes, commonly known as the Diels–Alder reaction after its discoverers, is an extraordinarily versatile reaction for the construction of six-membered carbocyclic compounds. Its preparative value lies in the high stereo- and regioselectivity with which the adducts are formed, and in the range of substituents which the reaction partners can bear. The reaction is thermally allowed according to the Woodward–Hoffmann rules, and the selectivity of the reaction can be accounted for, as with 1,3-dipolar addition, by means of frontier orbital theory.[26]

---

25 A. Padwa, G. E. Fryxell, and L. Zhi, *J. Org. Chem.*, 1988, **53**, 2875; A. Padwa, R. L. Chinn, and L. Zhi, *Tetrahedron Lett.*, 1989, **30**, 1491.

26 For a discussion of the mechanism of the Diels–Alder reaction see J. Sauer and R. Sustmann, *Angew. Chem. Int. Edn Engl.*, 1980, **19**, 779.

The reaction has been less widely used for the synthesis of six-membered heterocycles even though one of the first [4+2] cycloadditions gave the pyridazine 9 from cyclopentadiene and diethyl azodicarboxylate.[27] Nevertheless, a growing number of useful heterocyclic syntheses are based on this method.

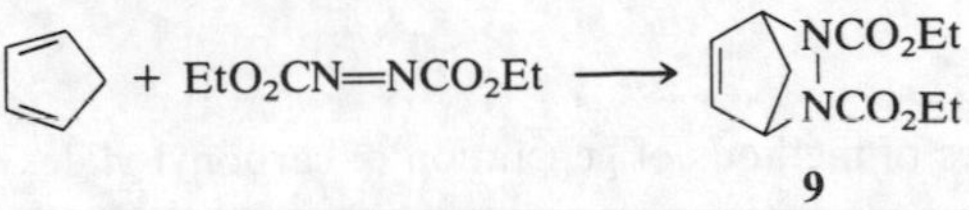

The reactions can be grouped on the basis of the component containing the heteroatoms. Common dienophiles that incorporate heteroatoms are shown in Table 4.19, and heterodienes are shown in Table 4.20.

The majority of all-carbon Diels–Alder reactions involve the interaction of a nucleophilic diene with an electrophilic dienophile. In terms of frontier orbitals, the dominant interaction is that between the HOMO of the diene and the LUMO of the dienophile. The reactions show high stereoselectivity, with products arising from *endo* transition states being kinetically favoured.

Table 4.19 Some compounds that act as heterodienophiles in the Diels–Alder reaction.[a]

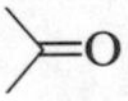 $\rangle{=}O$	Aldehydes; ketones
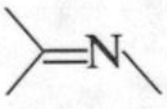 $\rangle{=}N\backslash$	Imines
$-C{\equiv}N$	Nitriles
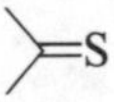 $\rangle{=}S$	Thioaldehydes; thioketones; thioesters; thiophosgene
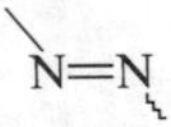 $N{=}N$	Azocarbonyl compounds
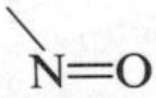 $N{=}O$	Nitroso compounds
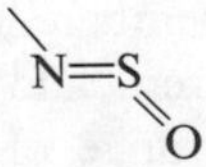 $N{=}S{\diagdown}_O$	Sulphinylamines

[a] Comprehensive review: D. L. Boger and S. M. Weinreb, *Hetero Diels–Alder Methodology in Organic Synthesis*, Academic Press, San Diego, 1987. See also G. W. Kirby, *Chem. Soc. Rev.*, 1977, **6**, 1 (nitroso compounds); C. J. Moody, *Adv. Heterocycl. Chem.*, 1982, **30**, 1 (azocarbonyl compounds); S. M. Weinreb, *Acc. Chem. Res.*, 1985, **18**, 16 (intramolecular cycloaddition of imines); T. Kametani and S. Hibino, *Adv. Heterocycl. Chem.*, 1987, **42**, 245 (applications in natural product synthesis).

27 O. Diels, J. H. Blom, and W. Koll, *Liebigs Ann. Chem.*, 1925, **443**, 242.

Table 4.20 Some compounds that act as heterodienes in the Diels–Alder reaction.[a]

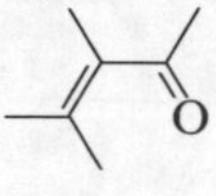

α,β-Unsaturated carbonyl compounds; quinone methides

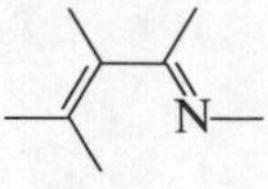

a,β-Unsaturated imines and dimethylhydrazones

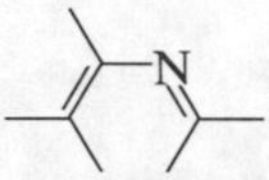

N-Vinyl- and N-arylimines; oxazoles; 1,2,4-triazines

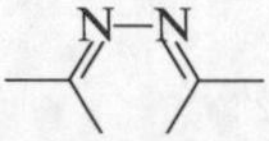

1,2,4,5-Tetrazines

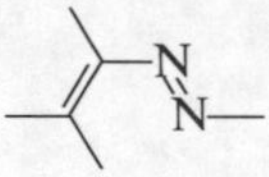

Azoalkenes

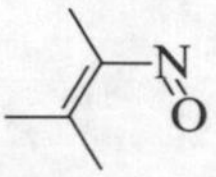

Nitrosoalkenes

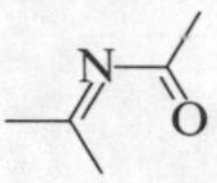

N-Acylimines

a Comprehensive review: D. L. Boger and S. M. Weinreb, *Hetero Diels–Alder Methodology in Organic Synthesis*, Academic Press, San Diego, 1987. See also R. R. Schmidt, *Synthesis*, 1972, 333 (N-acylimines); T. L. Gilchrist, *Chem. Soc. Rev.*, 1983, **12**, 53 (nitrosoalkenes); D. L. Boger, *Chem. Rev.*, 1986, **86**, 781 (heterocyclic azadienes as synthetic intermediates); T. Kametani and S. Hibino, *Adv. Heterocycl. Chem.*, 1987, **42**, 245 (applications in natural product synthesis).

The regioselectivities of the reactions can be accounted for on the basis of the dominant HOMO–LUMO interaction and the coefficients of these orbitals. Both the rates and the selectivities of the additions can often be enhanced by the use of Lewis acid catalysts which coordinate to the dienophile. Many of these features also apply to Diels–Alder reactions of dienes with heterodienophiles, although much less detailed mechanistic work has been done on these reactions.

On the other hand, Diels–Alder reactions of heterodienes often go most easily with electron-rich dienophiles. This is because the heterodienes, being more electron deficient than all-carbon dienes, can act as the electron acceptors in the transition state. The dominant frontier orbital interaction in such reactions is between the LUMO of the diene and the HOMO of the dienophile. Diels–Alder reactions in which this type of interaction is dominant are referred to as showing 'inverse electron demand'. Such Diels–Alder

Table 4.21 Examples of Diels–Alder cycloaddition with heterodienophiles.

Hererodienophile (conditions)	Diene	Product	Refs.
(i) 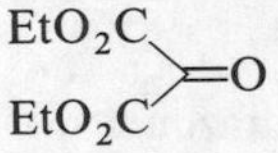 EtO₂C—C(=O)—CO₂Et (130°C, 4 h)		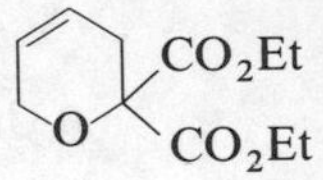 CO₂Et, CO₂Et (63%)	a
(ii) 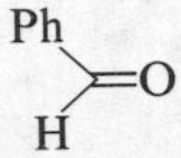 Ph—CH=O, H (ZnCl₂, r.t.)	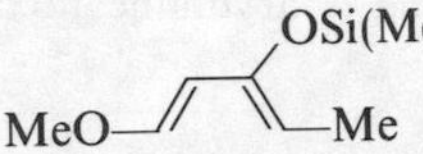 OSi(Me₂)Buᵗ, MeO—CH=CH—C—Me	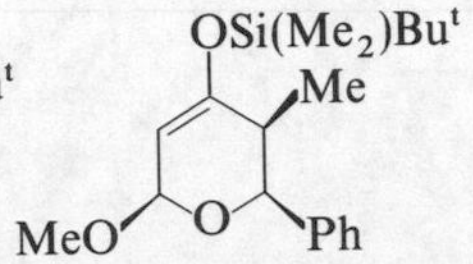 OSi(Me₂)Buᵗ, Me, MeO—O—Ph (42%)	b
(iii) 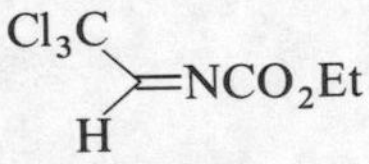 Cl₃C—CH=NCO₂Et, H (100°C, 80 h)	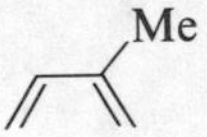 Me	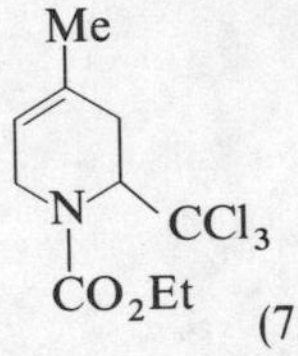 Me, CCl₃, CO₂Et (75%)	c
(iv) 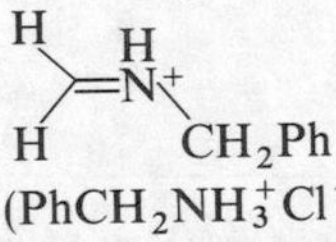 H₂C=N⁺H—CH₂Ph (PhCH₂NH₃⁺Cl⁻, HCHO aq., 35°C, 48 h)	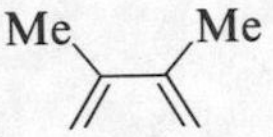 Me, Me	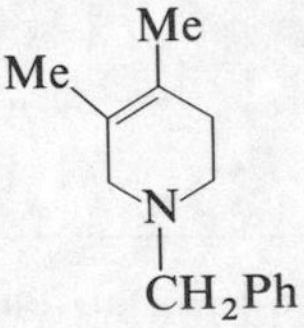 Me, Me, CH₂Ph (64%)	d
(v) 4-Me.C₆H₄SO₂C≡N (r.t., 30 min.)	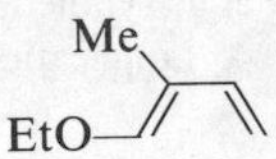 Me, EtO	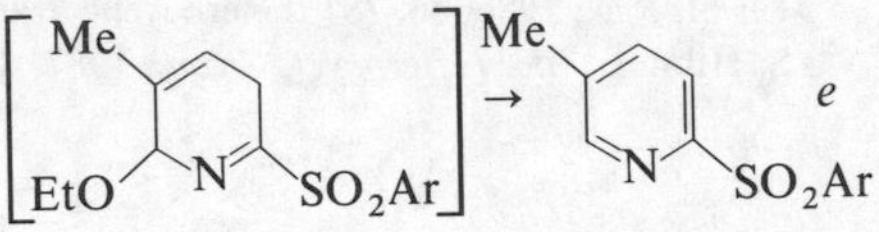 [Me, EtO—N—SO₂Ar] → Me, N—SO₂Ar (75%)	e
(vi) 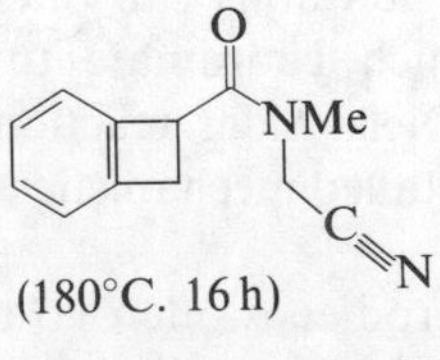 (180°C. 16 h)	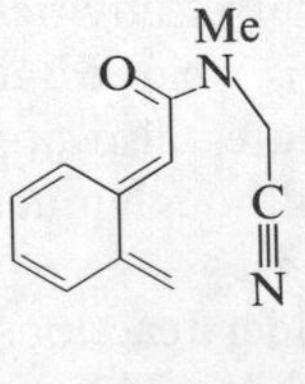 Me, N, C≡N	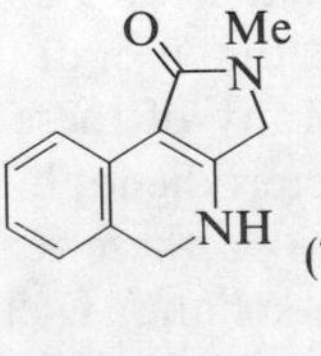 Me, O, NH (76%)	f
(vii) 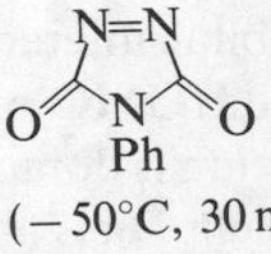 (−50°C, 30 min.)		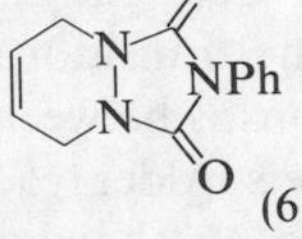 N, NPh, O, O (61%)	g

Table 4.21 *continued*

Hererodienophile (conditions)	Diene	Product	Refs.

(viii) PhCH$_2$O$_2$C — N=O

(PhCH$_2$O$_2$CNHOH, Et$_4$N$^+$IO$_4^-$, $-10°$C)

(72%)

h

a R. A. Ruden and R. Bonjouklian, *J. Am. Chem. Soc.*, 1975, **97**, 6892. *b* E. R. Larson and S. Danishefsky, *J. Am. Chem. Soc.*, 1982, **104**, 6458. *c* T. Imagawa, K. Sisidu, and M. Kawanisi, *Bull. Chem. Soc. Jpn.*, 1973, **46**, 2922. *d* S. D. Larsen and P. A. Grieco, *J. Am. Chem. Soc.*, 1985, **107**, 1768. *e* U. Rüffer and E. Breitmaier, *Synthesis*, 1989, 623. *f* W. Oppolzer, *Angew. Chem. Int. Edn Engl.*, 1972, **11**, 1031. *g* R. C. Cookson, S. S. H. Gilani, and I. D. R. Stevens, *J. Chem. Soc. (C)*, 1967, 1905. *h* G. W. Kirby, H. McGuigan, J. W. M. MacKinnon, D. McLean, and R. P. Sharma, *J. Chem. Soc., Perkin Trans. 1*, 1985, 1437.

reactions, in which electron-rich alkenes and even simple unactivated alkenes often participate under mild conditions, provide a very useful method of synthesis of many six-membered heterocycles.

A few of the dienophiles and dienes which incorporate heteroatoms and which participate in the Diels–Alder reaction are described briefly below. There are many other such species, not described here, which also provide routes to six-membered heterocycles: the review references in Tables 4.19 and 4.20 give information on their generation and reactions.

Heterodienophiles

Carbonyl compounds. The cycloaddition of dienes to carbonyl compounds provides a route to 4,5-dihydro-2*H*-pyrans. The reaction takes place readily between nucleophilic dienes and activated aldehydes such as trichloro-acetaldehyde and glyoxylates, RO$_2$CCHO. It is also well-established with highly electrophilic ketones such as dialkyl oxomalonates [Table 4.21 (i)] and vicinal tricarbonyl compounds. Simple aliphatic and aromatic aldehydes and ketones normally react poorly, with the exceptions mentioned below, and the reaction is virtually unknown with esters or amides.

Unactivated aldehydes can react with dienes substituted with one or more alkoxy groups under special conditions. 1-Methoxybutadiene adds to acetaldehyde, benzaldehyde and other simple aldehydes under high pressure (20 kbar or 20 000 atm). More nucleophilic dienes such as that in example (ii) of Table 4.21 undergo cycloaddition to aldehydes at normal pressure in the presence of a Lewis acid. The most common catalysts are zinc chloride, boron trifluoride, and mild Lewis acids such as the europium(III) complexes which are used in NMR as shift reagents. This is an important method of synthesis of oxygenated pyrans related to carbohydrates.

Imines. The Diels–Alder reaction of nucleophilic dienes with imines occurs readily if the imines are substituted, on carbon or on nitrogen, with electron-withdrawing groups. The imines can be cyclic or acyclic. A typical cycloaddition to an activated acyclic imine is shown in Table 4.21 (iii). Some imines are generated and trapped *in situ*, as illustrated in Fig. 4.36.[28] There are several useful intramolecular cycloadditions of activated imines: one such reaction is the synthesis of the alkaloid δ-coniceine described in Chapter 5, Fig. 5.6. Iminium cations can also undergo cycloaddition to nucleophilic dienes. One very versatile procedure is to generate iminium salts from formaldehyde and primary amines in the presence of dienes under conditions for the Mannich reaction. An example is shown in Table 4.21 (iv). Iminium salts can also be generated from imines by the use of Lewis acids such as zinc chloride.

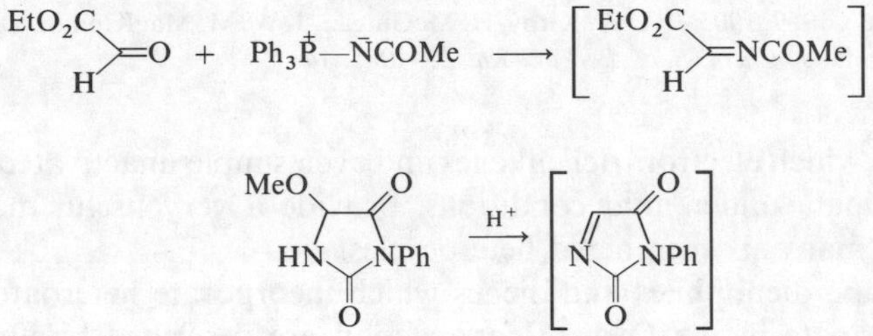

Fig. 4.36 Examples of acylimines generated *in situ*. These and related imines react well with dienes.

Nitriles. Intermolecular Diels–Alder reactions in which nitriles act as dienophiles are uncommon. A powerfully electrophilic nitrile is nearly always necessary, and even then the reaction can require forcing conditions. The most useful reactions of this type occur with arenesulphonyl nitriles: the conditions are mild and the yields can be good, as illustrated in Table 4.21 (v). Intramolecular cycloadditions to *o*-quinodimethanes take place even with unactivated nitriles [Table 4.21 (vi)].

Thiocarbonyl compounds. Thioketones are good dienophiles: they add to a range of dienes under mild conditions to give 5,6-dihydro-2*H*-thiopyrans. Thioaldehydes are unstable species but there are several methods available for generating them *in situ* and intercepting them in Diels–Alder reactions. Some of these methods are illustrated in Fig. 4.37. Thiophosgene ($Cl_2C{=}S$) and some activated thioesters are also good dienophiles.

28 See also J. Sisko and S. M. Weinreb, *Tetrahedron Lett.*, 1989, **30**, 3037; P. D. Bailey, R. D. Wilson, and G. R. Brown, *Tetrahedron Lett.*, 1989, **30**, 6781.

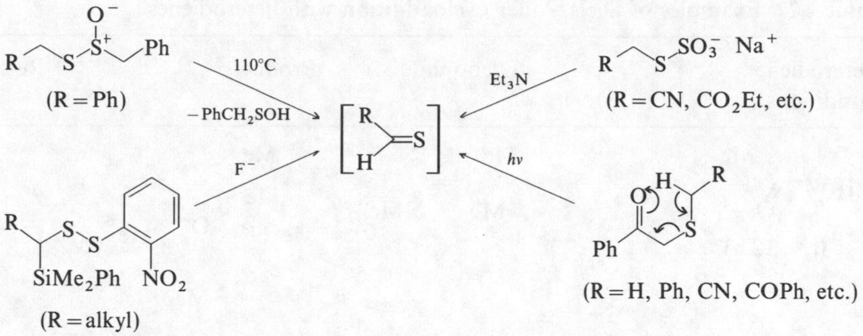

Fig. 4.37 Some methods of generation of thioaldehydes.

Azo compounds. Azobenzenes and azoalkanes rarely act as dienophiles. On the other hand, acyclic and cyclic azo compounds substituted by two electron-withdrawing groups are among the most reactive of the isolable dienophiles in the Diels–Alder reaction. The reaction of diethyl azodicarboxylate with cyclopentadiene, referred to above, was the first of many examples of cyclo-addition reactions of azodicarboxylic esters. Cyclic azocarbonyl compounds such as *N*-phenyltriazolinedione are even more reactive [Table 4.21 (vii)]. They have been widely used to intercept dienes at low temperature. The cycloaddition can also be used as a method of protection of dienes because the adducts can be degraded to regenerate the dienes.

Some acyclic azocarbonyl compounds are also capable of acting as electron-deficient *dienes* in the Diels–Alder reaction. An example of a cycloaddition of this type is shown in Table 4.22 (vi).

Nitroso compounds. Diels–Alder reactions of nitroso compounds are useful in heterocyclic synthesis mainly for two reasons. The first is that nitroso compounds are excellent dienophiles: even nitrosobenzenes undergo Diels–Alder reactions quite readily and electrophilic nitroso compounds such as nitrosocarbonyl compounds, which are generated and trapped *in situ*, are among the most reactive dienophiles known. Examples of this procedure are shown in Table 4.21 (viii) and in Fig. 5.22. The second reason is that the cycloadducts contain a relatively weak N—O bond which is easily cleaved by reduction, and this can provide an indirect route to other heterocycles. An illustration is the synthesis of a pyrrolidin-3-one shown in Fig. 4.38(a). α-Chloronitroso compounds such as 2-chloro-2-nitrosopropane have a particular use because the cycloadducts are easily hydrolysed [Fig. 4.38(b)]; the *N*-unsubstituted oxazines so formed can then be reduced to amino alcohols.

Compounds containing the N=S *group.* The most widely used dienophiles of this type are electrophilic *N*-sulphinyl compounds such as *N*-sulphinyl-*p*-toluenesulphonamide, TsN=S=O, which add to many nucleophilic dienes

Table 4.22 Examples of Diels–Alder cycloaddition with heterodienes.

Heterodiene (conditions)	Dienophile	Product	Refs.
(i) 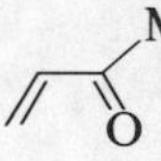 (r.t., 12 h)	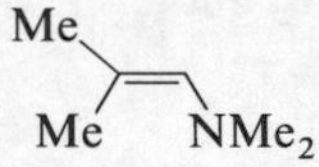	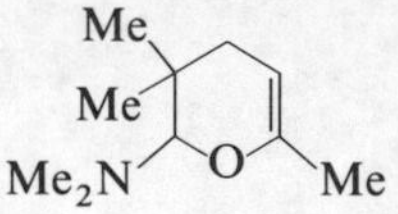 (60%)	a
(ii) 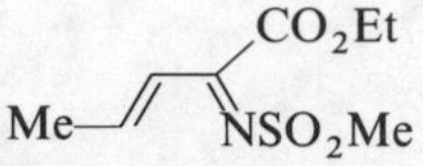 (25°C, 17-26 h)	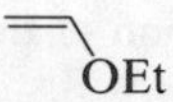	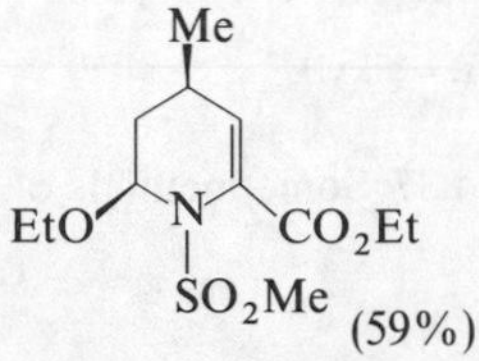 (59%)	b
(iii) 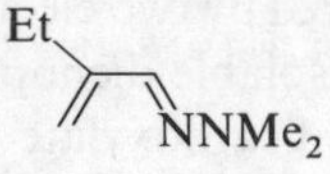 (55°C, 3 h) [R=C(Me)(CN)CHMe₂]	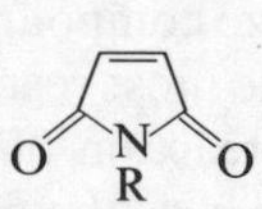	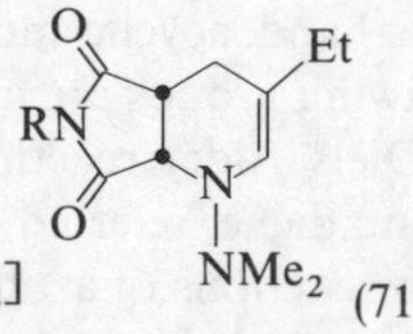 (71%)	c
(iv) 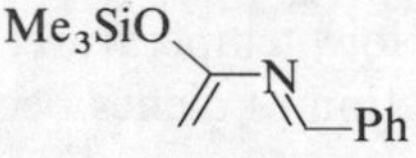 (20°C, MeOH)	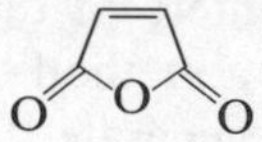	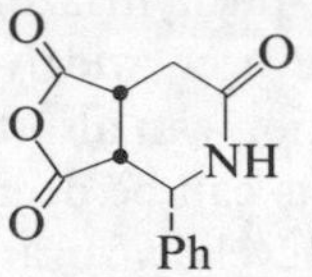 (92%)	d
(v) 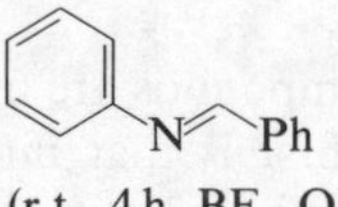 (r.t., 4 h, BF₃.OEt₂)		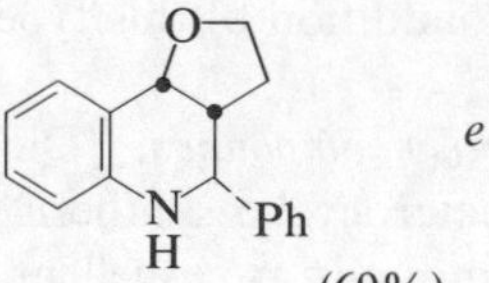 (69%)	e
(vi) 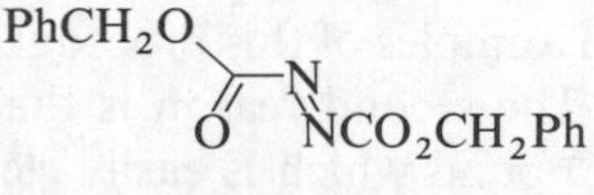 (35°C, 4 days, hv/350 nm)	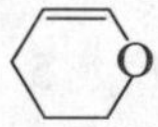	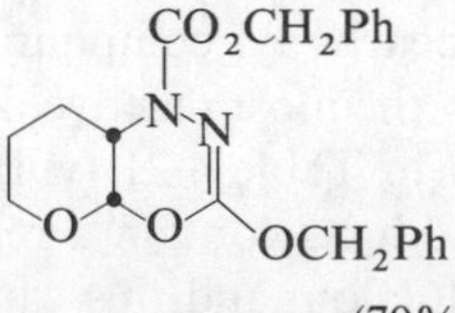 (70%)	f
(vii) 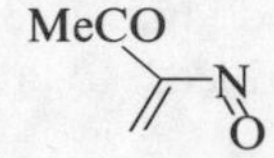 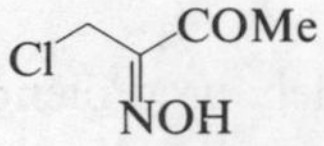 (Cl, Na₂CO₃, CH₂Cl₂, r.t., 24 h)	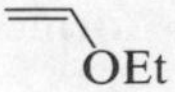	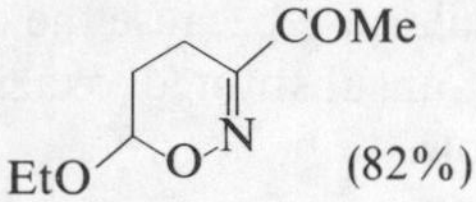 (82%)	g

at room temperature. Electrophilic sulphurdiimides, such as $TsN=S=NTs$, are also excellent dienophiles. Simple thionitroso compounds $RN=S$ are unstable and difficult to generate, but they also participate readily in the Diels–Alder reaction.[29]

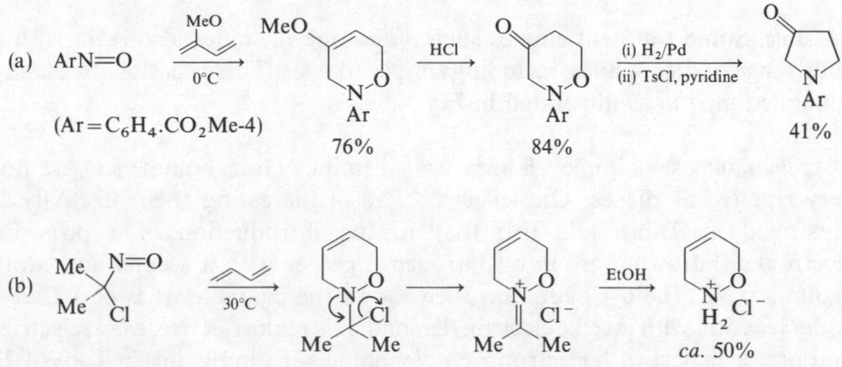

Fig. 4.38 Some uses of the Diels–Alder reactions of nitroso compounds. (a) Cycloaddition followed by reductive N—O bond cleavage; (b) cycloaddition of an α-chloronitroso compound followed by ethanolysis.

Heterodienes

αβ-Unsaturated carbonyl compounds. Cycloaddition reactions in which αβ-unsaturated carbonyl compounds (particularly aldehydes) act as dienes represent an important method of synthesis of dihydropyrans. Many of these reactions show the characteristics of 'inverse electron demand' Diels–Alder cycloadditions: that is, they go most easily with electron-rich dienophiles such as enamines. An example is shown in Table 4.22 (i). Enol ethers usually add to simple αβ-unsaturated carbonyl compounds only at elevated temperatures although Lewis acids can drastically reduce the activation energy for addition: this is presumably because of coordination of the Lewis acid to the carbonyl oxygen, so lowering the energy of the LUMO. The presence of an additional electron-withdrawing group at C-2 of the αβ-unsaturated carbonyl component makes cycloaddition to simple alkenes

---

29 M. R. Bryce and P. C. Taylor, *J. Chem. Soc., Perkin Trans. 1*, 1990, 3225.

Footnotes to Table 4.22
a I. Fleming and M. H. Karger, *J. Chem. Soc. (C)*, 1967, 226, *b* D. L. Boger, W. L. Corbett, and J. M. Wiggins, *J. Org. Chem.*, 1990, **55**, 2999. *c* K. Waldner, *Helv. Chim. Acta*, 1988, **71**, 493. *d* P. Bayard and L. Ghosez, *Tetrahedron Lett.*, 1988, **29**, 6115 (product isolated after hydrolysis of the cycloadduct). *e* T. Kametani, H. Takeda, Y. Suzuki, and T. Honda, *Synth. Commun.*, 1985, **15**, 499. *f* Y. Leblanc, B. J. Fitzsimmons, J. P. Springer, and J. Rokach, *J. Am. Chem. Soc.*, 1989, **111**, 2995 (irradiation is used to increase the concentration of the *cis* azoester, which is more reactive). *g* T. L. Gilchrist and T. G. Roberts, *J. Chem. Soc., Perkin Trans. 1*, 1983, 1283.

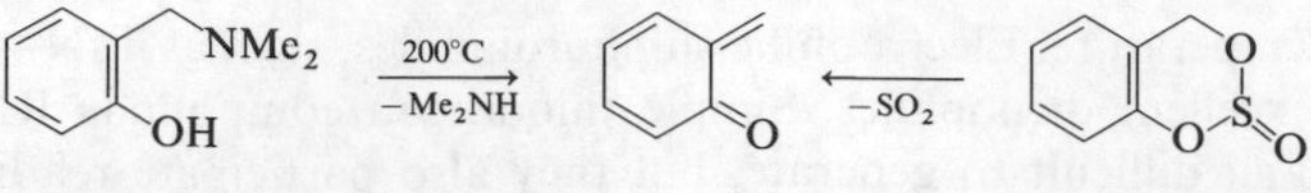

Fig. 4.39 Examples of methods of generation of *o*-quinone methide.

possible. Some transient enones such *o*-quinone methides also react with a wide range of dienophiles, including simple alkenes. These species are usually generated *in situ*, as illustrated in Fig. 4.39.

1-Azabutadienes. Simple $\alpha\beta$-unsaturated imines (1-azabutadienes) are not very reactive as dienes. One effective way of increasing their reactivity is illustrated in Table 4.22 (ii); that is, the introduction of a powerful electron-withdrawing group on nitrogen together with a second activating group at C-2. The cycloaddition then shows the characteristics of a Diels–Alder reaction with inverse electron demand: it is regioselective, *endo* selective, and proceeds best with electron-rich dienophiles. Example (iii) in Table 4.22 shows a 1-azadiene of the opposite type, in which the substituents on nitrogen and at C-3 are electron donating. This unsaturated hydrazone undergoes cycloaddition to electron-deficient dienophiles under mild conditions. It is thus possible to promote either 'inverse electron demand' or 'normal electron demand' Diels–Alder reactions by a careful choice of substituents.

2-Azabutadienes. 2-Azabutadienes can participate in the Diels–Alder reaction either as electron-rich or as electron-deficient components. The most widely used electron-rich 2-azadienes are oxazoles: cycloaddition of oxazoles is described in Chapter 8. There are also several types of acyclic 2-azadienes. Example (iv) in Table 4.22 shows the cycloaddition of an electron-rich acyclic azadiene which is activated by a trimethylsilyloxy group. Reactions which can be represented as Diels–Alder reactions with inverse electron demand are exemplified by the addition of Schiff bases, derived from aniline and aromatic aldehydes, to electron-rich dienophiles such as enol ethers and enamines in the presence of Lewis acid catalysts [Table 4.22 (v)]. Some examples of six-membered heterocycles (pyrimidines and 1,2,4-triazines) acting as electron-deficient 2-azadienes are given in Chapter 5, Table 5.1.

Azadienes with additional heteroatoms. Several other heterodienes are known to participate in the Diels–Alder reaction. These incorporate in the diene skeleton a nitrogen atom and one or more additional heteroatoms, usually nitrogen or oxygen. The presence of the additional heteroatoms usually makes these heterodienes more electron deficient so that their Diels–Alder reactions take place with electron-rich dienophiles. Two examples are shown in Table 4.22 [(vi) and (vii)]: these show reactions of an acyclic azocarbonyl compound and of a nitrosoalkene with vinyl ethers. 1,2,4,5-Tetrazines have been known for a long time to act as electron-deficient dienes in the Diels–Alder reaction: these reactions are described in Chapter 7.

4.3.4 [2 + 2] *Cycloaddition*

Several four-membered heterocycles can be prepared by [2 + 2] cycloaddition reactions but most of the reactions are not general. The range of substituents which can be incorporated is limited, and the inherent instability of many four-membered heterocycles means that side reactions often compete or predominate. The most important of these are cycloreversion and expansion to give six-membered rings (Fig. 4.40).

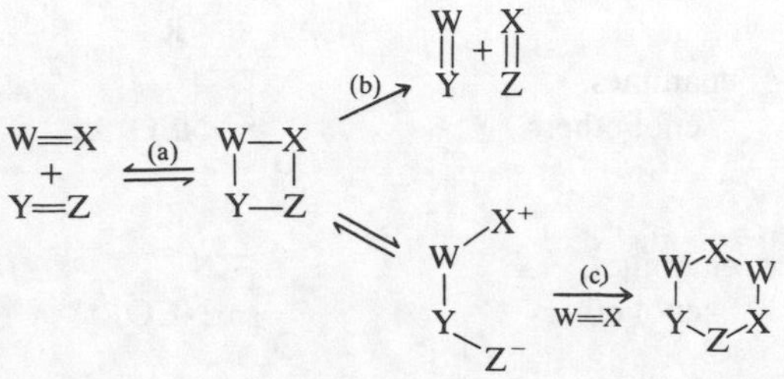

Fig. 4.40 Competing reactions in [2 + 2] additions. (a) Cycloreversion to starting materials; (b) cycloreversion with dismutation; (c) formation of 2:1 adducts.

The most useful reaction partners are ketenes and heterocumulenes such as isocyanates, isothiocyanates, and carbodiimides. Azocarbonyl compounds also undergo thermal [2 + 2] cycloadditions with a range of olefins. Some of the heterocycles which can be prepared in good yields are listed in Table 4.23. Chlorosulphonyl isocyanate [example (i)] is a particularly good reagent for the formation of β-lactams because it adds to a wide range of nucleophilic olefins, and because the chlorosulphonyl group is readily removed from the cycloadducts by hydrolysis. Additions to olefins are generally highly stereoselective, but the reaction rates are much greater in polar than in nonpolar solvents. The most likely mechanism for the addition is a stepwise one, in which a dipolar intermediate is formed but which rapidly collapses to give the azetidinone (Fig. 4.41). Although alternative, thermally allowed concerted $[_\pi 2_s + _\pi 2^a]$ mechanisms can be formulated for the cycloaddition of chlorosulphonyl isocyanate and of other heterocumulenes, stepwise mechanisms seem more likely in most cases. Dipolar intermediates have been detected or intercepted in some of the additions, such as that of ketenes to imines.

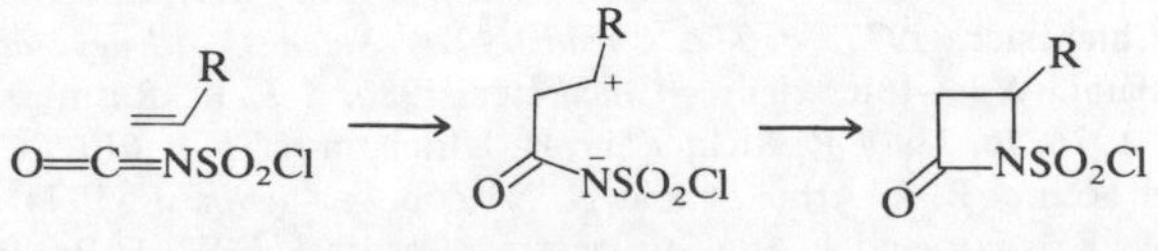

Fig. 4.41 Stepwise mechanism for chlorosulphonyl isocyanate addition.

Table 4.23 Reagents for the synthesis of four-membered heterocycles by [2+2] addition.[a,b,c]

Reagent	Typical reaction partners	Products	Refs. and notes
(i) $ClSO_2N{=}C{=}O$	alkenes, dienes, enol ethers	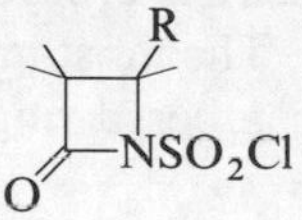 (β-lactam) R, NSO_2Cl, O	d, e
(ii) $ArSO_2N{=}C{=}O$	enamines, enol ethers	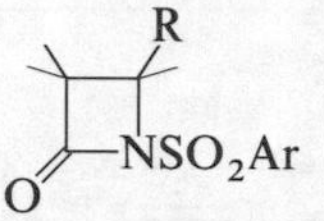 R, NSO_2Ar, O	a, e
(iii) $RCON{=}C{=}O$	enamines, enol ethers	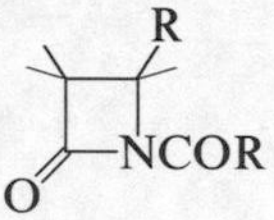 R, NCOR, O	f
(iv) $ArN{=}C{=}O$	imines	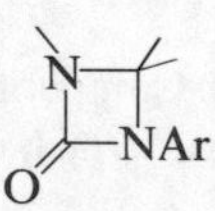 N, NAr, O	e
(v) $RSO_2N{=}C{=}S$	enamines	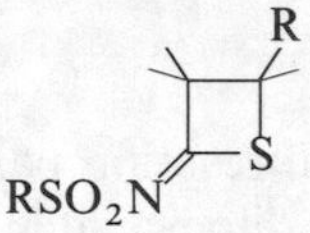 R, S, RSO_2N	g
(vi) $R_2C{=}C{=}O$	activated carbonyl compounds	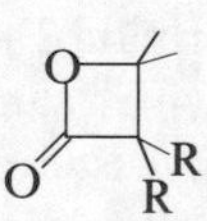 O, O, R, R	h
	imines	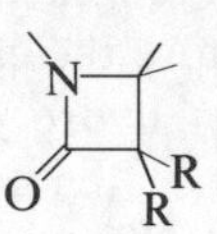 N, O, R, R	h
(vii) $RN{=}C{=}NR$	ketenes	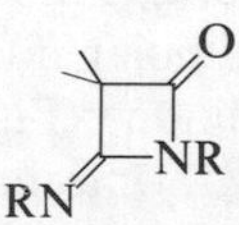 O, NR, RN	h, i
(viii) $RCON{=}NCOR$	cyclic alkenes, enol ethers	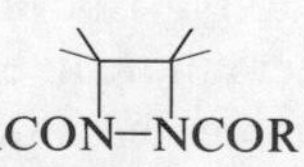 $RCON{-}NCOR$	j

a For a general survey see H. Ulrich, *Cycloaddition Reactions of Heterocumulenes*, Academic Press, New York, 1967. *b The Chemistry of Cyanates and their Thio Derivatives*, ed. S. Patai, Wiley-Interscience, Chichester, 1977. *c The Chemistry of Ketenes, Allenes, and Related Compounds*, ed. S. Patai, Wiley-Interscience, Chichester, 1980. *d* J. K. Rasmussen and A. Hassner, *Chem. Rev.*, 1976, **76**, 389. *e* R. Richter and H. Ulrich, in ref. *b*. p. 619. *f* Competitive [4+2] addition often occurs: B. A. Arbuzov and N. N. Zobova, *Synthesis*, 1974, 461; 1982, 433. *g* L. Drobnica, P. Kristián, and J. Augustín, in ref. *b*, p. 1003. *h* W. T. Brady, in ref. *c*, p. 279. *i* Y. Wolman, in ref. *c*, p. 721. *j* C. J. Moody, *Adv. Heterocycl. Chem.*, 1982, **30**, 1.

It is also possible to form four-membered rings by photochemical [2+2] addition. The most important such process in heterocyclic chemistry is the Paterno–Büchi reaction, which is the photoaddition of olefins to aldehydes or ketones.[30] Oxetanes are formed by the addition of the excited-state carbonyl compound (the singlet or triplet formed by n → π* excitation) to olefins. The reaction has been observed with simple alkenes, enol ethers, and electron-deficient olefins such as acrylonitrile. An example is shown in Fig. 4.42; the products in this case can be rationalized as being formed through diradical intermediates, with that from the more stable diradical predominating.

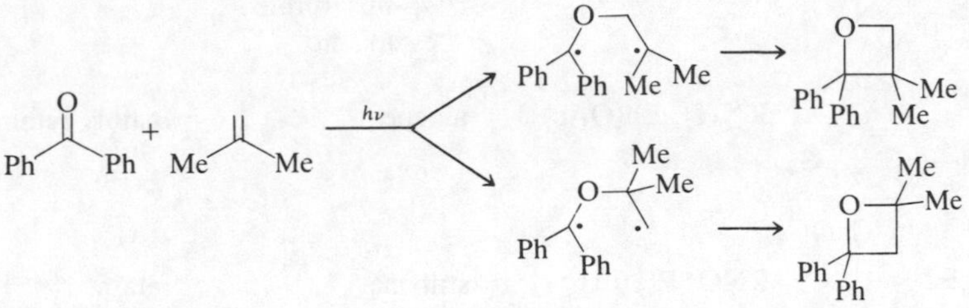

Fig. 4.42 The Paterno–Büchi reaction.

4.3.5 Cheletropic reactions

Reactions (d) and (e) of Fig. 4.20 represent types of cycloaddition process in which both new σ-bonds are made to the same atom. These so-called cheletropic reactions have a limited utility as methods of heterocyclic synthesis. Reactions of type (d) are represented by the cycloaddition of carbenes and nitrenes to double bonds. Many types of carbene add to olefins to give cyclopropanes but their addition to multiple bonds containing heteroatoms is not a general reaction. There are a few syntheses of aziridines and of thiiranes which are based on the addition of carbenes to C=N and C=S bonds.[31] The addition of nitrenes to olefins provides a good route to a limited range of *N*-substituted aziridines: some types of nitrene (for example, alkylnitrenes and simple arylnitrenes) do not add to olefins. Examples of the preparation of aziridines by nitrene cycloaddition are shown in Table 4.24.

Cheletropic reactions of type (e) in Fig. 4.20 are exemplified by the addition of sulphur dioxide to butadiene to give 2,5-dihydrothiophene-1,1-dioxide, **10**. The reaction is reversed at higher temperatures and compound

30 Review: H. Meier, in *Methoden der Organischen Chemie* (Houben Weyl), Vol. 4/5b, Georg Thieme Verlag, Stuttgart, 1975, p. 838.

31 Review: A. P. Marchand, in *The Chemistry of Double-bonded Functional Groups*, ed. S. Patai, Wiley-Interscience, London, 1977, p. 533.

Table 4.24 Synthesis of aziridines by the addition of nitrenes to olefins.[a]

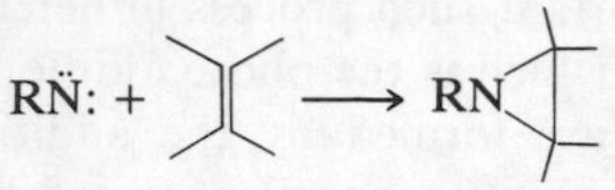

$$R\ddot{N}: + \;\;\;\;\longrightarrow\;\; RN$$

R	Precursor and conditions	Typical olefins	Stereochemistry[b]
(i) CO_2Et	RN_3, $h\nu$	alkenes, dienes	mainly retained
(ii) $COCMe_3$	RN_3, $h\nu$	alkenes	retained
(iii) 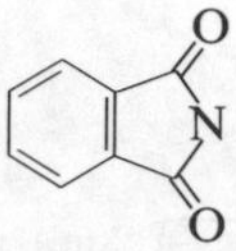 ![phthalimide structure]	RNH_2, $Pb(OAc)_4$	alkenes, dienes, $\alpha\beta$-unsaturated esters, etc.	retained
(iv) 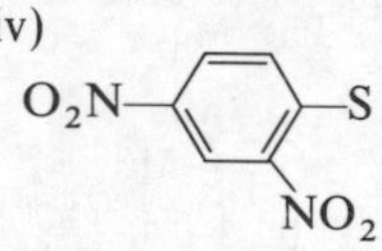 O_2N—⟨⟩—S, NO_2	RNH_2, $Pb(OAc)_4$	alkenes	mainly retained[c]
(v) C_6F_5	RNO, $P(OEt)_3$	stilbene	retained[d]

a For reviews and further examples, see J. A. Deyrup, in *Small Ring Heterocycles*, Part 1, ed. A. Hassner, Wiley-Interscience, New York, 1983, p. 1; W. Lwowski, *Nitrenes*, Wiley-Interscience, New York, 1970. *b* Refers to retention of olefin stereochemistry in the product. *c* R. S. Atkinson, B. D. Judkins, and N. Khan, *J. Chem. Soc., Perkin Trans. 1*, 1982, 2491. *d* R. A. Abramovitch, S. R. Challand, and Y. Yamada, *J. Org. Chem.*, 1975, **40**, 1541.

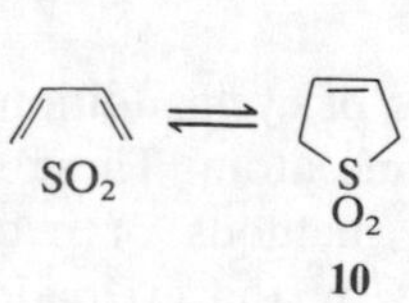

$$\text{(diene)} \;\; SO_2 \;\rightleftharpoons\; \underset{O_2}{\overset{}{S}}$$

10

10 can be used to generate butadiene. Other acyclic dienes react in a similar way and the stereochemistry of the diene is retained in the adduct.[32]

4.3.6   Ene reactions

The ene reaction is shown schematically in Fig. 4.43(a). It bears a close mechanistic relationship to [4 + 2] cycloaddition although the intermolecular ene reaction does not lead to the formation of a cyclic product. In an intramolecular ene reaction, however, the formation of the new σ-bond to the allylic carbon atom suffices to produce a new ring system. The reaction shows high stereoselectivity and has been used for the synthesis of several saturated five- and six-membered heterocycles. Some examples are given in Table 4.25.

32 Review: T. S. Chou and H. H. Tso, *Org. Prep. Proceed. Int.*, 1989, **21**, 257.

(a) 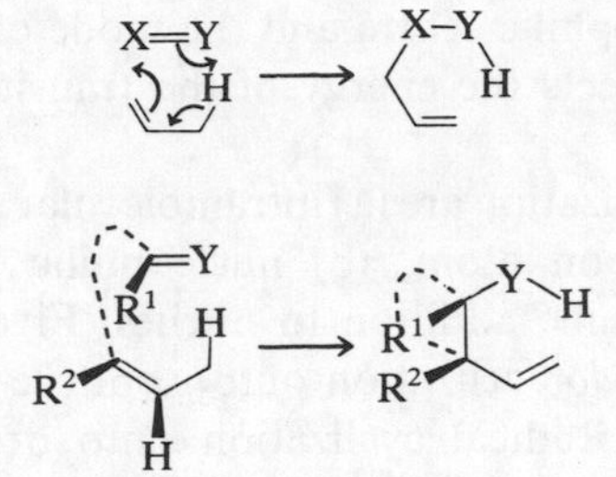
(b)

Fig. 4.43 (a) The ene reaction. (b) An intramolecular version, showing the stereochemistry of the process.

Table 4.25 Synthesis of heterocycles by intramolecular ene reactions.[a]

	Precursor	Product	Refs. and notes
(i)	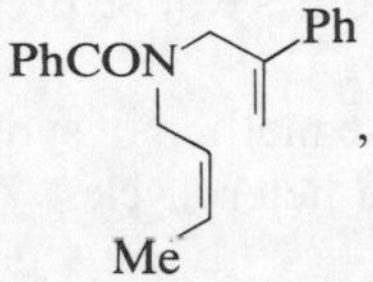, 230°C	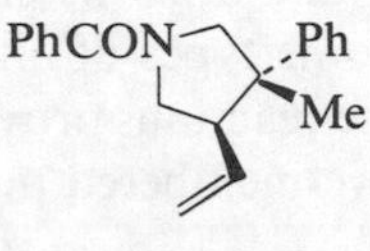	*b*
(ii)	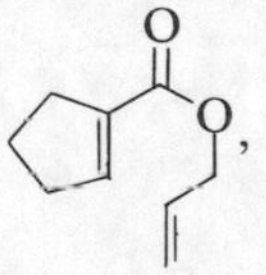, 300–400°C	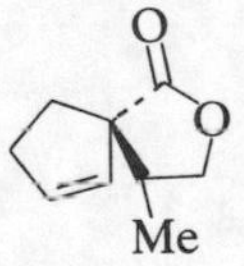	*c*
(iii)	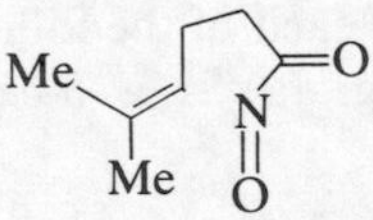, 80°C	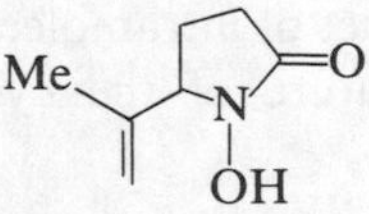	*d*

a For a review and further examples see W. Oppolzer and V. Snieckus, *Angew. Chem. Int. Edn Engl.*, 1978, **17,** 476. *b* W. Oppolzer, E. Pfenninger, and K. Keller, *Helv. Chim. Acta*, 1973, **56,** 1807. *c* M. Bortolussi, R. Bloch, and J. M. Conia, *Bull. Soc. Chim. Fr.*, 1975, 2727. *d* G. E. Keck and R. R. Webb, *Tetrahedron Lett.*, 1979, 1185; *J. Am. Chem. Soc.*, 1981, **103,** 3173.

Summary

1. Cyclization reactions are, in general, most favourable when they produce five- and six-membered rings and least favourable for the formation of eight- to eleven-membered rings.

2. The hybridization of the electrophilic centre and the mode of cyclization onto that centre (*exo* or *endo*) affects the energy of the transition state in cyclization reactions.

3. The most important types of cyclization are (a) intramolecular nucleophilic displacement at a saturated carbon atom, (b) nucleophilic addition to carbonyl groups, and (c) nucleophilic addition to nitriles. Five-membered heterocycles are also formed by *endo* cyclization onto isonitrile groups and by 1,5-dipolar electrocyclization. Radical cyclization onto double bonds generally takes place by *exo* addition.

4. 1,3-Dipolar cycloaddition is an important method of synthesis of five-membered heterocycles in which the 1,3-dipolar species add stereo-selectively to double and triple bonds. Frontier molecular orbital theory can be used to account for the stereoselectivity and the regioselectivity of the addition.

5. Heteroatoms can be incorporated into the diene or dienophile component of the Diels–Alder reaction, and the reaction provides a route to six-membered heterocycles which is generally regio- and stereoselective.

6. Four-membered heterocycles are formed by [2+2] cycloadditions of ketenes, isocyanates, and other heterocumulenes. Aziridines can be prepared by the addition of some types of nitrene to olefins.

7. Intramolecular ene reactions provide a stereoselective method of synthesis of some saturated five-membered nitrogen and oxygen heterocycles.

Problems

1. Predict the product of intramolecular cyclization of each of the following and analyse the nature of the cyclization step [e.g. (a) is a 5-*exo-dig* cyclization].

(a) 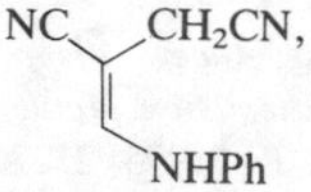 NC CH$_2$CN, KOEt
         NHPh

(b) 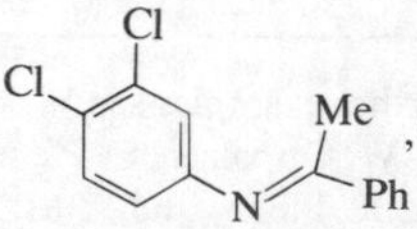 Cl, Cl ... Me, Ph, N , KOBu^t/NaNH$_2$

(c) 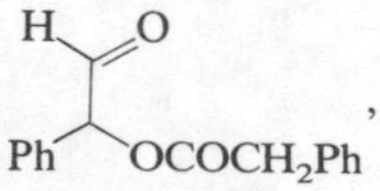 H O
     Ph OCOCH$_2$Ph , heat

(d) PhC≡C[CH(OH)]$_2$C≡CPh, NaOEt

(e) 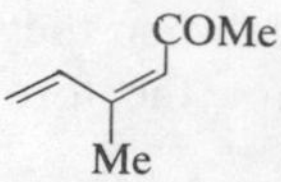 COMe, TsNHNH$_2$, H$^+$
     Me

(f) 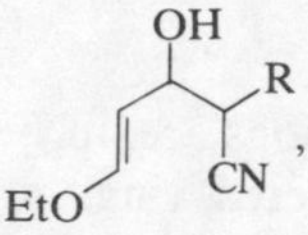 OH
     EtO R CN , HBr/MeCO$_2$H

2. Suggest syntheses of the following heterocycles from the reagents specified, together with any other suitable reagents.

(a) 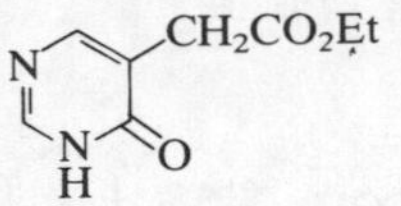   from   EtO₂CCH(CHO)CH₂CO₂Et

(b) 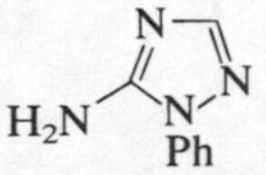   from   EtOCH=NCN

(c) 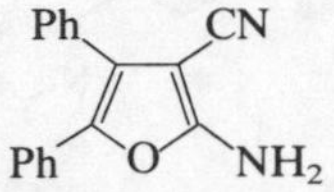   from   PhCOCH(OH)Ph

(d) 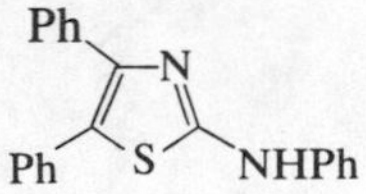   from   PhNHCSNH₂

(e) 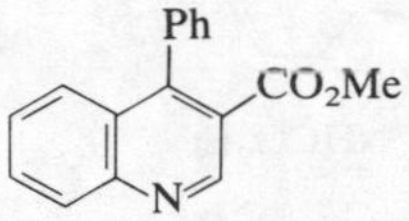   from

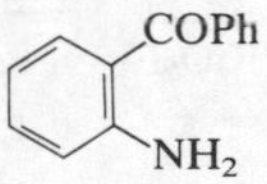

3. Write down the structures of 1,3-dipoles and dipolarophiles appropriate for the construction of each of the following.

(a) 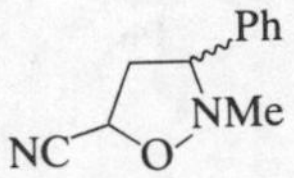   (b) 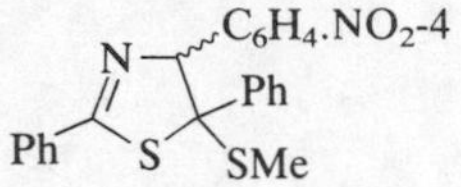   (c)

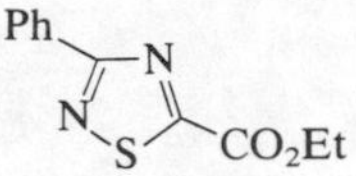

(d) 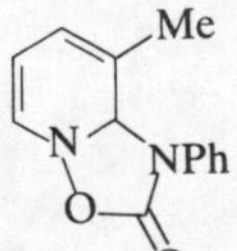   (e) 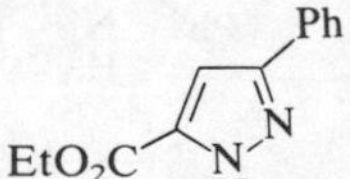   (f)

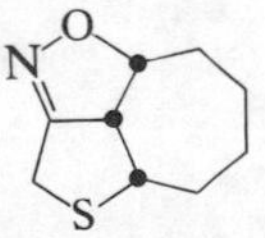

4. Suggest mechanistic pathways for each of the following.

(a)

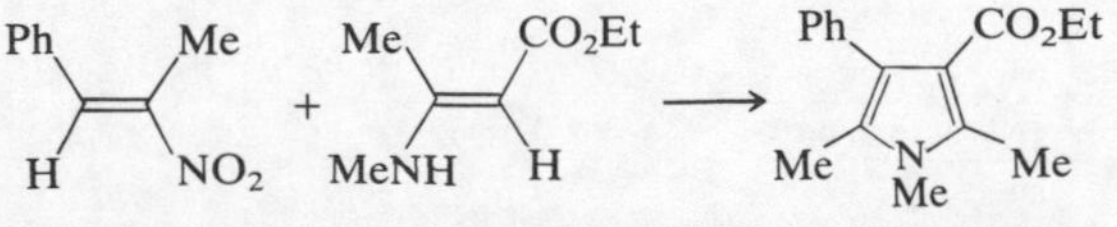

(b)

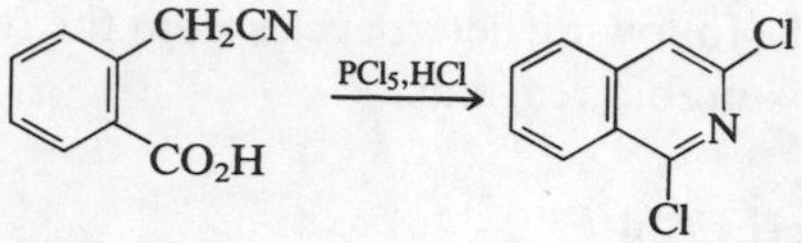

$$\xrightarrow{PCl_5, HCl}$$

(c)

$$\xrightarrow{heat}$$

(d)

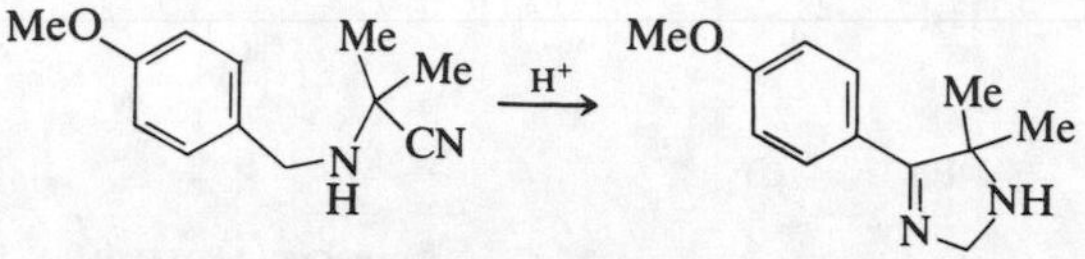

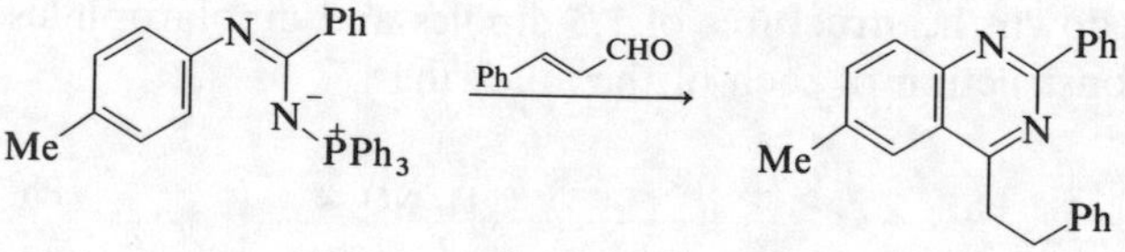

$$\xrightarrow{H^+}$$

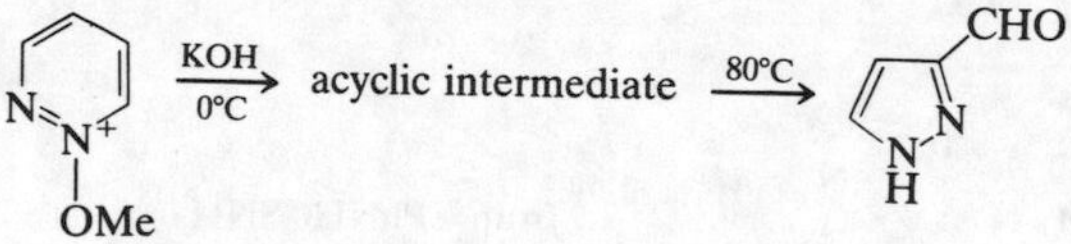

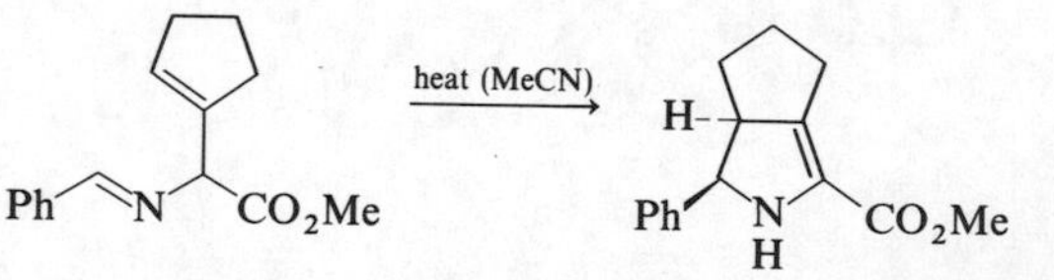

(e)

$$\xrightarrow[0°C]{KOH}$$ acyclic intermediate $$\xrightarrow{80°C}$$

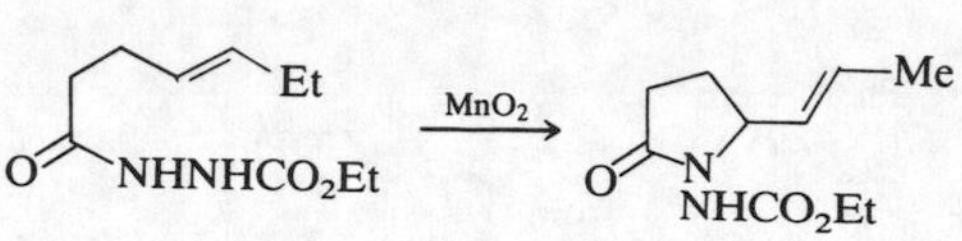

(f)

$$\xrightarrow{MnO_2}$$

(g)

$$\xrightarrow{Ph \diagup CHO}$$

(h)

$$\xrightarrow{heat \ (MeCN)}$$

(i)

$$\xrightarrow[AIBN, \ heat]{Bu_2SnH}$$

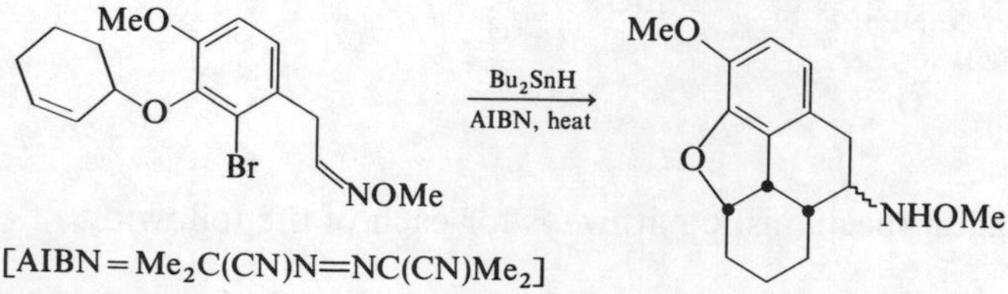

$[AIBN = Me_2C(CN)N{=}NC(CN)Me_2]$

(j)

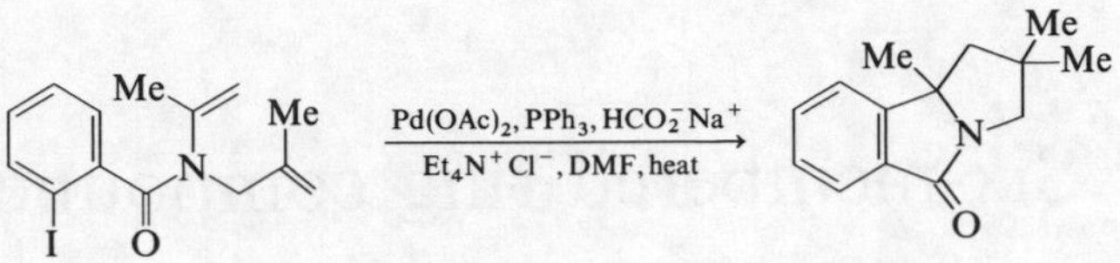

$$\xrightarrow[\text{Et}_4\text{N}^+\text{Cl}^-,\,\text{DMF, heat}]{\text{Pd(OAc)}_2,\,\text{PPh}_3,\,\text{HCO}_2^-\text{Na}^+}$$

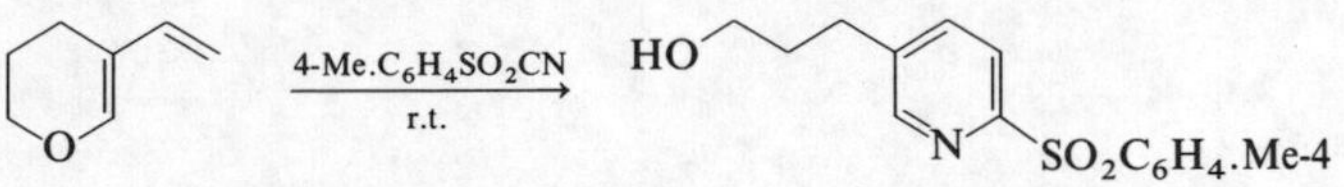

(k)

$$\xrightarrow[\text{r.t.}]{4\text{-Me.C}_6\text{H}_4\text{SO}_2\text{CN}}$$

5 Six-membered ring compounds with one heteroatom

5.1 Introduction

It was shown in Chapter 2 that a CH group of benzene could formally be replaced by a nitrogen atom, or by an atom of one of the other Group V elements, without substantially altering the character of the π-orbitals. The unsaturated six-membered heterocycles of this type, which are the subject of this chapter, can thus be regarded as analogues of benzene. Pyridine is one of the most abundant and best known of the aromatic heterocycles but the other group V 'heterobenzenes' are, as yet, laboratory research materials. Most of the chapter is concerned with the chemistry of pyridine and the two benzopyridines, quinoline and isoquinoline. Some other types of fused pyridine ring system, including quinolizines and indolizines in which the nitrogen is at a ring junction, are also described briefly.

It is also possible to incorporate oxygen or another Group VI element into a fully unsaturated six-membered ring, but the heteroatom bears a positive charge. The properties of the pyrylium cation are described, since this cation and the related pyrones occur widely in natural products derived from plants.

5.2 Pyridines[1]

5.2.1 Introduction

Pyridine and several methyl- and ethylpyridines are available on a large scale from the carbonization of coal. Coal tar contains about 0.2% of a mixture of pyridine bases, which are extracted with acid and then separated. Several large-scale synthetic routes to pyridine and methylpyridines are also available. For example, the vapour-phase reactions of acetaldehyde, formaldehyde and ammonia over a silica–alumina catalyst give pyridine and a mixture of methylpyridines. Acrolein and ammonia, and butadiene, formaldehyde and

---

1 Review: *Pyridine and its Derivatives*, Supplement, ed. R. A. Abramovitch, Wiley-Interscience, New York, 1974.

ammonia have also been used as starting materials. Several more highly substituted pyridines are also prepared commercially by cyclization reactions [see Fig. 5.4(f)].

Pyridine is a water-miscible liquid, b.p. 115°C, with an unpleasant odour. It is also an excellent polar solvent, a base (pK_a = 5.23), and a donor ligand in metal complexes. The three *methylpyridines* (formerly called picolines) are all obtained from coal tar and can be manufactured by ring synthesis. 3-Methylpyridine is a commercially important precursor to pyridine-3-carboxylic acid (nicotinic acid), one of the B group of vitamins. The di- and trimethylpyridines are also constituents of coal tar. The dimethylpyridines were formerly given the trivial name 'lutidines' and the trimethylpyridines, 'collidines'.

Pyridine-3-carboxamide (nicotinamide) occurs as a component of the structure of the important coenzymes nicotinamide adenine dinucleotide (NAD$^+$), **1** (R = H), and its phosphate (NADP$^+$), **1** [R = PO(OH)$_2$]. The latter coenzyme, one of the B$_2$ complex of vitamins, occurs in red blood corpuscles and participates in biochemical redox reactions (see Section 5.2.10).

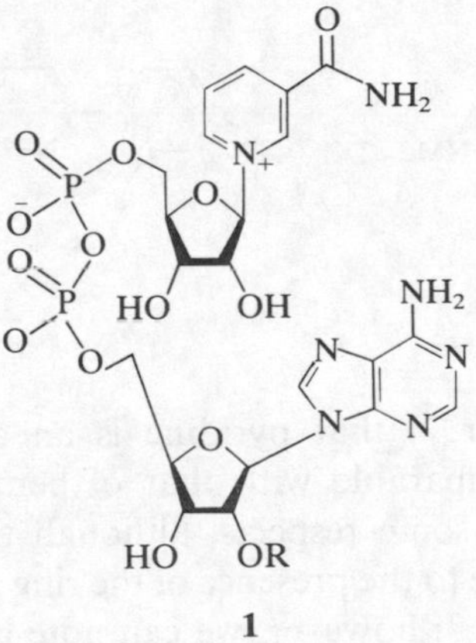

1

Pyridoxol (vitamin B$_6$), **2** (R = CH$_2$OH), occurs in yeast and wheatgerm, and is an important food additive. Related compounds with common trivial names are pyridoxal (R = CHO) and pyridoxamine (R = CH$_2$NH$_2$). The 5-phosphate of pyridoxal is a coenzyme in decarboxylation and trans-amination reactions of α-amino acids.

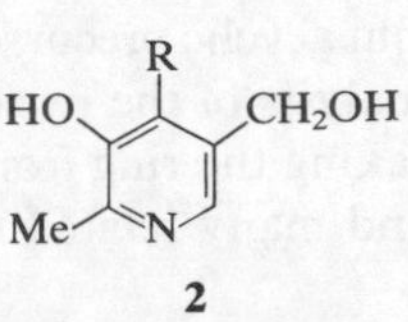

2

There are many other naturally occurring pyridines, the tobacco alkaloid nicotine (structure **6** in Chapter 6) being the best known.

Among the many pyridines which have been developed as pharmaceuticals are piroxicam, **3**, an anti-inflammatory, nifedipine, **4**, and amlodipine, **5**, which are effective for the treatment of angina, and pinacidil, **6**, used to treat hypertension. Several bipyridyls are effective herbicides, notably paraquat, **7** (X = halide or MeSO$_4$), and diquat, **8** (X = halide).

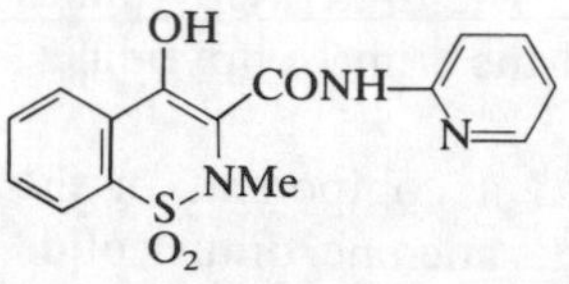

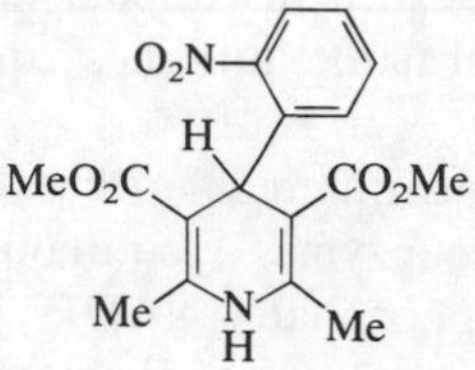

There is abundant evidence (Chapter 2) that pyridine is an aromatic molecule, with a resonance energy comparable with that of benzene. Its chemistry resembles that of benzene in some respects, although there are many important differences which are due to the presence of the ring nitrogen atom. These are discussed in Section 5.2.3; however, we can note here that electrophilic substitution reactions are generally more difficult than in the benzene series. One consequence of this is that complex pyridines are often made by ring synthesis rather than by substitution, and methods for synthesis of the pyridine ring are relatively more important than in the benzene series.

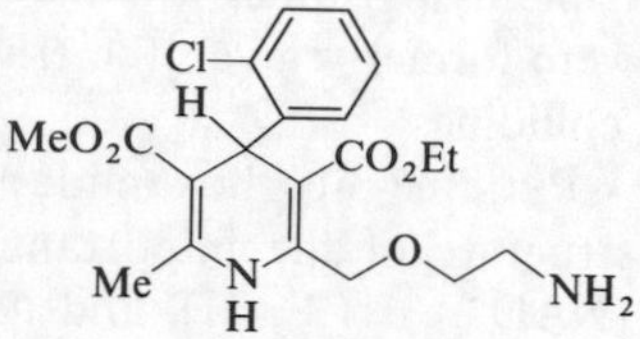

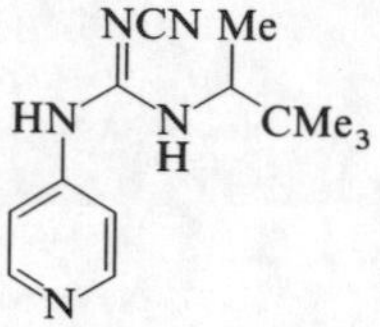

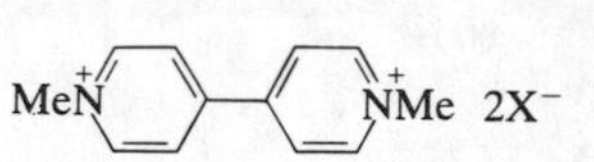

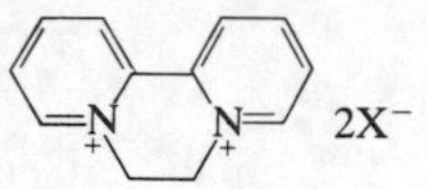

5.2.2   Ring synthesis

The classical methods of constructing pyridines from acyclic precursors make use of cyclization reactions. The retrosynthetic analysis of the pyridine ring system in Fig. 5.1 shows three possible ways of making the ring from readily available starting materials. All three methods, and many related methods, are used to prepare pyridines.

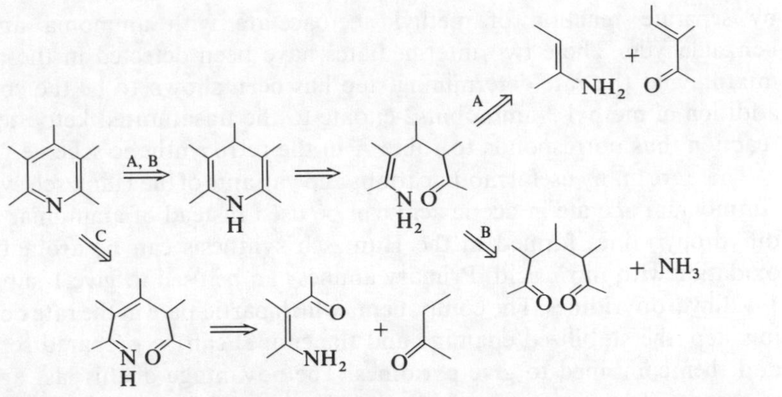

Fig. 5.1 Analysis of three cyclization routes to pyridines.

The Hantzsch synthesis and related reactions.[2] The reaction of 1,3-dicarbonyl compounds with aldehydes and ammonia was first described by Hantzsch over a century ago as a route to 1,4-dihydropyridines. The reaction has been widely used since as a method of synthesis of dihydropyridines and pyridines with symmetrical substitution patterns. The reaction has also come into prominence in recent years following the discovery that some products of the Hantzsch synthesis have useful properties. For example, nifedipine, **4**, can act as a calcium antagonist *in vivo* and has therapeutic value as an antihypertensive agent.[3]

The reaction and its mechanism are illustrated in Fig. 5.2. Four components participate in the reaction: in this example they are methyl acetoacetate (two moles), benzaldehyde, and ammonia. Two intermediates are produced

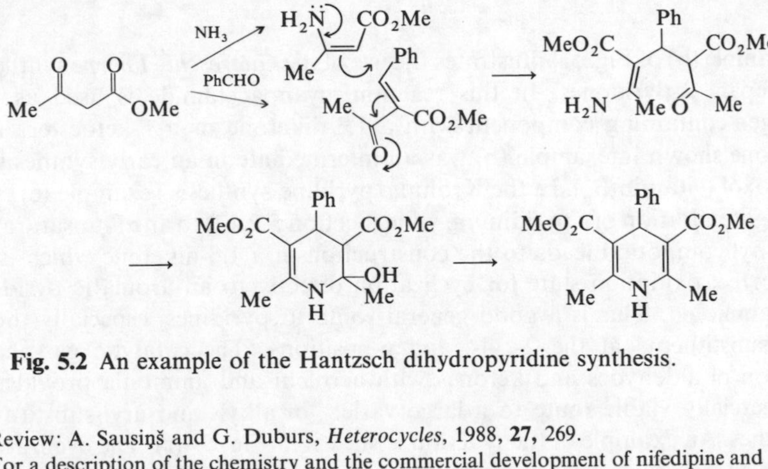

Fig. 5.2 An example of the Hantzsch dihydropyridine synthesis.

2 Review: A. Sausiņš and G. Duburs, *Heterocycles*, 1988, **27**, 269.

3 For a description of the chemistry and the commercial development of nifedipine and related compounds, see F. Bendall, *Dihydropyridines in Action* (DTI/Pfizer), Hobsons Scientific, Cambridge, 1989.

by separate reaction of methyl acetoacetate with ammonia and with benzaldehyde. These two intermediates have been detected in the reaction mixture and the rate determining step has been shown to be the conjugate addition of methyl 3-aminobut-2-enoate to the unsaturated ketoester.[4] The reaction thus corresponds to route A in the retrosynthetic scheme.

There are many useful modifications and variants of the Hantzsch synthesis. Ammonium acetate in acetic acid can be used instead of ammonia, and the dihydropyridines formed in the Hantzsch synthesis can be aromatized by oxidation with nitric acid. Primary amines can be used to give 1-substituted 1,4-dihydropyridines. The components which participate in the rate determining step (the stabilized enamine and the enone) can be prepared separately and then combined to give pyridines. The advantage of this last approach is that it allows unsymmetrically subsituted pyridines to be prepared; an example of a reaction of this type is shown in Fig. 5.4(a).

Route C of the retrosynthetic scheme of Fig. 5.1 is also well known as a method of preparation of pyridines: an example is shown in Fig. 5.3. Route B, the cyclization of 1,5-diketones with ammonia, is of rather limited use, but a modern variant, the Kröhnke synthesis, is described below. This and a few of the many other variants of the cyclization approach are illustrated in Fig. 5.4.

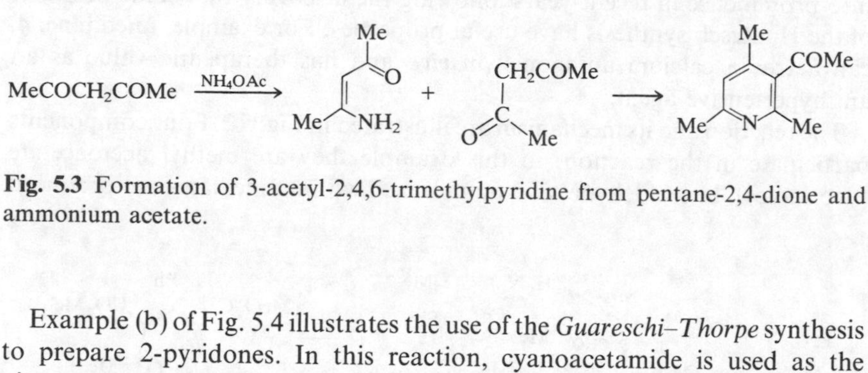

Fig. 5.3 Formation of 3-acetyl-2,4,6-trimethylpyridine from pentane-2,4-dione and ammonium acetate.

Example (b) of Fig. 5.4 illustrates the use of the *Guareschi–Thorpe* synthesis to prepare 2-pyridones. In this reaction, cyanoacetamide is used as the nitrogen-containing component with a 1,3-diketone or a β-ketoester. The pyridone shown in example (b) was an intermediate in an early synthesis of pyridoxol (vitamin B$_6$). In the Kröhnke pyridine synthesis [example (e)] the conjugate addition of a pyridinium ylide (Section 5.2.12) to an $\alpha\beta$-unsaturated carbonyl compound leads to the construction of a 1,5-diketone which is at the correct oxidation state for cyclization directly to an aromatic pyridine with ammonia. This is a good general route to pyridines, especially those with substituents at the 2-, 4-, and 6-positions. The catalytic gas-phase reaction of aldehydes and ketones with acrolein and ammonia provides a commercially viable route to a large variety of alkyl- and aryl-substituted pyridines. An example of the reaction is shown in Fig. 5.4(f). The cyclization

4 A. R. Katritzky, D. L. Ostercamp, and T. I. Yousaf, *Tetrahedron*, 1987, **43**, 5171.

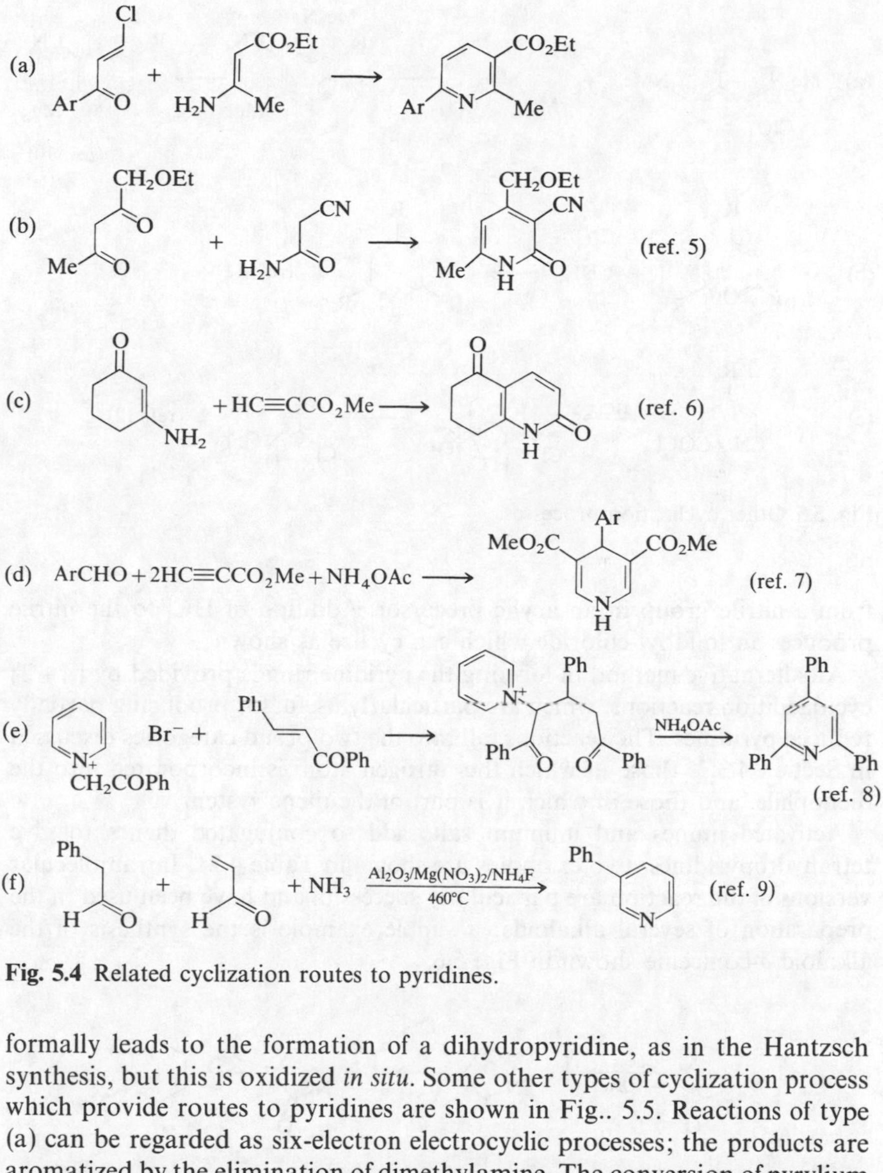

Fig. 5.4 Related cyclization routes to pyridines.

formally leads to the formation of a dihydropyridine, as in the Hantzsch synthesis, but this is oxidized *in situ*. Some other types of cyclization process which provide routes to pyridines are shown in Fig.. 5.5. Reactions of type (a) can be regarded as six-electron electrocyclic processes; the products are aromatized by the elimination of dimethylamine. The conversion of pyrylium salts to pyridinium salts shown in (b) is illustrated later, in Fig. 5.53. Reaction (c) is one of several 2-pyridone syntheses in which the ring nitrogen comes

5 S. A. Harris and K. Folkers, *J. Am. Chem. Soc.*, 1939, **61**, 1245.
6 M. A. T. Sluyter, U. K. Pandit, W. N. Speckamp, and H. O. Huisman, *Tetrahedron Lett.*, 1966, 87.
7 T. Chennat and U. Eisner, *J. Chem. Soc., Perkin Trans. 1*, 1975, 926.
8 Review: F. Kröhnke, *Synthesis*, 1976, 1.
9 Review: H. Beschke, *Aldrichimica Acta*, 1981, **14**, 13.

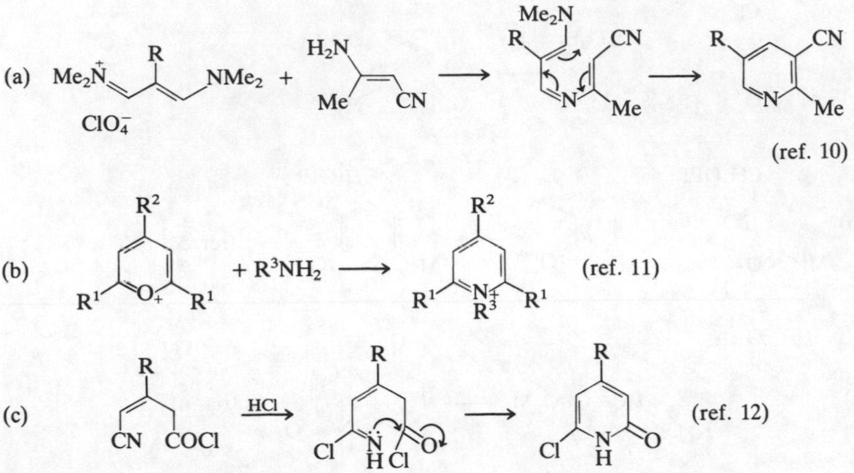

(a)

(ref. 10)

(b)

(ref. 11)

(c)

(ref. 12)

Fig. 5.5 Other cyclization processes.

from a nitrile group in an acylic precursor. Addition of HCl to the nitrile produces an imidoyl chloride which can cyclize as shown.

An alternative method of forming the pyridine ring is provided by [4+2] cycloaddition reactions, which are particularly useful for producing partially reduced pyridines. The reactions fall into the two broad categories discussed in Section 4.3.3: those in which the nitrogen atom is incorporated into the dienophile, and those in which it is part of the diene system.

Activated imines and iminium salts add to conjugated dienes to give tetrahydropyridines: two examples are shown in Table 4.21. Intramolecular versions of the reaction are particularly successful and have been used in the preparation of several alkaloids: a simple example is the synthesis of the alkaloid δ-coniceine shown in Fig. 5.6.

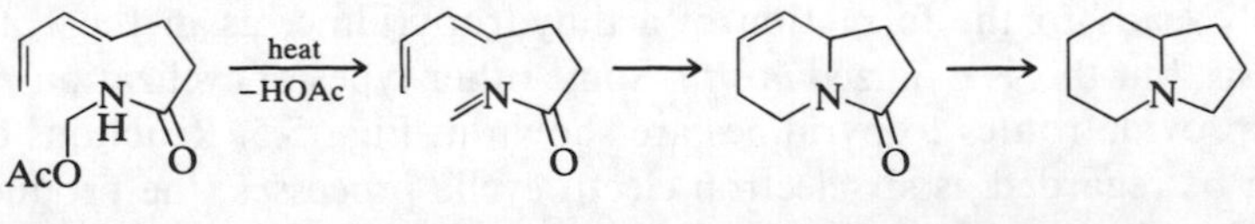

Fig. 5.6 Synthesis of δ-coniceine by intramolecular Diels–Alder addition. The intermediate imine was not isolated.

The addition of dienes to the C≡N bond should, in principle, provide a route to dihydropyridines, but, with a few exceptions, the conditions required

10 Review: J. C. Jutz, *Top. Curr. Chem.*, 1978, **73**, 125.
11 Review: A. R. Katritzky, *Tetrahedron*, 1980, **36**, 679.
12 Review: G. Simchen and G. Entenmann, *Angew. Chem. Int. Edn Engl.*, 1973, **12**, 119.

are so vigorous that the intermediate is aromatized. An example of the reaction is shown in Table 4.21.

Although several examples of addition of acyclic azadienes are shown in Table 4.22, most of the available azadienes are cyclic rather than acyclic.[13] Electron-rich alkenes or alkynes are usually required to react with them, and the initial cycloadditions are followed by further reactions to produce aromatic pyridines. Examples of such reactions are shown in Table 5.1. In all these examples, the cycloaddition is followed by a retro-Diels–Alder reaction in which a stable molecule is eliminated. In example (i), CO_2 is lost from the primary cycloadduct, in (ii) and (iii) HCN is eliminated from the adducts, and in (iv) nitrogen is extruded.

Table 5.1 Examples of pyridine synthesis by cycloaddition of alkenes and alkynes to heterocyclic dienes.[a]

	Diene	Dienophile	Conditions	Product[c]
(i)	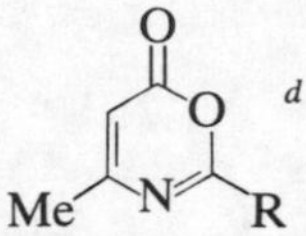	$Et_2NC≡CMe$	Et_2O, 20°C	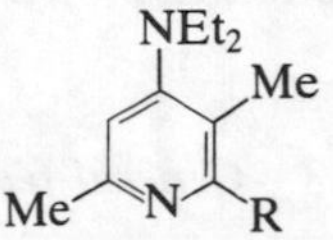
(ii)	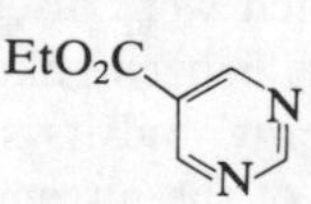	$Et_2NC≡CMe$	MeCN, 80°C	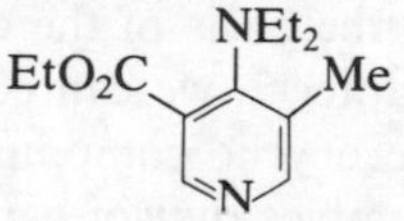
(iii)	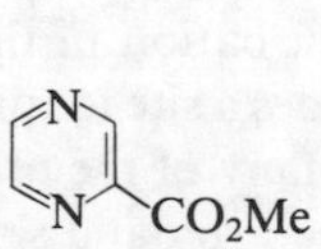	$Et_2NC≡CMe$	$CHCl_3$, 25°C	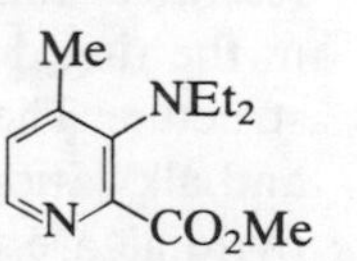
(iv)	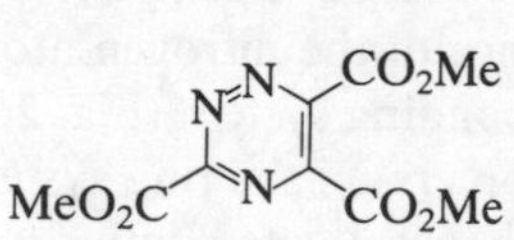	$H_2C=CHOAc$	dioxane, 100°C	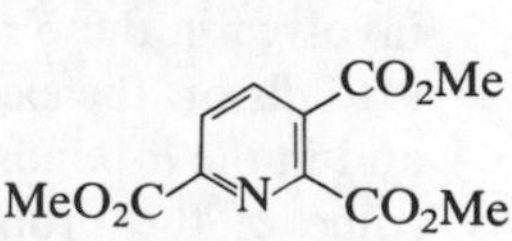

[a] For further examples and leading references see ref. 13. [b] See also the cycloaddition of oxazoles to alkenes (Section 8.5.3).

Pyridines can also be prepared by the combination of alkynes and nitriles on soluble organocobalt catalysts.[14] π-Cyclopentadienylcobalt(I) catalysts of various types can be used, or a catalyst can be generated *in situ* from cobalt(II) chloride and sodium borohydride. Two moles of acetylene or of

13 Review: D. L. Boger, *Chem. Rev.*, 1986, **86**, 781.
14 Review: H. Bönnemann and W. Brijoux, *Adv. Heterocycl. Chem.*, 1990, **48**, 177.

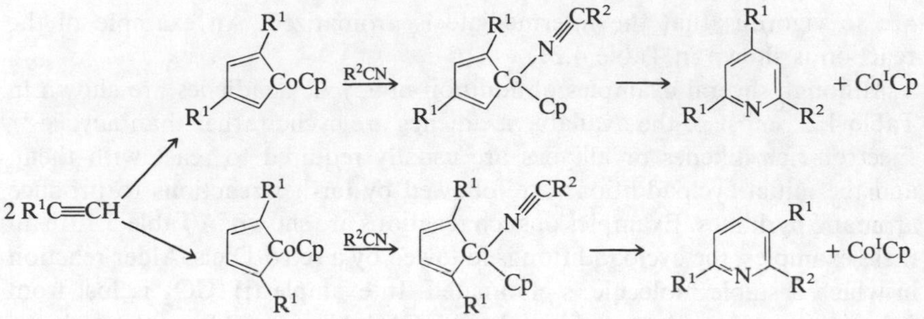

Fig. 5.7 Cobalt-catalysed formation of pyridines from alkynes and nitriles (Cp = π-cyclopentadienyl).

a substituted acetylene combine with a nitrile on heating at 120–130°C with the catalyst. The overall yields are generally good but isomeric mixtures are formed if monosubstituted acetylenes are used. A likely mechanism for the reaction is shown in Fig. 5.7.

5.2.3   General features of the chemistry of pyridines

On the basis of the criteria of aromatic character which were discussed in Chapter 2, we can conclude that pyridine is the most 'benzene-like' of the heterocyclic compounds. It has a high resonance energy and its structure resembles that of benzene quite closely. The presence of the nitrogen atom in the ring does, of course, represent a major perturbation of the benzene structure. The lone pair in the plane of the ring provides a site for protonation and alkylation which has no analogy in benzene. Many of the properties of pyridine are thus those of a tertiary amine and the aromatic sextet is not involved in these reactions. The other major influence of the nitrogen atom is to distort the electron distribution both in the π-bonding system (Fig. 2.4) and in the σ-bonds (by the inductive effect). This confers on the ring system some of the properties which are associated with conjugated imines or conjugated carbonyl compounds. The distortion of electron distribution is greater still in quaternary pyridinium salts.

We can thus expect to find reactions of pyridines and of pyridinium salts which show analogies to those of three types of model system:

(a) Tertiary amines: reactions at the nitrogen lone pair, including protonation, alkylation, acylation, N-oxide formation, and coordination to Lewis acids.
(b) Benzene: substitution reactions and resistance to addition and ring opening.
(c) Conjugated imines or carbonyl compounds: susceptibility to attack by nucleophiles at the α- and γ-carbon atoms.

This last analogy has important consequences for the nature and type of substitution reactions that pyridines undergo. The dominance of *electrophilic substitution* which typifies benzene chemistry is not apparent in the chemistry of pyridines. First, the kinetic product of the reaction of pyridine with an electrophile is that in which the electrophile is coordinated to the nitrogen lone pair. This makes electrophilic substitution at carbon even more difficult, because further reaction has to take place either on the pyridinium salt, or on the small amount of uncomplexed pyridine which may be present. Second, the attack of the electrophile on carbon is selective; it tends to go mainly at the 3- and 5-positions which have the greatest π-electron density (Fig. 2.1). Attack at these positions also leads to the formation of intermediates which are the least destabilized by the presence of the nitrogen atom. This is analogous to the preferred *meta*-substitution of deactivated benzene derivatives such as nitrobenzene by electrophiles. The resonance structures in Fig. 5.8 show the destabilizing influence of the nitrogen atom (nitrenium ion resonance forms, *) on the intermediates produced by electrophilic attack at the 2- and 4-positions.

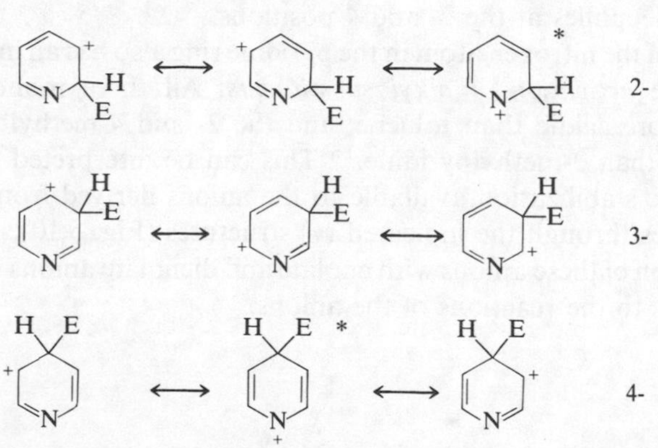

Fig. 5.8 Intermediates in the electrophilic substitution of pyridine.

These effects can be counterbalanced to some extent by the presence of cation-stabilizing substituents, such as alkoxy and amino groups, in the ring. Nevertheless, the general level of reactivity is lower than in the corresponding benzene derivatives.

Nucleophilic substitution is not a common process in benzene chemistry, but it is much easier in pyridines, particularly at the 2- and 4-positions which are activated by the nitrogen.

Nucleophilic displacement of a good leaving group by an addition–elimination mechanism occurs most readily for leaving groups at the 2- and 4-positions. The intermediates in such processes are shown in Fig. 5.9. Structures for anions in which the negative charge can be delocalized onto nitrogen (forms *) are stabilized relative to the others.

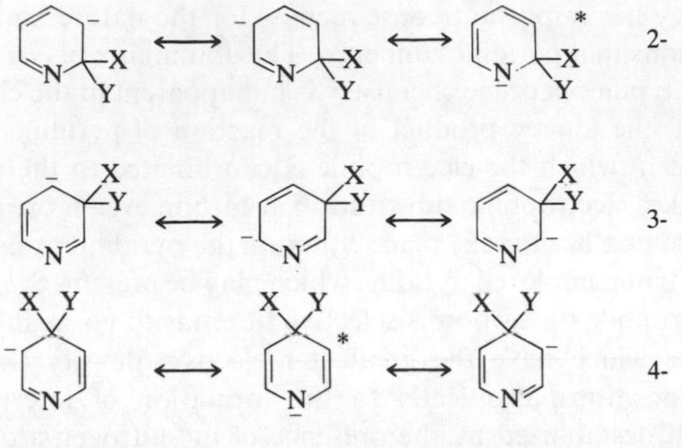

Fig. 5.9 Intermediates in the nucleophilic displacement of X^- by Y^- in X-substituted pyridines.

Nucleophilic displacement of leaving groups by the addition–elimination mechanism is easier still in *N*-alkylpyridinium salts; such salts are very readily attacked by nucleophiles at the 2- and 4-positions.

The presence of the nitrogen atom in the pyridine ring also has an important influence on the *activation of alkyl substituents*. All three monomethyl-pyridines are more acidic than toluene, and the 2- and 4-methylpyridines are more acidic than 3-methylpyridine.[15] This can be interpreted in terms of the mesomeric stabilization available to the anions derived from 2- and 4-methylpyridines through the indicated (∗) structures (Fig. 5.10).

The comparison of these anions with enolate and dienolate anions is a valid one, and extends to the reactions of the anions.

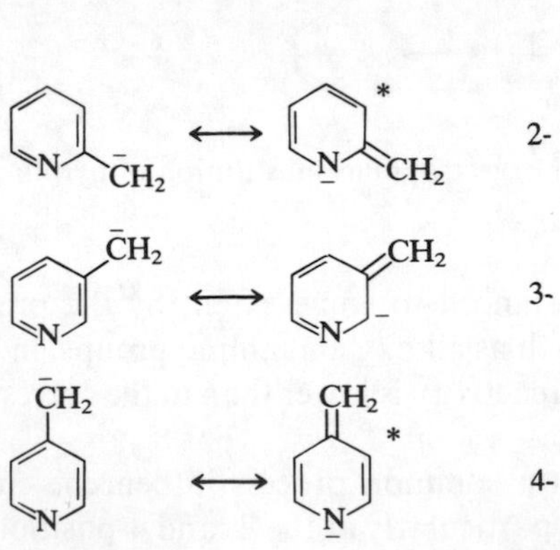

Fig. 5.10 Carbanions derived from methylpyridines.

15 Comparable pK_a values are: 2-methylpyridine, 29.5; 3-methylpyridine, 33.5; 4-methyl-pyridine, 26; toluene (estimated), *ca.* 42: G. Seconi, C. Eaborn, and A. Fischer, *J. Organometal. Chem.*, 1979, **177**, 129.

In summary, the chemistry of pyridine and substituted pyridines bears some resemblance to that of a deactivated benzene such as nitrobenzene. Substitution by electrophilic reagents is least disfavoured at the 3-position, whereas the 2- and 4-positions are activated to nucleophilic attack by the ring nitrogen atom. Alkyl substituents at the 2- and 4-positions are also activated.

5.2.4 Basicity

Ionization constants of some protonated pyridines are listed in Table 5.2.

Table 5.2 pK_a values (20°C, H_2O) of pyridines.[a]

Substituent	pK_a
None	5.23
2-NH_2	6.86
3-NH_2	5.98
4-NH_2	9.17
2-OMe	3.28
4-OMe	6.62
4-NO_2	1.61
2-CN	−0.26
4-CN	1.90

a From K. Schofield, *Hetero-aromatic Nitrogen Compounds*, Butterworths, London, 1967, p. 146.

Pyridine and simple alkylpyridines are weak bases which form salts with strong acids. The basicity is lowered by electron-withdrawing substituents, especially in the 2- and 6-positions. At the 4-position, groups capable of mesomeric stabilization of the cation increase the basicity. For example, 4-dimethylaminopyridine is protonated on the pyridine nitrogen to give a stabilized cation, **9**, with a pK_a value of 9.70.[16] 4-Aminopyridine is also a moderately strong base.

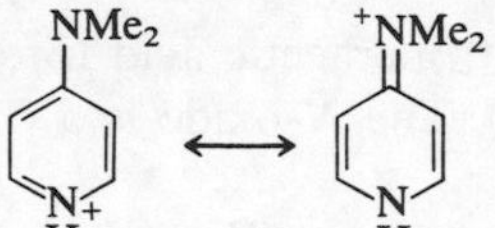

9

16 Review: G. Höfle, W. Steglich, and H. Vorbrüggen, *Angew. Chem. Int. Edn Engl.*, 1978, **17**, 569.

5.2.5 *Alkylation, acylation, and complexation on nitrogen*

Pyridine is a good nucleophile and can be alkylated with alkyl halides, tosylates, and similar compounds. *N*-Methylpyridinium iodide is a water-soluble crystalline solid. It also reacts readily, but reversibly, with acyl halides, anhydrides and similar acid derivatives and is often used as a catalyst in acylation reactions. The *N*-acylpyridinium salts **10** ($R^1 = H$) so formed are better acylating agents than acid anhydrides or acid chlorides. 4-Dimethylaminopyridine is a much superior catalyst for acylation, and will catalyse the acylation of sterically hindered and tertiary alcohols.[16] This is due to the much higher equilibrium concentration of the *N*-acylpyridinium salt **10** ($R^1 = NMe_2$) which is resonance stabilized. The nucleophile is then acylated by attack on the *N*-acylpyridinium salt (Fig. 5.11). An excess of the pyridine is usually used to neutralize the acid produced in the reaction.

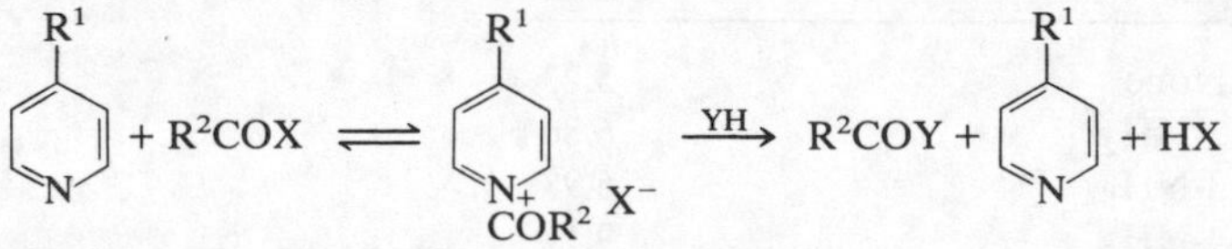

10

Fig. 5.11 Nucleophilic catalysis of acylation.

Pyridine reacts with a wide range of Lewis acids at nitrogen to give complexes, some of which are stable enough to be isolated. Examples of such compounds which are useful synthetic reagents include the pyridinium sulphonate **11**, a crystalline solid which is used to prepare sulphate esters and to sulphonate indoles,[17a] the borane complex **12**, a mild and selective reducing agent,[17b] *N*-nitropyridinium tetrafluoroborate **13**, a mild nitrating agent,[18] and *N*-fluoropyridinium triflate **14**, a useful fluorinating agent.[19]

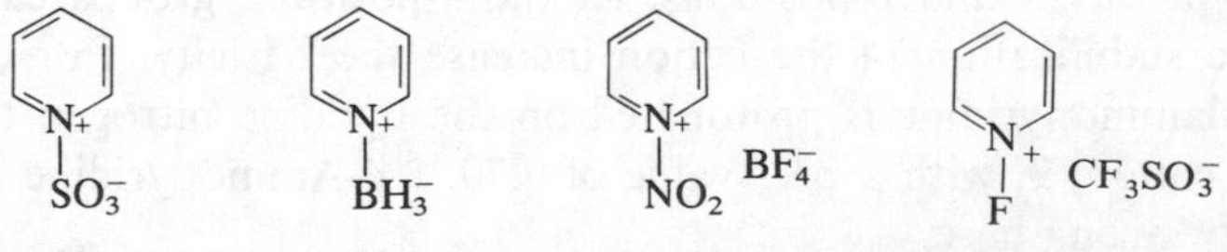

| **11** | **12** | **13** | **14** |

Pyridine can be *N*-oxidized by reaction with peracetic acid (prepared *in situ* from hydrogen peroxide and acetic acid). The *N*-oxide is a colourless

17 L. F. Fieser and M. Fieser, *Reagents for Organic Synthesis* (a) Vol. 1, p. 1127; Vol. 4, p. 473; (b) Vol. 1, p. 963; Wiley-Interscience, New York, 1967 and 1974.
18 C. A. Cupas and R. L. Pearson, *J. Am. Chem. Soc.*, 1968, **90**, 4742; G. A. Olah, S. C. Narang, R. L. Pearson, and C. A. Cupas, *Synthesis*, 1978, 452.
19 T. Umemoto, S. Fukami, G. Tomizawa, K. Harasawa, K. Kawada, and K. Tomita, *J. Am. Chem. Soc.*, 1990, **112**, 8563.

crystalline solid, m.p. 66°C, which is soluble in water. *N*-Aminopyridinium salts can be prepared by the reaction of electrophilic aminating agents (such as hydroxylamine-*O*-sulphonic acid) with pyridine. When pyridine is heated with acyl or sulphonyl azides, pyridinium *N*-imides, **15**, are formed, with the loss of nitrogen. The chemistry of the *N*-oxide and *N*-imides is discussed in section 5.2.12.

NX̄   (X = COR, SO₂R)

15

5.2.6 *Electrophilic substitution at carbon*

For the reasons discussed in Section 5.2.3, the electrophilic substitution of pyridine is a difficult process. Pyridine is itself about a million times less reactive than benzene. The pyridinium cation (which is often the major species present in the conditions required for electrophilic substitution) is very much less reactive still. Where such reactions do occur, the 3-position is usually attacked preferentially, The majority of electrophilic substitutions go by way of an addition–elimination mechanisms involving intermediates of the types shown in Fig. 5.8. One exception is the acid-catalysed hydrogen–deuterium exchange, which takes place selectively at the 2- and 6-positions. This reaction is believed to go by way of a transient pyridinium ylide intermediate, **16** (compare the analogous exchange process in imidazole discussed in Section 8.2.3).

16

Friedel–Crafts alkylation and acylation is unsuccessful with pyridine. It can be nitrated in poor yield (less than 5%) at the 3-position by reaction with fuming sulphuric acid and potassium and sodium nitrates at 300°C. 3-Chloropyridine is obtained in moderate yield by the reaction of pyridine with chlorine in the presence of 2 mol of aluminium chloride, and 3-bromopyridine can be prepared in high yield from bromine in fuming sulphuric acid at 130°C. Sulphonation of pyridine takes place with fuming sulphuric acid and a mercury(II) chloride catalyst: the 3-sulphonic acid (75–80%) is obtained at 265°C. All these reactions involve substitution of a pyridinium cation.

In view of the limitations of conventional electrophilic substitution procedures, a number of alternatives have been investigated. Pyridine N-oxide is nitrated by reaction with sulphuric acid and fuming nitric acid at 90°C to give 4-nitropyridine-1-oxide in high yield (85–90%). A very small amount of 2-nitropyridine-1-oxide is also formed. The reaction takes place on the unprotonated N-oxide, which is 4×10^{-6} times as reactive as benzene. Since the N-oxide can subsequently be deoxygenated by reaction with phosphorus trichloride or with nitrogen monoxide, this nitration provides a route to 4-nitropyridine (Fig. 5.12). As shown, the intermediate formed by attack at the 4-position is stabilized by electron release from the oxygen. 4-Nitropyridine has been obtained in good yield from pyridine, via the N-oxide, in a 'one pot' preparation.[20]

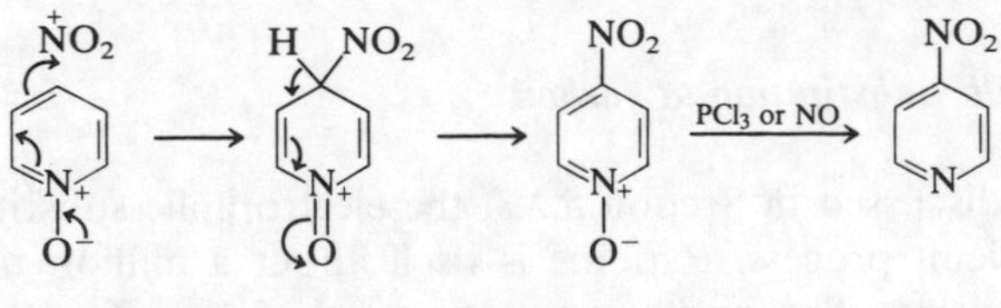

Fig. 5.12 Nitration of pyridine N-oxide.

Electrophilic substitution by way of the lithiated heterocycles is not easy because, in contrast to many of the five-membered heterocycles, pyridine tends to undergo nucleophilic addition reactions with alkyllithium reagents rather than hydrogen–lithium exchange (see Fig. 5.16). The lithiated pyridines can, however, be prepared from the bromopyridines at low temperatures and will react with electrophiles, an example of the reaction is shown in Fig. 5.13(a).[21]

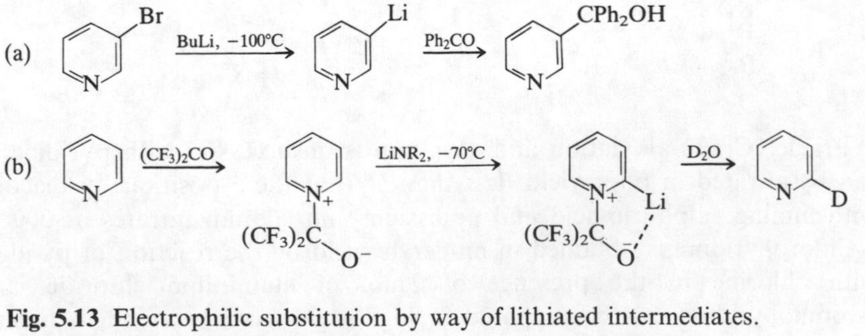

Fig. 5.13 Electrophilic substitution by way of lithiated intermediates.

20 F. Kröhnke and H. Schäfer, *Chem. Ber.*, 1962, **95**, 1098.
21 W. E. Parham and R. M. Piccirilli, *J. Org. Chem.*, 1977, **42**, 257.

Hexafluoroacetone complexes with pyridine and the complex undergoes selective lithiation at C-2 [Fig. 5.13(b)].[22]

Electrophilic substitution in the pyridine ring is more satisfactory if electron-releasing substituents are present. Methylpyridines can be halogenated in good yield by the methods used for pyridine. 2,6-Dimethylpyridine is nitrated in good yield (at the 3-position) by reaction with potassium nitrate and fuming sulphuric acid at 100°C. 2,6-Di-t-butylpyridine, **17**, which is sterically hindered at nitrogen, can be sulphonated at the 3-position in moderate yield by reaction with sulphur trioxide in liquid sulphur dioxide at −10°C. This may be because sulphur trioxide can no longer coordinate to nitrogen in the usual manner.[23]

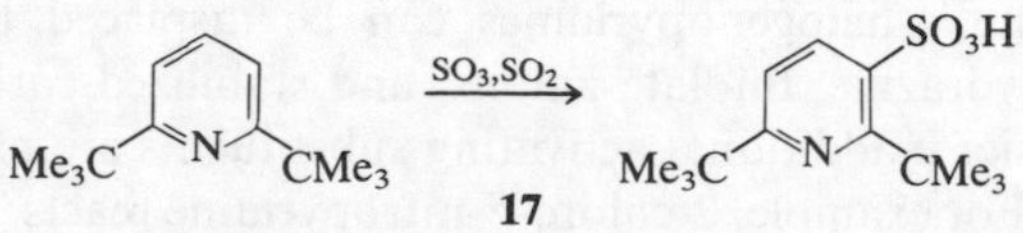

Alkoxypyridines and hydroxypyridines (or pyridones) undergo electrophilic substitution very readily. The *ortho–para* directing effect of the activating substituent is dominant: thus, 2- and 4-alkoxypyridines are substituted preferentially at the 3- and 5-positions, and 3-alkoxypyridines mainly at the 2-position, with some substitutions also taking place at the 6-position. For example, 2-pyridone is converted into the 3,5-dibromo derivative by reaction with bromine water or bromine in acetic acid.[24] It can be selectively nitrated at the 3- or 5-positions in high yield,[25] and under more forcing conditions it gives the 3,5-dinitro derivative in moderate yield. 3-Hydroxypyridine and 3-methylpyridine give the 2-nitro derivatives in moderate yield when nitrated with mixtures of nitric and sulphuric acids below 40°C.[26] Another method of substitution of alkoxypyridines is by lithiation: the alkoxy group stabilizes an *ortho*-lithio substituent and so facilitates lithiation. For example, 3-ethoxypyridine is lithiated at the 2-position by butyllithium at −40°C and the lithiated species **18** reacts with a wide range of electrophiles (Fig. 5.14).[27]

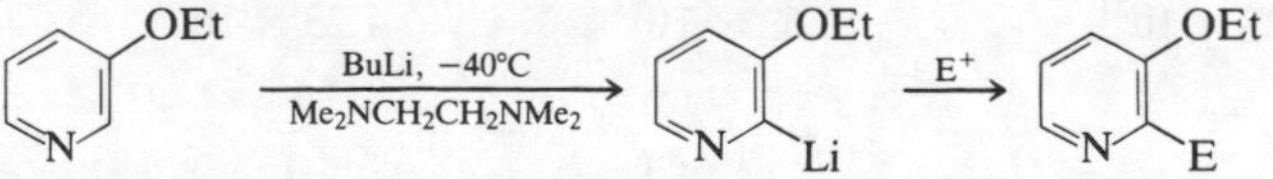

Fig. 5.14 Lithiation and substitution of 2-ethoxypyridine.

22 S. L. Taylor, D. Y. Lee, and J. C. Martin, *J. Org. Chem.*, 1983, **48**, 4156.

23 Review: B. Kanner, *Heterocycles*, 1982, **18**, 411.

24 O. S. Tee and M. Paventi, *J. Am. Chem. Soc.*, 1982, **104**, 4142.

25 A. G. Burton, P. J. Halls, and A. R. Katritzky, *J. Chem. Soc., Perkin Trans.* 2, 1972, 1953.

26 A. R. Katritzky, H. O. Tarhan, and S. Tarhan, *J. Chem. Soc. (B)*, 1970, 114.

27 F. Marsais, G. Le Nard, and G. Quéguiner, *Synthesis*, 1982, 235; F. Trécourt, M. Mallet, F. Marsais, and G. Quéguiner, *J. Org. Chem.*, 1988, **53**, 1367. For related *ortho*-lithiation see V. Snieckus, *Chem. Rev.*, 1990, **90**, 879.

5.2.7   Nucleophilic substitution

The discussion of the reactivity of pyridines in Section 5.2.3 indicated that nucleophilic substitution by the addition–elimination mechanism should be much easier than for the corresponding benzenes, specially at the 2- and 4-positions. This is borne out by the data for a specific reaction, the displacement of chloride by methoxide (Table 5.3). From these data it can be seen that the chloropyridines are all much more reactive than chlorobenzene, the order of reactivity being $4 > 2 > 3$. The order of reactivity is the same as for the chloronitrobenzenes, to which there is a close parallel. The pyridinium salts are very much more reactive but the rate of attack at position 3 is again the lowest.

The halide in 2- and 4-halogenopyridines can be displaced by good nucleophiles such as hydrazine, thiolate anions, and stabilized carbanions. These reactions are easier if additional activating substituents are present in appropriate positions. For example, 2-chloro-5-nitropyridine reacts 7.3×10^6 times faster than 2-chloropyridine with ethanol, the intermediate **19** gaining additional stabilization from the nitro group.[28]

Table 5.3 Rate factors for benzene and pyridine derivatives in the reaction $ArCl + MeO^- \rightarrow ArOMe + Cl^-$.[a]

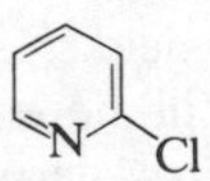 2.76×10^8	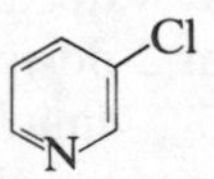 9.12×10^4	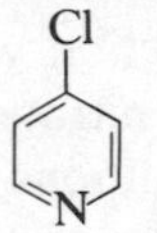 7.43×10^9
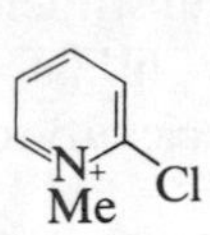 1.28×10^{21}	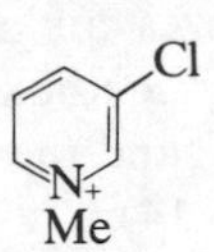 2.62×10^{13}	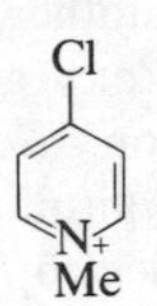 4.23×10^{19}
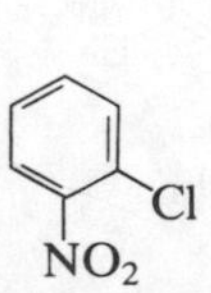 2.10×10^{10}	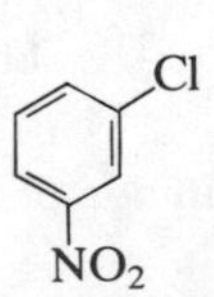 5.64×10^5	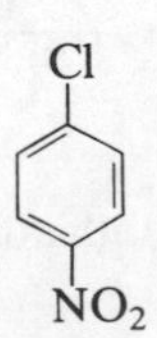 7.05×10^{10}

a Values are relative to chlorobenzene = 1; M. Liveris and J. Miller, *J. Chem. Soc.*, 1963, 3486.

28 Reviews: G. Illuminati, *Adv. Heterocycl. Chem.*, 1964, **3**, 285; R. G. Shepherd and J. L. Fedrick, *Adv. Heterocycl. Chem.*, 1965, **4**, 145.

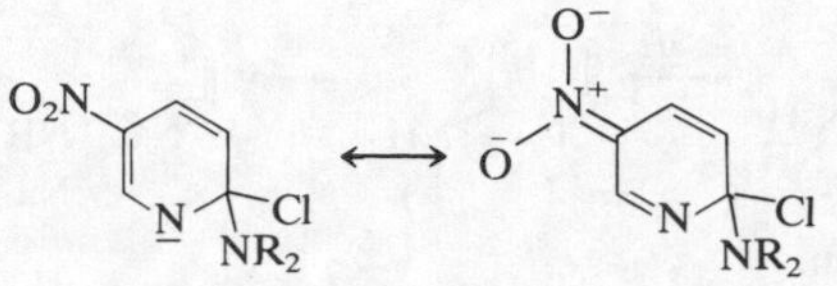

19

As illustrated in Table 5.3, halopyridinium salts are very much more reactive than neutral halopyridines. 2-Halo-*N*-methylpyridinium salts react with a wide range of nucleophiles. A good application of this reaction in synthesis is their conversion into 2-acyloxypyridinium salts, **20**, by reaction with carboxylic acids.[29] These salts are then susceptible to attack by nucleophiles with the loss of *N*-methyl-2-pyridone. Even poor nucleophiles react well; for example, acids can be converted into t-butyl esters in good yield (YH = Bu^{t}OH).

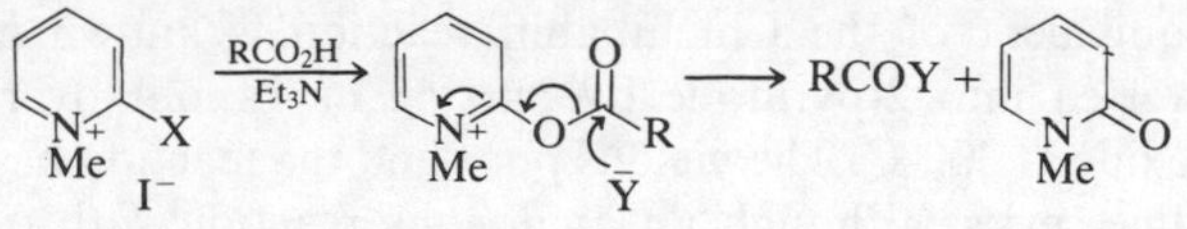

20

Nucleophilic displacement of halide in halopyridines can be catalysed by acids or Lewis acids, so that in these cases it is the pyridinium salt which is the reacting species. For example, 2- and 4-chloropyridines can be converted into the amino compounds by reaction with a zinc chloride–ammonia complex. The chloropyridine *N*-oxides are also more reactive then the corresponding chloropyridines towards nucleophiles.

2-Bromopyridine is usually more reactive than 2-chloropyridine in nucleophilic displacement of this type; 2-methylsulphonylpyridine also undergoes displacement readily. For example, the stabilized ylides **21** were prepared by reaction of 2-substituted pyridines **22** (X = Br or MeSO$_2$) with triphenylphosphonium methylide, but 2-chloropyridine failed to react.[30]

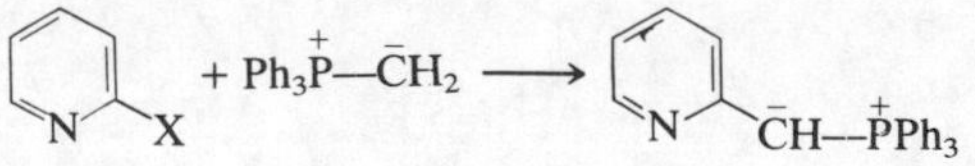

22   **21**

Nucleophilic displacement of hydride is an important reaction in pyridine chemistry. The *Chichibabin reaction*[31] is the amination of pyridine and related heterocycles by sodamide. The reaction is usually carried out by heating pyridine with powdered sodamide in an inert solvent such as *NN*-dimethylaniline. At 110°C, 2-aminopyridine is formed in good yield, and

29 Review: T. Mukaiyama, *Angew. Chem. Int. Edn Engl.*, 1979, **18**, 707.
30 E. C. Taylor and S. F. Martin, *J. Am. Chem. Soc.*, 1974, **96**, 8095.
31 Review: C. K. McGill and A. Rappa, *Adv. Heterocycl. Chem.*, 1988, **44**, 1.

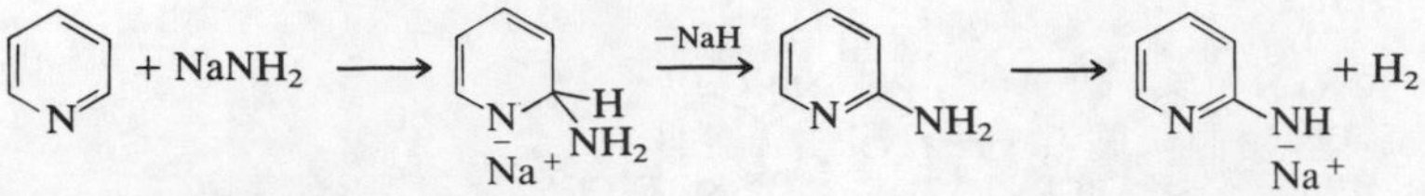

Fig. 5.15 The Chichibabin reaction.

hydrogen is evolved. 4-Aminopyridine is formed only in trace amounts. The reaction is overall one of the addition–elimination type (Fig. 5.15).

The preference for 2-substitution has been ascribed to the higher electrophilicity of the 2-position (which will be a controlling factor for a reaction with an early transition state), and to the coordination of sodium at the pyridine nitrogen atom. 3- and 4-methylpyridines are aminated at the 2-position, and 2-methylpyridine at the 6-position, but if both α-positions are blocked, as in 2,6-dimethylpyridine, amination takes place at the 4-position only in very poor yield.

An oxygen equivalent of the Chichibabin reaction is known in which pyridine is converted into 2-pyridone by passing its vapour over molten potassium hydroxide at 300°C. The yield is poor and the reaction has limited use in the pyridine series, although an analogous reaction with quinoline and isoquinoline (Section 5.3.6) is preparatively useful.

Pyridine can be alkylated or arylated at the 2- and 6-positions by reaction with organolithium reagents (Fig. 5.16). The intermediate σ-complexes **23** formed with phenyllithium and t-butyllithium have been isolated and characterized.[32]

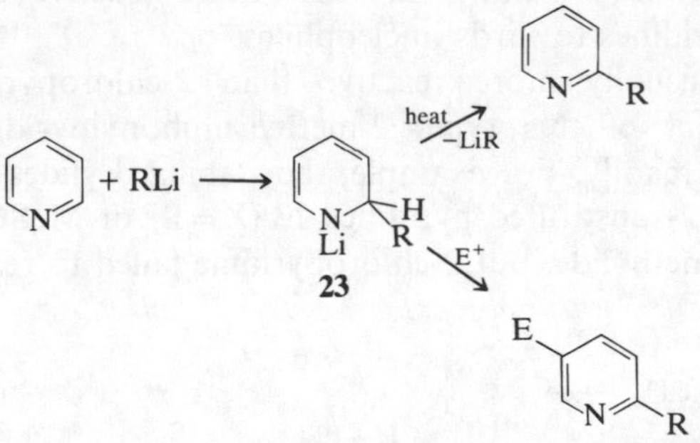

Fig. 5.16 Addition of organolithium reagents to pyridine and reaction of the adducts.

The intermediates are aromatized by heat, and further alkylation can take place at the 6-position if an excess of the organolithium reagent is present. Another useful property of the intermediates **23** is that they can be intercepted

32 C. S. Giam and J. L. Stout, *Chem. Commun.*, 1969, 142; R. F. Francis, W. Davis, and J. T. Wisener, *J. Org. Chem.*, 1974, **39**, 59.

by electrophiles which attack at the 5-position.[33] Thus, a combination of nucleophilic and electrophilic substitution leads to the formation of 2,5-disubstituted pyridines.

N-Alkylpyridinium salts are much more readily attacked by nucleophiles. In aqueous base an equilibrium is set up with a covalent adduct **24**. This 'pseudobase'[34] can be oxidized by oxidants such as potassium ferricyanide to give the 1-alkyl-2-pyridone.[35]

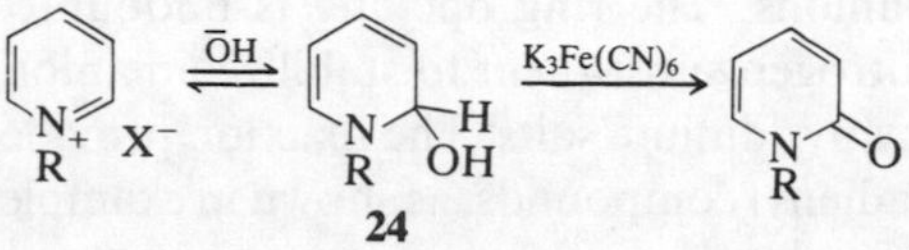

Another method by which the intermediates can be converted into stable products is by the loss of a leaving group from nitrogen. For example, pyridine *N*-oxide reacts with acetic anhydride to give 2-acetoxypyridine. This reaction can be formulated as shown in Fig. 5.17(a): a 1-acetoxypyridinium salt is formed which can be intercepted by acetate and which can then aromatize by elimination of acetic acid. The pyridinium salt in (b) has a good leaving group on nitrogen which is sufficiently bulky to inhibit the attack of nucleophiles at the 2-position. Thus, attack by alkylmagnesium halides, enolate anions, thiols, and other 'soft' nucleophiles is directed to the 4-position.[36]

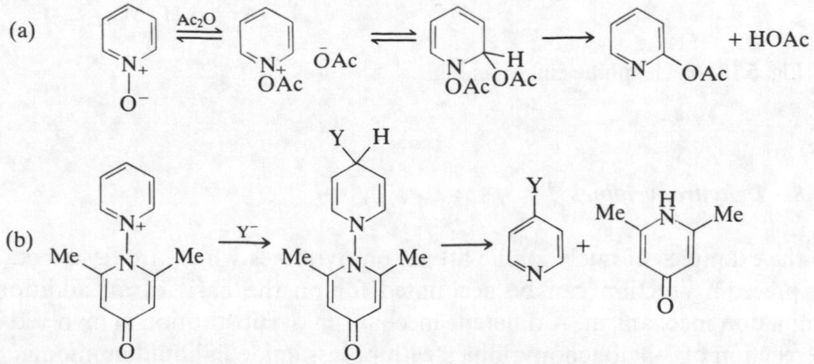

Fig. 5.17 Selective nucleophilic attack on pyridinium salts.

33 Review: D. L. Comins and S. O'Connor, *Adv. Heterocycl. Chem.*, 1988, **44**, 199.
34 Review: J. W. Bunting, *Adv. Heterocycl. Chem.*, 1979, **25**, 1.
35 Review: H. Weber, *Adv. Heterocycl. Chem.*, 1987, **41**, 275.
36 A. R. Katritzky, H. Beltrami, J. G. Keay, D. N. Rogers, M. P. Sammes, C. W. F. Leung, and C. M. Lee, *Angew. Chem. Int. Edn Engl.*, 1979, **18**, 792.

Reactions of nucleophiles with pyridine *N*-oxide and with pyridinium salts can also lead to products of ring cleavage. Some examples are shown in Fig. 5.18. In reaction (a) the addition of phenylmagnesium bromide to pyridine *N*-oxide results in the formation of an open-chain oxime. This can be rationalized as the result of electrocyclic ring opening of the initial adduct.[37] A similar sequence can account for the alkaline hydrolysis of the *N*-methoxy-pyridinium salt shown in (b). *N*-2,4-Dinitrophenylpyridinium chloride is cleaved by several types of nucleophile including primary and secondary amines and stabilized carbanions. The ring opening is undoubtedly made easier by the ability of the nitrogen substituent to stabilize an anion, because it does not occur with *N*-alkylpyridinium salts. The reaction provides a useful route to functionalized pentadienyl compounds, as shown in example (c).[38]

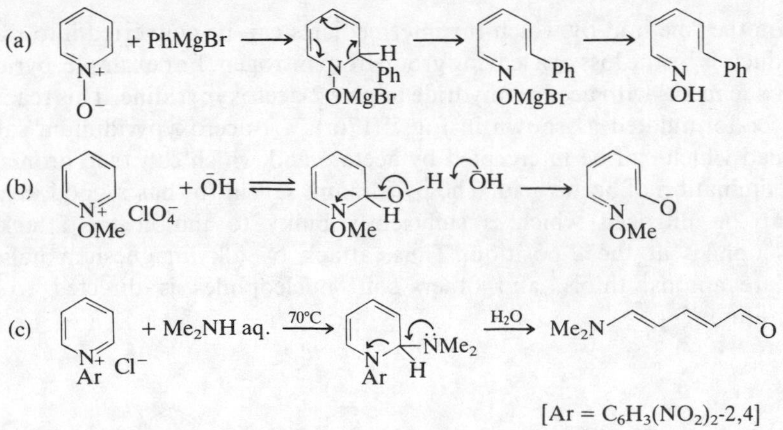

$$[Ar = C_6H_3(NO_2)_2\text{-}2,4]$$

Fig. 5.18 Nucleophilic ring opening.

5.2.8   Dehydropyridines[39]

All the examples of nucleophilic attack on pyridines which are described in the preceding section can be accounted for on the basis of an addition–elimination mechanism. A different mechanism of substitution is involved in the reaction of 3-halogenopyridines with potassamide in liquid ammonia, or with potassium t-butoxide. The halide at the 3-position is relatively unreactive towards direct displacement, so a competitive reaction, in which the proton is removed from C-4, takes place. The anions then give 3,4-dehydropyridine

37 P. Schiess, C. Monnier, P. Ringele, and E. Sendi, *Helv. Chim. Acta*, 1974, **57**, 1676.
38 S. S. Malhotra and M. C. Whiting, *J. Chem. Soc.*, 1960, 3812.
39 Review: H. C. van der Plas and F. Roeterdink, in *The Chemistry of Triple-bonded Functional Groups*, Part 1, ed. S. Patai and Z. Rappoport, Wiley-Interscience, Chichester, 1983, p. 421; see also C. May and C. J. Moody, *Tetrahedron Lett.*, 1985, **26**, 2123.

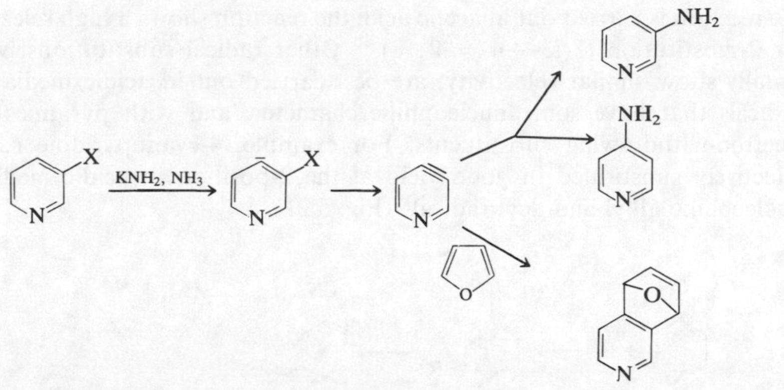

Fig. 5.19 Generation of 3,4-dehydropyridine.

by loss of halide. This can be intercepted by amide anions or ammonia to give a mixture of 3- and 4-aminopyridine. In the presence of furan the cycloadduct is also formed (Fig. 5.19).

The ratio of 3- and 4-aminopyridines (35:65) is independent of the nature of the leaving group X^-. No 2-aminopyridine is formed and none of the isomeric 2,3-dehydropyridine intermediate is generated in these reactions. This intermediate can, however, be generated by other methods commonly used to produce arynes. For example, 3-bromo-2-chloropyridine is dehalogenated by butyllithium in furan and the expected adduct is formed as shown in Fig. 5.20.[40]

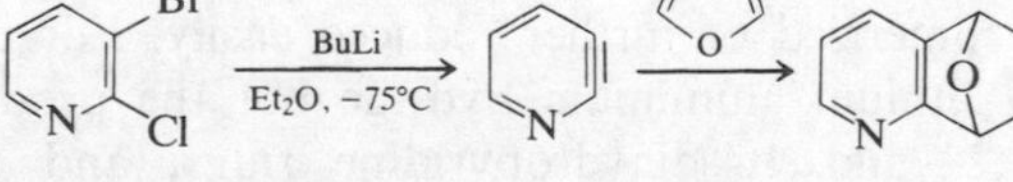

Fig. 5.20 Generation of 2,3-dehydropyridine.

5.2.9 *Radical substitution*[41]

Pyridines undergo homolytic substitution (replacement of H˙ by R˙) by several different types of radical species. Radical substitution reactions generally suffer from the major disadvantage, as synthetic methods, that they are unselective. In pyridines, however, substitution usually takes place preferentially at the 2-position, and if the reaction is carried out in an acid medium, the selectivity is high. For example, the phenylation of pyridine by phenyl radicals, produced by the decomposition of benzoyl peroxide in the absence of acid, gives 2-, 3-, and 4-phenylpyridine in a ratio of 54:32:14, but when

40 J. D. Cook and B. J. Wakefield, *J. Chem. Soc.* (*C*), 1969, 1973.
41 Reviews: F. Minisci and O. Porta, *Adv. Heterocycl. Chem.*, 1974, **16**, 123; F. Minisci, E. Vismara, and F. Fontana, *Heterocycles*, 1989, **28**, 489.

the reaction is carried out in acetic acid, the reaction shows a high selectivity for 2-substitution (2-:3-+4-=82:18).[42] Other radical substitutions, which usually show similar selectivity, are best carried out in acidic media with radicals that have some nucleophilic character, and with pyridines with electron-withdrawing substituents. For example, 4-cyanopyridine can be selectively substituted in good yield at the 2-position in acidic media by nucleophilic alkyl and acyl radicals (Fig. 5.21).

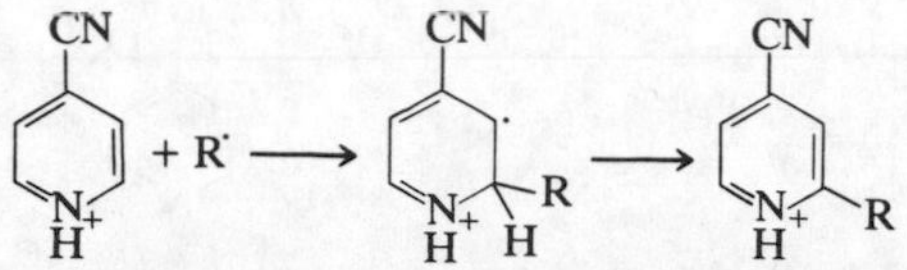

(R = Et, PhCH$_2$, EtCO, Me$_2$NCO)

Fig. 5.21 Selective homolytic substitution of 4-cyanopyridine.

The photochemical halogenation of pyridine leads to the formation of 2-chloropyridine in good yield; this is also probably a radical substitution.

5.2.10   *Reduction of pyridines and pyridinium salts: dihydropyridines*[43]

The reaction of pyridines and pyridinium salts with reducing agents of the complex hydride type is essentially a nucleophilic addition. It is therefore much easier to reduce pyridines than benzenes. In simple pyridines it is, however, difficult to achieve selective reduction. The initial reduction products may be unstable and isomerized or further reduced easily. For example, pyridine is reduced by lithium aluminium hydride but the product is a complex containing 1,2- and 1,4-dihydropyridine rings, and the free dihydropyridines cannot be isolated from it. Pyridine is not reduced by sodium borohydride but pyridines bearing electron-withdrawing substituents, and pyridinium salts, are reduced. A good example of a synthetically useful application of the reaction is provided by the reduction of pyridine in the presence of chloroformate esters (Fig. 5.22).[44] Both 1,2- and 1,4-dihydro-pyridines can be formed in these reactions but the reaction conditions can be controlled to give the 1,2-isomers **25** almost exclusively.[44] These 1,2-dihydropyridines are useful intermediates for further synthetic trans-formations. They undergo selective electrophilic substitution,[33] an example

42 H. J. M. Dou and B. M. Lynch, *Bull. Soc. Chim. Fr.*, 1966, 3815.
43 Reviews: D. M. Stout and A. I. Meyers, *Chem. Rev.*, 1982, **82**, 223; A. Sausiņš and G. Duburs, *Heterocycles*, 1988, **27**, 291 (dihydropyridines); J. E. Toomey, *Adv. Heterocycl. Chem.*, 1984, **37**, 167 (electrochemical reduction); J. E. Keay, *Adv. Heterocycl. Chem.*, 1986, **39**, 1 (complex hydride reduction).
44 F. W. Fowler, *J. Org. Chem.*, 1972, **37**, 1321.
33 Review: D. L. Comins and S. O'Connor, *Adv. Heterocycl. Chem.*, 1988, **44**, 199.

being the formylation shown in Fig. 5.22[45] (this reaction is analogous to the substitution of the dihydropyridine **23** shown earlier in Fig. 5.16). They are also good dienes, and Diels–Alder reactions of these 1,2-dihydropyridines have provided the starting point for the synthesis of several alkaloids and other natural products. An example of a Diels–Alder reaction is also shown in Fig. 5.22. The adduct **26** and related acylnitroso adducts have been used to prepare aminosugars.[46]

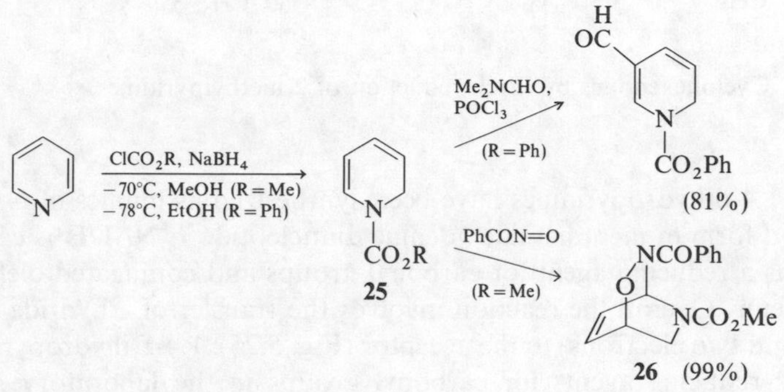

Fig. 5.22 Examples of formation and reactions of 1,2-dihydropyridine-1-carboxylic esters.

Under the conditions of the Birch reduction, pyridines can be reduced readily. The products are dependent upon the substituents and the reaction conditions. The radical anion intermediate formed in the first step of the reaction may dimerize, and, if unsubstituted at C-4, the dimer can be oxidized to a 4,4-bipyridyl. Further reduction in the presence of a proton source, such as ethanol, leads to the formation of monomeric 1,4-dihydropyridines (Fig. 5.23).

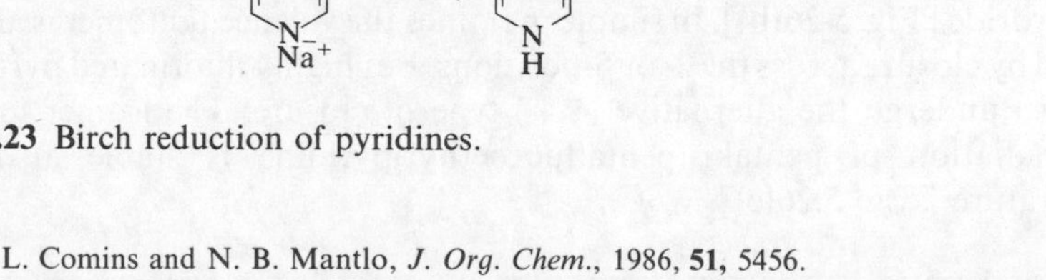

Fig. 5.23 Birch reduction of pyridines.

45 D. L. Comins and N. B. Mantlo, *J. Org. Chem.*, 1986, **51**, 5456.
46 A. Defoin, H. Fritz, C. Schmidlin, and J. Streith, *Helv. Chim. Acta*, 1987, **70**, 554.

The Birch reduction provides a method of using 2-methylpyridines as equivalents of cyclohexenones in synthesis. The stable pyridine unit can be carried through several steps of a synthetic sequence and then 'unmasked' by Birch reduction to reveal a cyclohexenone unit, as shown in Fig. 5.24. The reaction has been used in steroid synthesis.

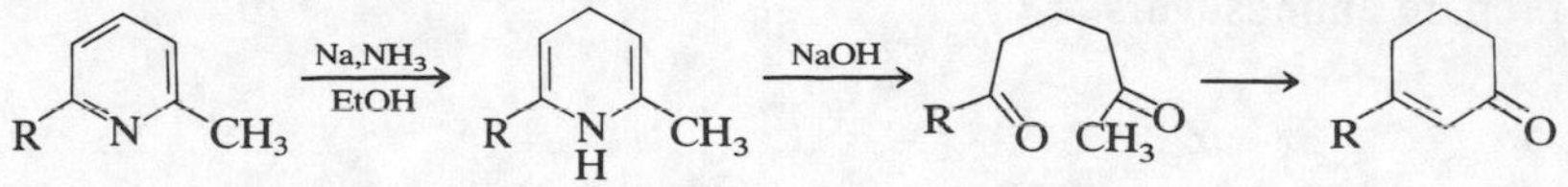

Fig. 5.24 Cyclohexenones by Birch reduction of 2-methylpyridines.

Several 1,4-dihydropyridines have been synthesized as mimics of NADH, the reduced form of nicotinamide adenine dinucleotide **1**. NADH is capable of acting as a reducing agent for carbonyl groups and conjugated olefins in living systems. Overall the reaction involves the transfer of a hydride (or of a proton and two electrons) to the acceptor (Fig. 5.25). 1,4-Dihydropyridines can act as reducing agents for carbonyl groups in the laboratory; metal cations such as Zn^{2+} and Mg^{2+} enhance the rate of the reaction.

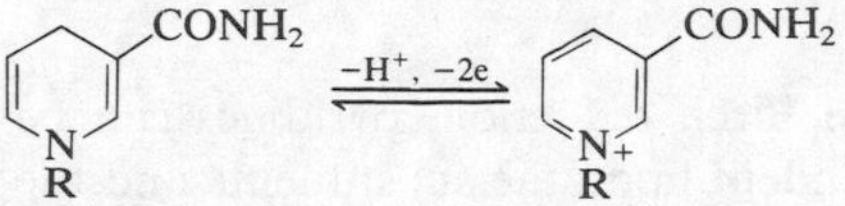

Fig. 5.25 NADH as a reducing agent.

5.2.11   Photochemical isomerization[47]

1,2-Dihydropyridines can undergo a light-induced disrotatory electro-cyclization to give stable azabicyclohexenes [Fig. 5.26(a)]. The same type of cyclization can take place when aromatic pyridines are irradiated, but the valence isomers, the azabicyclohexadienes ('Dewar pyridines') are, in general, very unstable. The parent compound has a halflife of 2.5 min at 25°C but it can be intercepted, for example, by reduction with sodium borohydride [Fig. 5.26(b)]. In simple pyridines the valence tautomer is always formed by closure across the 2- or 5-positions, but highly fluorinated pyridines can also undergo the alternative (1,4-) type of closure. The isomer formed by irradiation of pentakis(pentafluoroethyl)pyridine is stable at room temperature [Fig. 5.26(c)].

47 Review: Y. Kobayashi and I. Kumadaki, *Adv. Heterocycl. Chem.*, 1982, **31**, 169.

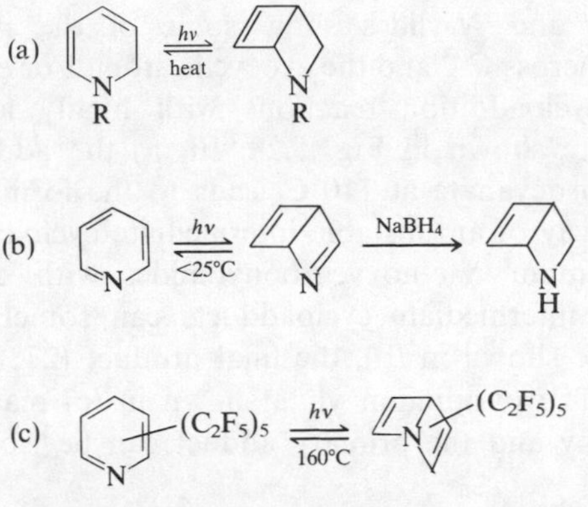

Fig. 5.26 Photoisomerization of 1,2-dihydropyridines and pyridines.

5.2.12 *Pyridine* N-*oxides,* N-*imides, and* N-*ylides*

Some of the properties of pyridine *N*-oxide have been described in earlier sections. Pyridine, like other tertiary amines, is also capable of forming zwitterionic imides, **27**, and ylides, **28**,[48] which are stable isolable solids when conjugative groups R^1 and R^2 are present to delocalize the negative charge. They can be prepared by amination and alkylation of pyridine, followed by treatment with an appropriate base (Fig. 5.27). Alternatively, they can be formed from azides or diazo compounds and pyridine if good stabilizing substituents (R^1, $R^2 = COR$, SO_2R, etc.) are incorporated.

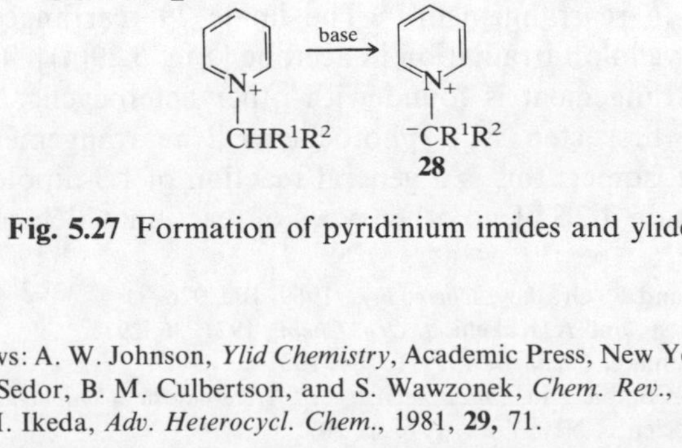

Fig. 5.27 Formation of pyridinium imides and ylides.

48 Reviews: A. W. Johnson, *Ylid Chemistry*, Academic Press, New York, 1966; W. J. McKillip, E. A. Sedor, B. M. Culbertson, and S. Wawzonek, *Chem. Rev.*, 1973, **73**, 255; Y. Tamura and M. Ikeda, *Adv. Heterocycl. Chem.*, 1981, **29**, 71.

The *N*-oxides, *N*-imides and *N*-ylides show some of the reactions characteristic of 1,3-dipoles across C-2 and the exocyclic atom. For example, all three types undergo cycloaddition reactions with highly activated dipolarophiles. Examples are shown in Fig. 5.28. In (a) the addition of pyridine-1-oxide to phenyl isocyanate at 110°C leads to the formation of 2-phenylaminopyridine by way of an unstable intermediate cycloadduct.[49] In the reactions of pyridinium *N*-ethoxycarbonylimides with dimethyl acetylenedicarboxylate, the intermediate cycloadducts can sometimes be isolated, but, in the example shown in (b), the final product (27%) is the pyrazolo[1,5-*a*]pyridine.[50] The pyridinium ylide shown in (c) reacts with acrylonitrile in a similar way and the primary adduct can be isolated in moderate yield.[51]

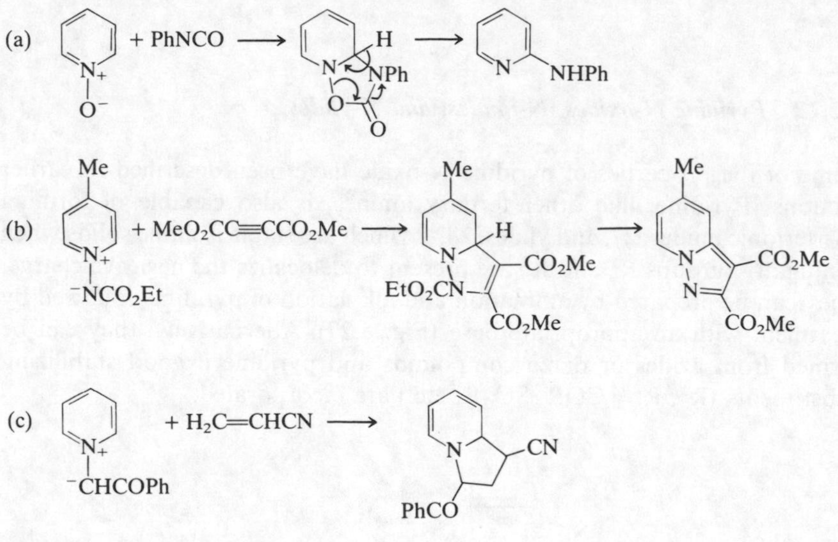

Fig. 5.28 Cycloaddition reactions.

The photolysis of these compounds can result either in cleavage of the exocyclic group, or in rearrangement.[52] The imide **29** rearranges to the diazepine **30** in high yield on irradiation in acetone [Fig. 5.29(a)]. This type of light-induced rearrangement is found with other heterocyclic *N*-oxides and *N*-imides. The first step is a photochemical rearrangement to a three-membered ring isomer; this is a general reaction of 1,3-dipoles which was discussed in Section 4.2.8. Two other types of reaction which can occur

49 H. Seidl, R. Huisgen, and R. Grashey, *Chem. Ber.*, 1969, **102**, 926.
50 T. Sasaki, K. Kanematsu, and A. Kakehi, *J. Org. Chem.*, 1971, **36**, 2978.
51 J. Fröhlich and F. Kröhnke, *Chem. Ber.*, 1971, **104**, 1621.
52 Review: A. Lablache-Combier, in *Photochemistry of Heterocyclic Compounds*, ed. O. Buchardt, Wiley-Interscience, New York, 1976, p. 207.

on irradiation are shown for the ylide in Fig. 5.29(b). The exocyclic group is lost as a carbene which can be intercepted, and a competing ring contraction also takes place. Analogous reactions to these also occur in low yield with pyridine *N*-oxide.

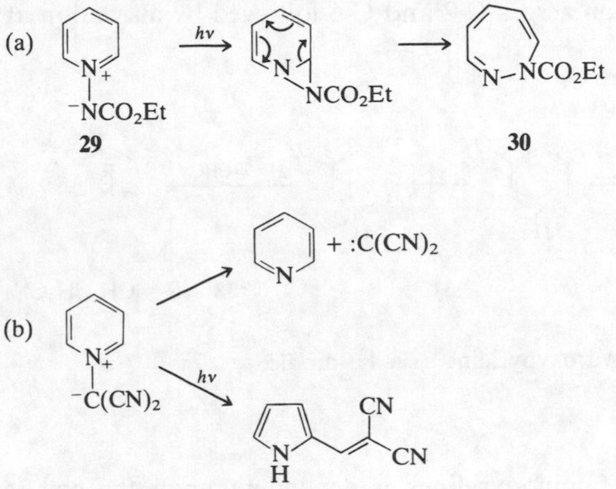

Fig. 5.29 Photochemical reactions.

5.2.13 Hydroxy- and aminopyridines

The tautomeric equilibria in the 2- and 4-substituted derivatives of these types were discussed in Chapter 2, Section 2.5. The 2- and 4- hydroxypyridines exist in the pyridone tautomeric forms in solution. 2-Pyridone is a water-soluble solid, m.p. 107°C. It is alkylated on nitrogen by iodomethane, and it can be converted into 2-methoxypyridine by way of 2-chloropyridine, as shown in Fig. 5.30. Its reaction with phosphorus pentachloride is analogous to that of an amide.

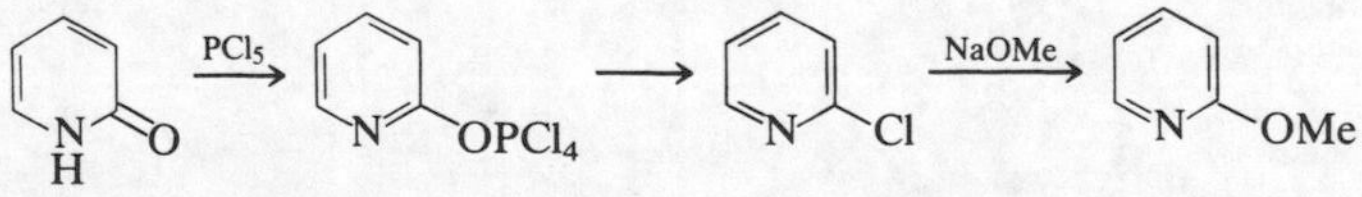

Fig. 5.30 2-Methoxypyridine from 2-pyridone.

N-Methyl-2-pyridinone shows some dienic character; it forms Diels–Alder adducts with such dienes as maleic anhydride and (in poor yield) methyl acrylate and acrylonitrile.[53]

53 H. Tomisawa, H. Hongo, H. Kato, T. Naraki, and R. Fujita, *Chem. Pharm. Bull.*, 1979, **27**, 670.

3-Hydroxypyridine cannot exist in a pyridone form but the zwitterionic tautomer **31** is present in solution. Indeed, 3-hydroxypyridine and 1-substituted derivatives undergo cycloaddition reactions, as 1,3-dipoles, quite readily.[54] Thus, 3-hydroxypyridine reacts with acrylonitrile on heating to give the adducts **32** (*exo* and *endo*) in high yield; these are formed by 1,3-dipolar addition across C-2 and C-6 followed by alkylation at nitrogen (Fig. 5.31).

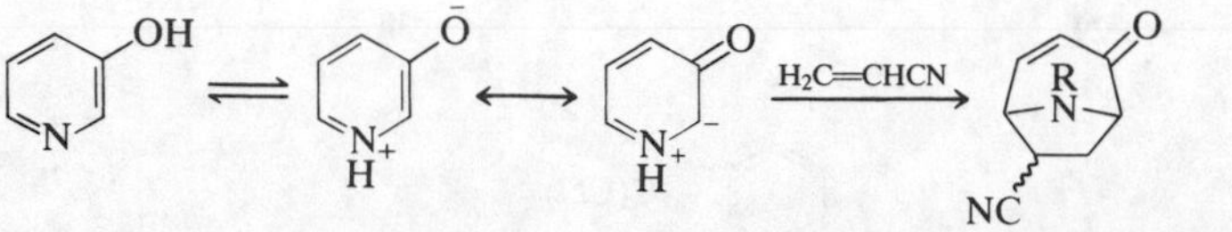

31   **32** (R = CH$_2$CH$_2$CN)

Fig. 5.31 3-Hydroxypyridine as a 1,3-dipole.

The basicity of aminopyridines is dependent upon the position of the substituent (Table 5.2); all are protonated preferentially on the ring nitrogen. 3-Aminopyridine is the least basic of the three, and the most like an 'aromatic' amine: for example, it can be diazotized and the diazonium salts can be coupled in the normal way. 2- and 4-Pyridinediazonium salts, on the other hand, are activated by the ring nitrogen atom, and when the amines are diazotized in the dilute acid, the corresponding pyridones are obtained by nucleophilic displacement of nitrogen by water. In concentrated hydrochloric acid, chloropyridines are formed in a similar way. Diazonium salts derived from the corresponding *N*-oxides are more stable, as shown for the 4-substituted cation **33**.

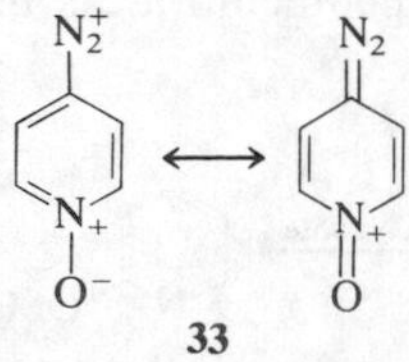

33

5.2.14   *Alkyl- and alkenylpyridines*

Alkyl groups in the 2- and 4-positions of pyridines and pyridinium cations are activated by the ring nitrogen atom, as discussed in Section 5.2.3, and the methyl or methylene groups attached to the ring are appreciably acidic.

---

54 Review: A. R. Katritzky and N. Dennis, *Chem. Rev.*, 1989, **89**, 827.

This is an important feature of pyridine chemistry because it allows a wide range of functional groups to be introduced into the readily available alkylpyridines: they can undergo alkylation and acylation reactions, aldol condensations, and conjugate additions. Aldol reactions are usually best carried out in the presence of acetic anhydride or a Lewis acid, such as zinc chloride, which can coordinate to the ring nitrogen atom and increase the acidity of the alkyl group. Examples of aldol and acylation reactions of 2-methylpyridine which go in good yield are shown in Fig. 5.32.

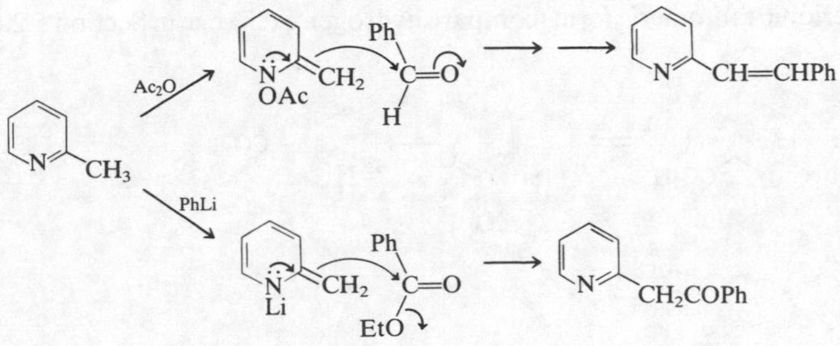

Fig. 5.32 Substitution of the methyl group of 2-methylpyridine.

3-Methylpyridine is less acidic and will not, for example, undergo the catalysed aldol reaction as 2-methylpyridine does. It can, however, be deprotonated by very strong bases such as sodamide in liquid ammonia, and the anions can then be alkylated. 2-Methyl groups can be selectively functionalized in the presence of 3-methyl groups: an example is the selective formation of the sulphide **34** in high yield from 2,3-dimethylpyridine, butyllithium, and diphenyl disulphide.[55]

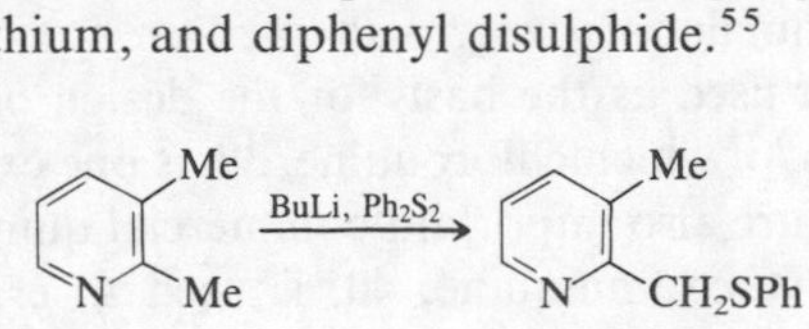

Alkenyl groups at the 2- and 4-positions of the pyridine ring are activated by the nitrogen atom, and as a result they are susceptible to nucleophilic attack. For example, both 2- and 4-vinylpyridine react with secondary amines in the presence of acid to give addition products. The products **36** formed from 2-vinylpyridine **35** are shown.

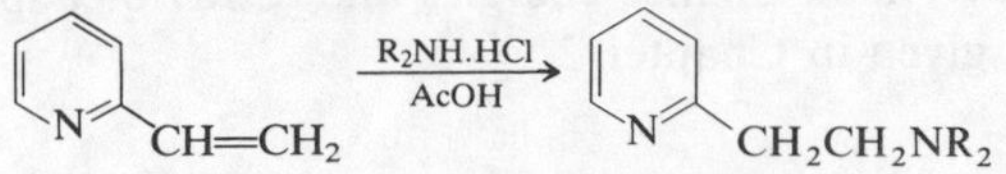

55 E. Ghera, Y. Ben-David, and H. Rapoport, *J. Org. Chem.*, 1981, **36**, 2059.

5.2.15   *Pyridinecarboxylic acids*

The three monocarboxylic acids have trivial names which are often found in the literature: the 2-, 3- and 4-carboxylic acids are called picolinic, nicotinic, and isonicotinic acid, respectively. All are weaker acids than benzoic acid and exist mainly as zwitterions, such as **37**. They can be esterified in acidic conditions in the normal way but in base they can also be alkylated on nitrogen. All are decarboxylated readily, the ease of decarboxylation being in the order $2 > 4 > 3$. Decarboxylation probably takes place in the zwitterionic tautomeric form (compare hydrogen exchange in Section 5.2.6).

37

5.3   Quinolines and isoquinolines[56]

5.3.1   *Introduction*

Quinoline and isoquinoline are the two heterocycles in which a benzene ring and a pyridine ring are fused through carbon. Both ring systems occur naturally and both were originally isolated from coal tar. The alkaloid quinine, **38**, is a traditional antimalarial drug, also used in tonics. The quinoline skeleton has since been used as the basis for the design of many synthetic antimalarial compounds, of which chloroquine, **39**, is one example. The cyanine dyes (Section 5.3.11) are also important commercial quinolines. The tetrahydroquinoline derivative oxamniquine, **40**, is used to eradicate blood flukes (*Schistosoma mansoni*), which are a major cause of disease in tropical regions. Papaverine, **41**, is an opium alkaloid which is a nonspecific smooth muscle relaxant and a vasodilator. There are several other families of isoquinoline alkaloids, which occur widely in the plant kingdom; all are derived in nature from the amino acid tyrosine.

Quinoline is a stable liquid, b.p. 237°C, which is often used in the laboratory as a high boiling basic solvent. Isoquinoline has a m.p. of 26.5°C when pure and a b.p. of 243°C. The resonance energies and ultraviolet spectra of the two compounds are given in Chapter 2.

---

56 Reviews: G. Jones, in *Quinolines*, ed. G. Jones, Wiley-Interscience, London, 1977, p. 1; S. F. Dyke and R. G. Kinsman, in *Isoquinolines*, ed. G. Grethe, Wiley-Interscience, New York, 1981, p. 1.

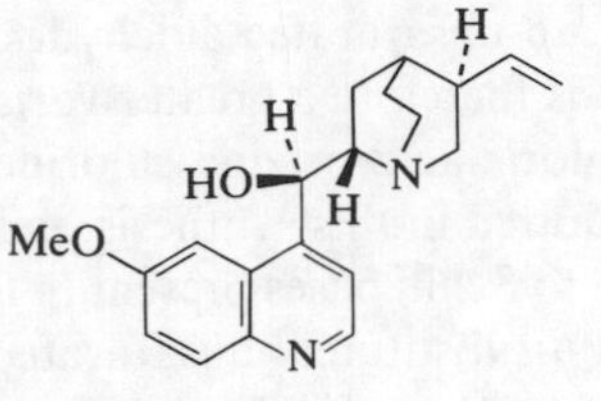

38

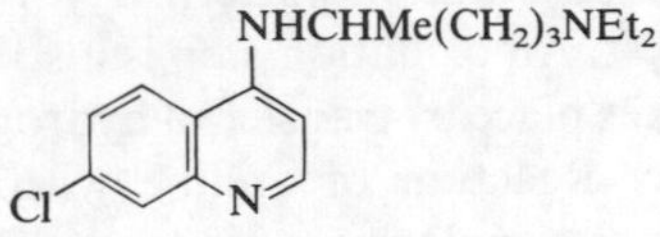

39

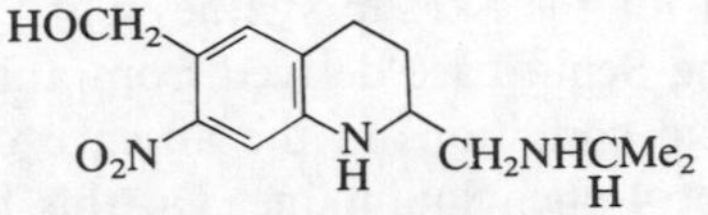

40

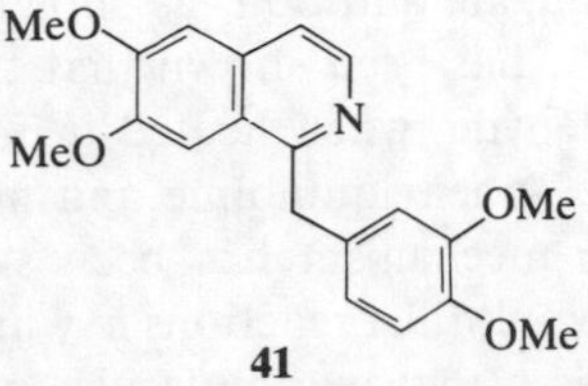

41

5.3.2 Synthesis of quinolines

There are many methods of synthesis of this ring system. The classical methods are cyclization processes in which substituted benzene derivatives are the starting materials and the heterocyclic rings are constructed in the synthesis. In a few more recent syntheses, a carbocyclic ring has been built up onto a substituted pyridine. Those derived from a benzenoid starting material are of two types. The more important type is that in which a side chain is constructed and cyclized onto a free *ortho* position of the benzene ring. Such reactions include the Skraup, Doebner–von Miller, and Combes syntheses.

The *Skraup* synthesis consists of heating an aniline with glycerol and sulphuric acid, which acts as a dehydrating agent and an acid catalyst. The nitrobenzene corresponding to the aniline used is added as a dehydrogenating agent. Since the reaction can be violently exothermic, a moderator such as iron(II) sulphate is also usually added. The details of the reaction sequence have not all been established, but it seems most likely that glycerol is dehydrated to acrolein. This may then react with the aniline by conjugate addition. This intermediate is then cyclized, oxidized, and dehydrated (probably in that order) to give the quinoline. The sequence for quinoline itself is shown in Fig. 5.33.

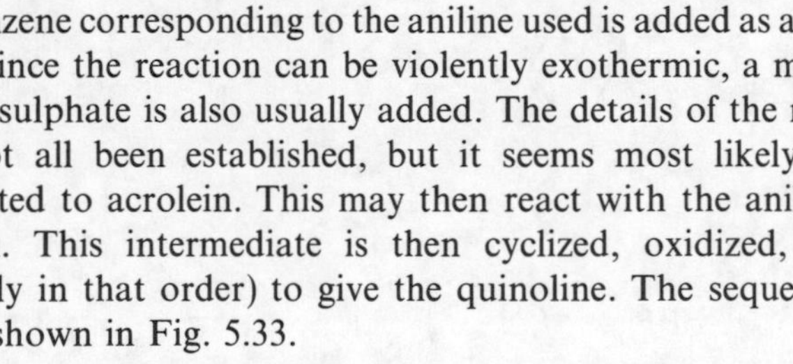

Fig. 5.33 The Skraup synthesis.

In the *Doebner–von Miller* variation, $\alpha\beta$-unsaturated aldehydes or ketones are used in place of glycerol, and there is therefore a greater variation in the possible substitution pattern. Hydrochloric acid or zinc chloride is used as a catalyst. An oxidation step is also required in this synthesis; dehydrogenation takes place by transfer of hydrogen to Schiff bases present in the reaction mixture. Reaction of aniline with a β-substituted unsaturated aldehyde such as crotonaldehyde leads to the formation of a 2-substituted quinoline and not a 4-substituted quinoline [Fig. 5.34(a)]. This is consistent with an initial conjugate addition, as shown for the Skraup synthesis in Fig. 5.33. However, it has been shown that the Schiff base derived from aniline and cinnamaldehyde, when heated in acid conditions in the absence of water, also gives 2-phenylquinoline and not 4-phenylquinoline. On this basis, an alternative mechanism has been suggested for both the Skraup and the Doebner–von Miller reactions in which Schiff bases are the key intermediates. It is proposed that two molecules of the protonated Schiff base combine to give a 1,3-diazetinium ion [Fig. 5.34(b)].[57]

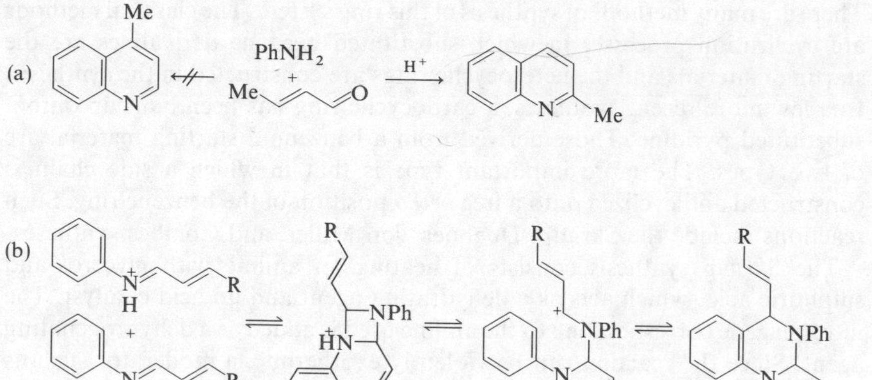

Fig. 5.34 (a) Doebner–von Miller synthesis of 2-methylquinoline from aniline and crotonaldehyde; (b) a possible alternative mechanism for the Skraup and Doebner-von Miller syntheses.

57 J. J. Eisch and T. Dłuzniewski, *J. Org. Chem.*, 1989, **54**, 1269.

In the *Combes* synthesis [Table 4.4 (v)] the aniline is reacted with a 1,3-diketone in the presence of acid. After formation of a Schiff base, cyclization probably takes place by way of the diprotonated form and the intermediate is then dehydrated.

When a β-ketoester is used in place of a 1,3-diketone, two products can be formed. For example, ethyl acetoacetate can give 2-methyl-4-quinolone, **42**, or 4-methyl-2-quinolone, **43**. Attack on the ketone carbonyl is kinetically favoured and so compound **43** is formed by carrying out the initial reaction at room temperature, followed by thermal cyclization. Attack on the ester function, which is thermodynamically favoured, occurs when the reaction is carried out at 110–140°C.

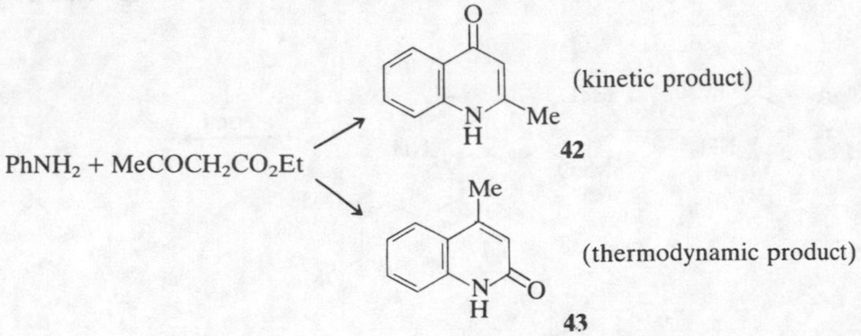

$$PhNH_2 + MeCOCH_2CO_2Et$$

(kinetic product) **42**

(thermodynamic product) **43**

All these methods suffer from the disadvantage that if the aniline bears a *meta*-substituent, there are two different *ortho* positions available for cyclization. In some cases the ratio of products favours one isomer (for example, 3-methoxyaniline gives mainly 7-methoxyquinoline rather than 5-methoxyquinoline in the Skraup synthesis) but often both possible isomers are formed. By starting with an *ortho*-disubstituted benzene as the basis of the construction of the quinoline ring, this problem is avoided.

The *Friedländer* synthesis is an example of such a reaction, in which an o-aminobenzaldehyde or an o-aminoketone is cyclized by reaction with an α-methyleneketone in the presence of base. Its synthetic use is limited by the difficulty in preparing the o-aminocarbonyl compounds. A useful modification is the reaction of an o-nitrocarbonyl compound with the activated methylene and subsequent reduction of the nitro group of the product (Fig. 5.35).

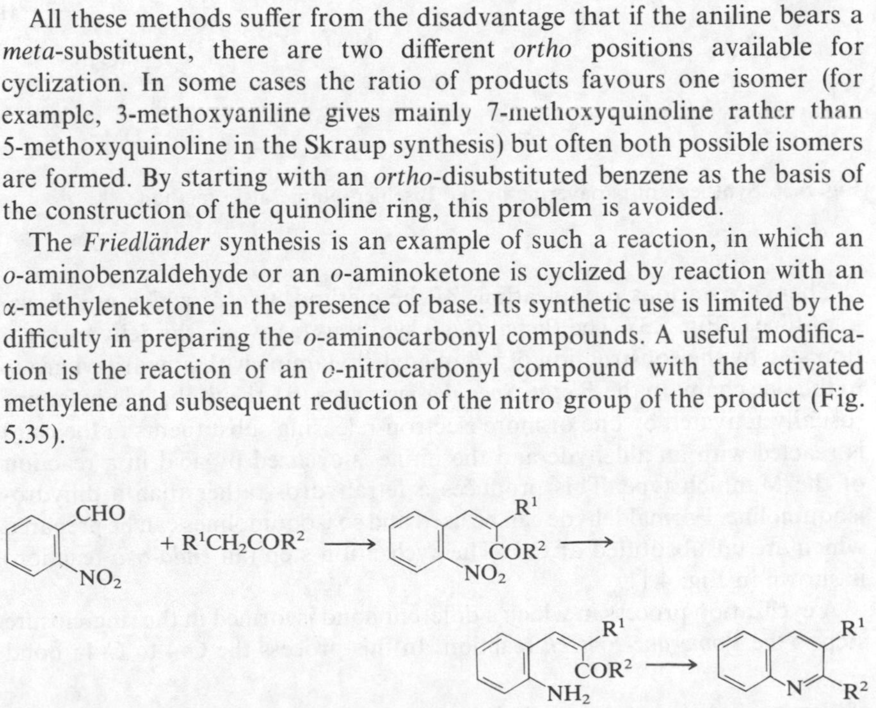

Fig. 5.35. A modified Friedländer synthesis.

5.3.3   Synthesis of isoquinolines[58]

The isoquinoline ring system is constructed in nature from tyrosine by incorporation of an additional carbon atom at the *ortho* position to complete the skeleton. Most of the classical laboratory syntheses use the same approach: a β-phenylethylamine or a similar compound is the usual starting material.

In the *Bischler–Napieralski* reaction the β-phenylethylamine is acylated and then cyclodehydrated by reaction with phosphoryl chloride, phosphorus pentoxide, or other Lewis acids. This gives a 1-substituted-3,4-dihydro-isoquinoline which is dehydrogenated by heating with palladium. An example is provided by a synthesis of the alkaloid papaverine, **41** (Fig. 5.36).

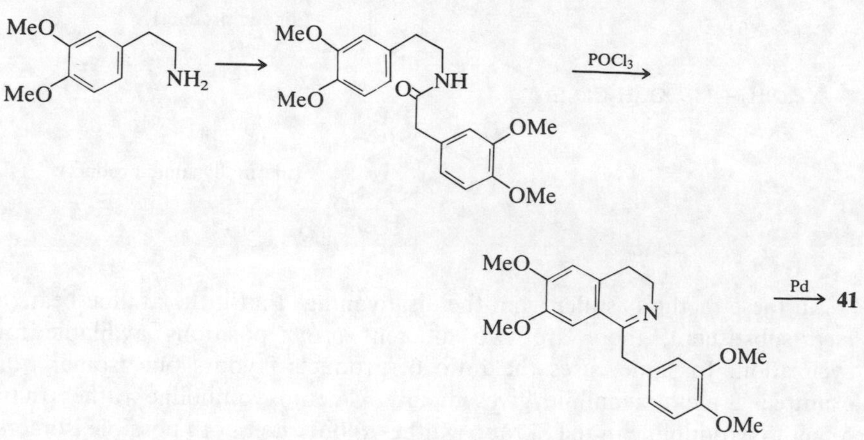

Fig. 5.36 Synthesis of papaverine by the Bischler–Napieralski method.

There are various modifications of the method, two of which are shown in outline in Fig. 5.37. The *Pictet–Gams* synthesis (a) avoids the dehydrogena-tion step by the construction of a β-phenylethylamine with a hydroxyl group in the side chain. In the *Pictet–Spengler* synthesis (b), the β-phenylethylamine (usually activated by one or more electron-releasing substituents in the ring) is reacted with an aldehyde and the imine is cyclized by acid in a reaction of the Mannich type. This produces a tetrahydro- rather than a dihydro-isoquinoline. Formaldehyde can be used and so isoquinolines can be prepared which are unsubstituted at C-1. The cyclization step (an *endo-trig* reaction) is shown in Fig. 4.11.

A cyclization process in which a different bond is formed in the ring-closure step is the *Pomeranz–Fritsch* reaction. In this process the C-4 to C-4a bond

58 Review: T. J. Kametani and K. Fukumoto, in *Isoquinolines*, ed. G. Grethe, Wiley-Interscience, New York, 1981, p. 139.

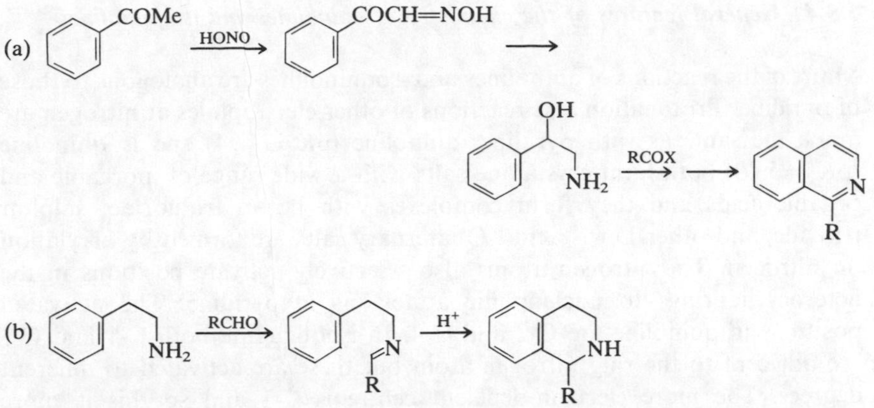

Fig. 5.37 (a) The Pictet–Gams synthesis and (b) the Pictet–Spengler synthesis.

is formed by intramolecular electrophilic substitution [Fig. 5.38(a)] and electron-releasing substituents in the benzene ring therefore make the cyclization easier. The cyclization intermediate is constructed from a benzaldehyde and aminoacetaldehyde diethyl acetal. The cyclization step is carried out in sulphuric acid but the yields are variable, and very sensitive to changes in acid strength. A modification (b) which gives tetrahydro-isoquinolines in good yield is to reduce the intermediate acetal before cyclization. Since aromatic ketones do not form Schiff bases easily, the Pomeranz–Fritsch method is not a good one for 1-substituted isoquinolines, but a variation, shown in (c), enables these derivatives to be prepared.

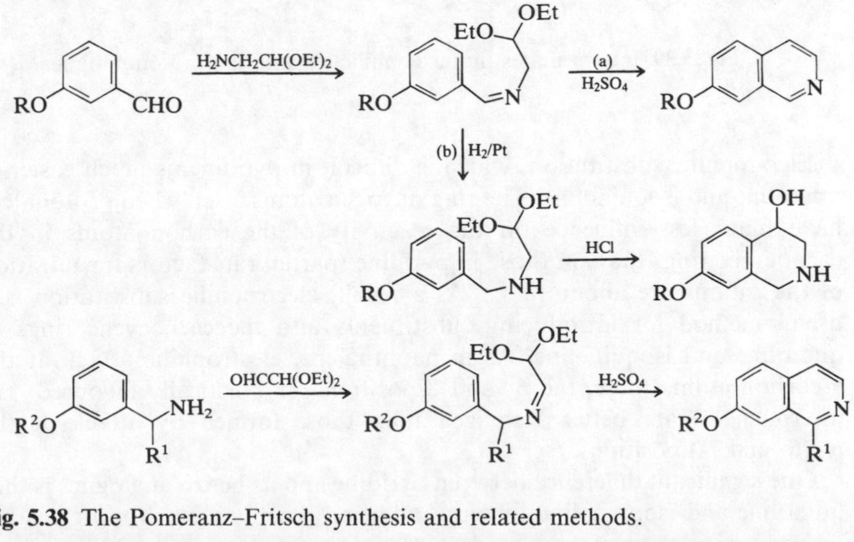

Fig. 5.38 The Pomeranz–Fritsch synthesis and related methods.

5.3.4   General features of the chemistry of quinolines and isoquinolines

Many of the reactions of quinolines and isoquinolines are analogous to those of pyridine. Protonation and reactions of other electrophiles at nitrogen are much the same as with pyridine: quinoline ($pK_a = 4.94$) and isoquinoline ($pK_a = 5.40$) both form crystalline salts with a wide range of inorganic and organic acids and they form complexes with boron trifluoride, sulphur trioxide, and other Lewis acids. Quaternary salts are formed by alkylation on nitrogen. The nitrogen atoms also selectively activate positions in the heterocyclic rings to nucleophilic attack, as in pyridine. The activated positions in quinoline are C-2 and C-4. In isoquinoline both C-1 and C-3 are adjacent to the ring nitrogen atom but these are activated to different degrees. The more electron-deficient centre is C-1 and so this is more susceptible to attack by 'hard' nucleophiles. The intermediates formed in addition–elimination reactions of nucleophiles at C-1 are more stable than those formed at C-3 (Fig. 5.39), since the negative charge cannot be accepted by the nitrogen atom without involving the π-electrons of the benzene ring.

Alkyl, amino, and hydroxy substituents at these 'activated' positions (2- and 4- in quinoline, 1- in isoquinoline) can be expected to show properties similar to those in the analogous pyridines, with respect to tautomerism, substitution of hydrogen, and displacement.

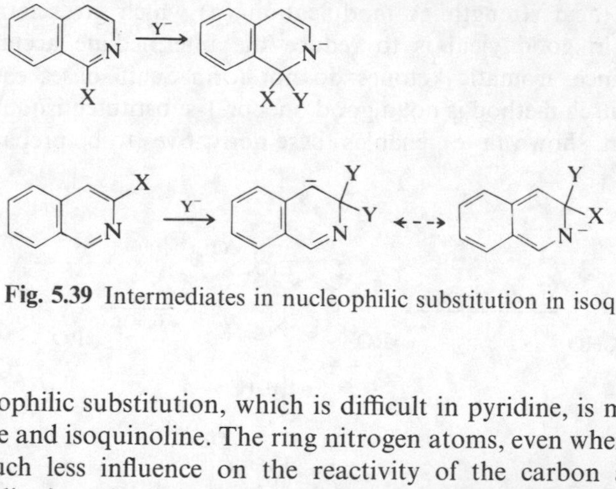

Fig. 5.39 Intermediates in nucleophilic substitution in isoquinoline.

Electrophilic substitution, which is difficult in pyridine, is much easier in quinoline and isoquinoline. The ring nitrogen atoms, even when protonated, have much less influence on the reactivity of the carbon atoms in the carbocyclic rings than on those in pyridine (partial rate factors for nitration of the cations are about 10^{-7}). As a result, electrophilic substitution is a useful method for introducing substituents into the carbocyclic rings of quinoline and isoquinoline. As in naphthalene, electrophilic attack at the α-carbon atoms (here, the 5- and 8-positions) is kinetically favoured; the intermediates are better stabilized than those formed by attack at the β- (6- and 7-) positions.

One significant difference between pyridine and its benzo analogues is that quinoline and isoquinoline undergo addition reactions much more readily in the nitrogen-containing ring. The initial coordination of an electrophile at nitrogen is quite frequently followed by the addition of a nucleophile at an adjacent carbon atom, as shown in Fig. 5.40. Nucleophilic attack on the heterocycle also sometimes results in the formation of addition products rather than substitution products.

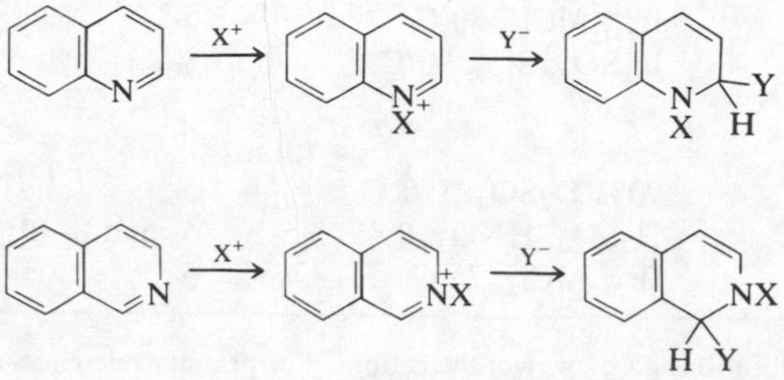

Fig. 5.40 Addition reactions initiated by electrophilic attack at nitrogen.

Such intermediates can clearly distort the results of attempted substitution reactions on these heterocycles: for example, the intermediate formed from isoquinoline is likely to undergo further electrophilic attack at C-4, rather than in the benzene ring, because the intermediate is an enamine. This tendency of the bicyclic systems to undergo addition reactions more readily than pyridine is analogous to the behaviour of naphthalene in comparison with benzene. In these bicyclic compounds, much of the resonance stabilization is retained in the remaining benzene ring of the dihydro species, whereas resonance stabilization is lost in dihydro derivatives of the monocyclic systems.

5.3.5 *Electrophilic substitution*

Under strongly acidic conditions, electrophilic substitution of both quinoline and isoquinoline takes place preferentially in the carbocyclic rings, and usually at the 5- or 8-positions, or both. Examples of such reactions are given in Table 5.4. These reactions go by way of the quinolinium and isoquinolinium cations and the preferential substitution at positions 5 and 8 is in accord with an addition–elimination mechanism of the usual type.

There are, however, several useful substitution reactions both of quinolines and of isoquinolines which do not follow this pattern, and where different mechanisms are probably involved. For example, the major product of nitration of quinoline by nitric acid in acetic anhydride is 3-nitroquinoline; isoquinoline gives 4-nitroisoquinoline in low yield with the same reagent. Both quinoline and isoquinoline can be brominated in high yield by heating

Table 5.4 Electrophilic substitution of quinolines and isoquinolines.[a]

Electrophile	Reagents and conditions	Major products
Quinoline		
D$^+$	70% D$_2$SO$_4$, 150°C	8-
NO$_2^+$	H$_2$SO$_4$, HNO$_3$, 0°C	5- and 8- (1:1)
Br$^+$	Br$_2$, AlCl$_3$, 80°C	5-[b]
SO$_3$H$^+$	H$_2$SO$_4$, SO$_3$, 90°C	8-[c]
Isoquinoline		
D$^+$	90% D$_2$SO$_4$, 180°C	5-
NO$_2^+$	H$_2$SO$_4$, HNO$_3$, 0°C	5- and 8- (9:1)
Br$^+$	Br$_2$, AlCl$_3$, 75°C	5-[d]

a All these substitutions go by way of the cations. For primary references see ref. 56. b 8-Bromoquinoline is also formed. Excess bromine gives 5,8-dibromoquinoline (86%). c The 5-sulphonic acid is also formed at higher temperatures (up to 220°C) but at 300°C both rearrange to the 6-sulphonic acid (see text). d 5-Bromoisoquinoline is isolated in 78% yield. Bromine (2 mol) gives 5,8-dibromoisoquinoline.

their hydrochlorides with bromine in nitrobenzene: the products are 3-bromoquinoline and 4-bromoisoquinoline, respectively.[59]

Such reactions can be rationalized by assuming that the reagents can reversibly add to the heterocycles across the 1,2-bonds: substitution then takes place in the intermediate dihydro species. This is illustrated for the bromination of quinoline in Fig. 5.41.

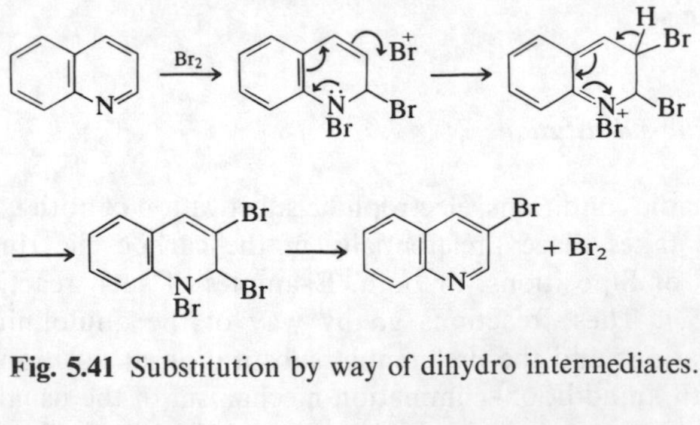

Fig. 5.41 Substitution by way of dihydro intermediates.

A different type of 'abnormal' substitution is exemplified by the formation of quinoline-6-sulphonic acid in the sulphonation of quinoline at 300°C. This is the thermodynamic product of the reaction because the 5- and 8-derivatives

59 T. J. Kress and S. M. Constantino, *J. Heterocycl. Chem.*, 1973, **10**, 409.

which are formed at lower temperatures are destabilized by *peri* interaction, and rearrange to the 6-sulphonic acid at 300°C (Fig. 5.42). The same type of rearrangement takes place with naphthalene-1-sulphonic acid, for the same reason.

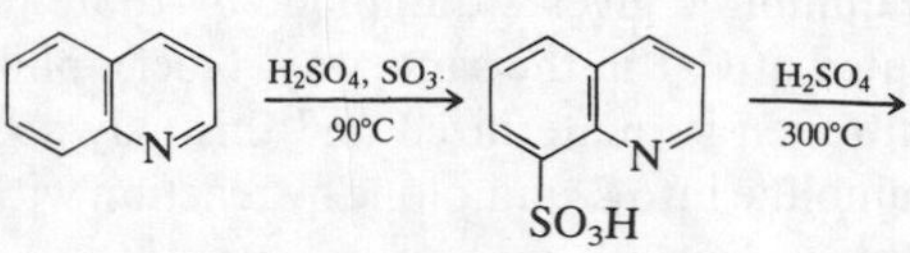

Fig. 5.42 Sulphonation of quinoline.

5.3.6 *Nucleophilic substitution*

The displacement of chloride from the 2- and 4-positions of quinoline and from the 1-position of isoquinoline by nucleophiles such as alkoxides, thiophenoxide, and secondary amines takes place readily by an addition–elimination mechanism. Two examples of selective displacement of chloride by carbanions are shown in Fig. 5.43. It is possible to displace chloride from the 3-position of isoquinolines under more vigorous conditions.

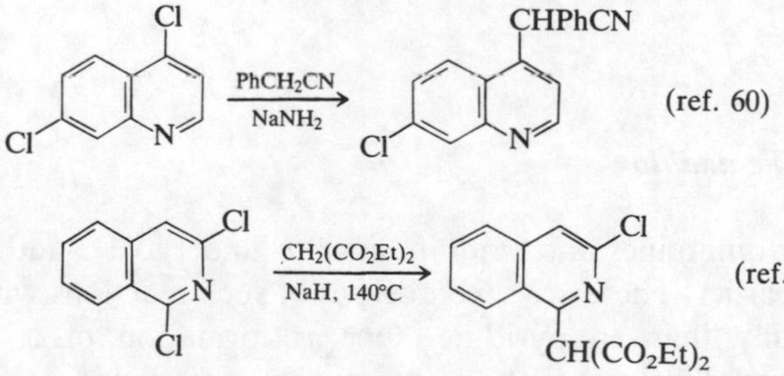

Fig. 5.43 Selective displacement of chloride.

Formal displacement of hydride by nucleophiles takes place in the same manner as in pyridine. Organolithium compounds and Grignard reagents can alkylate or arylate quinoline in the 2-position and isoquinoline in the 1-position. The Chichibabin reaction goes very readily with isoquinoline; it is aminated in high yield exclusively at the 1-position either by heating with sodamide in dimethylaniline or with potassamide in liquid ammonia. When the 1-position is blocked, the isoquinoline is not aminated at the 3-position.

60 R. A. Cutler, A. R. Surrey, and J. B. Cloke, *J. Am. Chem. Soc.*, 1949, **71**, 3375.
61 T. Kametani, K. Kigasawa, and M. Hiiragi, *Chem. Pharm. Bull.*, 1967, **15**, 704.

Quinoline gives 2-aminoquinoline in poor yield when heated with sodamide in dimethylaniline; it can be aminated in good yield by barium amide in liquid ammonia. 2-Substituted quinolines generally undergo amination in the 4-position more readily than the corresponding pyridines.

Quinoline and isoquinoline can be hydroxylated by heating with dry potassium hydroxide: thus, quinoline gives 2-quinolone by route (a) in Fig. 5.44. Isoquinoline is substituted at C-1 in the same way. Nucleophilic attack is greatly facilitated if the nitrogen is quaternized and this may account for much easier conversion of quinoline into 2-quinolone by reaction with sodium hypochlorite, as shown in (b).

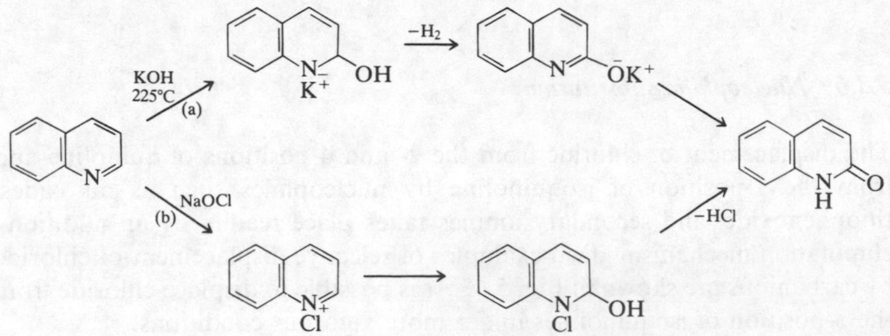

Fig. 5.44 Routes to 2-quinolone from quinoline.

5.3.7   *Nucleophilic addition*

The tendency for quinolines and isoquinolines to undergo 1,2-addition has been referred to earlier. There are several types of such reactions which take place more readily than in pyridine. One is formation of a covalent 'pseudobase' in equilibrium with a quaternary hydroxide such as 1-methylquinolinium hydroxide [Fig. 5.45(a)].[34] Although the equilibrium concentrations of the covalent form of this species and of the corresponding isoquinolinium salt are small ($K \simeq 10^{-6}$ for both), they are appreciably higher than for the pyridine species **24**. Another example is the formation of so-called Reissert compounds by the reaction of quinoline or isoquinoline with an acyl halide and potassium cyanide in aqueous dichloromethane.[62] The formation of such an adduct, **44**, with quinoline is shown in (b). These compounds, which are isolable solids, undergo several useful transformations; one which has been used in the synthesis of isoquinoline alkaloids[63] is the

34 Review: J. W. Bunting, *Adv. Heterocycl. Chem.*, 1979, **25**, 1.
62 Review: F. D. Popp, *Adv. Heterocycl. Chem.*, 1968, **9**, 1; 1979, **24**, 187.
63 Review: F. D. Popp, *Heterocycles*, 1973, **1**, 165.

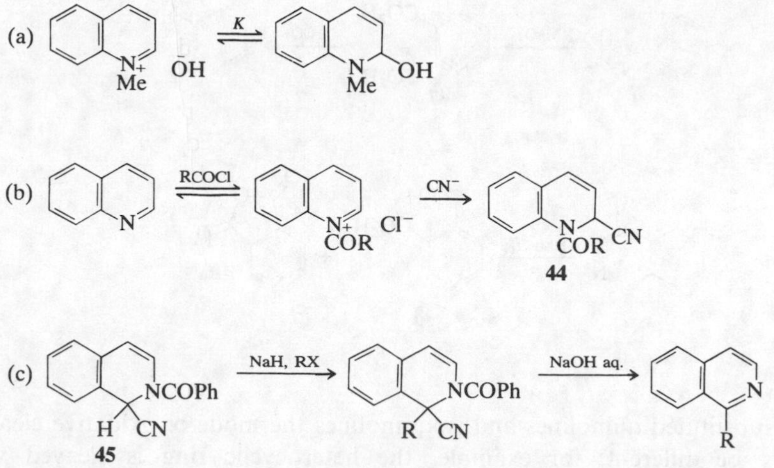

Fig. 5.45 (a) Pseudobase formation; (b) and (c) Reissert compounds.

alkylation of the isoquinoline adduct **45** to give 1-substituted isoquinolines by the route shown in (c).

Quinolines can be selectively reduced at the 1,2-bond by reaction with lithium aluminium hydride or diisobutylaluminium hydride but the 1,2-dihydroquinolines are unstable and disproportionate easily. The product of diisobutylaluminium hydride reduction can, however, be intercepted before hydrolysis by ethyl chloroformate to give the 1-ethoxycarbonyl compounds **46** in good yield.[64] Quinoline can be reduced to 1,2,3,4-tetrahydroquinoline by catalytic hydrogenation or by tin and hydrochloric acid. Isoquinolines are reduced in a similar way to tetrahydroisoquinolines.

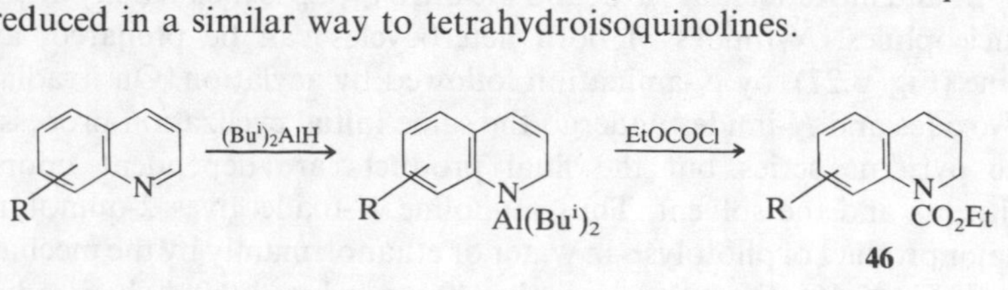

5.3.8   Oxidative cleavage

The carbocyclic ring of quinoline can be oxidatively cleaved to give pyridine-2,3-dicarboxylic acid, **47**. This can be isolated in good yield by electrochemical oxidation but the usual product isolated from oxidation by alkaline permanganate and other chemical oxidants is nicotinic acid, derived from it by decarboxylation. Isoquinoline similarly gives pyridine-3,4-dicarboxylic acid, **48**, together with the corresponding anhydride, on oxidation with alkaline potassium permanganate.

64 D. E. Minter and P. L. Stotter, *J. Org. Chem.*, 1981, **46**, 3965.

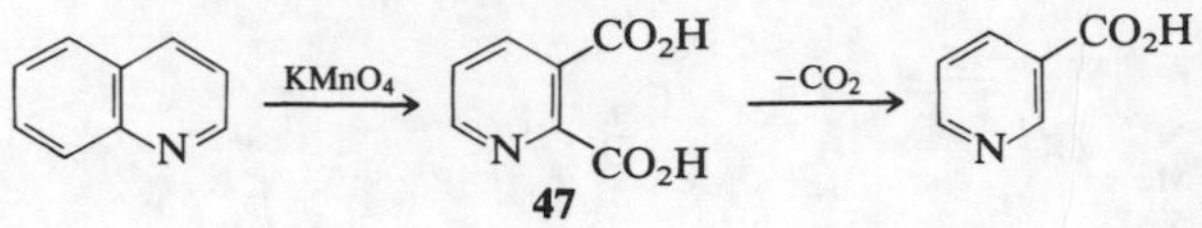

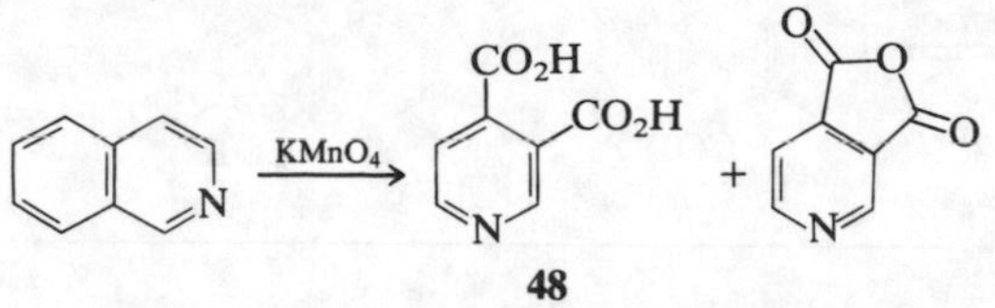

In substituted quinolines and isoquinolines the mode of oxidative cleavage may be different; for example, the heterocyclic ring is cleaved when 2-alkylquinolines are oxidized by acidic permanganate. However, a series of substituted pyridinedicarboxylic acids has been prepared by ozonolysis of the corresponding quinolines.[65]

5.3.9 N-*Oxides and* N-*imides*

Quinoline and isoquinoline both form N-oxides when treated with hydrogen peroxide in acetic acid, or with organic peracids. Quinoline N-oxides[66] are useful synthetic intermediates in much the same way as pyridine N-oxides; for example, quinoline N-oxide can be nitrated at C-4 by reaction with sulphuric and nitric acids at 70°C, and the nitro group is then readily displaced by nucleophiles. N-Imides of both heterocycles can be prepared, as for pyridine (Fig. 5.27), by N-amination followed by acylation. On irradiation, the N-oxides and N-imides undergo the same initial cyclization processes as in the pyridine series but the final products are dependent upon the substituents and the solvent. Thus quinoline N-oxide gives 2-quinolone as the major product of photolysis in water or ethanol, mainly by the mechanism shown in Fig. 5.46(a), but the oxazepine **49** is produced (60%) by irradiation of the N-oxide in cyclohexane.[67] The pathway (b) shown in Fig. 5.46 can account for the formation of this product: two thermal rearrangements, a [1, 5] sigmatropic migration of the bridging oxygen function and an electrocyclic ring opening, give the final product, which is thermodynamically favoured because it relieves both the strain in the three-membered ring and the *ortho*-quinonoid character of its precursor.

65 C. O'Murchu, *Synthesis*, 1989, 880.
66 Review: G. Jones and J. D. Baty, in *Quinolines*, Part II, ed. G. Jones, Wiley-Interscience, Chichester, 1982, p. 377.
67 A. Albini, G. F. Bettinetti, and G. Minoli, *Tetrahedron Lett.*, 1979, 2761.

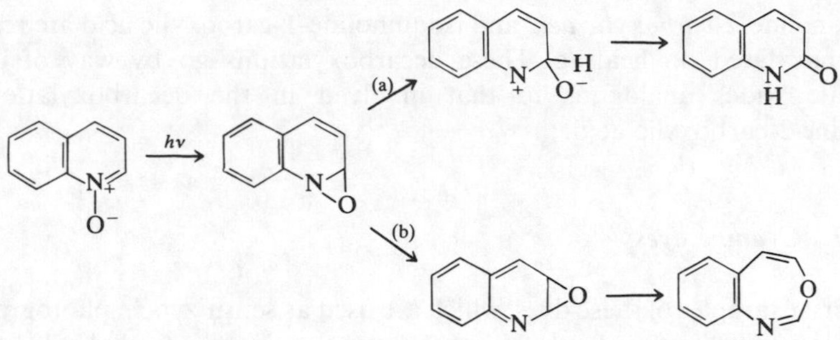

Fig. 5.46 Photoisomerization of quinoline *N*-oxide.

5.3.10 *Properties of substituents*

Quinolines substituted at the 2- and 4-positions by *alkyl* groups are, like those in pyridine, readily deprotonated and they can be alkylated and acylated. As in the pyridine series, aldol condensations can be carried out in the presence of acids or Lewis acids such as zinc chloride. The 1-alkylisoquinolines are activated in the same way but 3-alkylisoquinolines are much less activated. For example, the zinc chloride catalysed aldol condensation of 3-methylisoquinoline with benzaldehyde goes only in poor yield, even after heating the reagents for a week at 160°C.[68]

2- and 4-*Hydroxy*quinolines exist in the quinolone form (shown in Fig. 5.44) and 1-hydroxyisoquinoline in the isoquinolone form. The trivial names 'carbostyril' and 'isocarbostyril' are sometimes seen in the literature for 2-quinolone and 1-isoquinolone, respectively. In 3-hydroxyisoquinoline, however, the equilibrium between hydroxy and keto forms is finely balanced because the keto form **50** has isoquinonoid character. 8-Hydroxyquinoline **51** can form an intramolecular hydrogen bond. It is also a good chelating agent for metals.

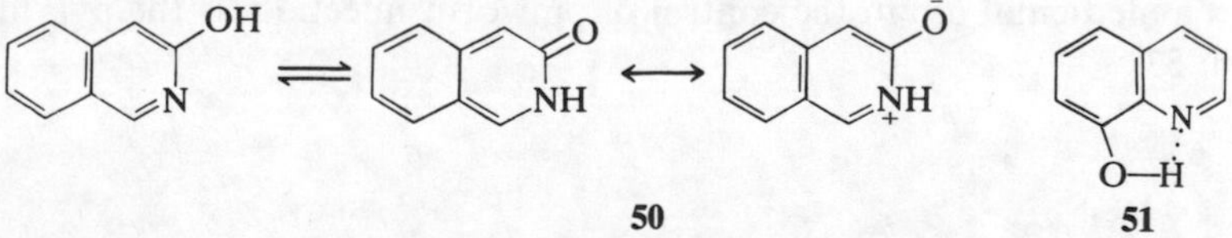

All the *amino*quinolines, with the exception of the 2- and 4-isomers, act as normal aromatic amines and can, for example, be diazotized and coupled. Similarly, both 1- and 3-aminoisoquinoline give the corresponding isoquinolones when diazotized in aqueous solution but the other amino-isoquinolines give normal aromatic diazonium salts.

68 H. Erlenmeyer, H. Baumann, and E. Sorkin, *Helv. Chim. Acta*, 1948, **31**, 1978.

Quinoline-2-*carboxylic acid* and isoquinoline-1-carboxylic acid are readily decarboxylated on heating. These decarboxylations go by way of intermediate ylides analogous to that involved in the decarboxylation of pyridine-2-carboxylic acid.

5.3.11   Cyanine dyes[69]

The first examples of these dyes, which are used as sensitizers in photographic emulsions, were formed by the base-catalysed reaction of a methyl-substituted quinolinium salt with *N*-ethylquinolinium iodide. An example of the formation of a cyanine dye, the purple compound **52**, is shown. The reaction makes use of two of the characteristics of the chemistry of quinolinium salts which we have seen earlier: the activation of the 4-position to nucleophilic attack, and the increased acidity of a methyl group at the 2- or 4-positions.

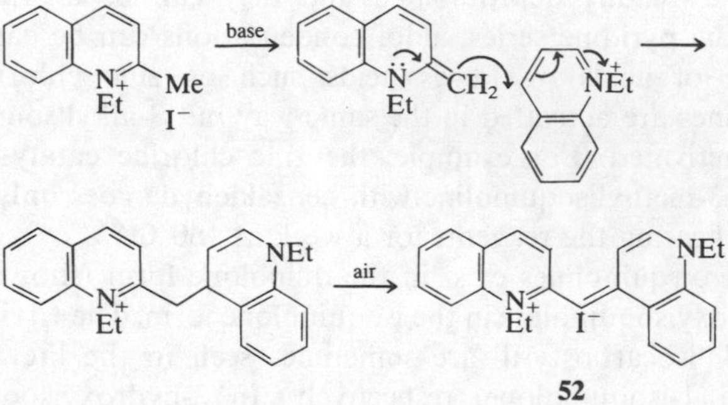

52

This type of reaction is not limited to quinoline components and has been extended to a wide range of other heterocycles; the linking methine chain can also be extended. An example of a mixed cyanine dye, which has also found a medicinal use in the control of pinworm infection, is the pyrvinium cation **53**.

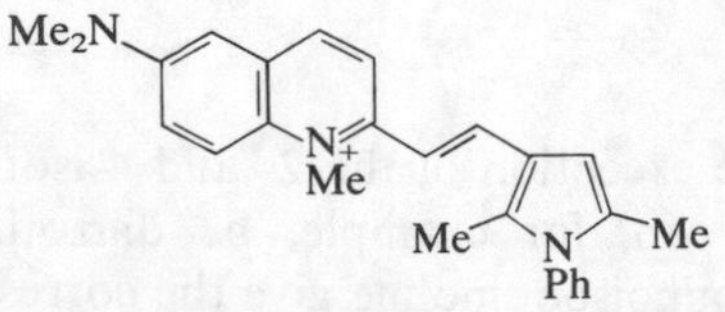

53

69 Review: D. M. Sturmer, in *Special Topics in Heterocyclic Chemistry*, ed. A. Weissberger and E. C. Taylor, Wiley-Interscience, New York, 1977, p. 441.

5.4   Other fused pyridines

5.4.1   Quinolizines[70]

The quinolizinium salts such as **54** and its derivatives are stable, water-soluble crystalline solids which can be regarded as aromatic. The ultraviolet spectrum of the iodide in water is very similar to that of quinoline. The salts can be prepared by several routes starting from 2-substituted pyridines; one such route, in which the appropriate side chain is introduced by a Grignard reaction, is shown in Fig. 5.47.[71]

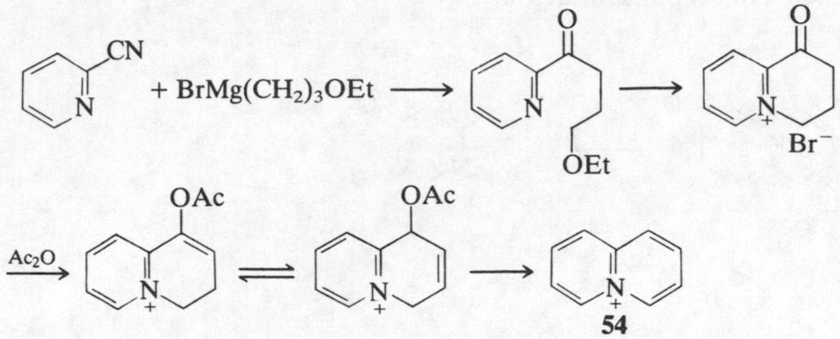

Fig. 5.47 A route to quinolizinium salts.

The ring system occurs naturally, usually in a partially or fully reduced form.[72] Lupinine **55** is one of the quinolizidine alkaloids which are present in plants of the lupin genus.

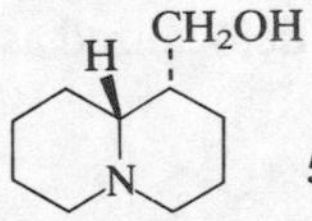

Quinolizinium salts retain many of the properties of pyridinium salts: they are resistant to electrophilic attack but are susceptible to nucleophilic attack, particularly at the 4-position. The ring system is easily cleaved by nucleophiles. Hydride is not displaced from C-4 because the intermediate **56** can undergo electrocyclic ring opening, but chloride can be lost from 4-chloroquinolizinium perchlorate (Fig. 5.48).[73]

Hydroxy, amino, and alkyl substituents at C-2 and C-4 are activated in the same way as in pyridinium salts: for example, 2-hydroxyquinolizinium salts are easily deprotonated to 2-quinolizones, **57**.

70 Reviews: P. Claret, in *Comprehensive Organic Chemistry* Vol. 4, ed. P. G. Sammes, Pergamon, Oxford, 1979, p. 233; G. Jones, *Adv. Heterocycl. Chem.*, 1982, **31**, 1.
71 E. E. Glover and G. Jones, *J. Chem. Soc.*, 1958, 3021.
72 Review: D. R. Dalton, *The Alkaloids*, Marcel Dekker, New York, 1979, p. 122.
73 D. F. Farquhar, T. T. Gough, and D. Leaver, *J. Chem. Soc., Perkin Trans. 1*, 1976, 341.

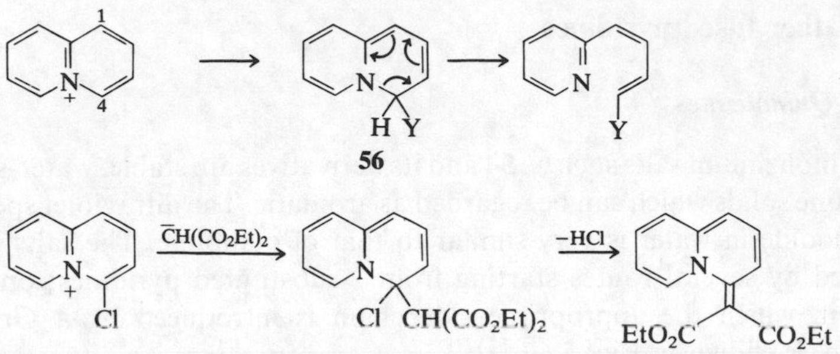

Fig. 5.48 Nucleophilic attack at C-4.

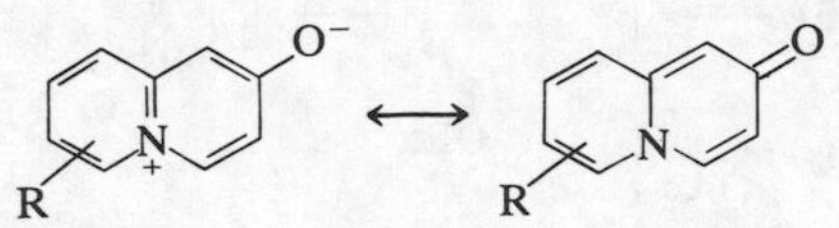

57

5.4.2 Acridines[74,75]

Acridine is a linear tricyclic aromatic heterocycle which is similar in structure to anthracene. Some aminoacridines, particularly 9-aminoacridine and 3,6-diaminoacridine (proflavine), have antiseptic properties. Other amino-acridines are used as dyestuffs: the cation **58** (Acridine Orange L) is an example. In this cation the positive charge is extensively delocalized as shown.

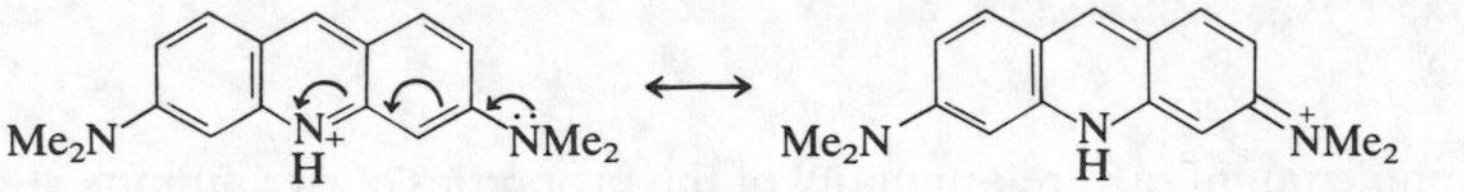

58

A general method of synthesis of the ring system is the cyclization of 2-(phenylamino)benzoic acids (Fig. 5.49).

9-Chloroacridine obtained in this way can be converted into other 9-substituted acridines by nucleophilic displacement. For example, 9-amino-acridine is obtained by heating it with ammonium carbonate in phenol. Acridine itself can be obtained from 9-chloroacridine by reduction with hydrogen and Raney nickel to give 9,10-dihydroacridine, followed by oxidation with chromic acid.

74 Review: N. Campbell, in *Rodd's Chemistry of Carbon Compounds* Vol. IVG, ed. S. Coffey, Elsevier, Amsterdam, 1978, p. 1.
75 Review: R. M. Acheson, *Acridines*, Wiley-Interscience, New York, 1973.

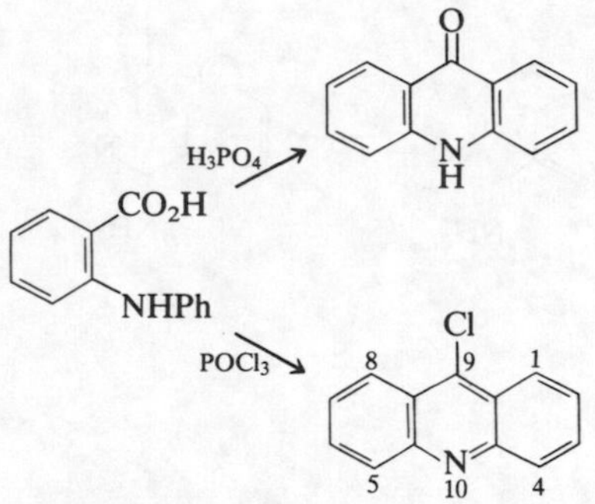

Fig. 5.49 9-Acridone and 9-chloroacridine by cyclization.

Acridine, which occurs in coal tar, is a crystalline solid, m.p. 110°C, which is a base ($pK_a = 5.60$) of comparable strength with pyridine. The 9-position is activated by the ring nitrogen atom (as in 9-chloroacridine above) and, as in anthracene, addition reactions to the 9- and 10-positions take place very readily. Acridine is easily reduced to 9,10-dihydroacridine, and with potassium cyanide it gives 9-cyano-9,10-dihydroacridine. It can also be aminated at C-9 by sodamide in *NN*-dimethylaniline. A methyl group at C-9 is sufficiently activated to couple with diazonium salts, and, when the nitrogen atom is quaternized, a methide **59** is easily formed by reaction with alkali.

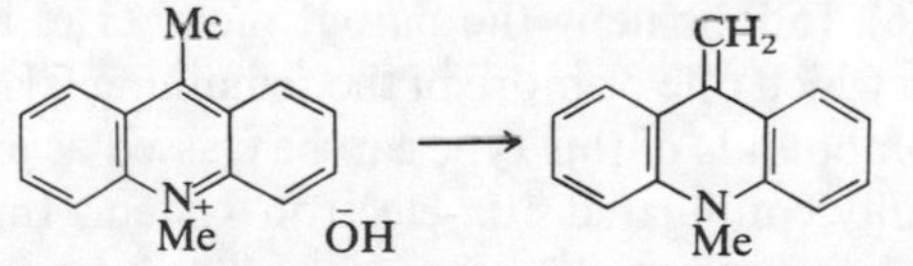

5.4.3 *Phenanthridines*[74,76]

Phenanthridine, **60**, is a heterocyclic analogue of phenanthrene. It is a solid, m.p. 108°C, which is weakly basic ($pK_a = 4.52$). It can be *N*-alkylated and *N*-oxidized like the other fused pyridines. The ring system can be prepared by a number of cyclization reactions. A good route to phenanthridine and several derivatives is that shown in Fig. 5.50, involving an aryne intermediate.[77]

The activated position of the phenanthridine system is C-6, and the properties of substituents at C-6 differ from those at other positions. Chloride at C-6 is easily displaced by nucleophiles, the diazonium salt

76 Review B. R. T. Keene and P. Tissington, *Adv. Heterocycl. Chem.*, 1971, **13**, 315.
77 Reviews: S. Kessar, *Accounts Chem. Res.*, 1978, **11**, 283.

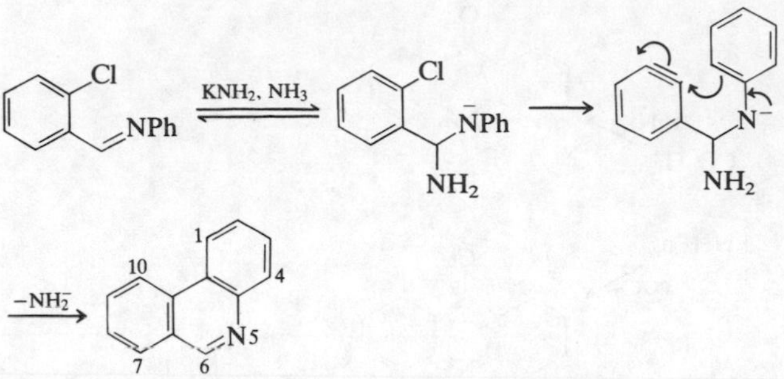

Fig. 5.50 Phenanthridine by aryne cyclization.

is unstable and easily transformed to 6-phenanthridone, and 6-methyl-phenanthridine can be lithiated and alkylated at the methyl group, for example. Phenanthridine is reduced to the 5,6-dihydro derivative by reaction with lithium tetrahydroaluminate.

5.4.4   Indolizines[78]

The structure of indolizine differs from those of the other heterocycles discussed so far in that a five-membered ring is fused across the 1,2-bond of pyridine. The ring system occurs naturally in a fully reduced form, as the alkaloid δ-coniceine (Fig. 5.6). Indolizine is the parent member of a family of heterocyclic compounds in which one or more of the remaining CH groups is replaced by nitrogen.[79] Compounds of this type can be classed as aromatic because the periphery is a fully conjugated 10π-electron system. Indolizine has significant resonance energy (Section 2.3.2) and its ultraviolet spectrum (λ_{max} 237, 281, and 337 nm) resembles that of naphthalene.

There are several synthetic routes to indolizines and most start with a pyridine derivative. The Chichibabin synthesis is an example; in this, a pyridine with an activated methylene substituent at C-2 is alkylated with an α-halocarbonyl compound. An intramolecular aldol reaction is then used to generate the five-membered ring (Fig. 5.51). The ring system can also be produced by 1,3-dipolar addition reactions of pyridinium ylides; an example is shown in Fig. 5.28.

Indolizine is an air-sensitive solid, m.p. 75°C, which shows some properties similar to those of pyrrole. For example, it is a weak bease ($pK_a = 3.94$) and the most stable cation **61** is formed by protonation at C-3, which is equivalent

78 Reviews: H. L. Blewitt, in *Special Topics in Heterocyclic Chemistry*, ed. A. Weissberger and E. C. Taylor, Wiley-Interscience, New York, 1977, p. 117; F. J. Swinbourne, J. H. Hunt, and G. Klinkert, *Adv. Heterocycl. Chem.*, 1978, **23**, 103.
79 For a survey of these 'aza-indolizines' see M. Tisler, *Pure Appl. Chem.*, 1980, **52**, 1611.

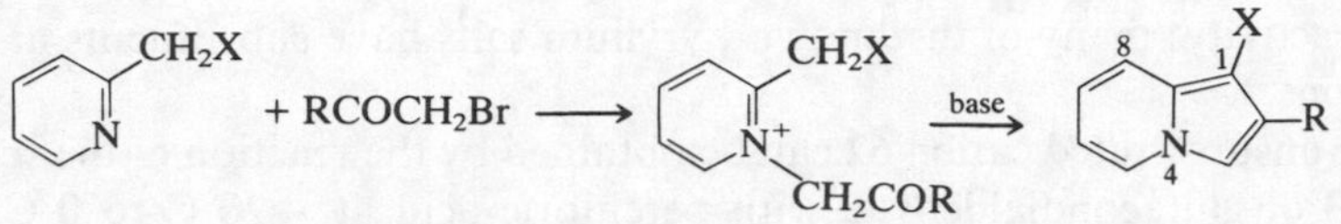

Fig. 5.51 The Chichibabin synthesis of indolizines.

to the 2-position of pyrrole. Indolizines similarly undergo electrophilic substitution preferentially at C-3, and, if that position is blocked, at C-1. Acetylation can be carried out in the absence of a catalyst, and the Mannich and Vilsmeier reactions go readily. The ring system is preferentially hydrogenated in the six-membered ring but if the reduction is carried out in acid, the cation **61** is selectively hydrogenated at the C1–C2 double bond.

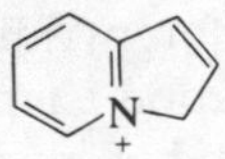

61

5.5 Ring systems containing oxygen

5.5.1 Pyrylium salts[80,81,82a]

The pyrylium cation **62** is a fully unsaturated six-membered heterocycle with one oxygen atom bearing a positive charge. Oxonium cations are, in general, very reactive and susceptible to nucleophilic attack. By comparison with typical oxonium ions, the pyrylium cation is unreactive: for example, it is stable even in nucleophilic solvents when the reaction medium is strongly acidic. It seems reasonable to ascribe the lower reactivity to the aromatic character of the ring system. The ultraviolet spectra of pyrylium cations are similar to those of the corresponding *N*-alkylpyridinium cations, and the ^{1}H NMR spectrum of pyrylium perchlorate is like that of *N*-ethylpyridinium perchlorate, the signals appearing in the δ 8–10 region.[83] Pyrylium salts are, however, much more reactive towards nucleophiles than the corresponding pyridinium salts, as might be expected because of the higher electronegativity of oxygen. Alkyl or aryl groups at the 2-, 4-, and 6-positions help to moderate

80 Review: A. T. Balaban, in *New Trends in Heterocyclic Chemistry*, ed. R. B. Mitra *et al.*, Elsevier, Amsterdam, 1979, p. 79.
81 Review: R. Livingstone, in *Rodd's Chemistry of Carbon Compounds* Vol. IVE, ed. S. Coffey, Elsevier, Amsterdam, 1977, p. 1.
82 Reviews: J. Staunton, in *Comprehensive Organic Chemistry* Vol. 4, ed. P. G. Sammes, Pergamon, Oxford, 1979, (a) p. 607, (b) p. 629, (c) p. 659.
83 S. Yoneda, T. Sugimoto, and Z. Yoshida, *Tetrahedron*, 1973, **29**, 2009.

this reactivity; many of the known pyrylium salts have substituents at these positions.

The unsubstituted cation **62** can be obtained by the reaction of the sodium salt **63** of glutacondialdehyde with perchloric acid at $-20°C$ to $0°C$. The salt is obtained from the pyridine–sulphur trioxide complex, **11**, by alkaline hydrolysis.

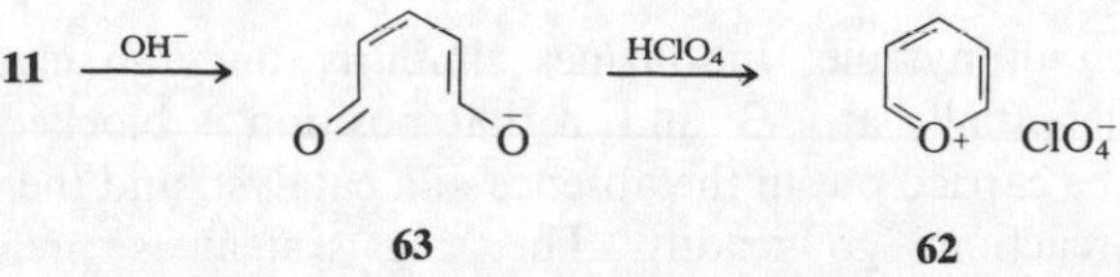

Substituted pyrylium salts can be prepared by a variety of cyclization methods, many of which are analogous to those used for pyridines and discussed in Section 5.2.2. In most cases the cyclization step is the acid-catalysed dehydration of a 1,5-diketone, but the method of construction of the diketone and its oxidation level can vary. Two examples (which lead to the formation of tetrafluoroborate salts rather than the explosive perchlorates) are outlined in Fig. 5.52.[84] When saturated 1,5-diketones are the precursors, an oxidizing agent such as iron(III) chloride is usually added to the medium.

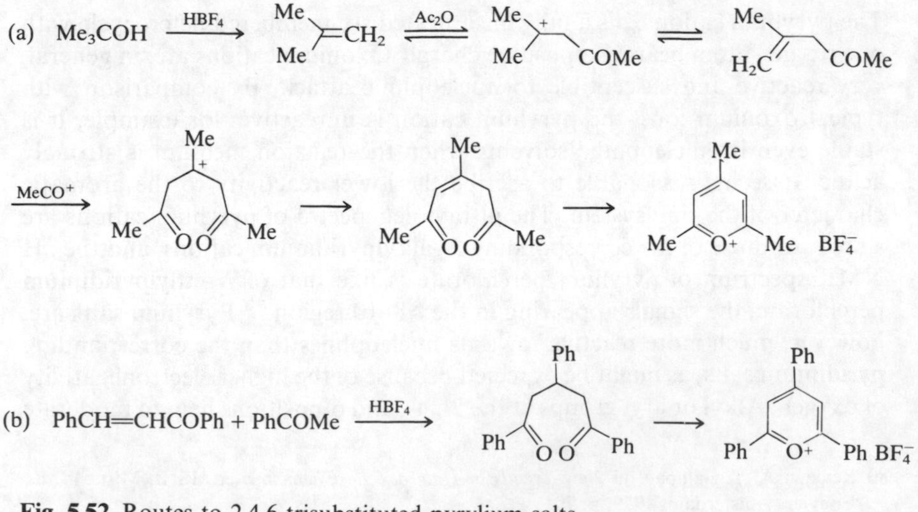

Fig. 5.52 Routes to 2,4,6-trisubstituted pyrylium salts.

84 (a) A. T. Balaban and A. J. Boulton, *Org. Synth.*, *Coll. Vol.* V, 1973, 1112; (b) K. Dimroth, C. Reichardt, and K. Vogel, *Org. Synth.*, *Coll. Vol.* V, 1973, 1135.

These trisubstituted pyrylium salts are useful intermediates for the formation of a range of other carbocyclic and heterocyclic compounds. An initial attack by a nucleophile at C-2 produces a 2*H*-pyran, which can exist in equilibrium with an open valence tautomer **64**. Compounds of the general structure **64** can be isolated from reactions of 2,4,6-trimethylpyrylium salts with several nucleophiles, including phenyllithium (Y = Ph), sodium cyanide (Y = CN), and sodium borohydride (Y = H).

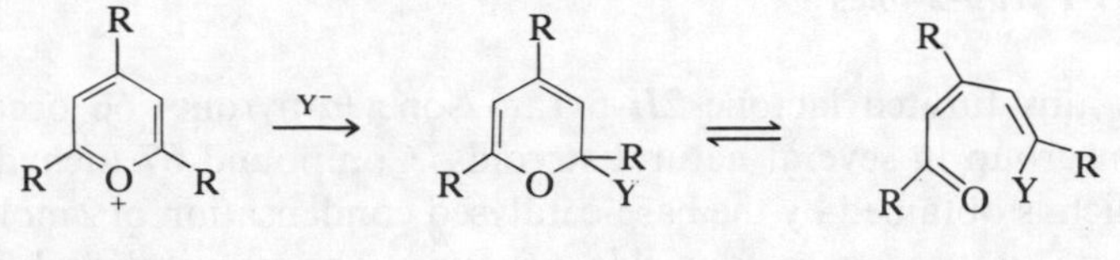

64

In other reactions of this type these intermediates can react further and may recyclize. Some examples of the conversion of the 2,4,6-trimethyl-pyrylium cation into other cyclic compounds are shown in Fig. 5.53.

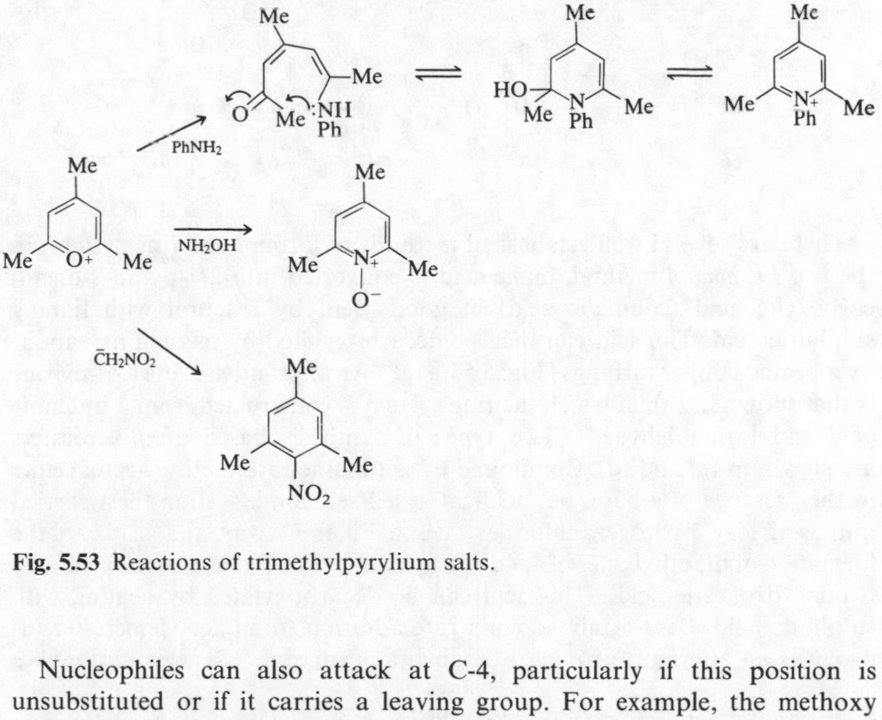

Fig. 5.53 Reactions of trimethylpyrylium salts.

Nucleophiles can also attack at C-4, particularly if this position is unsubstituted or if it carries a leaving group. For example, the methoxy group in the cation **65** is replaced when the cation is reacted with secondary amines, thiols, and other nucleophiles.

65

5.5.2 2H-*Pyran-2-ones*[81,82b]

The fully unsaturated lactone 2H-pyran-2-one (α-pyrone) **66** occurs as a side-chain group in several natural steroids. Compound **67** (dehydroacetic acid), which is obtained by the base-catalysed condensation of 2 mol of ethyl acetoacetate, is used as a fungicide on fruit. The physical and chemical properties of the 2H-pyran-2-one ring system are consistent with its representation as an unsaturated lactone rather than as an aromatic zwitterion. Its ^{1}H NMR spectrum shows no evidence for the existence of a ring current, and its carbonyl stretching frequency ($1739 \, cm^{-1}$ in $CHCl_3$) is that of an unsaturated lactone. Its resonance energy, calculated by the method described in Section 2.2.6, is negligible.[85]

66   **67**

There are several well-established methods of forming the ring system. In the *von Pechmann* method, malic acid is converted into 2H-pyran-2-one-4-carboxylic acid (coumalic acid) in good yield by reaction with fuming sulphuric acid. The acid can then be decarboxylated by passing its vapour over heated copper turnings [Fig. 5.54(a)].[86] An alternative route to α-pyrone is that shown in (b), in which the ring system is constructed from 3-butenoic acid and formaldehyde.[87] Two types of synthesis based on β-ketoesters are shown in (c) and (d). Compound **67** is obtained from ethyl acetoacetate in the presence of a base, but acid-catalysed self-condensation (best carried out using dry hydrogen chloride) goes a different way and leads to the formation of the ethyl ester of 4,6-dimethyl-2H-pyran-2-one-5-carboxylic acid (isodehydroacetic acid). This acid can be decarboxylated by heating with sulphuric acid. Base-catalysed conjugate addition to an acetylenic ester (d) provides an alternative route to trisubstituted pyrones. Reaction (e) makes

85 J. Aihara, *J. Am. Chem. Soc.*, 1976, **98**, 2750.
86 H. E. Zimmerman, G. L. Grunewald, and R. M. Paufler, *Org. Synth., Coll. Vol.* V, 1973, 982.
87 M. Nakagawa *et al.*, *Org. Synth.*, 1977, **56**, 49.

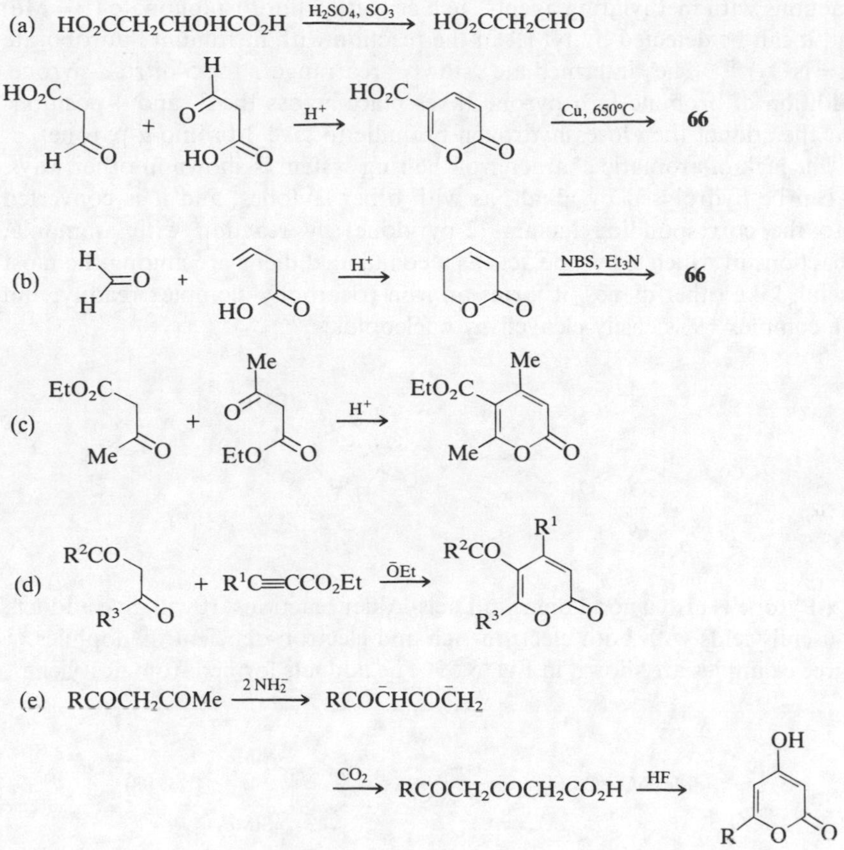

(a) $HO_2CCH_2CHOHCO_2H$ $\xrightarrow{H_2SO_4,\ SO_3}$ HO_2CCH_2CHO

(b)

(c)

(d)

(e) $RCOCH_2COMe$ $\xrightarrow{2\,NH_2^-}$ $RCO\bar{C}HCO\bar{C}H_2$

$\xrightarrow{CO_2}$ $RCOCH_2COCH_2CO_2H$ $\xrightarrow{HF}$

Fig. 5.54 Routes to α-pyrones.

use of the property of dianions of 1,3-diketones to be acylated preferentially at the terminal methylene group.[88] Carboxylation of the anions, followed by ring closure, gives 4-hydroxy derivatives.

α-Pyrone can be substituted by electrophiles at the 3- and 5-positions, which are those activated by the ring oxygen atom. The initial attack of an electrophile may, however, take place at the carbonyl oxygen atom to give an intermediate pyrylium cation, **68**. Such a cation can be isolated from

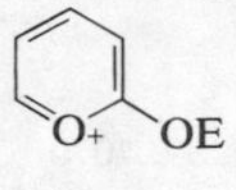

68

88 Review: T. M. Harris, C. M. Harris, and K. B. Hindley, *Progr. Chem. Org. Nat. Prod.*, 1974, **31**, 217.

reactions with methylating agents such as methyl fluorosulphonate (E = Me) and it can be detected by NMR in the reaction with nitronium fluoroborate (E = NO$_2$).[89] The intermediate slowly rearranges to 5-nitro-α-pyrone. Addition of bromine to α-pyrone takes place across the 3- and 4-positions but the adduct then loses hydrogen bromide to give 3-bromo-α-pyrone.

The lack of aromatic character of the ring system is shown in other ways. It can be hydrolysed by alkali, as with other lactones, and it is converted into the corresponding lactam (2-pyridone) by reaction with ammonia. Reactions in which α-pyrone acts as a conjugated diene are among the most useful. Like other dienes, it forms an iron tricarbonyl complex readily, and the complex **69** is easily cleaved by nucleophiles.[90]

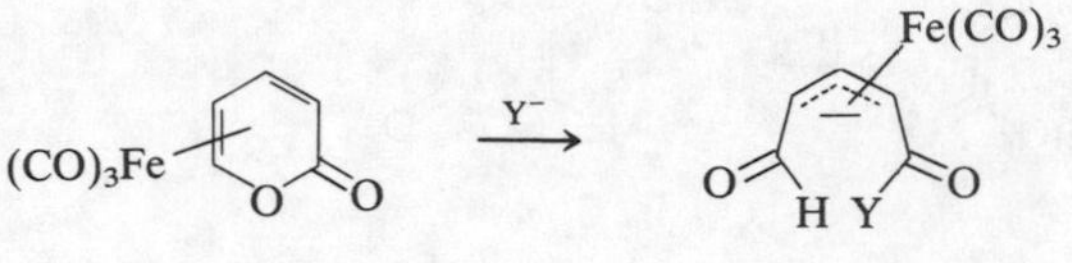

69

α-Pyrone is also a good diene in Diels–Alder reactions. It will give adducts in useful yields with both electron-rich and electron-deficient dienophiles.[91] Three examples are shown in Fig. 5.55. The adducts formed from acetylenes,

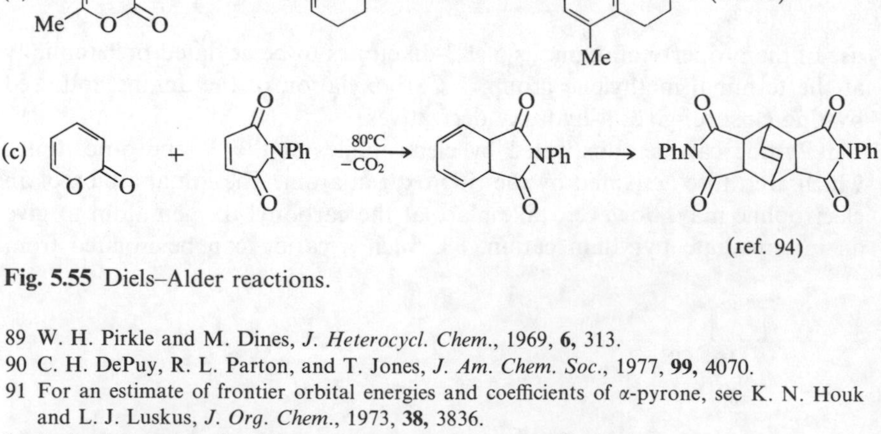

Fig. 5.55 Diels–Alder reactions.

89 W. H. Pirkle and M. Dines, *J. Heterocycl. Chem.*, 1969, **6**, 313.
90 C. H. DePuy, R. L. Parton, and T. Jones, *J. Am. Chem. Soc.*, 1977, **99**, 4070.
91 For an estimate of frontier orbital energies and coefficients of α-pyrone, see K. N. Houk and L. J. Luskus, *J. Org. Chem.*, 1973, **38**, 3836.
92 D. Seyferth and D. L. White, *J. Organometal. Chem.*, 1972, **34**, 119.
93 G. Märkl and R. Fuchs, *Tetrahedron Lett.*, 1972, 4695.
94 N. P. Shusherina and V. S. Pilipenko, *Zh. Org. Khim.*, 1978, **14**, 895.

as in example (a), are benzene derivatives formed by the loss of carbon dioxide from the primary products of the cycloaddition. This is a good route to substituted benzenes. Enamines can be used in place of acetylenes, as in (b). Example (c) is typical of the reactions of α-pyrone with electron-deficient olefins: 2:1 adducts can be formed by addition of a second mole of the dienophile after decarboxylation of the 1:1 adducts.

Another useful rection which is typical of *cisoid* dienes is the light-induced electrocyclic ring closure of α-pyrone to the bicyclic lactone **70**. This lactone can be formed in good yield by irradiating α-pyrone at low temperature in pyrex.[95] The cyclization is in competition with the reversible ring opening to the ketene **71**, which can be intercepted if the photolysis is carried out in methanol. The lactone **70** can be decarboxylated when it is irradiated in quartz (Fig. 5.56) and this has been used as a method for the generation of cyclobutadiene.

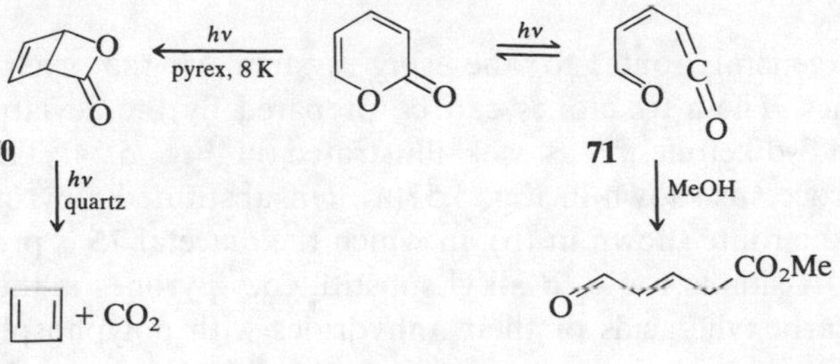

Fig. 5.56 Photochemistry of α-pyrone.

5.5.3   *4H-Pyran-4-ones*[81,82c]

The 4*H*-pyran-4-one, or γ-pyrone, ring system has several simple naturally occurring derivatives including maltol **72**, which is present in pine needles, and kojic acid **73**, which is produced by mould of the genus *Aspergillus*. Like the isomeric α-pyrone, the ring system shows little evidence of aromatic character. Its resonance energy is very small[85] and it is best regarded as a

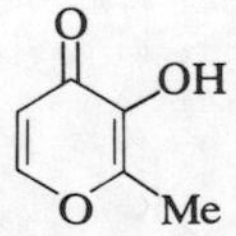

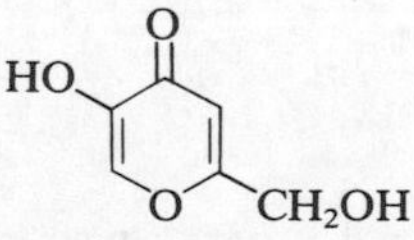

95 O. L. Chapman, C. L. McIntosh, and J. Pacansky, *J. Am. Chem. Soc.*, 1973, **95**, 614.
81 Review: R. Livingstone, in *Rodd's Chemistry of Carbon Compounds* Vol. IVE, ed. S. Coffey, Elsevier, Amsterdam, 1977, p. 1.
82 Reviews: J. Staunton, in *Comprehensive Organic Chemistry* Vol. 4, ed. P. G. Sammes, Pergamon, Oxford, 1979, (a) p. 607, (b) p. 629, (c) p. 659.
85 J. Aihara, *J. Am. Chem. Soc.*, 1976, **98**, 2750.

vinylogous lactone. As with α-pyrone, it can be methylated on the exocyclic (carbonyl) oxygen atom, and 2,6-dimethyl-γ-pyrone forms a crystalline hydrochloride 74 with hydrogen chloride. The pK_a of this cation is 0.4, which is unusually high for an oxonium salt; this can be related to the increased aromatic character of the cation, which is a pyrylium derivative. Thus, although the γ-pyrone system is not aromatic, it can be aromatized by the attack of electrophiles at the exocyclic oxygen atom.

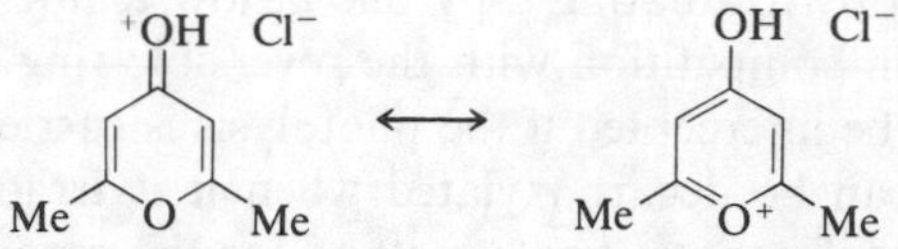

74

The best general route to the ring system is the cyclization of 1,3,5-triketones. These triketones can be prepared by the acylation of the dianions of 1,3-diketones,[88] as was illustrated in Fig. 5.54. The general cyclization process is shown in Fig. 5.57(a). Unsubstituted γ-pyrone can be prepared by the route shown in (b), in which the diacetal 75 is probably an intermediate. A simple route to alkyl-substituted γ-pyrones is the reaction of aliphatic carboxylic acids or their anhydrides with polyphosphoric acid at 200°C.[96] The course of this reaction is outlined in (c).

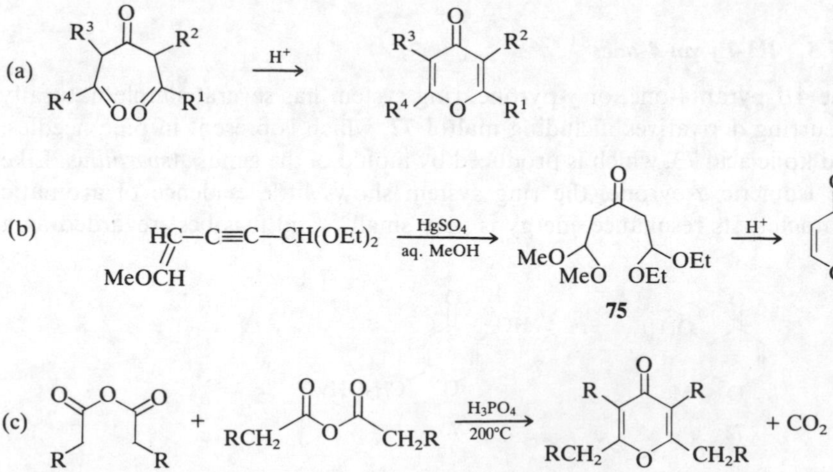

Fig. 5.57 Routes to γ-pyrones.

96 E. B. Mullock and H. Suschitzky, *J. Chem. Soc.* (*C*), 1967, 828.

As indicated earlier, γ-pyrones are best regarded as vinylogous lactones rather than as aromatic heterocycles. The carbonyl group does not show 'ketonic' properties and neither oximes nor hydrazones can be formed. Nucleophiles attack the ring system at the 2- and 6-positions and this often leads to ring cleavage. Alkaline hydrolysis, for example, is essentially the reverse of the reaction in Fig. 5.57(a). The products may, however, react further; an example is the intramolecular aldol condensation of the intermediate **76**, which leads to the formation of a phenol.

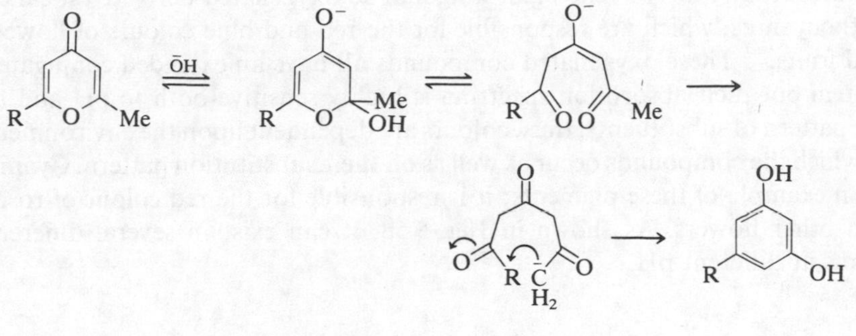

76

Grignard reagents and other 'hard' nucleophiles can react at C-4 rather than at C-2. Nucleophiles can also attack C-4 in the presence of Lewis acid catalysts: this type of reaction probably involves the displacement of a leaving group from a pyrylium cation. For example, the reaction of 2,6-dimethyl-γ-pyrone with malononitrile in the presence of acetic anhydride may go by way of the cation **77**.

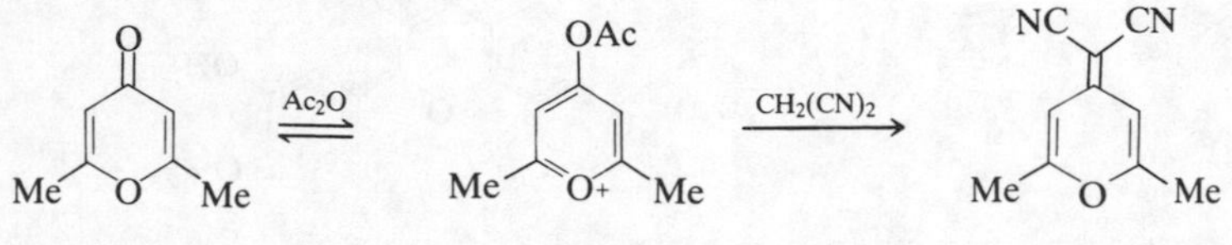

77

The carbonyl group has a predictable effect on the ring substituents; for example, methyl groups at the 2- and 6-positions are activated and will undergo aldol and other substitution reactions. Hydroxyl groups at both the 2- and 3-positions are acidic; kojic acid, **73**, for example, has a pK_a of 7.9.

5.5.4 *Benzo-fused systems*[81,82]

Benzo-fused analogues of the pyrylium cation and the pyrones represent the ring systems of several important groups of natural products. The ring systems are the benzopyrylium cation **78**, coumarin **79**, and chromone **80**.

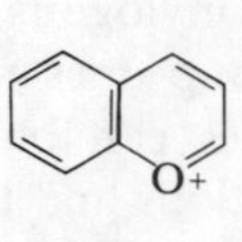

78 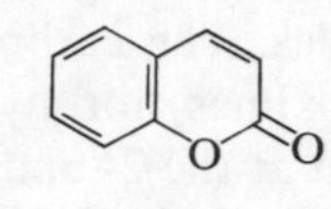   **79** 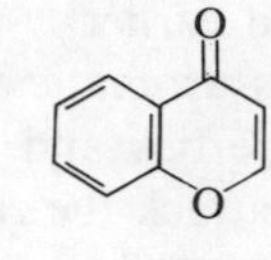   **80**

Benzopyrylium cations and chromones with a 2-phenyl substituent have the trivial names 'flavylium cations' and 'flavones', respectively. Flavylium cations occur widely in the plant kingdom as oxygenated derivatives, called anthocyanins, which are responsible for the red and blue colours of flowers and fruits.[97] These oxygenated compounds all have an extended conjugated system but their absorption spectrum is highly sensitive both to pH and to the pattern of substituents; thus, colours are dependent upon the environment in which the compounds occur as well as on their substitution pattern. Cyanin is an example of these pigments; it is responsible for the red colour of roses and other flowers. As shown in Fig. 5.58, it can exist in several different forms at different pH.

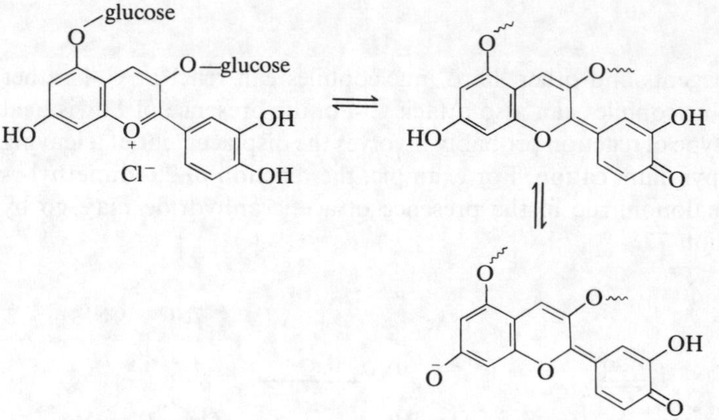

Fig. 5.58 Acid–base equilibria in cyanin.

The flavylium cations of this type clearly show indicator properties but they are too unstable to be used as dyestuffs. Flavones, which are also widely distributed in plants, are considerably more stable. Compounds having the structure **81** (R = H, OH, or O-sugar) are yellow pigments which have been used as natural dyestuffs.

Coumarin **79** occurs naturally and many natural derivatives are known. 4-Hydroxycoumarins have anticoagulant activity; a naturally occurring toxin present in clover is a 4-hydroxycoumarin derivative. The property has been

97 Review: G. A. Iacobucci and J. G. Sweeny, *Tetrahedron*, 1983, **39**, 3005.

exploited in pharmaceuticals which are used in the treatment of cardio-vascular disease. Warfarin **82**, originally developed as a rodenticide, is a useful anticoagulant drug. Some important drugs are also based on the chromone system: an example is the antiallergic agent nedocromil sodium **83**.[98]

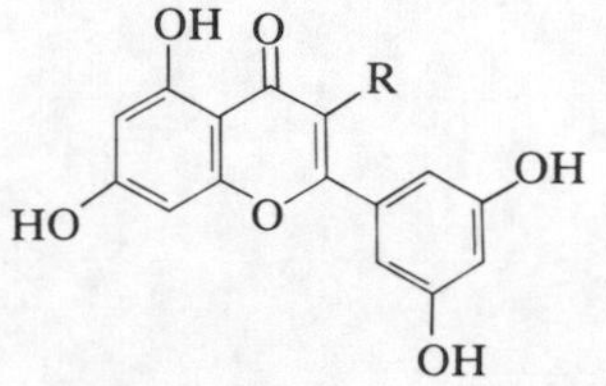

81

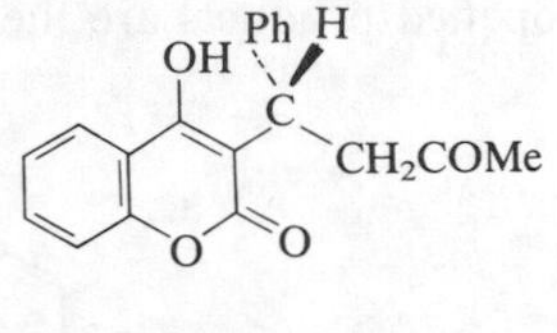

82

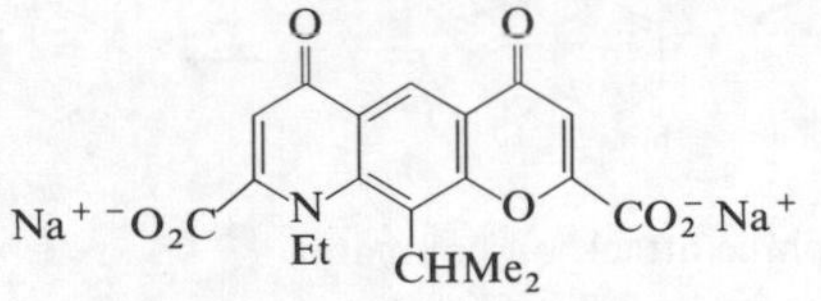

83

Coumarins and chromones can be synthesized by cyclization reactions of monosubstituted or *o*-disubstituted benzenoid compounds: three examples are shown in Fig. 5.59.

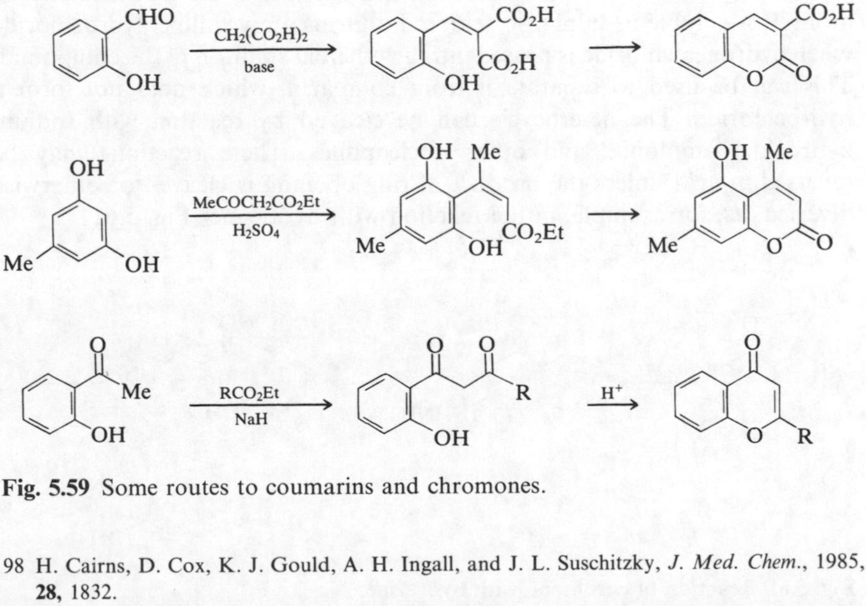

Fig. 5.59 Some routes to coumarins and chromones.

98 H. Cairns, D. Cox, K. J. Gould, A. H. Ingall, and J. L. Suschitzky, *J. Med. Chem.*, 1985, **28**, 1832.

Many of the reactions of the benzo-fused heterocycles are analogous to those of the monocyclic compounds. Thus, coumarin is attacked by nucleophiles at C-4 and at C-2. Attack at C-4 by 'soft' nucleophiles such as CN^- leads to the formation of addition products. Opening of the ring by attack at C-2 is not followed by recyclization, as it often is in the monocyclic series, because the benzenoid ring is not susceptible to nucleophilic attack. Ring-opened products are therefore isolated (Fig. 5.60).

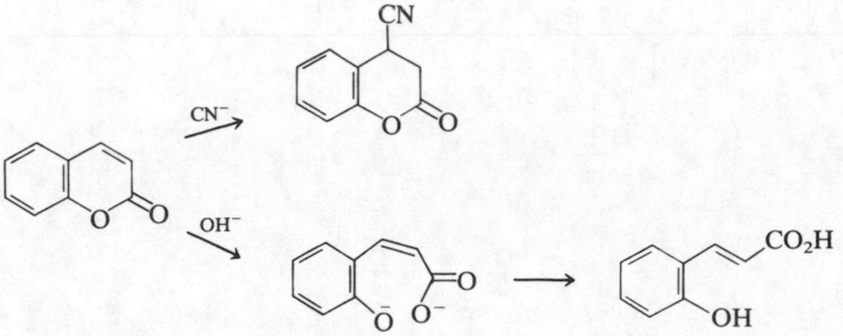

Fig. 5.60 Nucleophilic attack on coumarin.

Coumarin can be brominated at C-3, in the same way as α-pyrone, by addition of bromine across the 3,4-bond followed by the loss of hydrogen bromide. It can be chloromethylated at C-3, probably by a similar mechanism, but it is nitrated and sulphonated in the benzene ring, at C-6.

Like γ-pyrone, chromone reacts with electrophiles at the exocyclic oxygen atom. It is soluble in sulphuric acid and it forms a crystalline hydrochloride when hydrogen chloride is passed into an ethereal solution of the compound. This can be used to separate it from coumarin, which does not form a hydrochloride. The heterocycle can be cleaved by reaction with sodium hydroxide, ammonia, and other nucleophiles. These reactions may be reversed in acid unless the product of ring opening is cleaved or otherwise diverted, as, for example, in the reaction with hydrazine (Fig. 5.61).

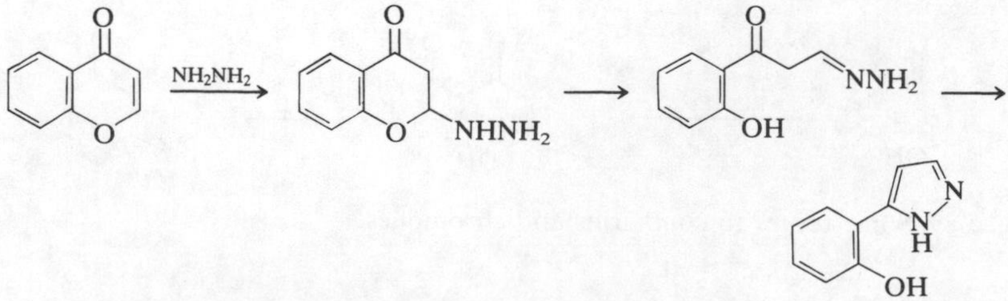

Fig. 5.61 Reaction of chromone with hydrazine.

Summary

1. The chemistry of pyridine can be compared with that of three types of model compound: tertiary amines, benzene, and conjugated imines or carbonyl compounds. Pyridine is a base and a good nucleophile; the nitrogen lone pair is the primary site of attack of protons and other electrophiles. Electrophilic substitution at carbon is therefore difficult: in strongly acidic media it is the pyridinium cation which is substituted. The 3- and 5-positions are the least deactivated, and the most 'benzene-like' in character. Electrophilic substitution at C-4 can, however, be achieved by way of pyridine *N*-oxide.

Nucleophilic substitution, in contrast, occurs at the 2-, 4-, and 6-positions which are activated by the ring nitrogen atom. Reactions at these positions resemble those in 2- or 4-substituted nitrobenzenes but not those in benzene itself. Substitution occurs even more readily in pyridinium salts. Substitution reactions which have no direct parallel in benzene chemistry include amination (the Chichibabin reaction) and substitution by organolithium reagents; both take place preferentially at C-2.

Substituents at the 3- and 5-positions show properties broadly similar to substituents in benzene whereas those at the 2-, 4-, and 6-positions are activated by the ring nitrogen atom. Hydroxy substituents at these positions exist preferentially in the amide (pyridone) tautomeric forms, diazonium salts are very easily hydrolysed, and methyl groups are readily deprotonated. The methyl groups at these positions can therefore be alkylated and acylated preferentially.

Thus, in comparison with benzene, reactions involving positions marked → show some similarities, but reactions at positions marked * are generally quite different because the influence of the ring nitrogen atom is dominant. Similar properties are found at the corresponding ring positions of other heterocycles of this type.

2. Quinoline and isoquinoline show properties which are in many ways similar to those of pyridine. The most important differences are:

(a) Electrophilic substitution is much easier than in pyridine and it takes place preferentially in the carbocyclic rings, usually at positions 5 and 8.

(b) Attack by nucleophiles on the heterocyclic ring can lead to the formation of addition products more easily than with pyridine; the formation of Reissert compounds, for example, occurs readily with quinoline and isoquinoline.

(c) Position 1 in isoquinoline is activated much more strongly than position 3.

3. The properties of other fused pyridines can generally be extrapolated from those of pyridine and quinoline, with the exception of indolizine, which is an activated electron-rich heterocycle of the pyrrole type.

4. Pyrylium cations are unusually stable oxonium ions with aromatic character and undergo some reactions analogous to those of pyridinium cations. The 2-, 4-, and 6-positions are strongly activated to nucleophilic attack and these reactions often result in ring opening unless the reaction medium is strongly acidic.

5. 2*H*-Pyran-2-ones are nonaromatic unsaturated lactones which can be cleaved by nucleophiles and undergo cyclization and cycloaddition reactions typical of conjugated dienes. The 4*H*-pyran-4-one system is also nonaromatic and shows the properties expected of a vinylogous lactone; it can, however, be protonated by mineral acids on the exocyclic oxygen atom to give a pyrylium cation.

Problems

1. Compound **A** was prepared from 2-amino-4-methylpyridine by the route shown.

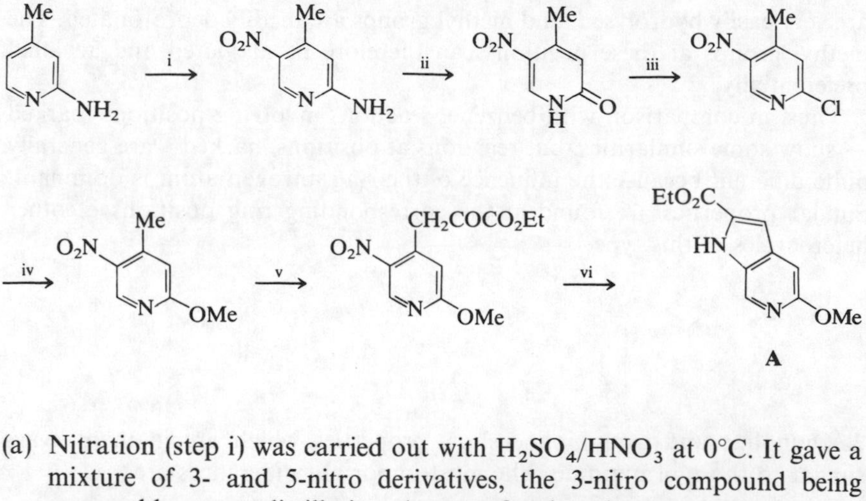

(a) Nitration (step i) was carried out with H_2SO_4/HNO_3 at 0°C. It gave a mixture of 3- and 5-nitro derivatives, the 3-nitro compound being removed by steam distillation. Account for the selective nitration at C-3 and C-5 and explain the use of steam distillation to remove the 3-nitro compound.

(b) Suggest reagents and conditions for carrying out each of the steps ii to v and write mechanisms for the reactions in each step.

(c) Step vi was carried out by catalytic hydrogenation, compound **A** being formed in high yield. Account for its formation.

2. Devise a synthetic route to nifedipine (structure **4** in Section 5.2).

3. Suggest a synthetic route to the antibacterial agent nalidixic acid **B** starting from 2-amino-6-methylpyridine.

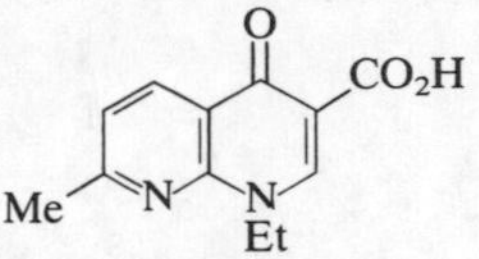

B

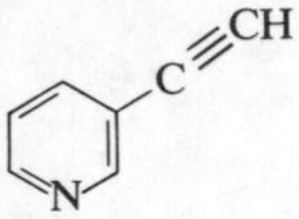

C

4. 3-Ethynylpyridine, **C**, was heated with sodamide and gave a crystalline product $C_7H_6N_2$ (54%) which did not contain a triple bond. Suggest a structure for it.

5. Give two possible mechanisms for the substitution of 3-bromo-4-pyridone, **D**, shown. Which do you favour, and what experiments can you suggest to investigate the mechanism?

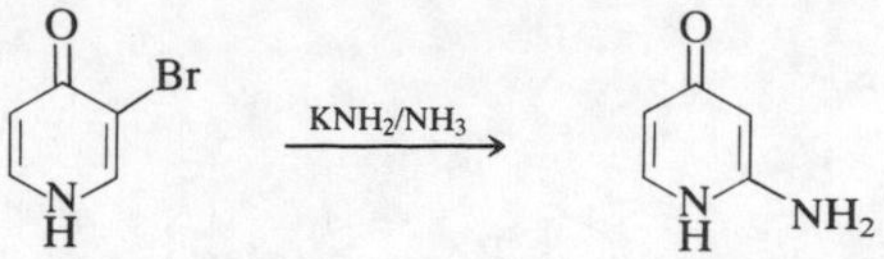

D

6. Explain the differences in the pK_a values of the following:

(a) 1-Aminoisoquinoline, $pK_a = 7.62$; 3-amino-, $pK_a = 5.0$.
(b) 3-Aminoacridine, $pK_a = 8.04$; 4-amino-, $pK_a = 4.40$; 9-amino-, $pK_a = 9.99$.

7. The pyrylium salt **E** on treatment with triethylamine gave a red crystalline product $C_{15}H_{17}O_3^+$ ClO_4^-. Propose a structure for it and explain its formation.

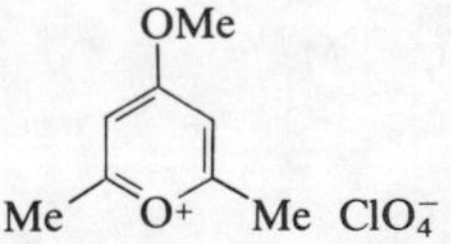

E

8. Suggest explanations for each of the following:

(a)

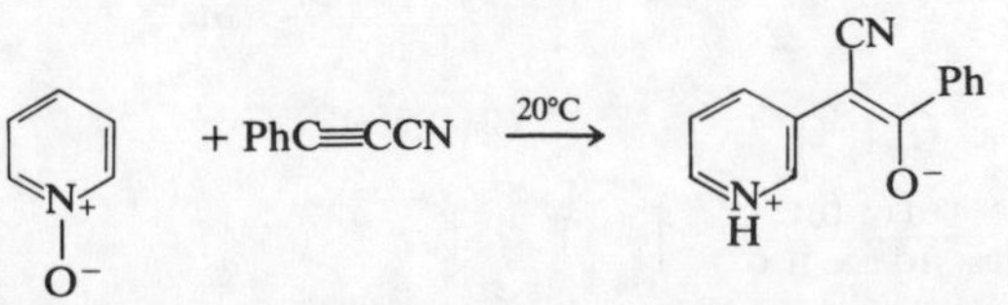

$+ PhC{\equiv}CCN \xrightarrow{20°C}$

(61%)

(b)

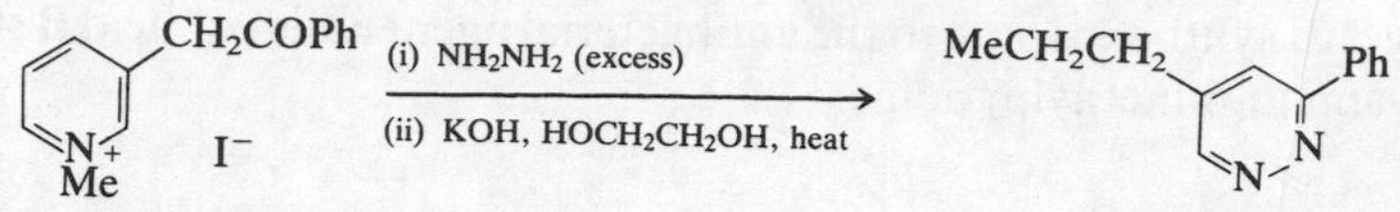
CH2COPh
N+
Me
I−
(i) NH2NH2 (excess)
(ii) KOH, HOCH2CH2OH, heat
MeCH2CH2
Ph
N
N

(c)

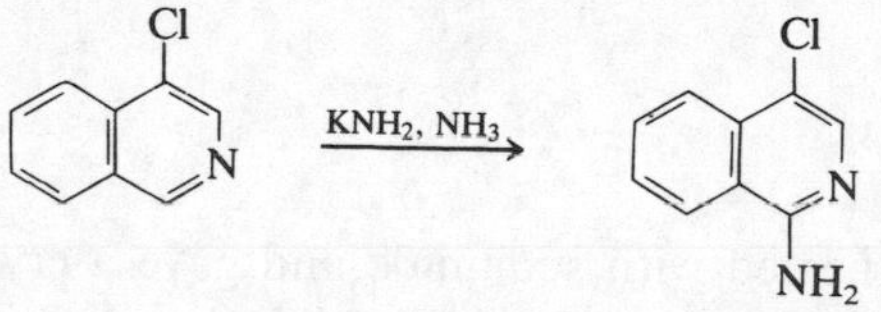
Cl
N
KNH2, NH3
Cl
N
NH2

(86%)

(d)

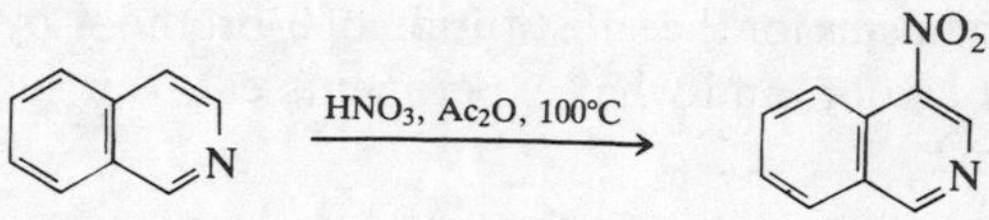
N
HNO3, Ac2O, 100°C
NO2
N

(14%)

(e)

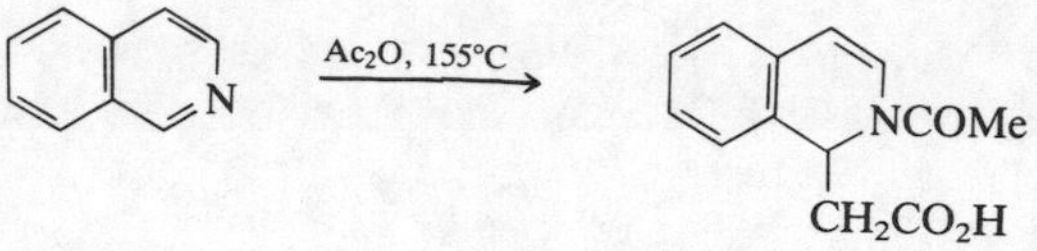
N
Ac2O, 155°C
NCOMe
CH2CO2H

(43%)

(f)

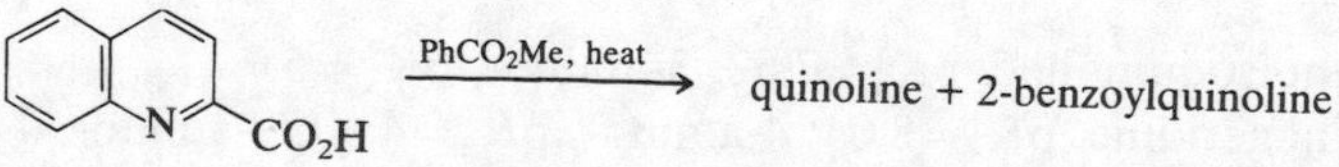
N
CO2H
PhCO2Me, heat
quinoline + 2-benzoylquinoline

(g)

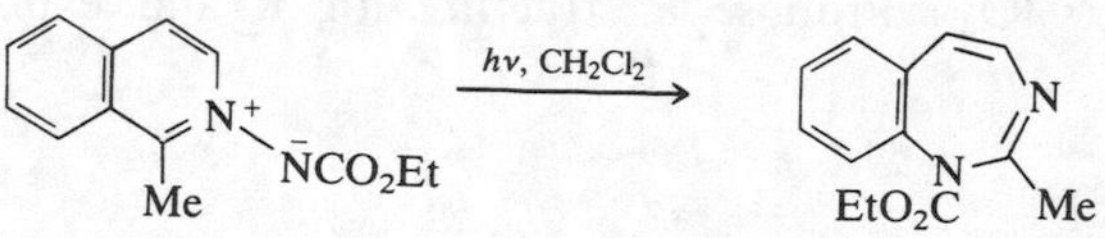
N+
−
NCO2Et
Me
hν, CH2Cl2
N
N
EtO2C
Me

(20%)

(h)

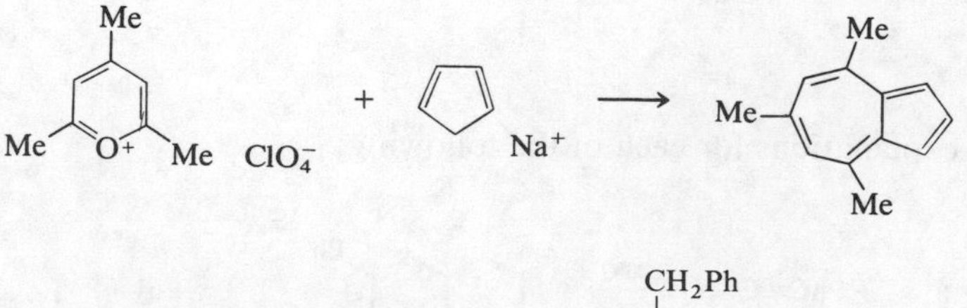
Me
Me
O+
Me
ClO4−
+
Na+
Me
Me
Me

(43%)

(i)

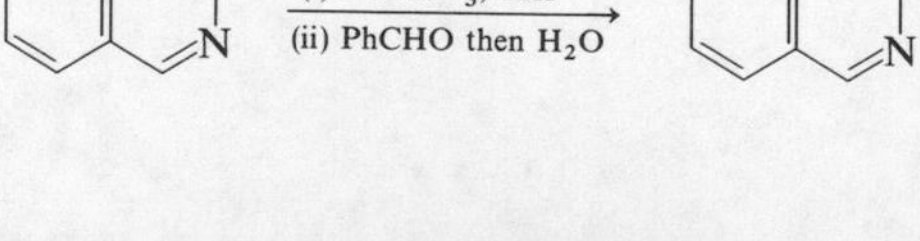
N
(i) NaBHEt3, THF
(ii) PhCHO then H2O
CH2Ph
N

(65%)

(j)

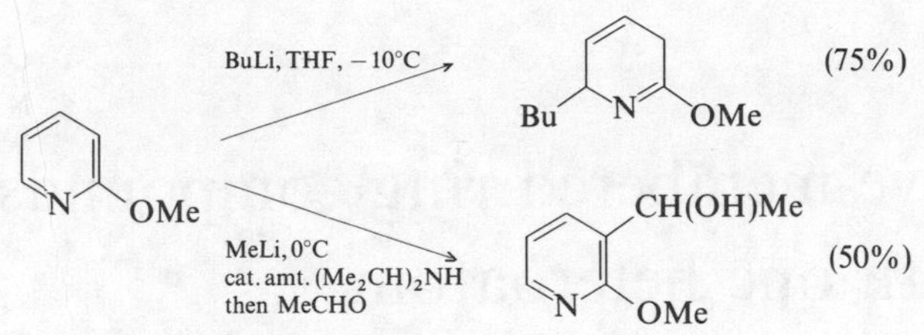

BuLi, THF, −10°C

(75%)

MeLi, 0°C
cat. amt. (Me₂CH)₂NH
then MeCHO

(50%)

(k)

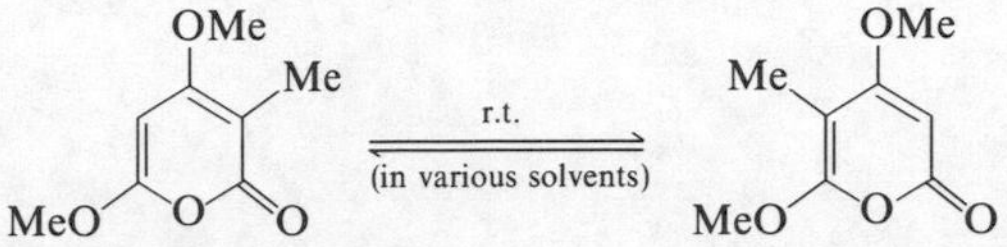

r.t.
(in various solvents)

9. Reaction of pyridine (2 mol) with phosgene (COCl₂) (1 mol) in pentane gave a solid 2:1 adduct in 92% yield. This had ν_{max} (KBr) 1752 cm⁻¹; ¹H NMR (−55°C) δ 9.54 (2 H, d), 8.43 (1 H, t), 8.03 (2 H, dd) 7.51 (1 H, d), 7.46 (1 H, d), 7.33 (1 H, dd), 5.61 (1 H, dd), and 5.55 (1 H, dd). The low temperature ¹³C NMR spectrum included a signal at 64.3 ppm. The compound readily reverted to pyridine and phosgene in solution. Suggest a structure for the compound and mechanism for its formation.

6 Five-membered ring compounds with one heteroatom

6.1 Introduction

The chemistry of one of the most important groups of heterocycles, five-membered compounds containing a single heteroatom in the ring system, is dealt with in this chapter. The commonest of these, and the ones that will be discussed in detail, are pyrrole, furan, thiophene, and their benzo derivatives. It was shown in Chapter 2 that these heterocycles can be regarded as aromatic, to a greater or lesser degree, on the basis of their physical properties and their resonance energies. All these heterocycles have an excess of π-electrons since six electrons are distributed over five atoms. Because of this, their chemistry has some parallels with that of nucleophilic benzenoid aromatics such as aniline and phenol. At the other extreme, an appropriate model system with which to compare their chemistry would be a conjugated diene substituted with the appropriate heteroatom at the 1-position. Both types of model system are nucleophilic and thus react preferentially with electrophiles, but the products are of different kinds, resulting from substitution in the aromatic model and addition in the nonaromatic model. Both types of reaction occur with all these heterocycles but the predominant type correlates approximately with the resonance energy of the system. The parallel with aniline and phenol is closest with pyrrole, which has appreciable resonance energy and which undergoes electrophilic substitution reactions very readily. On the other hand, furan has very low resonance energy and its chemistry is much closer to that of a nucleophilic diene than of a benzenoid aromatic compound. Thiophene is also appreciably aromatic and undergoes substitution reactions with electrophiles, but much less readily than pyrrole.

The chemistry of each of these ring systems is dealt with separately in the following sections, but there are many broad similarities and many reactions that are common to them all. It is helpful to keep in mind the chemistry of the appropriate 'aromatic' and 'nonaromatic' model systems when reading these sections, and to make a mental comparison of the chemistry of one ring system with that of the others.

6.2 Pyrroles[1]

6.2.1 Introduction

Apart from its inherent theoretical interest as an aromatic system, pyrrole is an important heterocycle because its structure is incorporated into that of haem, chlorophyll, and related natural products such as vitamin B_{12} and the bile pigments. These compounds provided the impetus for most of the early work on the synthesis and reactions of pyrroles, and the synthesis of pyrroles related to these naturally occurring complex molecules continues to be a very active area of research. The chemistry of pyrrole has also been explored in order to understand its relationship to the benzenoid aromatics.

The structures and chemistry of haem, chlorophyll and other porphyrins are outlined in Section 6.2.8. The structures of some important monopyrrolic compounds are shown below. Porphobilinogen, **1**, is the biosynthetic precursor of the natural porphyrins and of vitamin B_{12}. Pyrrolnitrin, **2** and pyoluteorin, **3**, are naturally occurring pyrroles which have some antibiotic activity, and the simple ester **4** is an insect pheromone. Of the reduced pyrroles the most important naturally occurring compound is the amino acid (S)-proline, **5**. The alkaloid nicotine, **6**, contains a pyrrolidine as well as a pyridine ring.

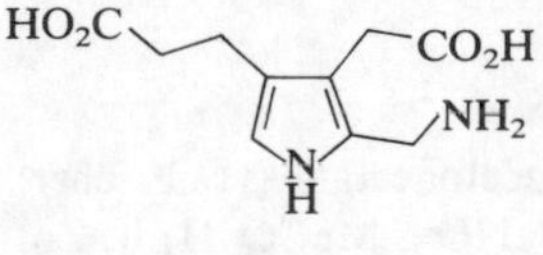

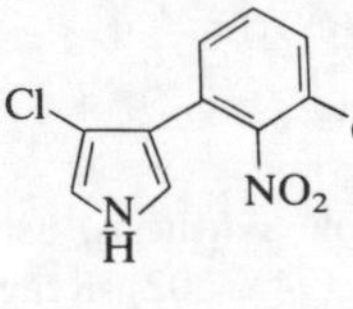

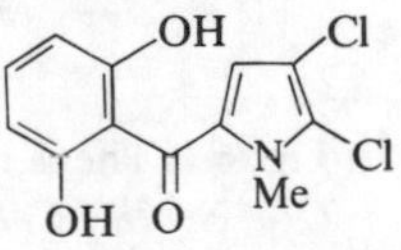

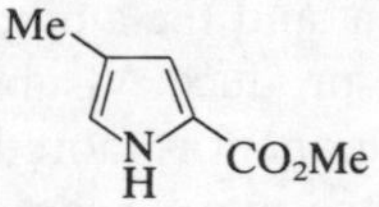

The parent compound, pyrrole, is a liquid, b.p. 129°C, which is liable to darken on exposure to air and light. The molecules are associated in the liquid phase. Its structure and many of its physical properties are discussed in Chapter 2. It is a polar molecule with a dipole moment which is directed from nitrogen to carbon; its magnitude ranges from 1.55 D to 3.0 D, depending upon the solvent and its ability to form hydrogen bonds.

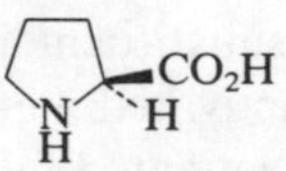

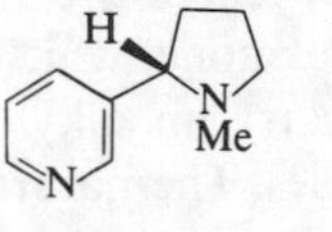

1 Reviews: (a) R. A. Jones and G. P. Bean, *The Chemistry of Pyrroles*, Academic Press, London, 1977; (b) A. Gossauer, *Die Chemie der Pyrrole*, Springer-Verlag, Berlin, 1974; (c) *Pyrroles* Part 1, ed. R. A. Jones, Wiley-Interscience, New York, 1990.

6.2.2   Ring synthesis

The most widely used methods of constructing the pyrrole ring system are all cyclization reactions. Three of these are outlined in Chapter 4: the Paal–Knorr synthesis (Section 4.2.1), the Knorr synthesis (Table 4.4), and isonitrile cyclization (Table 4.8).

The *Knorr* synthesis is the most important and widely used method. The classical example of the reaction is shown in Fig. 6.1. Two moles of ethyl acetoacetate are used in the reaction, one being nitrosated and reduced *in situ* to give an aminoketone which reacts with the second mole. The pyrrole diester **7** is thus available on a large scale, and many other pyrroles, particularly those used in porphyrin synthesis, have been derived from it by modification of the substituents. To this end, benzyl and t-butyl esters have been used instead of ethyl esters in the Knorr synthesis, in order to facilitate the subsequent removal of the alkoxycarbonyl groups.

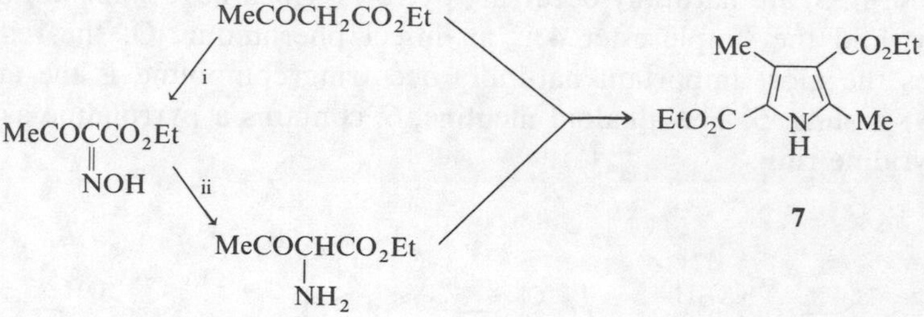

$$\text{MeCOCH}_2\text{CO}_2\text{Et}$$

$$\text{MeCOCCO}_2\text{Et}$$
$$\underset{\text{NOH}}{\|}$$

$$\text{MeCOCHCO}_2\text{Et}$$
$$\underset{\text{NH}_2}{|}$$

(i, ii)

Me, CO$_2$Et / EtO$_2$C, N(H), Me **7**

Fig. 6.1 The Knorr pyrrole synthesis, using ethyl acetoacetate (H. Fischer, *Org. Synth., Coll. Vol. 2*, 1943, 202). Reagents: i, NaNO$_2$, MeCO$_2$H; ii, Zn, MeCO$_2$H.

The reaction can also be carried out using an independently prepared aminoketone, which allows some structural variation, and the amino group can carry an alkyl or aryl substituent in order to produce *N*-substituted pyrroles. The nature of the activated methylene component is more limited, however: it must be doubly activated in order that the pyrroles are formed in reasonable yields.

The *Paal*–Knorr synthesis (the preparation of pyrroles from 1,4-dicarbonyl compounds) is a very good method when the appropriate dicarbonyl compounds are readily available. Ammonia, primary amines, hydroxylamines, and hydrazines can be used as the nitrogen component. The reaction has been used extensively for the preparation of 2,5-dimethylpyrroles from hexane-2,5-dione. In fact the reaction is so straightforward that it has been used as a method for the protection of primary amines;[2] the amines RNH$_2$ can be regenerated from the dimethylpyrroles **8** by reaction with hydroxylamine hydrochloride. It is also a good method for the preparation of

2 S. P. Breukelman, G. D. Meakins, and M. D. Tirel, *J. Chem. Soc., Chem. Commun.*, 1982, 800.

alkyl-bridged and fused pyrroles, such as compounds of the types **9**[3] and **10**.[4] A variant of the method which is applicable to the synthesis of 1-substituted pyrroles is the reaction of amines with the cyclic acetal 2,5-dimethoxytetrahydrofuran, **11**[5] (see Section 6.3.6).

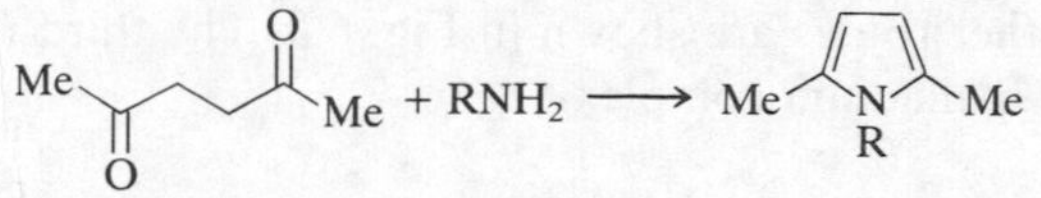

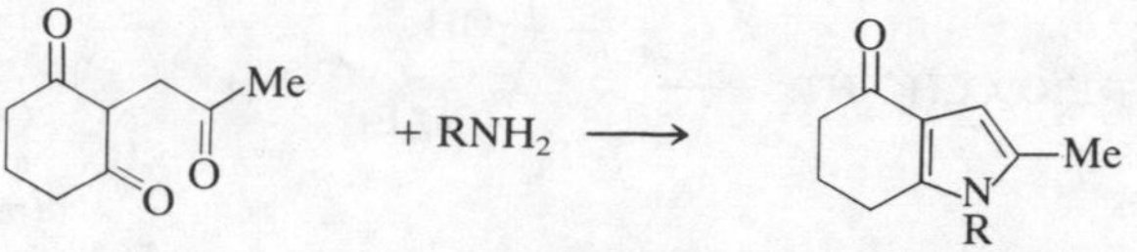

The *Hantzsch* synthesis is the reaction of an α-haloketone with a β-ketoester and ammonia or a primary amine (Fig. 6.2). It has been used to prepare some 2,5-dialkylpyrrole-3-carboxylates, although it should be noted that α-haloketones can react directly with β-ketoesters to give furans [the Fiest–Benary furan synthesis; see example (iii) in Table 4.3] and this can be a troublesome side reaction.

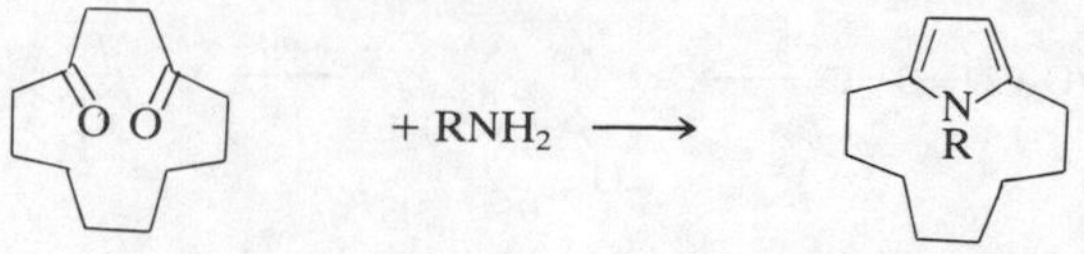

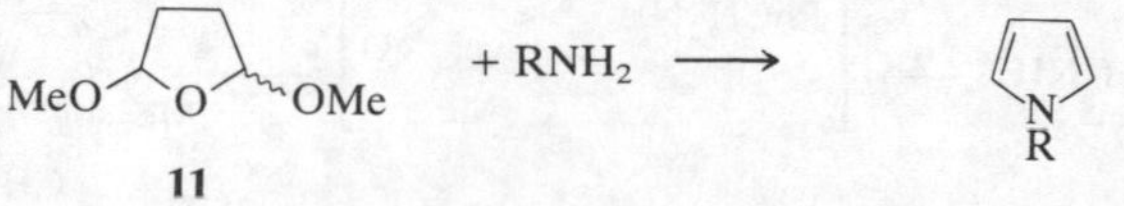

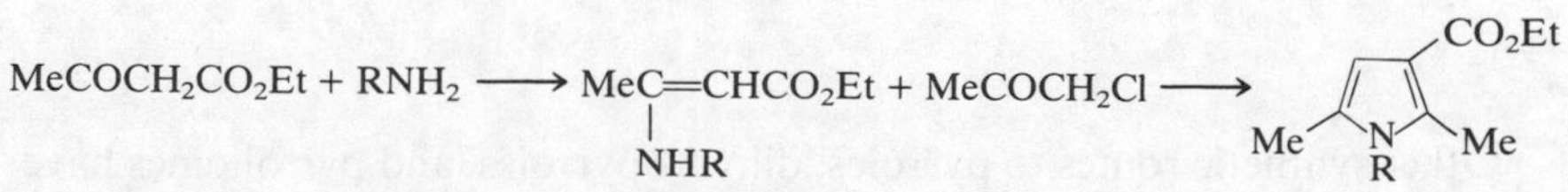

Fig. 6.2 The Hantzsch pyrrole synthesis.

3 H. Stetter and R. Lauterbach, *Liebigs Ann. Chem.*, 1962, **655**, 20.
4 H. Nozaki, T. Koyama, and T. Mori, *Tetrahedron*, 1969, **25**, 5357.
5 N. Elming and N. Clauson-Kaas, *Acta Chem. Scand.*, 1952, **6**, 867; T. H. Chan and S. D. Lee, *J. Org. Chem.*, 1983, **48**, 3059.

Other methods

One limitation of these classical methods of pyrrole synthesis is that they do not generally provide a direct route to 2- or 2,5-unsubstituted pyrroles. Some of the more recent synthetic methods have the advantage of overcoming this limitation to some extent. The isonitrile cyclization shown in Table 4.8 is one example; other routes are shown in Fig. 6.3. The third reaction goes by way of a π-allylpalladium complex.

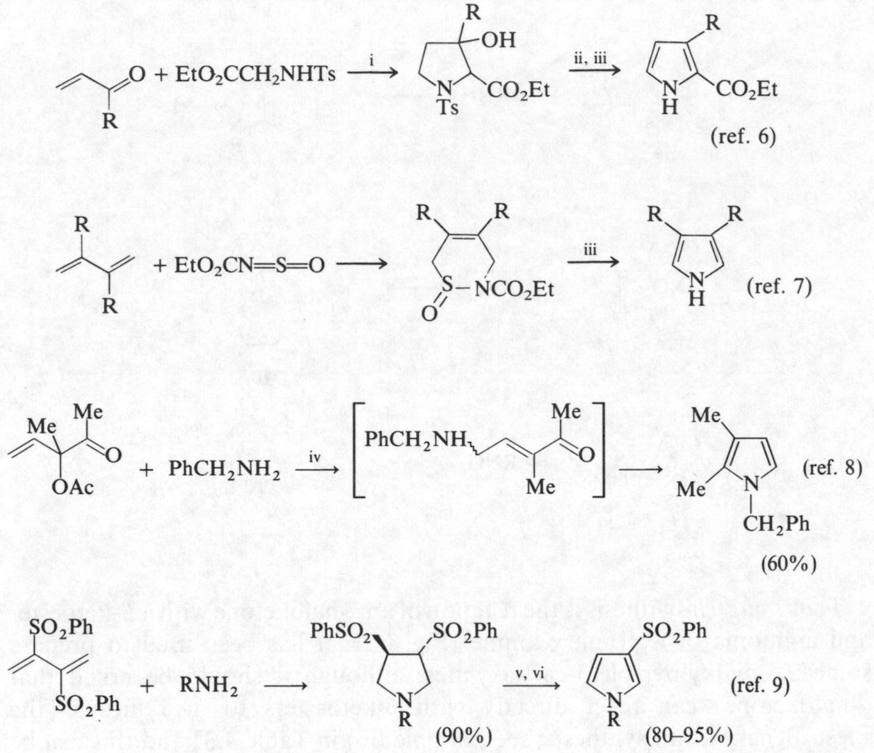

Fig. 6.3 Routes to pyrroles with a free α-position. Reagents: i, KOBu^t; ii, POCl$_3$/pyridine; iii, NaOR; iv, cat. Pd(PPh$_3$)$_4$; v, NaOMe; vi, DDQ.

Other synthetic routes to pyrroles, dihydropyrroles, and pyrrolidines have been outlined earlier (see Tables 4.6, 4.9, 4.11, 4.17, 4.21, and 4.25, and Figs. 4.15 and 4.38).

6 W. G. Terry, A. H. Jackson, G. W. Kenner, and G. Kornis, *J. Chem. Soc.*, 1965, 4389.
7 K. Ichimura, S. Ichikawa, and K. Imamura, *Bull. Chem. Soc. Jpn*, 1976, **49**, 1157.
8 B. M. Trost and E. Keinan, *J. Org. Chem.*, 1980, **45**, 2741.
9 A. Padwa and B. H. Norman, *J. Org. Chem.*, 1990, **55**, 4801.

6.2.3 Acidity and metallation reactions

Pyrrole is a weak acid, of about the same strength as simple alcohols; its pK_a in aqueous solution is 17.5. The pK_a is considerably reduced by the presence of conjugative electron-withdrawing groups in the 2- and 5-positions (2-nitropyrrole, for example, has a pK_a of 10.6). Pyrrole can be converted into its sodium or potassium salt by reaction with the metal amide in liquid ammonia, or with the metal in an inert solvent. The magnesium derivative **12** can be prepared by the reaction of pyrrole with ethylmagnesium bromide in ether; alkyllithium reagents similarly produce 1-lithiopyrrole. The *N*-metallated pyrroles are very useful for achieving controlled substitution reactions with electrophiles (see the following sections). Their properties are determined by the degree of covalent character in the nitrogen-to-metal bond. This depends upon the nature of the metal and upon the ability of the solvent to stabilize the metal cation (it is helpful to compare the influence of the metal cation and the solvent on the properties of enolate anions; the general trends are very similar). Thus, the sodium and potassium salts are ionic species whereas the lithium and magnesium derivatives are largely covalent, except in the presence of dipolar aprotic solvents such as hexamethylphosphoramide.

N-Substituted pyrroles can be selectively metallated at C-2 by reaction with alkyllithium reagents.[10] These 2-lithiated pyrroles are useful intermediates for further substitution (Section 6.2.5).

6.2.4 Substitution at nitrogen

Pyrrole undergoes very rapid proton exchange at nitrogen in neutral, basic, or acidic media, but exchange at carbon is much slower. In acid, kinetic protonation occurs at nitrogen to give the cation **13**, probably associated with one or more molecules of water, but this is unstable and rapidly reverts to starting materials. *N*-Substitution products are normally isolated only from reaction of pyrryl anions with electrophiles.

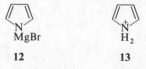

12   **13**

N-Substitution is promoted (a) by the use of the sodium or potassium salt, (b) by a dipolar aprotic solvent or cosolvent, and (c) by the use of a 'hard' electrophile. Pyrrole can be acylated, sulphonated and silylated specifically at the 1-position in this way, using the potassium salt. Pyrrole can also be acetylated at the 1-position under mildly basic conditions, for

10 Review: H. Gschwend and H. R. Rodriguez, *Org. Reactions*, 1979, **26**, 21.

example in the presence of imidazole as a weakly basic catalyst. Alkylation of the potassium salt normally occurs predominantly at the 1-position. Simple alkylation can be achieved under mild conditions by the use of potassium t-butoxide with 18-crown-6 as a catalyst.[11] Conjugate addition of the pyrryl anion to αβ-unsaturated ketones and nitriles also results in the formation of 1-substituted pyrroles. N-Lithiopyrrole and the magnesium derivative **12** are normally alkylated predominantly at carbon, however, unless a solvent is used which strongly solvates the metal cations. Pyrroles with electron-withdrawing groups on carbon undergo substitution at nitrogen much more easily than pyrrole itself.

6.2.5   Substitution at carbon

Because of its electron-rich character pyrrole reacts very readily with a wide range of electrophiles. Substitution takes place most readily at the 2-position but products of 3-substitution or of polysubstitution are also commonly formed.

The attack of an electrophile E^+ on pyrrole will lead to the formation of 2- and 3-substitution products by way of the cations **14** and **15**. The transition state for substitution might resemble one of these intermediates. In this case, 2-substitution should be favoured because the positive charge in the intermediate **14** is more extensively delocalized than it is in structure **15**, so it is of lower energy. The resonance forms shown in Fig. 6.4 illustrate the extent of delocalization in the two cations. Note that attack on nitrogen would give a cation analogous to **13** in which the charge is effectively localized on nitrogen (see Section 2.4 and Fig. 2.19).

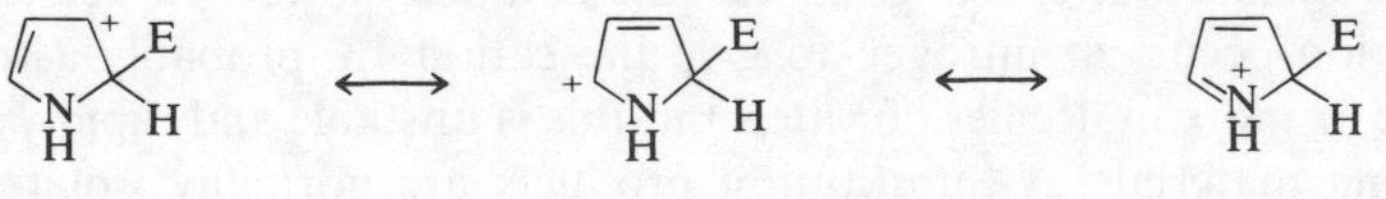

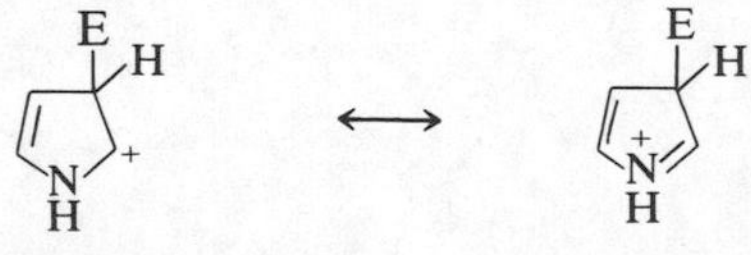

Fig. 6.4 Intermediates in electrophilic substitution of pyrrole.

If the transition state occurs early, it will resemble the reactants rather than these intermediates. The 'frontier orbitals' should then be important in determining the site of initial interaction of the electrophile with the π system.

11 W. C. Guida and D. J. Mathre, *J. Org. Chem.*, 1980, **45**, 3172.

The highest occupied molecular orbital of pyrrole has the largest coefficients centred at C-2 and C-5, as shown in Fig. 2.7, so this should also lead to predominant substitution at C-2 rather than at C-3.

The preference for electrophilic attack at the 2-position is illustrated by the protonation of pyrrole in strongly acidic media. The major protonated species present in these conditions is the cation **16**, which has a pK_a value of -3.60. The pK_a of the alternative cation **17** formed by protonation at C-3 is estimated to be -5.9, so there is very little of this present at equilibrium. In practice the treatment of simple pyrroles with strong acids often leads to the formation of polymers, which are formed through C-3 protonated species of type **17**. Although this cation is present only in low concentration, it is much more reactive than cation **16** and acts as an electrophile towards a molecule of unprotonated pyrrole. This reaction produces a new electrophilic species which can react in turn with another molecule of pyrrole; eventually a polymer results from a sequence of such reactions.

16   **17**

In other electrophilic substitutions, pyrrole shows the characteristics of a very nucleophilic aromatic compound. Thus, it undergoes acetylation without a catalyst, it gives 2,3,4,5-tetrachloropyrrole under mild conditions, and it reacts with diazonium salts. In this respect it resembles phenol or an aromatic amine; pyrrole has been shown to be more reactive than *NN*-dimethylaniline in azo coupling reactions.

These and some other examples of electrophilic substitution of pyrrole are given in Table 6.1. Reagents must be used that avoid the exposure of pyrrole to strong acids, which would cause it to polymerize.

Formylation is a particularly important reaction in pyrrole chemistry because of the use of pyrrole-2-carboxaldehydes in porphyrin synthesis. The Vilsmeier reaction provides an effective method for the introduction of the formyl group. A closely related reaction is the Houben–Hoesch acylation process, in which a nitrile RCN and hydrogen chloride combine to produce the electrophile $RC\equiv\overset{+}{N}H$. This electrophile can attack a wide range of pyrroles, including those with a deactivating substituent, to give acylated pyrroles after hydrolysis. A related reaction, in which a preformed nitrilium salt is used as the electrophile, is shown in Table 6.1.

Another useful type of electrophilic substitution occurs with aldehydes and ketones in the presence of an acid catalyst. With pyrrole itself this often leads to the formation of polymers, although the Ehrlich reaction (Fig. 6.5) applies to pyrrole as well as to substituted pyrroles and it illustrates the

Table 6.1 Electrophilic substitution of pyrrole.

Functional group introduced	Reagents and conditions	Products	Refs. and notes
NO_2	HNO_3, $(MeCO)_2O$, 20°C	2- and 3- (14:1)	a, b
Cl	SO_2Cl_2, ether	2- and 2,5-	a
Br	N-Bromosuccinimide	2-	c
CHO	Me_2NCHO, $POCl_3$	2-	a, d
COMe	$MeC\equiv\overset{+}{N}Me$, BF_4^- then H_2O	2-	e
$COCH_2Cl$	Me_2NCOCH_2Cl, POCl	2-	a, f
CH_2CH_2COMe	$H_2C{=}CHCOMe$, BF_3	2- and 2.5-	a
CH_2NMe_2	CH_2O, Me_2NH, H^+	2-	a, g
SO_3H	SO_3–pyridine, 100°C	2-	a
MeS	MeSCl, K_2CO_3	2- and 2,5-	a
N=NPh	$PhN_2^+Cl^-$	2-	a

a See ref. 1a for primary reference. b J. H. Boyer, *Nitroazoles*, VCH, Deerfield Beach, Florida, 1986. c H. M. Gilow and D. E. Burton, *J. Org. Chem.*, 1981, **46**, 2221. d The Vilsmeier reaction; the electrophile is $CHCl{=}\overset{+}{N}Me_2$ and the formyl group is produced by hydrolysis. e S. C. Eyley, R. G. Giles, and H. Heaney, *Tetrahedron Lett.*, 1985, **26**, 4649. f P. Dalla Croce, C. La Rosa, and A. Ritieni, *Synthesis*, 1989, 783. g The Mannich reaction; the electrophile is $H_2C{=}\overset{+}{N}Me_2$.

general course of these substitutions. The reaction of 4-dimethylamino-benzaldehyde and hydrochloric acid with pyrrole (or an alkylpyrrole with a free 2- or 3-position) produces a purple colouration which is due to the formation of the cation **18**. With simpler aldehydes and ketones this type of cation is also formed but it then acts as an electrophile towards a second molecule of the pyrrole. A succession of such reactions can eventually lead to the formation of a polymer unless the 5-position of the pyrrole is substituted. In some cases, cyclic tetramers can be isolated from the reaction of pyrrole with aldehydes and ketones; for example, pyrrole reacts with

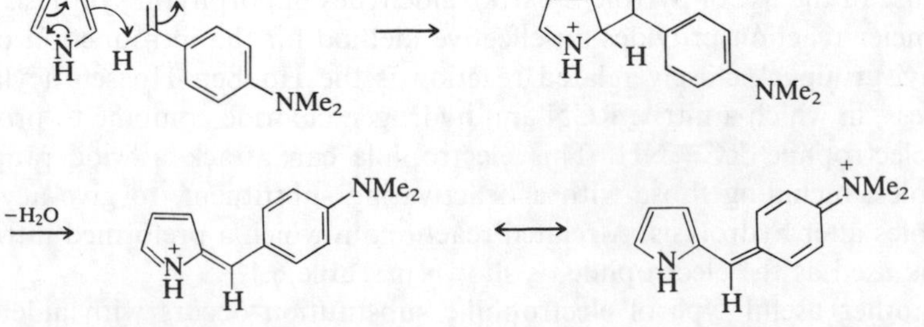

18

Fig. 6.5 The Ehrlich test for pyrroles.

benzaldehyde and acid in the presence of air to give tetraphenylporphyrin, **19**, in low yield.[12]

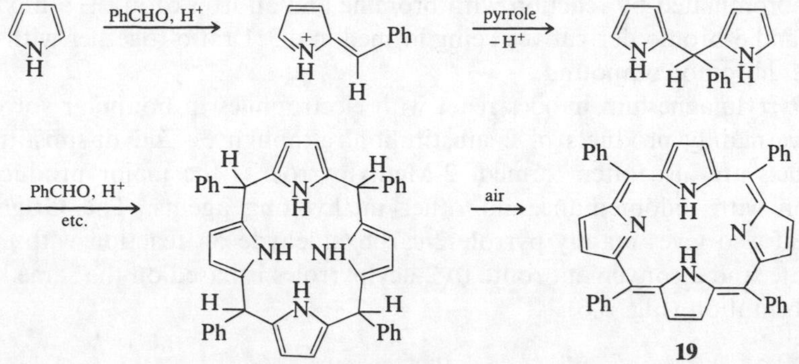

The effect of substituents on the reactivity and on the position of substitution by electrophiles follows a predictable pattern. An electron-releasing substituent at position 2 tends to direct an incoming electrophile to the 3- or 5-position, whereas an electron-withdrawing group directs it mainly to the 4-position and to a lesser extent to the 5-position. Substituents on nitrogen have less effect unless they are very bulky or strongly electron-withdrawing; in both cases, 3-substitution becomes favoured.[13] For example, 1-t-butylpyrrole undergoes Vilsmeier formylation mainly at the 3-position, and Friedel–Crafts benzoylation of 1-benzenesulphonylpyrrole (which requires an aluminium trichloride catalyst) goes in the 3-position. This is a useful reaction because the 1-benzenesulphonyl group is easily removed from the product. The silylated pyrrole **20** is particularly useful because it can be cleanly mono- and dibrominated at the β-positions.[14] The bromopyrroles are useful for preparing other β-substituted pyrroles by way of bromine–lithium exchange, and the silyl group can easily be removed.

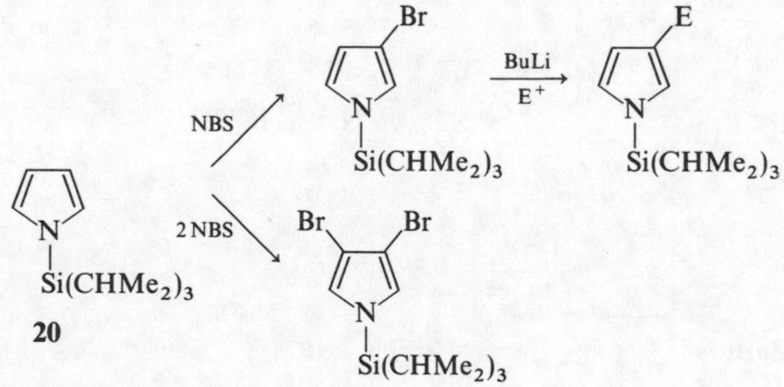

12 A. D. Adler, F. R. Longo, J. D. Finarelli, J. Goldmacher, J. Assour, and L. Korsakoff, *J. Org. Chem.*, 1967, **32**, 476.

13 Review: H. J. Anderson and C. Loader, *Synthesis*, 1985, 353.

14 B. L. Bray, P. H. Matthies, R. Naef, D. R. Solas, T. T. Tidwell, D. R. Artis and J. M. Muchowski, *J. Org. Chem.*, 1990, **55**, 6317.

The presence of a conjugatively electron-withdrawing group on the ring reduces the reactivity of the pyrrole towards electrophiles and allows more controlled substitution. For example, methyl pyrrole-2-carboxylate is mainly monobrominated by reaction with bromine and an iron catalyst, a mixture of 4- and 5-bromo derivatives being formed in a 3:1 ratio together with 5% of the dibromo compound.

1-Pyrrylmagnesium halides react with electrophiles in nonpolar solvents to give mainly products of 2-substitution, although 3- and di-substituted products are also often formed. 2-Methylpyrrole is the major product of rection with iodomethane and other methylating agents. The Grignard reagent also gives mainly pyrrole-2-carboxaldehyde on reaction with ethyl formate, and a convenient route to 2-acylpyrroles is based on the same kind of substitution (Fig. 6.6).

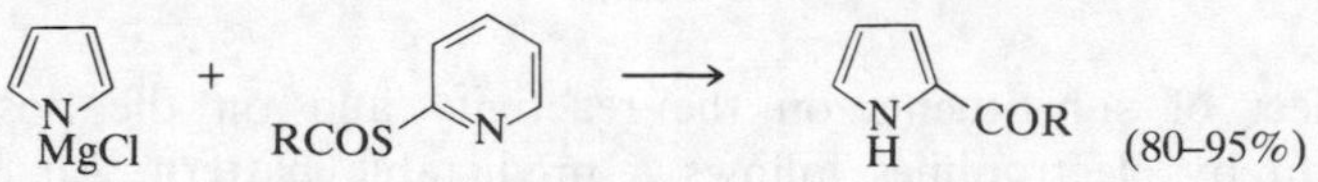

Fig. 6.6 Acylation of 1-pyrrylmagnesium chloride (K. C. Nicolaou, D. A. Claremon, and D. P. Papahatjis, *Tetrahedron Lett.*, 1981, **22**, 4647).

The lithiation of *N*-substituted pyrroles at the 2-position provides a means of achieving substitution at this position in a specific and controlled manner.[15] The 2-lithio derivative of 1-(dimethylamino)pyrrole can be alkylated in good yield by reaction with aldehydes. The 2-lithio derivative of 1-methylpyrrole has been specifically carboxylated at the 2-position by reaction with carbon dioxide, and it can be coupled and cross-coupled to provide 2-arylpyrroles. Examples of these reactions are shown in Fig. 6.7.

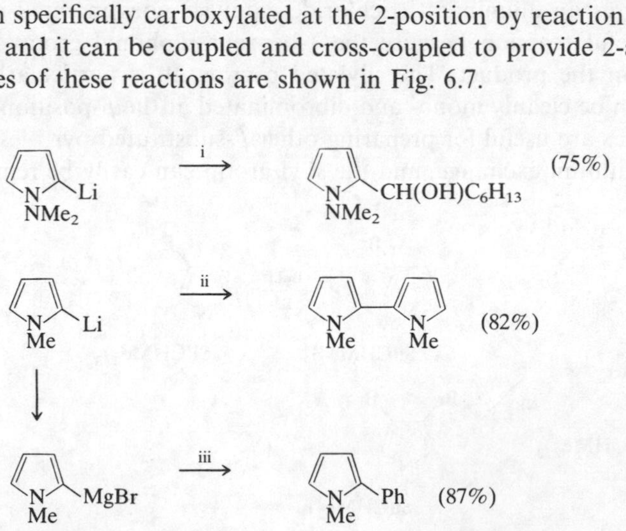

Fig. 6.7 Use of 2-lithiated pyrroles for specific substitution. Reagents: i, heptanal; ii, $NiCl_2$; iii, PhBr, $PdCl_2$, $Ph_2P(CH_2)_4PPh_2$.

15 B. J. Wakefield, *Organolithium Methods*, Academic Press, London, 1988; W. Chen and M. P. Cava, *Tetrahedron Lett.* 1987, **28**, 6025.

6.2.6   *Addition and cycloaddition reactions*

As an electron-rich heterocycle, pyrrole is not easily reduced. It is resistant to Birch reduction by alkali metals in liquid ammonia, and catalytic hydrogenation to pyrrolidines normally requires high temperatures and pressures (typical conditions are 150–200°C, 70 atm., and a Raney nickel catalyst; *N*-t-butoxycarbonylpyrroles are an exception in being hydrogenated at room temperature and atmospheric pressure[16]). In contrast, oxidation reactions are easy. Pyrrole and simple alkylpyrroles form insoluble polymers when treated with hydrogen peroxide in acidic media and with several other common oxidizing agents. Anodic oxidation of pyrrole in dilute acid also causes polymerization and the polymers may well be formed by way of a radical cation species. Simple pyrroles are also notoriously susceptible to degradation by molecular oxygen in the presence of light, and this may be the result of a radical-chain reaction initiated by the addition of peroxy radicals to the pyrrole.

Pyrroles do not normally undergo Diels–Alder reactions; they preferentially react with dienophiles by conjugate addition (Table 6.1). A few exceptions are known; for example, *N*-ethoxycarbonylpyrrole, **21**, reacts with dimethyl acetylenedicarboxylate to give the cycloadduct **22**. The *N*-aminopyrrole **23** and related compounds also undergo cycloaddition with dimethyl acetylenedicarboxylate.[17]

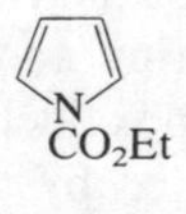

21 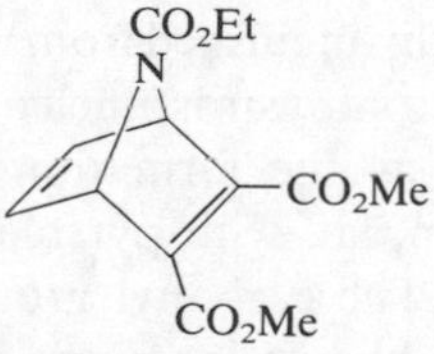   **22** 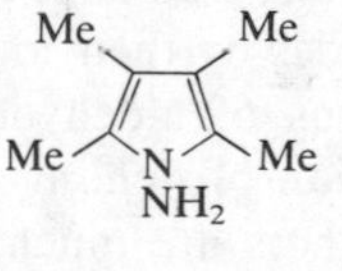   **23**

Pyrroles can also participate in photochemical [2+2] cycloadditions. Reaction of 1-methylpyrrole with a dialkyl ketone leads to the formation of a 3-substituted carbinol, probably by way of an oxetane intermediate (Fig.. 6.8).[18] This is a useful alternative to electrophilic substitution for the introduction of a 3-substituent.

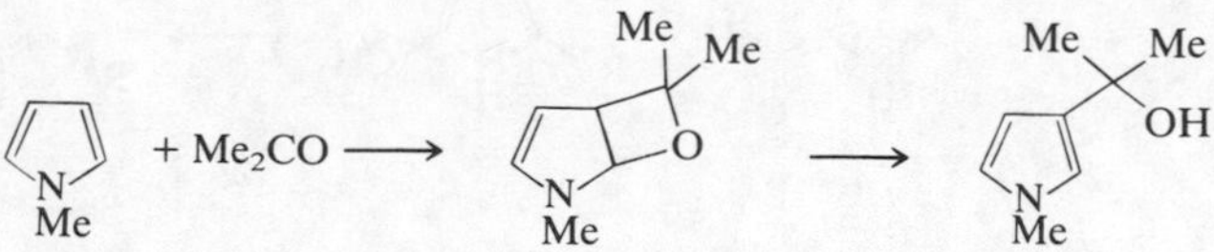

Fig. 6.8 Photoaddition of ketones to 1-methylpyrrole.

16 H.-P. Kaiser and J. M. Muchowski, *J. Org. Chem.*, 1984, **49**, 4203.
17 A. G. Schultz, M. Shen, and R. Ravichandran, *Tetrahedron Lett.*, 1981, 1767.
18 G. Jones, H. M. Gilow, and J. Low, *J. Org. Chem.*, 1979, **44**, 2949.

Pyrroles can also undergo cycloaddition reactions with carbenes across the 2,3-bond, but the adducts are unstable and react further. For example, addition of dichlorocarbene to 2,5-dimethylpyrrole in aprotic conditions gives 3-chloro-2,6-dimethylpyridine in good yield (Fig. 6.9).[19] Similar reactions of pyrrole and methylpyrroles can be carried out in the gas phase by heating with chloroform vapour.[20]

Pyrrole and N-alkylpyrroles are cleaved by reaction with hydroxylamine in acid media (see, for example, ref. 2). The reaction probably occurs by nucleophilic attack of hydroxylamine on the cation **17**.

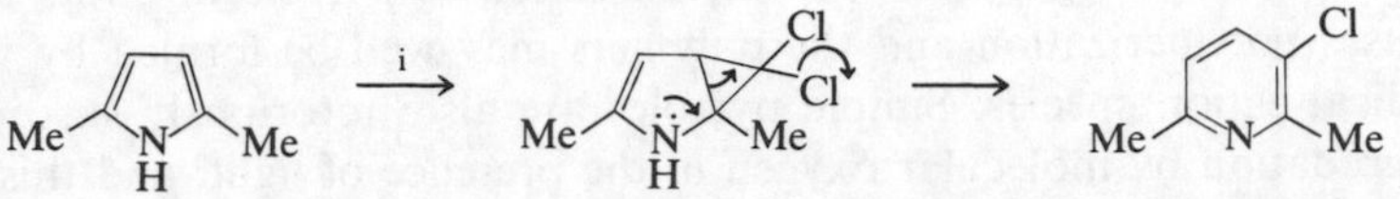

Fig. 6.9 Reaction of 2,5-dimethylpyrrole with dichlorocarbene. Reagent: i, $Cl_3CCO_2^- Na^+$, heat.

6.2.7   *Properties of substituted pyrroles*

The pyrrole ring system is an excellent π-electron donor and this influences the properties of substituents attached to the ring carbon atoms.

Pyrrolecarboxaldehydes are distinguished from most aromatic and aliphatic aldehydes by their low reactivity towards nucleophiles. Some of the typical reactions of aldehydes, such as the formation of cyanohydrins and the reduction of Fehling's solution, fail with pyrrole-2- and 3-carboxaldehyde, and others are much slower. The carbonyl group is deactivated by the π electron system of the ring, as shown for pyrrole-2-carboxaldehyde, **24**. As would be expected on this basis there is a significant barrier to rotation of the formyl group (it is about 11 kcal mol^{-1} (46 kJ mol^{-1}) in 1-alkylpyrrole-2-carboxaldehydes)[21] and the carbonyl stretching frequency is low (1667 cm^{-1} for pyrrole-2-carboxaldehyde). The *syn* conformer shown is the more stable.

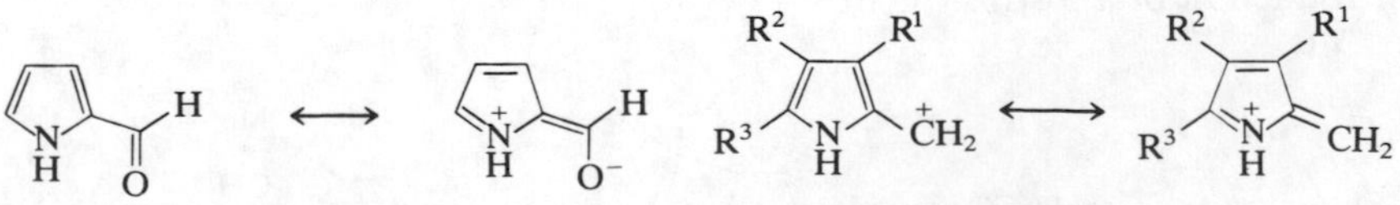

24   25

19 R. L. Jones and C. W. Rees, *J. Chem. Soc.* (*C*), 1969, 2249.
20 R. E. Busby, M. Iqbal, M. A. Khan, J. Parrick, and C. J. G. Shaw, *J. Chem. Soc., Perkin Trans. 1*, 1979, 1578.
21 C. Jaureguiberry, M. C. Fournie-Zaluski, and S. Combrisson, *Org. Magnetic Res.*, 1973, **5**, 165.

2-*Hydroxymethylpyrroles* are unstable in acidic media because of the formation of the cations **25**. This is then attacked by another molecule of the pyrrole and eventually a polymer is formed unless the pyrrole is fully substituted. Similarly, the great reactivity of 2-acetoxymethyl-, 2-chloromethyl-, and 2-bromomethylpyrroles towards nucleophiles can be ascribed to the formation of the same cation. This type of cation is an intermediate in coupling reactions of pyrroles which are used in porphyrin synthesis (Fig.. 6.10).[22]

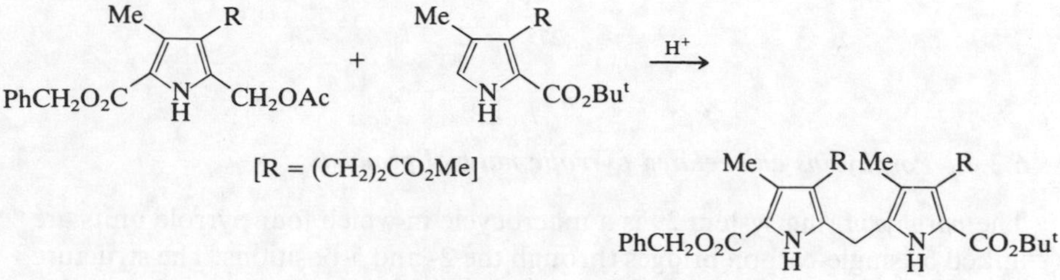

Fig. 6.10 Coupling of pyrroles through an acetoxymethyl derivative.

2-*Methylpyrroles* are easily halogenated in the side chain and can be acetoxylated by reaction with lead(IV) acetate, whereas 3-methylpyrroles are unreactive. These reactions have proved useful for the selective functionalization of 2-methyl groups in readily available pyrroles, such as those produced in the Knorr synthesis.

Pyrrolecarboxylic acids are readily decarboxylated, particularly those with electron-releasing substituents. Many such acids decarboxylate at their melting points. Bromination of a pyrrole-2-carboxylic acid can also cause decarboxylation of the acid and replacement of the carboxyl group by bromine. Some t-butyl esters of pyrrolecarboxylic acids can be cleaved under acid conditions. A possible mechanism for the acid-catalysed decomposition of pyrrole-2-carboxylic acid is shown in Fig. 6.11.

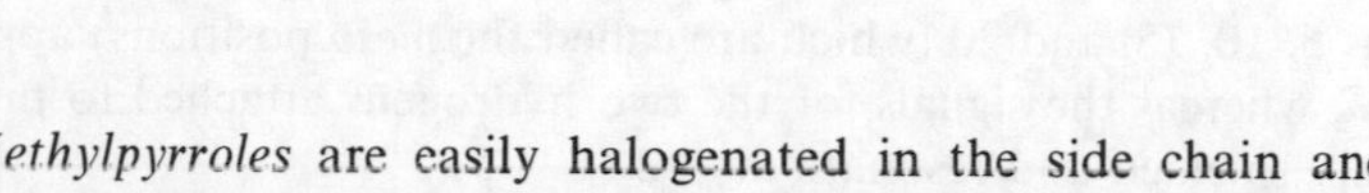

Fig. 6.11 Decarboxylation of pyrrole-2-carboxylic acid.

Hydroxypyrroles are well known but they are not analogues of phenols. They exist predominantly in keto tautomeric forms (Section 2.5). The tautomer **26** with the double bond in the 3,5-position is favoured, unless

22 J. A. P. B. Almeida, G. W. Kenner, J. Rimmer, and K. M. Smith, *Tetrahedron*, 1976, **32**, 1793.

there is a 4-acyl group, in which case the tautomer **27** (which allows conjugation between the nitrogen lone pair and the acyl function) is preferred. Hydroxy and amino tautomers are preferred when there is an adjacent substituent that can form intramolecular hydrogen bonds, such as in structures of the type **28** and example (i) in Table 4.6.

6.2.8   *Porphyrins and related pyrrolic natural products*[23]

The *porphyrin* ring system **29** is a macrocycle in which four pyrrole units are linked by single-carbon bridges through the 2- and 5-positions. The structure is a fully unsaturated one which shows evidence for the existence of a ring current in the NMR spectrum. The signals for the hydrogen atoms attached to carbons 5, 10, 15, and 20 (which are called the *meso* positions) appear at about $10\,\delta$, whereas the signals for the two hydrogens attached to nitrogen are strongly shielded and resonate at -2 to $-5\,\delta$.

The four nitrogen atoms are ideally arranged to chelate to metal cations, and porphyrins form complexes easily with many metals. Porphyrins also occur naturally as metal complexes, the most important being hemin (haem), **30**, which is the oxygen-carrying component of haemoglobin. Chlorophyll *a*, **31**, is a closely related structure with one saturated bond in ring D (this dihydroporphyrin ring system is called a *chlorin*). These compounds, of fundamental importance in animals and plants, have as a common biosynthetic precursor the cyclic tetrapyrrole uroporphyrinogen-III, **32**, which in turn is derived from four molecules of porphobilinogen, **1**. An interesting feature of the structure of compound **32** is that whereas three of the four pyrrolic units have peripheral substituents with alternating 'short' and 'long' carbon chains, the order is reversed in the fourth unit. It proved to be a major task to unravel the sequence of events by which this ring system is constructed in nature. It was finally established that the four porphobilinogen units are joined together in a regular fashion to produce a linear tetrapyrrole, and that the reversal of the fourth pyrrole unit takes place in the cyclization step.[24]

Bile pigments are oxidative degradation products of haem. Bilirubin-IXα, **33**, is one of these: it is a yellow pigment which is responsible for the characteristic yellow skin coloration in jaundice. The *corrin* ring system is related to that of the porphyrins but has one less carbon atom in the ring.

23 Review: *Porphyrins and Metalloporphyrins*, ed. K. M. Smith, Elsevier, Amsterdam, 1975.
24 A. R. Battersby and E. McDonald, *Accounts Chem. Res.*, 1979, **12**, 14.

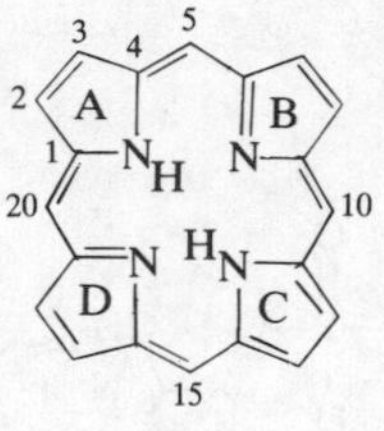

29

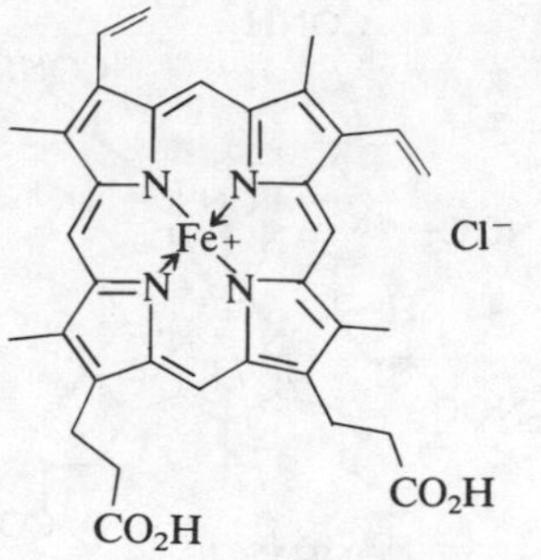

30

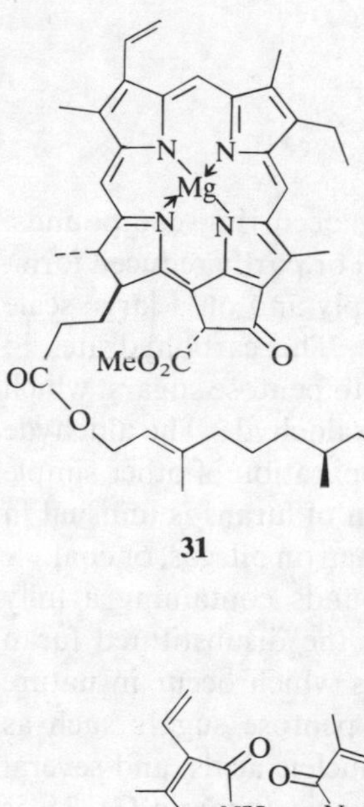

31

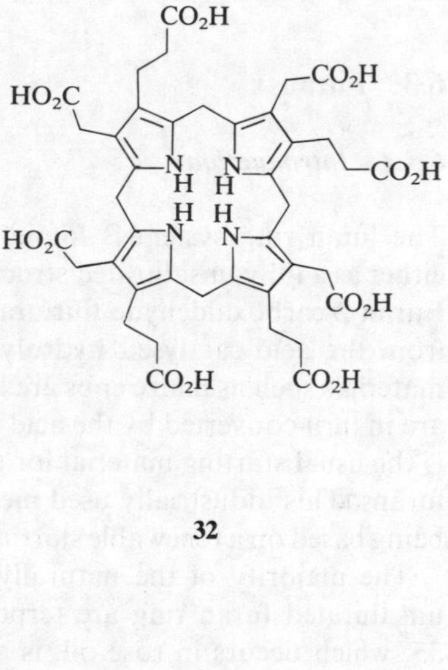

32

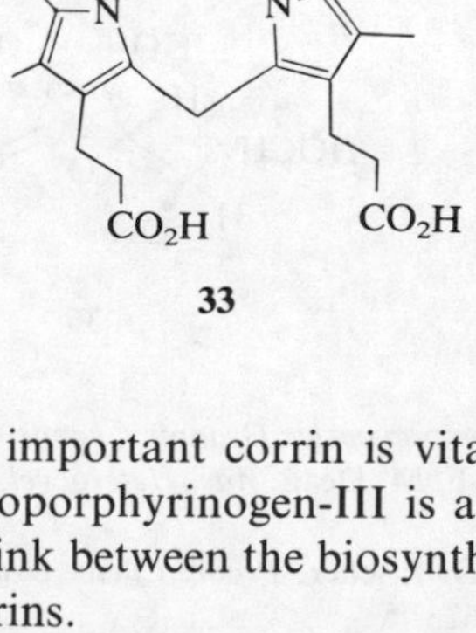

33

The most important corrin is vitamin B_{12}, **34**, the antipernicious-anaemia factor. Uroporphyrinogen-III is also a progenitor of vitamin B_{12}, so there is a clear link between the biosynthesis of the important porphyrins and that of the corrins.

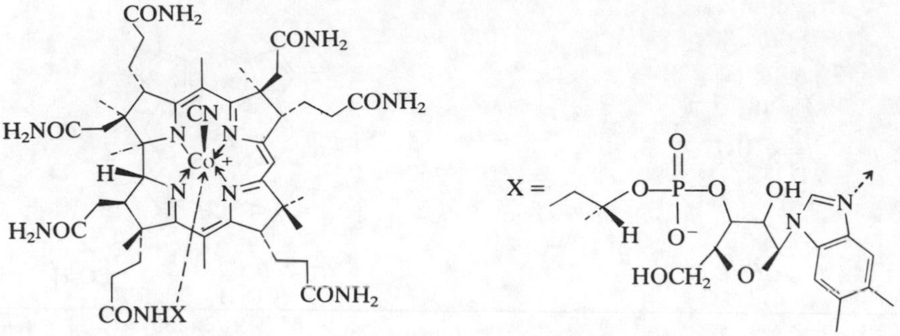

34

6.3 Furans[25]

6.3.1 Introduction

The furan ring system is found in many naturally occurring compounds, either as a fully unsaturated structure or in a reduced or partly reduced form. Furan-2-carboxaldehyde (furfural) is available cheaply and on a large scale from the acid-catalysed hydrolysis of cereal waste. The carbohydrates in materials such as maize cobs are hydrolysed by acid to pentose sugars, which are in turn converted by the acid into furan-2-carboxaldehyde. The aldehyde is the usual starting material for the commercial preparation of other simple furans. This industrially used method of preparation of furans is unusual in being based on a renewable starting material rather than on oil, gas, or coal.

The majority of the naturally occurring compounds containing a fully unsaturated furan ring are terpenoid in character; the disubstituted furan **35**, which occurs in rose oil, is an example. Furans which occur in nature in a reduced or otherwise modified form include pentose sugars such as ribose and deoxyribose, which are components of nucleic acids, and several types of unsaturated γ-lactone,[26] of which ascorbic acid (vitamin C), **36**, is

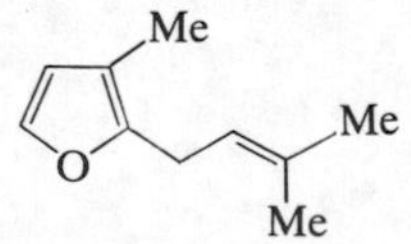

35 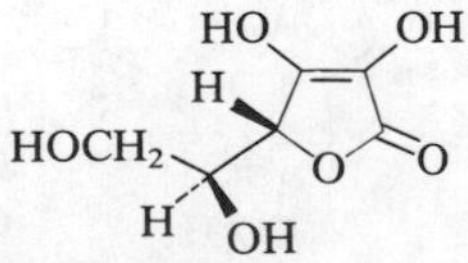   **36**

25 Reviews: M. V. Sargent and T. M. Cresp, in *Comprehensive Organic Chemistry*, Vol. 4, ed. P. G. Sammes, Pergamon, Oxford, 1979, p. 693; F. M. Dean, *Adv. Heterocycl. Chem.*, 1982, **30**, 167; 1982, **31**, 237.

26 Reviews: N. H. Fischer, E. J. Olivier, and H. D. Fischer, *Progr. Chem. Org. Nat. Prod.*, 1979, **38**, 47; Y. S. Rao, *Chem. Rev.*, 1976, **76**, 625.

one example. Some furan derivatives are useful chemotherapeutic agents. The semicarbazone of 5-nitrofuran-2-carboxaldehyde is a bactericide. The development of antiulcer compounds is mentioned in Chapter 1 and the furan derivative ranitidine, is a chemotherapeutic agent of this type.

Furan is a liquid, b.p. 31°C, which is slightly soluble in water. The dipole moment (0.72 D) is in the 'normal' direction with the heteroatom at the negative end of the dipole, unlike that of pyrrole. The structure and some other physical properties of furan are described in Chapter 2.

6.3.2 Ring synthesis

Because furan-2-carboxaldehyde is readily available, it is the starting point for the preparation of several other simple furans. It can be decarbonylated by heating in the gas phase and it can be oxidized, reduced, and further substituted. However, 3- and 4-substituted furans and other multisubstituted furans may require synthesis from acyclic precursors. The two classical methods of furan ring synthesis are the cyclization of 1,4-dicarbonyl compounds (the *Paal–Knorr* synthesis) and the reaction of α-haloketones with 1,3-dicarbonyl compounds (the *Feist–Benary* synthesis). The first of these is closely related to the Paal–Knorr pyrrole synthesis discussed in Section 6.2.2 and the second is outlined in Table 4.3 [reaction (iii)]. The Paal–Knorr method involves the use of an acid catalyst to bring about the cyclization (Fig. 6.12); catalysts such as sulphuric acid, phosphorus(V) oxide, zinc chloride, and an acidic ion-exchange resin (Amberlyst 15)[27] have been used. The limitation of the method, as with the corresponding pyrrole synthesis, is the availability of suitably substituted 1,4-dicarbonyl components. The Feist–Benary synthesis is a good method for the preparation of esters of furan-3-carboxylic acids, and an α-haloaldehyde such as chloroacetaldehyde can be used as the halocarbonyl component. A modern alternative to the Feist–Benary method, which is also shown in Table 4.3 [reaction (iv)], is the interaction of allenylsulphonium salts with 1,3-dicarbonyl compounds.

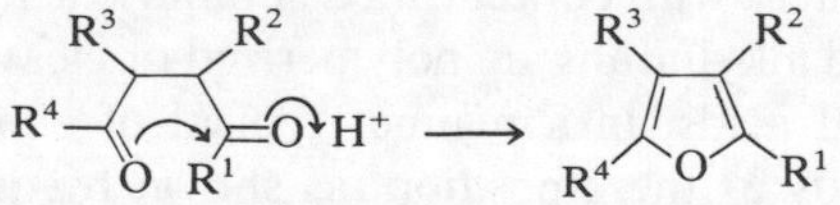

Fig. 6.12 The Paal–Knorr furan synthesis.

A good method for the preparation of 3-substituted furans is the oxidative cyclization of *Z*-butene-1,4-diols, **37**, by pyridinium chlorochromate.[28] The diols **37** are obtained by cuprate addition to dimethyl acetylenedicarboxylate, followed by reduction of the ester groups.

27 L. T. Scott and J. O. Naples, *Synthesis*, 1973, 209.
28 H. Nishiyama, M. Sasaki, and K. Itoh, *Chem. Lett.*, 1981, 1363.

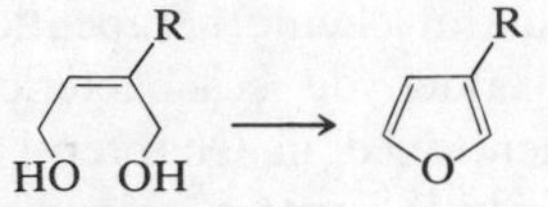

37

The furan ring system can also be formed by radical cyclization reactions (as illustrated in Fig. 4.14) and by 1,3-dipolar cycloaddition of carbonyl ylides (Section 4.3.2). Other useful routes to furans are described in later sections. One involves the Diels–Alder reactions of simpler furans (Section 6.3.5) and the other, oxazole cycloaddition (Section 8.5).

6.3.3   Electrophilic substitution reactions

Furan is an electron-rich heterocycle, and its reactions with electrophiles are in some respects similar to those of pyrrole described in Section 6.2.5. The major differences are as follows:

(a) Furan is considerably less aromatic than pyrrole, and there is therefore a much greater tendency for it to react with electrophiles to give addition products rather than substitution products. Even in those cases where substitution products are obtained, addition products can sometimes be shown to be intermediates.

(b) 2-Substitution is favoured over 3-substitution, as with pyrrole, but to a much greater extent. For example, both ring systems undergo trifluoro-acetylation (CF_3CO_2COMe, 75°C) but whereas for pyrrole the ratio of 2- and 3-substituted products is 6:1, it is 6000:1 for furan.[29]

(c) Furan is generally less reactive than pyrrole towards electrophiles (by a factor of about 10^5) although it is still very much more reactive than benzene.

Furan polymerizes when treated with concentrated sulphuric acid or with aluminium chloride. Furan and alkylfurans are polymerized or cleaved when warmed with aqueous mineral acids, in a manner typical of enol ethers. Furan is protonated most easily at the 2-position, as shown by deuterium exchange reactions, and indeed there is no evidence for deuterium exchange at the 3-position.[30] Thus, as with pyrrole, the cation **38** is more stable than the cation **39**, and is formed more readily; any ring opening may take place by the attack of water on the cation **39**.

29 S. Clementi, F. Fringuelli, P. Linda, G. Marino, G. Savelli, and A. Taticchi, *J. Chem. Soc., Perkin Trans. 2*, 1973, 2097.

30 P. Salomaa, A. Kankaanpera, E. Nikander, K. Kaipainen, and R. Aaltonen, *Acta Chem. Scand.*, 1973, **27**, 153.

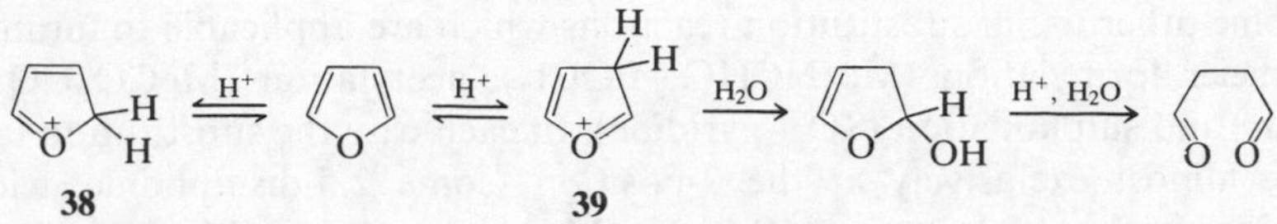

38 **39**

Furan can be converted into 2-bromofuran in good yield by reaction with bromine–dioxan complex at $-5°C$. If furan is brominated at $-50°C$ in carbon disulphide, however, a mixture of bromine addition products can be detected.[31] This indicates that the bromination may go by way of an addition–elimination mechanism, and certainly there is a tendency for the cation **40** to be intercepted by nucleophiles rather than to lose a proton. If the bromination of furan is carried out in methanol at $-10°C$, the acetal **41** is formed in a moderate yield. This is again the result of an addition process, of the type to be expected of a nonaromatic system.

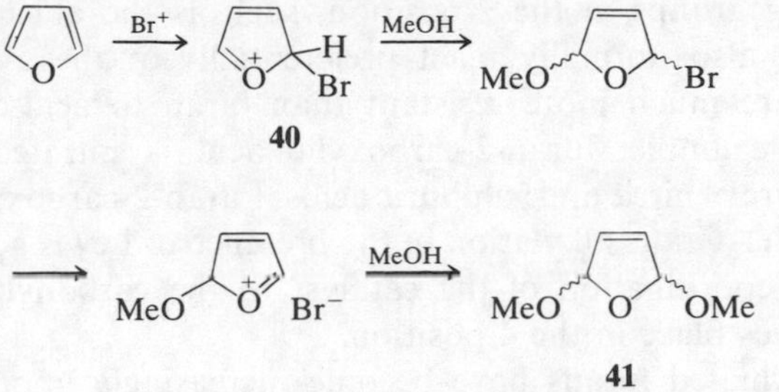

40

41

In general, therefore, if furan is subjected to electrophilic attack in the presence of a nucleophile, addition reactions are to be expected, especially at low temperatures, but the addition products may then undergo elimination to give substitution products. In the absence of an effective nucleophile the intermediate cation is deprotonated to give the substitution product in the normal way. This is further illustrated by the nitration of furan, in which both addition–elimination and normal electrophilic substitution mechanisms can operate, depending upon the choice of reagent (Fig. 6.13).

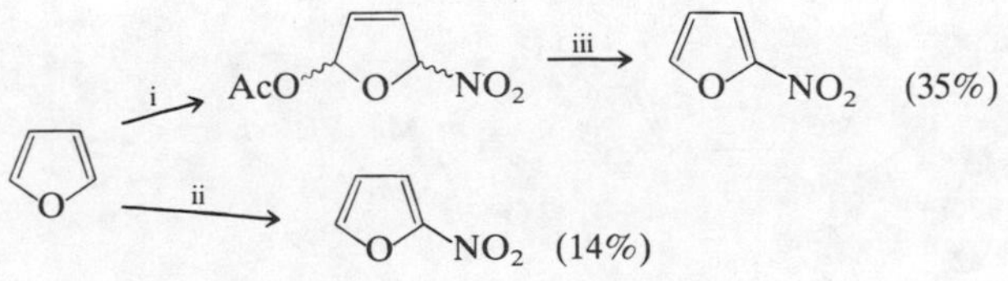

Fig. 6.13 Nitration of furan. Reagents: i, $NO_2^+ \bar{O}Ac$, $-5°C$; ii, $NO_2^+ BF_4^-$; iii, pyridine.

31 E. Baciocchi, S. Clementi, and G. V. Sebastiani, *J. Chem. Soc., Chem. Commun.*, 1975, 875.

Some other useful substitution reactions which are applicable to furan are Vilsmeier formylation (Me_2NCHO, $POCl_3$), acetylation ($MeCO_2COMe$, $SnCl_4$) and sulphonation (SO_3–pyridine). In each case the substitution takes place almost exclusively at the 2-position (some 2,5-disulphonic acid is obtained in the sulphonation). Both the Mannich reaction (Me_2NH, H_2CO, H^+) and azo coupling fail with furan, showing its lower reactivity compared with pyrrole, although alkylfurans with a free 2- or 5-position will undergo the Mannich reaction. The Mannich reaction can be achieved for furan itself by using a preformed iminium salt $Me_2\overset{+}{N}{=}CH_2$ Cl^-.[32] Furan also undergoes acid-catalysed conjugate addition to $\alpha\beta$-unsaturated aldehydes and ketones, in preference to cycloaddition (for an example see Fig. 6.18).

The effect of substituents upon the position of attack by electrophiles can be predicted on the basis of the electronic effect of the substituent. 2-Alkylfurans are substituted predominantly in the 5-position (in contrast to 2-alkylpyrroles, in which 4-substitution is usually more favourable). 2,5-Dialkylfurans undergo substitution at the 3-position. Furans with electron-withdrawing groups in the 2-position, such as the aldehyde and the carboxylic acid, also normally react preferentially at the 5-position. These compounds are much more resistant than furan to acid-catalysed decomposition; for example, furan-2-carboxylic acid is nitrated in the 5-position by a mixture of nitric and sulphuric acids. Furan-2-carboxaldehyde also undergoes Friedel–Crafts alkylation in the presence of Lewis acids, but, perhaps because of coordination of the catalyst to the carbonyl oxygen, substitution now takes place in the 4-position.

In recent years lithiated furans have become increasingly important in furan chemistry because they can be formed selectively and also react selectively, in mild conditions, to give substitution products.[10] Furan is lithiated at the 2-position by butyllithium, and it can be 2,5-dilithiated by careful control of reaction conditions.[33] Considerable selectivity is also possible in the lithiation of substituted furans; for example, furan-3-carboxylic acid can be selectively alkylated and deuterated at the 2-position. 3-Furyllithium can be prepared from 3-bromofuran by bromine–lithium exchange. It is also possible to achieve hydrogen–lithium exchange at the 3-position by using a powerful directing group such as the dimethyloxazolinyl group in structure **42**.[34] The organolithium reagent (*sec*-butyllithium)

42

32 H. Heaney, G. Papageorgiou, and R. F. Wilkins, *Tetrahedron Lett.*, 1988, **29**, 2377.

33 D. J. Chadwick and C. Willbe, *J. Chem. Soc., Perkin Trans. 1*, 1977, 887.

34 D. J. Chadwick, M. V. McKnight, and R. I. Ngochindo, *J. Chem. Soc., Perkin Trans. 1*, 1982, 1343.

coordinates to the nitrogen atom of this substituent, and lithiation is directed to the 3-position in preference to the more acidic 5-position. The alkylation of 3-furyllithiums has proved to be a synthetically useful reaction because many naturally occurring furans are 3-alkyl derivatives.

Furan and alkylfurans can also be mercurated by reaction with mercury(II) chloride and sodium acetate in aqueous ethanol; the $HgCl^+$ group is introduced preferentially at the 2-position.

6.3.4 Nucleophilic substitution

Some furans bearing electron-withdrawing substituents can undergo nucleophilic substitution reactions and they are generally more reactive than the corresponding benzene derivatives. Typical examples of such substitution reactions are shown in Fig. 6.14. In 5-nitrofuran-2-carboxaldehyde the nitro group can be displaced by azide and other nucleophiles.[35]

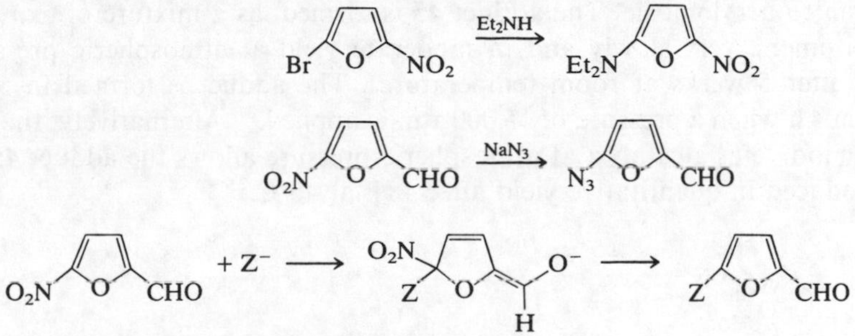

Fig. 6.14 Nucleophilic substitution in furans.

6.3.5 Cycloaddition reactions

Furans participate in several different types of cycloaddition process. By far the most important of these is the Diels–Alder reaction in which furan acts as the diene. Furan adds readily to maleic anhydride at or below room temperature to give the adduct **43** which results from *exo*-addition. It has been shown by NMR that the adduct initially formed is the *endo* adduct **44**, but that this is converted to its isomer **43** by retroaddition followed by readdition in the alternative orientation.[36] The *endo* adduct **44** is therefore the kinetic product and the *exo* adduct **43** the thermodynamic product. In the analogous addition of maleimide to furan at room temperature the *endo* adduct can be isolated but rearranges to the *exo* adduct on heating.

35 F. Lieb and K. Eiter, *Liebigs Ann. Chem.*, 1972, **761**, 130.
36 M. W. Lee and W. C. Herndon, *J. Org. Chem.*, 1978, **43**, 518.

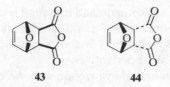

43   44

Yields of cycloadducts are generally unsatisfactory with dienophiles of lower reactivity, and the yields are rarely improved by carrying out the reactions at higher temperature because the reverse reaction becomes increasingly favourable. One method which has been used to overcome the problem is to carry out the reactions at very high pressures.[37] Since the transition state for cycloaddition is smaller than the reactants, the forward reaction is favoured by high pressure. Pressures in the range 8000–20 000 atm are needed to achieve a significant improvement in yield. The other useful practical modification is to add a Lewis acid catalyst, which increases the electrophilicity of the dienophile and so accelerates the reaction. An example in which both techniques have been used independently is the cycloaddition of furan to acrylonitrile. The adduct **45** is formed, as a mixture of *exo* and *endo* isomers, very slowly and in moderate yield at atmospheric pressure (39% after 5 weeks at room temperature). The adduct is formed in 55% yield in 4 h when a pressure of 15 000 atm is applied.[38] Alternatively, the use of zinc iodide as a catalyst at atmospheric pressure allows the adduct **45** to be produced in quantitative yield after 48 h at 40°C.[39]

45

Furan and alkylfurans also undergo cycloaddition to activated acetylenes. The adducts, such as that (**46**) formed from furan and dimethyl acetylene-dicarboxylate, can be converted into phenols by reaction with acid. They can also be hydrogenated preferentially at the less substituted double bond. The reduced adduct is thermally unstable and a new furan is formed from it by a retro-Diels–Alder reaction (Fig. 6.15). This is a good method of preparing esters of furan-3,4-dicarboxylic acids.

A second type of cycloaddition process of furans and alkylfurans takes place with oxallyl cations, **47**. These intermediates are generated from $\alpha\alpha'$-dibromoketones by reaction with iron carbonyls[40] or with sodium iodide[41] and they undergo a [4 + 3] cycloaddition reaction with furans which

37 Review: N. S. Isaacs, *Tetrahedron*, 1991, **47**, 8463.
38 W. G. Dauben and H. O. Krabbenhoft, *J. Am. Chem. Soc.*, 1976, **98**, 1992.
39 F. Brion, *Tetrahedron Lett.*, 1982, **23**, 5299.
40 H. Takaya, S. Makino, Y. Hayakawa, and R. Noyori, *J. Am. Chem. Soc.*, 1978, **100**, 1965.
41 D. I. Rawson, B. K. Carpenter, and H. M. R. Hoffmann, *J. Am. Chem. Soc.*, 1979, **101**, 1786.

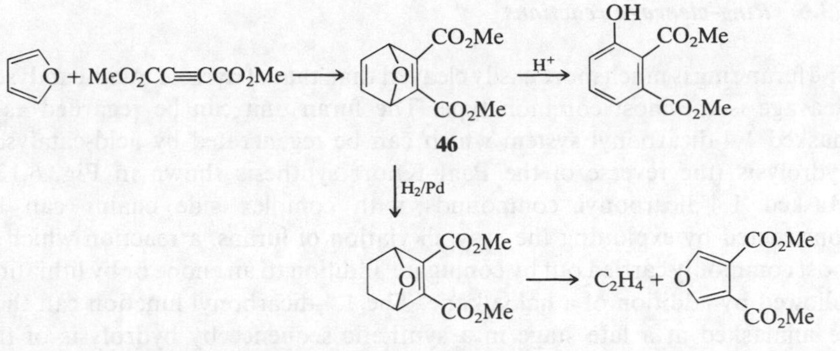

Fig. 6.15 Formation and reactions of the adduct **46**.

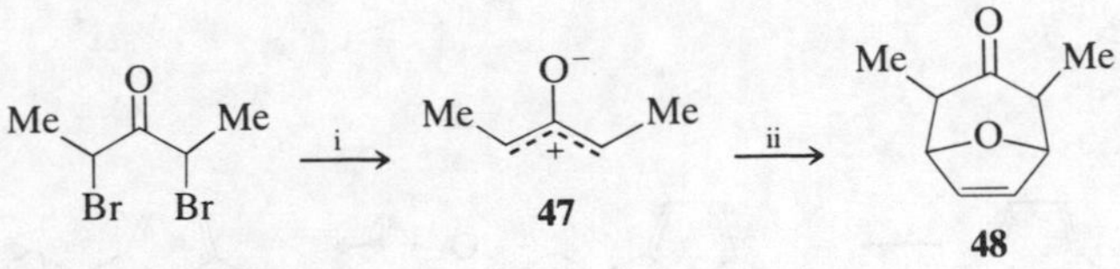

Fig. 6.16 Cycloaddition of furan to an oxallyl cation. Reagents: i, $Fe_2(CO)_9$ or NaI/Cu; ii, furan.

leads to the formation of bridged seven-membered cyclic ketones such as compound **48** (Fig. 6.16).

Furans also occasionally act as electron-rich 2π components in cycloaddition reactions. Some examples are shown in Fig. 6.17; one involves addition to a 1,3-dipole (a nitrile oxide)[42] and another to a heterodiene (an azo-olefin).[43] Furans can also undergo the Paterno–Büchi reaction with ketones to give isolable oxetanes, and can form cyclopropane derivatives with carbenes.[44]

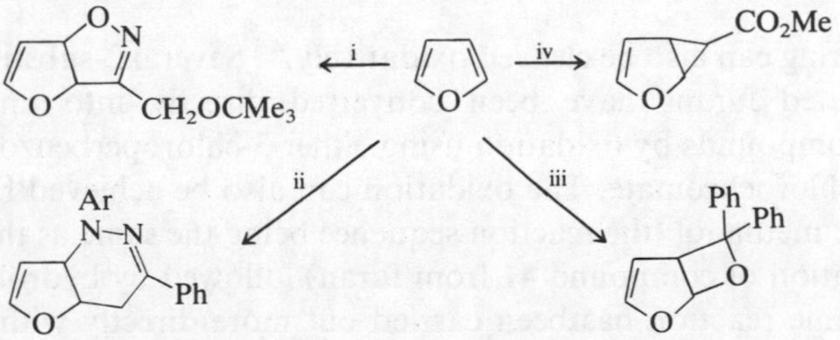

Fig. 6.17 Furan as a 2π component in cycloaddition. Reagents: i, $Me_3COCH_2C\overset{+}{N}\overset{-}{O}$; ii, $CH_2{=}C(Ph)N{=}NC_6H_3(NO_2)_2$-2,4; iii, Ph_2CO, hv; iv, N_2CHCO_2Me, hv.

42 I. Müller and V. Jäger, *Tetrahedron Lett.*, 1982, **23**, 4777.
43 R. Faragher and T. L. Gilchrist, *J. Chem. Soc., Perkin Trans. 1*, 1979, 249.
44 G. O. Schenck and R. Steinmetz, *Liebigs Ann. Chem.*, 1963, **668**, 19.

6.3.6   Ring-cleavage reactions

The furan ring is much more easily cleaved than that of pyrrole. Acid-catalysed cleavage is the most common type. The furan unit can be regarded as a masked 1,4-dicarbonyl system which can be regenerated by acid-catalysed hydrolysis (the reverse of the Paal–Knorr synthesis shown in Fig. 6.12). Masked 1,4-dicarbonyl compounds with complex side chains can be constructed by exploiting the easy alkylation of furans, a reaction which is most commonly carried out by conjugate addition to an enone or by lithiation followed by addition of a haloalkane. The 1,4-dicarbonyl function can then be unmasked at a late stage in a synthetic sequence by hydrolysis of the furan ring. An example is provided by the synthesis of *cis*-jasmone from 2-methylfuran shown in Fig. 6.18; two methods of alkylation of the furan are illustrated.[45]

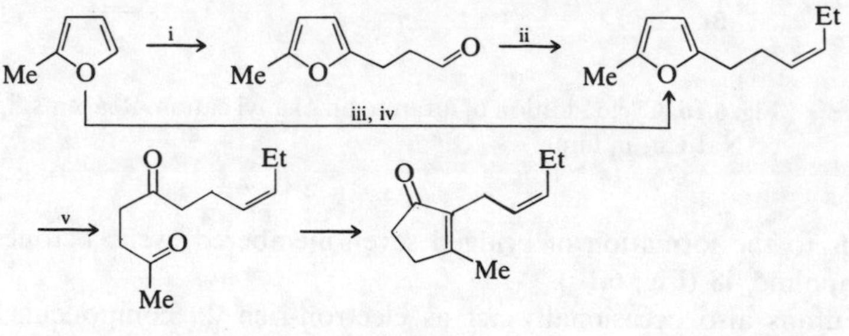

Fig. 6.18 Syntheses of *cis*-jasmone from 2-methylfuran. Reagents: i, CH_2=CHCHO, H^+; ii, $Ph_3\overset{+}{P}\overset{-}{C}HEt$, iii, BuLi; iv, $Br(CH_2)_2CH$=CHEt; v, H_2SO_4, $MeCO_2H$.

The furan ring can also be cleaved oxidatively.[46] Several 2-substituted and 2,5-disubstituted furans have been converted directly into unsaturated dicarbonyl compounds by oxidation using either 3-chloroperbenzoic acid or pyridinium chlorochromate. The oxidation can also be achieved by the use of bromine in methanol (the reaction sequence being the same as that shown for the formation of compound **41** from furan) followed by hydrolysis (Fig. 6.19). The same reaction has been carried out more directly with bromine in aqueous acetone at $-20°C$.[47]

45 G. Büchi and H. Wüest, *J. Org. Chem.*, 1966, **31**, 977; L. Crombie, P. Hemesley, and G. Pattenden, *J. Chem. Soc. (C)*, 1969, 1024,
46 Review: B. H. Lipshutz, *Chem. Rev.*, 1986, **86**, 795.
47 J. Jurczak and S. Pikul, *Tetrahedron Lett.*, 1985, **26**, 3039.

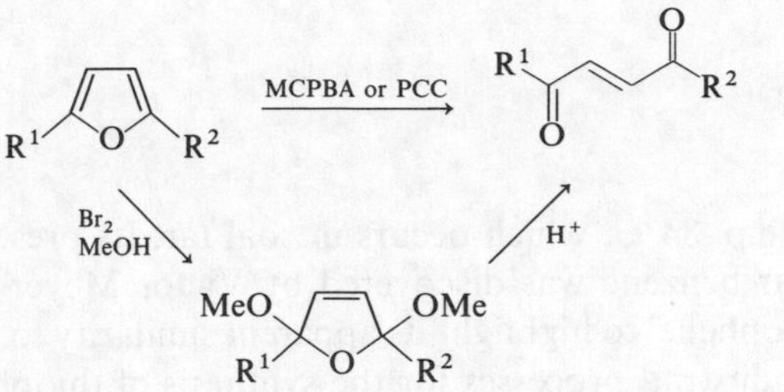

Fig. 6.19 Oxidative cleavage of furans to enediones.

6.3.7 Some properties of substituted furans

The chemistry of substituents at the 2-position of furan is influenced by the electronic properties of the ring system. Furan is a σ-electron acceptor but a π-electron donor. The π-electron-donating properties have much the same kind of influence on the substituents as in the analogous pyrroles: thus, cations are stabilized at the α-carbon atom, and nucleophilic attack on the carbonyl group of furan-2-carboxaldehydes and on 2-furyl ketones is slower than for aliphatic aldehydes and ketones. 2-(Chloromethyl)furan is unstable and difficult to handle, much more so than (chloromethyl)benzene; however, it can be stabilized by the introduction of a trimethylsilyl group at the 5-position.[48] The σ inductive effect of the furan oxygen can be seen by comparing the pK_a values of furan-2-carboxylic acid ($pK_a = 3.2$) and benzoic acid ($pK_a = 4.2$): furan-2-carboxylic acid is appreciably stronger.

Hydroxyfurans exist in the keto tautomeric forms, as γ-lactones, but alkoxy and silyloxyfurans are known. The presence of these electron-releasing substituents on the furan facilitates Diels–Alder reactions with electron-deficient dienophiles. For example, Diels–Alder reactions of 3-methoxyfuran, 3,4-dimethoxyfuran,[49] and 2,5-bis(trimethylsilyloxy)furans, **49**,[50] take place readily. The silyloxyfurans **49** are readily available from the corresponding succinic anhydrides.

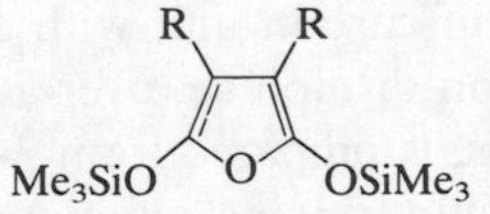

49

48 K. Takanishi, H. Urabe, and I. Kuwajima, *Tetrahedron Lett.*, 1987, **28**, 2281.
49 P. X. Iten, A. A. Hofmann, and C. H. Eugster, *Helv. Chim. Acta*, 1979, **62**, 2202.
50 P. Brownbridge and T.-H. Chan, *Tetrahedron Lett.*, *1980*, **21**, 3423.

6.4   Thiophenes[51]

6.4.1   Introduction

Thiophene is a liquid, b.p. 84°C, which occurs in coal tar. Its presence as a contaminant of coal-tar benzene was discovered by Victor Meyer in 1882; he coined the name 'thiophene' to highlight its apparent similarity to benzene. There are now good industrial processes for the synthesis of thiophene and alkylthiophenes which use aliphatic starting materials (see Section 6.4.2).

The structure of thiophene (Fig. 2.9) and the other properties discussed in Chapter 2 lead to its classification as an electron-rich aromatic compound. Its resonance energy is about the same as that of pyrrole, about half that of benzene, but much greater than that of furan. Like other members of this group, the ring system exerts a π-electron-donating and a σ-electron-withdrawing effect on substituents in the 2- and 5-positions. The d-orbitals on sulphur seem to play little part in its ground-state properties. The chemistry of thiophenes can thus reasonably be compared with that of pyrroles and furans.

The ring system occurs naturally in some plant products but it is of greater importance as a component of synthetic pharmaceuticals and dyestuffs. Biotin (vitamin H), **50**, contains a tetrahydrothiophene ring and it is the most important naturally occurring derivative of the ring system. It occurs in yeast and in eggs.

50

6.4.2   Ring synthesis

The important industrial routes to thiophenes involve the combination of an aliphatic compound containing a linear four-carbon unit with a source of sulphur, such as elemental sulphur or carbon disulphide, over a catalyst at 200–700°C. For example, 3-methylthiophene is prepared from 3-methyl-1-butanol and carbon disulphide over a chromium trioxide–alumina catalyst at 500°C. Thiophene itself can be made from butanol and carbon disulphide, or from butane and sulphur. Several good laboratory methods, based on cyclization reactions, are also available.[51] Three of the many different routes

51 Review: *Thiophene and its Derivatives*, ed. S. Gronowitz, Parts 1–3, Wiley-Interscience, New York, 1985 and 1986.

are illustrated in Fig. 6.20. The *Paal–Knorr* synthesis (a) is the reaction of 1,4-dicarbonyl compounds with phosphorus pentasulphide. The yields in the reaction have been significantly increased by the use of Lawesson's reagent in place of phosphorus pentasulphide.[52] A modification of the reaction makes use of hydrogen sulphide and hydrogen chloride to introduce the sulphur. The *Hinsberg* synthesis (b) is a good route to 3,4-disubstituted thiophenes because the initial products can be successively decarboxylated. The mechanism shown has been established by isotopic labelling experiments. The *Gewald* synthesis (c) is an example of the general method of synthesis of *C*-amino heterocycles by cyclization onto nitriles which was discussed in Chapter 4 and illustrated in Fig. 4.12. Instead of the α-mercaptocarbonyl compound a simple carbonyl compound with an α-methylene group can be used in the presence of elemental sulphur. This ring closure is potentially reversible and some 2-aminothiophenes have been cleaved by reaction with bases.[53]

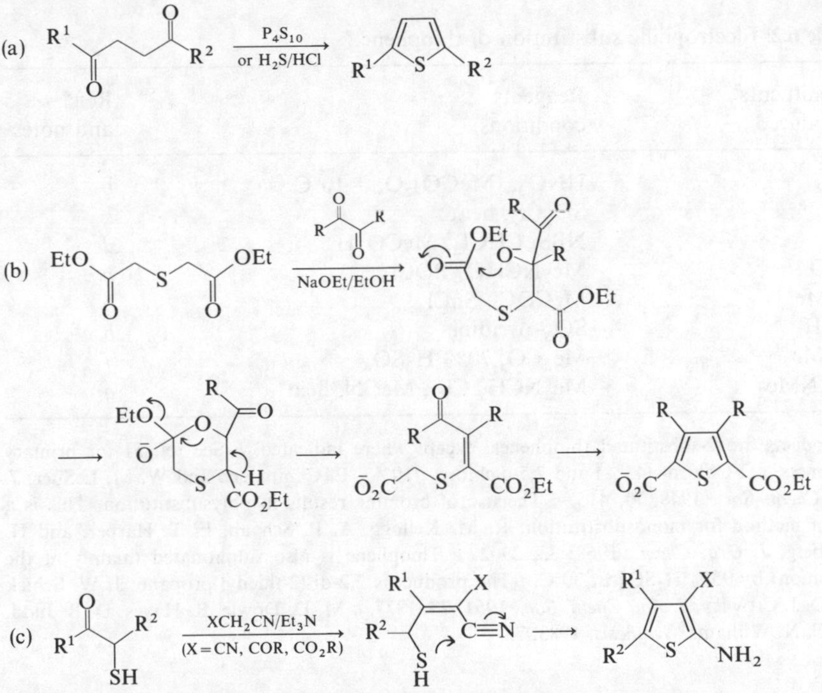

Fig. 6.20 Some methods of thiophene ring synthesis.

6.4.3 *Electrophilic substitution*

Thiophene is somewhat less reactive than furan toward electrophiles, and much less reactive than pyrrole, but it is still much more reactive than

52 D. R. Sridhar, M. Jogibhutka, P. S. Rao, and V. K. Handa, *Synthesis*, 1982, 1061.
53 W. J. Middleton, V. A. Engelhardt, and B. S. Fisher, *J. Am. Chem. Soc.*, 1958, **80**, 2822.

benzene, by a factor of 10^3-10^5. The electrophilic substitution of thiophenes is therefore a favourable process. Because of its greater aromatic character, thiophene tends to undergo substitution rather than addition reactions, and it is not so readily cleaved by acids as is furan. Thiophene is stable to aqueous mineral acids but not to 100% sulphuric acid or to strong Lewis acids such as aluminium(III) chloride. As with the other five-membered heterocycles, substitution in the 2-position is strongly favoured: substitution in the 3-position is negligible (less than 1%) with most electrophiles. The ratio of 3- to 2-substitution increases slightly with increase of reaction temperature. Nitration of thiophene (by benzoyl nitrate at 0°C) is exceptional in that the ratio of 2- or 3-substitution is only 6.2.[54] Some electrophilic substitution reactions of thiophene which go in good yield to give the 2-substituted thiophenes are shown in Table 6.2.

Table 6.2 Electrophilic substitution of thiophene.[a]

Substituents introduced	Reagents conditions	Refs. and notes
NO_2	HNO_3, $(MeCO)_2O$, $-10°C$	b
Cl	SO_2Cl_2, heat	c
Br	NBS, $CHCl_3$, $MeCO_2H$	d
CHO	Me_2NCHO, $POCl_3$	b
COMe	$MeCOCl$, $SnCl_4$	b
SO_3H	SO_3–pyridine	b,e
$—CMe_2—$[f]	Me_2CO, 70% H_2SO_4	f
CH_2NMe_2	$Me_2NCH_2^+Cl^-$, MeCN, heat	g

a Products are 2-substituted thiophenes, except where indicated. *b* See ref. 51 for primary references. *c* 2-Chloro- (43%) and 2,5-dichloro- (10%); E. Campaigne and W. M. LeSuer, *J. Am. Chem. Soc.*, 1948, **70**, 415. *d* The use of bromine results in polysubstitution. This is a milder method for monosubstitution; R. M. Kellogg, A. P. Schaap, E. T. Harper, and H. Wynberg, *J. Org. Chem.*, 1968, **33**, 2902. *e* Thiophene is also sulphonated (mainly at the 2-position) by 95% H_2SO_4 at 30°C. *f* The product is 2,2-di-(2-thienyl)propane; J. W. Schick and D. J. Crowley, *J. Am. Chem. Soc.*, 1951, **73**, 1377. *g* M. D. Dowle, R. Hayes, D. B. Judd, and C. N. Williams, *Synthesis*, 1983, 73.

Thiophenes with electron-releasing groups in the 2-position undergo electrophilic attack mainly or exclusively at the 5-position. With electron-withdrawing 2-substituents electrophilic substitution still occurs readily, but

---

54 For a comparative survey of electrophilic substitution in pyrrole, furan, and thiophene, see G. Marino, *Adv. Heterocycl. Chem.*, 1971, **13**, 235.

less selectively: both 4- and 5-substitution normally take place. For example, nitration of thiophene-2-carboxaldehyde gives the 4- and 5-nitro derivatives in a 3:1 ratio. As with the corresponding furans, electrophilic attack on 2-acylthiophenes in the presence of a Lewis acid catalyst occurs mainly at the 4-position, probably because of coordination of the Lewis acid to the oxygen of the carbonyl group.[55] 2-Substitution of thiophenes sometimes occurs at the site of an existing substituent, such as —Br or —CO_2H, which is subsequently expelled. An intramolecular example of such a reaction is the acid-catalysed cyclization of the thiophene **51** in which acylation occurs at the 2-position, followed by rearrangement.[56]

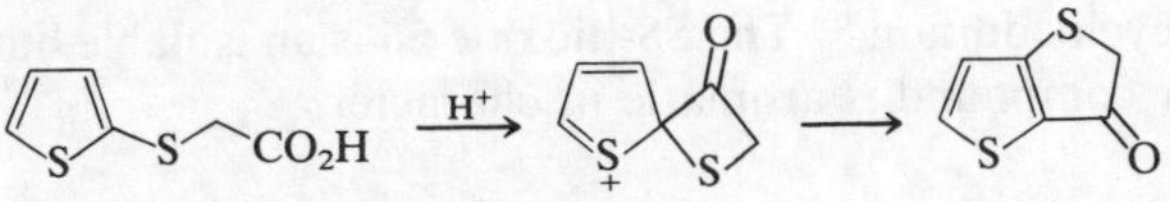

51

The electrophilic substitution of metallated thiophenes is a particularly useful reaction; thiophenes are easily and selectively metallated to give intermediates which react with a wide range of electrophiles. Thienyllithium compounds are most often used.[57] Simple thiophenes undergo hydrogen–lithium exchange selectively at the 2-position by reaction with butyllithium in ether or tetrahydrofuran at room temperature. Lithium can be introduced in other positions if the bromo or iodo compounds are available; these undergo rapid halogen–metal exchange at low temperatures with butyllithium. Directed lithiation, analogous to that described for the furan **42**, is also an effective way of introducing lithium into the 3-position of the thiophene ring.[57,58] 3-Lithiated thiophenes, **52**, are useful for the synthesis of

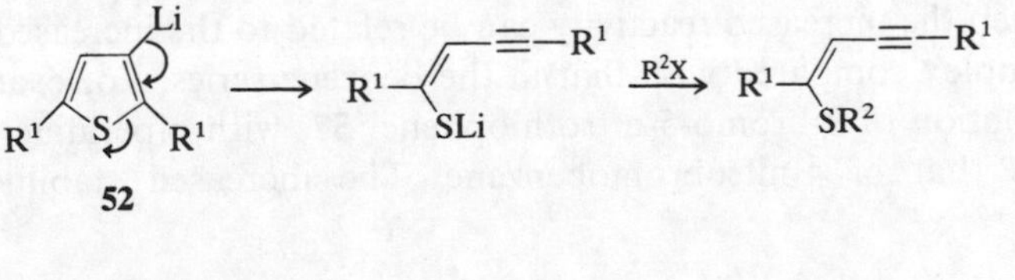

52

55 (a) M. Valenta and I. Koubek, *Coll. Czech. Chem. Commun.*, 1976, **41**, 78; (b) review: J. L. Goldfarb, J. B. Volkenstein, and L. I. Belenkij, *Angew. Chem. Int. Edn Engl.*, 1968, **7**, 519.
56 S. Gronowitz and P. Moses, *Acta Chem. Scand.*, 1962, **16**, 155.
57 Review: B. J. Wakefield, *Organolithium Methods*, Academic Press, London, 1988.
58 Review: V. Snieckus, *Chem. Rev.*, 1990, **90**, 879.

3-substituted thiophenes, although their reactions with electrophiles are best carried out at low temperature because they are thermally unstable. Their thermal ring opening provides a route to unsaturated sulphides.[59]

Electrophilic attack at the sulphur atom of thiophenes is not a common reaction. The sulphur can be alkylated using a 'hard' electrophile such as methyl fluorosulphonate,[60] but the sulphur atom in the product **53** is tetrahedrally substituted and the compound is, presumably, not aromatic. The rhodium(II)-catalysed addition of dimethyl diazomalonate to thiophene leads to the formation of a surprisingly stable ylide **54** in which the configuration about sulphur is pyramidal.[61] Thiophenes give related but unstable sulphur–nitrogen ylides when reacted with ethyl azidoformate, N_3CO_2Et.[62] Thiophene is oxidized by peroxides and peroxyacids to a sulphoxide **55** which is too unstable to be isolated but which can be intercepted by Diels–Alder cycloaddition.[63] The *SS*-dioxide **56** is an isolable but reactive diene.[63] Neither compound is aromatic in character.

| **53** | **54** | **55** | **56** |

6.4.4 Nucleophilic and radical substitution

Thiophenes which are substituted with conjugative electron-withdrawing groups (especially nitro) react much more readily with nucleophiles than do the corresponding benzenes. If we assume that the transition state resembles the Meisenheimer complex formed by bonding of the nucleophile to the ring system, then the increased reactivity can be related to the increased stability of the complex compared with that in the benzene series. For example, the rate of reaction of 2-bromo-5-nitrothiophene, **57**, with piperidine (25°C) is 150 times that of 4-nitrobromobenzene. The increased stability of the

59 Review: T. L. Gilchrist, *Adv. Heterocycl. Chem.*, 1987, **41**, 41.
60 R. F. Heldeweg and H. Hogeveen, *Tetrahedron Lett.*, 1974, 75.
61 Review: A. E. A. Porter, *Adv. Heterocycl. Chem.*, 1989, **45**, 151.
62 O. Meth-Cohn and G. van Vuuren, *Tetrahedron Lett.*, 1986, **27**, 1105.
63 Review: P. H. Benders, D. N. Reinhoudt, and W. P. Trompenaars, in ref. 51, Part 1, p. 671.

Meisenheimer complex, **58**, can be ascribed, in part, to the well-known ability of sulphur to stabilize an adjacent carbanion.[64]

These substitution reactions do not always follow the obvious course, and there are cases of *cine*-substitution (introduction of the nucleophile at the atom adjacent to the one bearing the leaving group) and *tele*-substitution (introduction of the nucleophile at a more remote atom). Nucleophiles can also cause ring cleavage. Examples of these processes are (a) the *cine*-substitution of PhS⁻ in 3,4-dinitrothiophene, **59**,[65] and (b) the ring cleavage of 2-nitrothiophene, **60**, by secondary amines.[66]

Bromide and iodide can be displaced by nucleophiles in unactivated thiophenes. The reactions may be catalysed by copper metal or by copper(I) salts and are not simple nucleophilic displacement processes. Useful examples are the displacement of bromide by chloride in bromothiophene by means of copper(I) chloride in dimethylformamide[67] and the displacement of bromide by acetoacetate anion in the carboxylic acid **61**.[68] 2-Bromothiophene reacts with potassamide in the absence of a catalyst to give 3-aminothiophene. This *cine*-substitution is neither an addition–elimination process nor an elimination–addition (aryne) process, but involves a complex series of steps which the carbanion **62** initiates by abstracting bromine from another molecule of 2-bromothiophene.[69]

64 It is not necessary to ascribe this stabilization to the use of sulphur d-orbitals: see J.-M. Lehn and G. Wipff, *J. Am. Chem. Soc.*, 1976, **98**, 7498.

65 C. Dell'Erba, D. Spinelli, and G. Leandri, *Gazz. Chim. Ital.*, 1969, **99**, 535.

66 G. Guanti, C. Dell'Erba, and S. Thea, *J. Chem. Soc., Perkin. Trans. I*, 1974, 2357.

67 S. Conde, C. Corral, R. Madronero, and A. S. Alvarez-Insua, *Synthesis*, 1976, 412.

68 D. E. Ames and O. Ribeiro, *J. Chem. Soc., Perkin Trans. 1*, 1975, 1390.

69 H. C. van der Plas, D. A. de Bie, G. Geurtsen, M. G. Reinecke, and H. W. Adickes, *Rec. Trav. Chim.*, 1974, **93**, 33. For a critical discussion of reactions purporting to go through 'dehydrothiophene' intermediates, see M. G. Reinecke, in *Reactive Intermediates*, Vol. 2, ed. R. A. Abramovitch, Plenum, New York, 1982, p. 367.

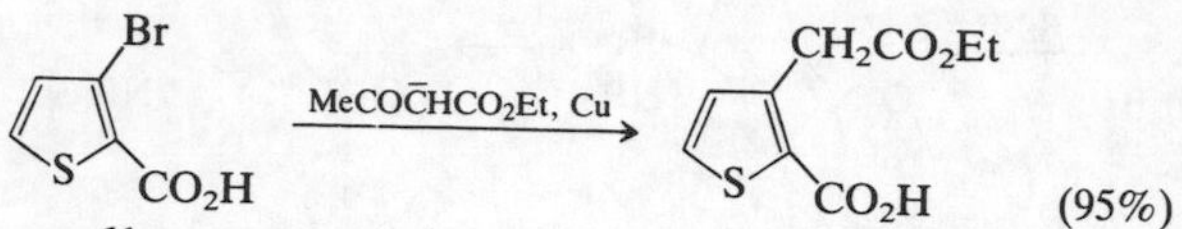

Thiophene can be phenylated, mainly in the 2-position, by reaction with aryl radicals generated from aniline and pentyl nitrite. Other 2-arylthiophenes can be made in the same way.[70]

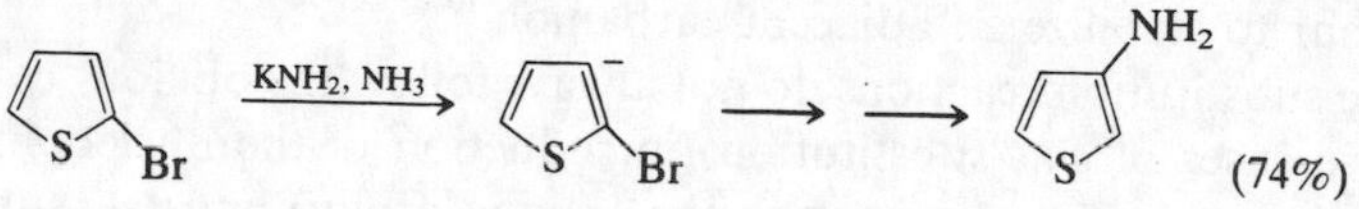

6.4.5 Addition and cycloaddition reactions

Thiophene can be induced to undergo the Diels–Alder reaction with highly activated dienophiles, but it is a poor diene.[63] Thus, in contrast to furan, it reacts with maleic anhydride only at 100°C and under high pressure.[71] It reacts with activated acetylenes and with benzyne but the adducts are unstable and extrude sulphur. Thus, benzyne, when generated thermally from the precursor 63 in the presence of thiophene, gives naphthalene as the major product.

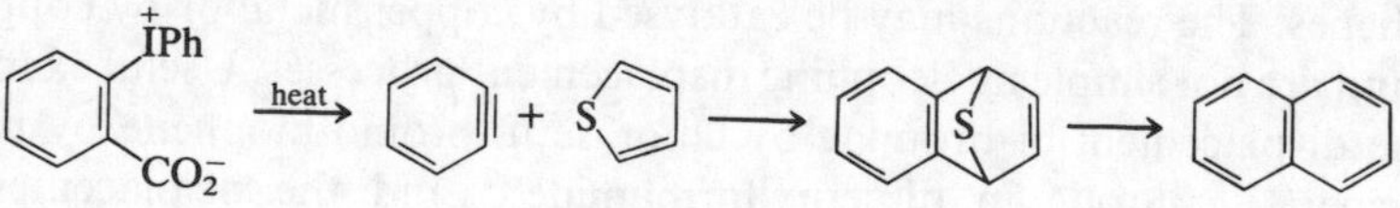

Some other examples of thermal cycloaddition reactions of thiophenes are shown in Table 6.3. Photochemical [2 + 2] addition of 2,5-dimethylthiophene to ketones, analogous to that observed with furans (Fig. 6.17), is also known.

6.4.6 Reductive desulphurization[72]

It is difficult to reduce thiophenes by means of hydrogen and noble metal catalysts because the catalysts are poisoned. Thiophenes are, however,

70 G. Vernin, J. Metzger, and C. Parkanyi, *J. Org. Chem.*, 1975, **40**, 3183.
71 H. Kotsuki, H. Nishizawa, S. Kitagawa, M. Ochi, N. Yamasaki, K. Matsuoka, and T. Tokoroyama, *Bull. Chem. Soc. Jpn.*, 1979, **52**, 544.
72 Review: L. I. Belen'kii and Ya. L. Gol'dfarb, in ref. 51, Part 1, p. 457.

Table 6.3 Some cycloaddition reactions of thiophenes.[a]

Thiophene substituents	Reagents	Conditions	Products	Refs. and notes
None	PhC≡CPh	200–280°C, 44 h	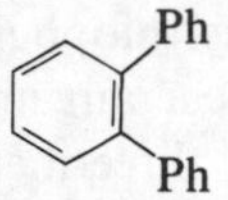	b
None (also with 2,5-Me$_2$)	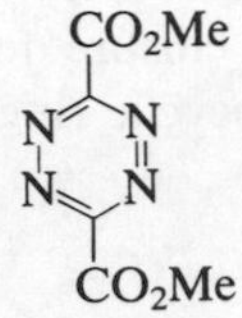	Room temperature	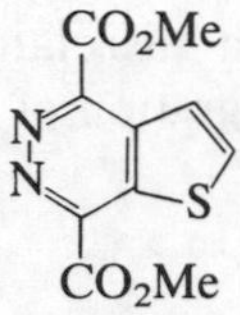	c
None	N$_2$CHCO$_2$Bu	Rh(II), 85°C	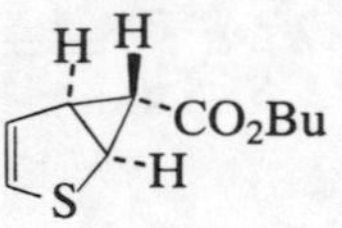	d
2,3,4,5-Me$_4$	NCC≡CCN	AlCl$_3$, 0°C	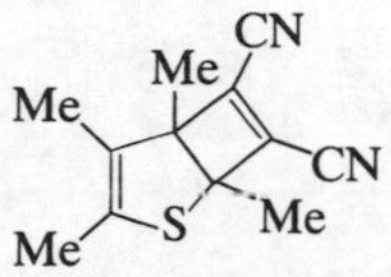	e

a Review: B. Iddon, *Heterocycles*, 1983, **20**, 1127; see also ref. 63. *b* Sulphur is lost from the primary cycloadduct: H. J. Kuhn and K. Gollnick, *Chem. Ber.*, 1973, **106**, 674. *c* G. Seitz and T. Kämpchen, *Arch. Pharm.*, 1978, **311**, 728. *d* R. J. Gillespie and A. E. A. Porter, *J. Chem. Soc., Perkin Trans. 1*, 1979, 2624. *e* R. H. Hall, H. J. den Hertog, and D. N. Reinhoudt, *J. Org. Chem.*, 1982, **47**, 967.

reductively cleaved by Raney nickel in an inert solvent, and the sulphur is removed from the molecule. Because thiophenes with a variety of substitution patterns are readily available, this has proved to be a very useful method of synthesizing saturated carbon compounds with specifically placed functional groups (Fig. 6.21). An application to the synthesis of a large ring ketone is shown.

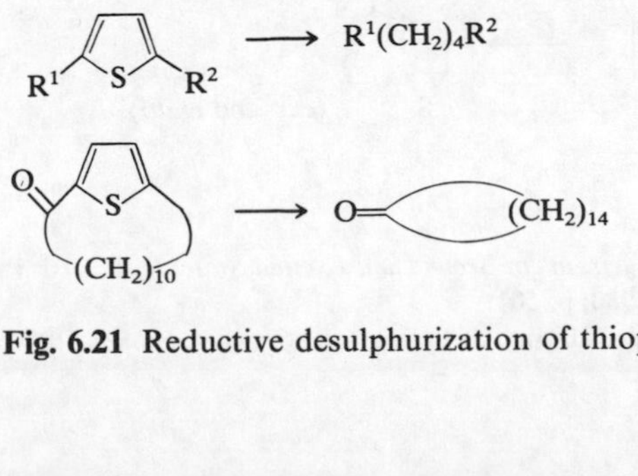

Fig. 6.21 Reductive desulphurization of thiophenes.

6.4.7   *Photochemical isomerization*

Thiophenes, in common with some furans, pyrroles, and other five-membered ring heterocycles, are photolabile and can rearrange when irradiated.[73] For example, 2-phenylthiophene can be converted by irradiation into a mixture of 2- and 3-phenylthiophene. Two of the mechanisms that can be involved in this type of rearrangement are shown in Fig. 6.22. Mechanism A is a contraction to a three-membered ring, followed by a ring-expansion. An analogous reaction, leading to a cyclopropane, occurs with 2-cyanofuran. Mechanism B is a disrotatory electrocyclic ring closure followed by a sigmatropic 1,3-migration of sulphur and an electrocyclic ring opening.

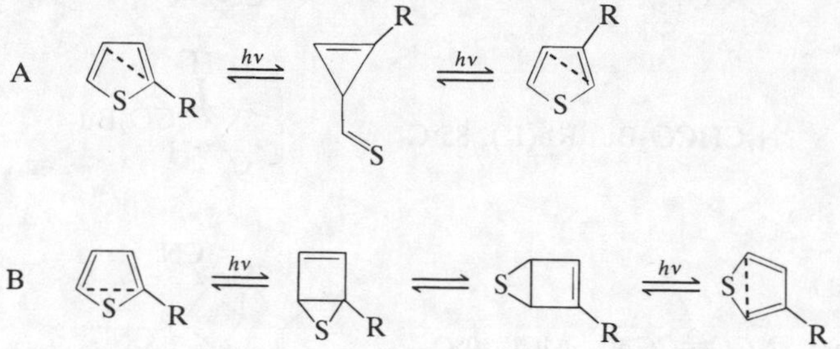

Fig. 6.22 Alternative methods of photoisomerization of thiophenes.

There is evidence for both types of intermediate in the irradiation of various thiophenes. The thioaldehyde intermediates of route A have been intercepted in low yield by reaction with amines, and the bicyclic sulphides of route B have been trapped in cycloaddition reactions. Thus, when thiophene-3-carbonitrile was briefly irradiated in furan, the Diels–Alder adducts of the valence isomer **64** and furan could be isolated in good yield. The identical adducts were also obtained in lower yield, by irradiating thiophene-2-carbonitrile in furan.[74]

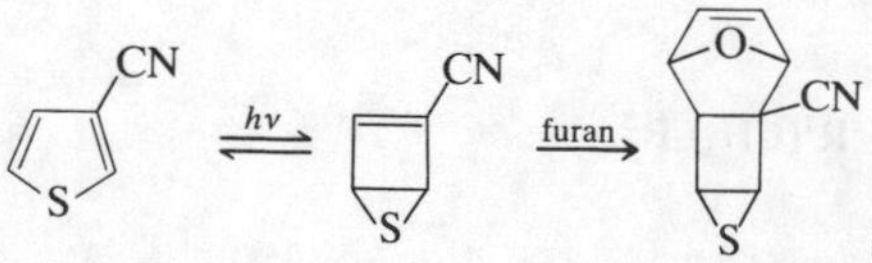

73 Review: A. Padwa, in *Rearrangements in Ground and Excited States* Vol. 3, ed. P. de Mayo, Academic Press, New York, 1980, p. 501.

74 J. A. Barltrop, A. C. Day, and E. Irving, *J. Chem. Soc., Chem. Commun.*, 1979, 881, 966.

6.4.8 Properties of some substituted thiophenes

As thiophene is much more aromatic in character than furan, and not so strongly π-electron-releasing as pyrrole, substituents on the thiophene ring have more of the character of their benzenoid analogues than in the other two heterocycles. Thus, thiophene-2-carboxaldehyde resembles benzaldehyde in most of its chemistry, and thiophene-2-carboxylic acid resembles benzoic acid, although its is somewhat stronger as an acid ($pK_a = 3.53$). 2-(Chloromethyl)thiophene, although more reactive than benzyl chloride, is not so unstable as the corresponding pyrrole and furan; it forms a Grignard derivative and undergoes a variety of nucleophilic displacement reactions. Simple hydroxy and amino derivatives are highly reactive and easily oxidized, as in the pyrrole and furan series. 2-Hydroxy compounds exist as keto tautomers but simple 3-hydroxythiophenes exist mainly in the enolic form (Section 2.5). If an adjacent electron-withdrawing group is present in the ring, these compounds are much more stable and, for example, the amino compounds can then be diazotized, and the diazonium salt coupled, as for benzenoid aromatic amines.

6.5 Indoles and related compounds[75]

6.5.1 Introduction

The investigation of the chemistry of indoles has been, and continues to be, one of the most active areas of heterocyclic chemistry. The indole unit occurs naturally in a wide variety of structures—there are well over a thousand indole alkaloids known—and many of these naturally occurring compounds have important physiological activity.[76] Some natural indoles are simple monosubstituted derivatives, such as indole-3-acetic acid, which is a plant growth regulator. Most indole alkaloids are derived from the amino acid (S)-tryptophan, **65**. Some related naturally occurring indoles are tryptamine, **66**, serotonin, **67**, and the NN-dimethylamines **68–70**, all of which are hallucinogenic alkaloids. These compounds are discussed in Chapter 1.

65; $R^1 = CO_2H$, $R^2 = H$
66; $R^1 = R^2 = H$
67; $R^1 = H$, $R^2 = OH$

68; $R^1 = H$, $R^2 = OH$
69; $R^1 = OH$, $R^2 = H$
70; $R^1 = H$, $R^2 = OMe$

75 Reviews: R. J. Sundberg, *The Chemistry of Indoles*, Academic Press, New York, 1970; *Indoles*, ed. W. J. Houlihan, Parts I–III, Wiley-Interscience, New York, 1972, 1979.

76 For a description of the main classes of indole alkaloids and their biosynthesis see (a) D. R. Dalton, *The Alkaloids: the Fundamental Chemistry*, Marcel Dekker, New York, 1979; (b) G. A. Cordell, *Introduction to Alkaloids*, Wiley-Interscience, New York, 1981.

Most of the important indole alkaloids have much more complex structures than these simple tryptamine derivatives, but the aminoethyl side chain of tryptophan is still discernible in their structure. The ergot alkaloids provide an illustration. Ergotamine, **71**, is a tetracyclic indole alkaloid with a complex peptide-based side-chain unit. This alkaloid is extracted from a fungus that infects rye, and it was the cause of sporadic outbreaks of poisoning in the Middle Ages and earlier periods. It is a potent vasoconstrictor and is used, as its tartrate salt, to treat migraine. On hydrolysis it gives lysergic acid (R = OH in structure **71**). The potent hallucinogen LSD is the *NN*-diethylamide of lysergic acid (**71**; R = NEt$_2$). It is a synthetic derivative.

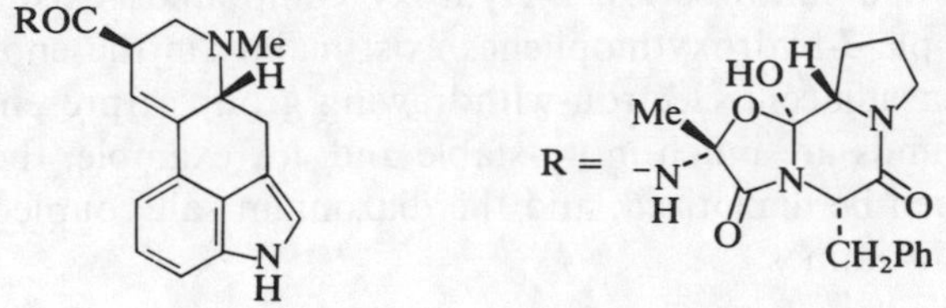

71

A group of alkaloids with complex bis-indole structures have inspired a great deal of research work because of their antileukaemic properties. These alkaloids occur in the plant *Catharanthus roseus*, a native of Madagascar which is now widely cultivated. The most important is probably leucocristine, **72**, which is used to treat leukaemia in children, and other cancers. Several other types of indole alkaloid have useful biological activity. These include the mitomycin antibiotics (see structure **1** in Chapter 9), reserpine, strychnine, and yohimbine.[76]

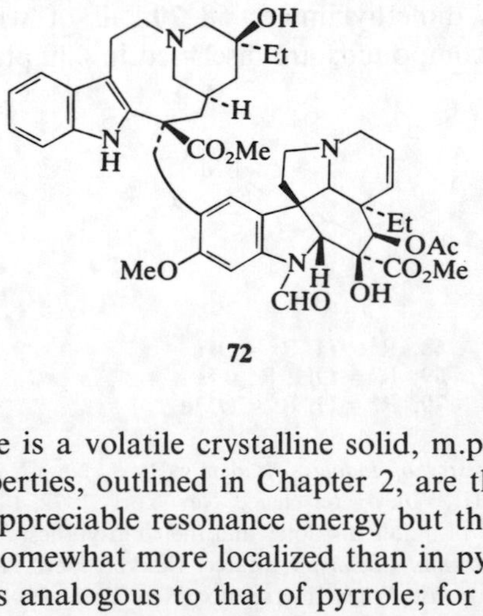

72

Indole is a volatile crystalline solid, m.p. 52°C, with a persistent odour. Its properties, outlined in Chapter 2, are those of an aromatic compound. It has appreciable resonance energy but the bonding in the five-membered ring is somewhat more localized than in pyrrole. Much of the chemistry of indole is analogous to that of pyrrole; for example, the compound is very weakly basic and the most stable cation is formed by protonation at carbon. Electrophiles normally attack the five-membered ring rather than the six-membered ring. The most important difference between the two ring systems is that, because of the greater localization of the N—C—C unit in the five-membered ring of indole, electrophiles attack at C-3 rather than at C-2. This point is developed further in Section 6.5.4.

6.5.2 Ring synthesis

Simple 3-substituted indoles are usually most easily prepared by substitution reactions on an existing indole nucleus, but for indoles with other substitution patterns it is often necessary to synthesize the ring system. All the important methods for doing this are cyclization reactions. As with other benzo-fused heterocycles (Section 4.2.1) there are two general approaches to their synthesis, one based on a benzenoid precursor with a single (nitrogen) substituent and a free *ortho* position, and the other on a precursor with adjacent carbon and nitrogen substituents. Several examples of indole ring synthesis of the second type are outlined in Chapter 4; these are intramolecular condensation (Table 4.4) and routes based on palladium catalysed cyclization (Fig. 4.10), isonitrile cyclization (Table 4.8) and nitrene cyclization (Table 4.10). Indole itself can be prepared commercially by high-temperature cyclization and dehydrogenation of 2-aminoethylbenzene or 2-aminostyrene. The most widely used indole syntheses are, however, of the first type, in which the ring is formed by the cylization of the side chain of a nitrogen-substituted benzene derivative onto a free *ortho* position. The Fischer indole synthesis, first described over a century ago, is still the most important and versatile method of this type.

Fischer synthesis[77]

This synthesis is the cyclization of arylhydrazones by heating, usually with an acid or Lewis acid catalyst. The reaction is illustrated in Fig. 6.23 for the cyclization of ketone phenylhydrazones; it is also possible to carry out the reaction with substituted phenylhydrazones, and with arylhydrazones of aldehydes, ketoacids, ketoesters, and diketones.

The most commonly used catalyst for the reaction is zinc chloride, which is often used in stoichiometric or greater amounts in a solvent at 150°C or higher temperatures, but the choice of catalyst, solvent, and temperature for optimum yields of indoles depends very much on the structure of the substrate. With phosphorus(III) chloride as catalyst, ketone phenylhydrazones can be converted into 2,3-disubstituted indoles at room temperature.[78] Indole itself is not isolated from the reaction of acetaldehyde phenylhydrazone and zinc

77 Review: B. Robinson, *The Fischer Indole Synthesis*, Wiley, Chichester, 1982.
78 G. Baccolini and P. E. Todesco, *J. Chem. Soc.*, *Chem. Commun.*, 1981, 563.

chloride in solution, but it can be obtained in moderate yield by passing the vapour over glass beads coated with the catalyst.[79] The Fischer synthesis can be carried out by heating arylhydrazones in the absence of a catalyst;[77] some arylhydrazones are even converted into indoles on attempted vapour phase chromatographic analysis.[79]

One disadvantage of the method is that unsymmetrical ketones can give mixtures of indoles if the substituent R^2 also has an α-methylene group. Arylhydrazones of methyl alkyl ketones (R^1 = alkyl, R^2 = methyl) normally give 2-methylindoles; that is, cyclization takes place at the more substituted alkyl group. In other cases the direction of cyclization can often be controlled by the choice of catalyst.

The synthesis is a simple and practicable one but the mechanism is not so straightforward, since one of the two nitrogen atoms of the hydrazone is lost. It has been established that the nitrogen atom which is retained is the one attached to the aryl group in the hydrazone. After extensive work, the sequence shown in Fig. 6.23 for the acid-catalysed reaction has considerable experimental support. The key substitution step can be regarded as an internal electrophilic substitution, or, perhaps better, as a [3,3] sigmatropic shift in which a weak (N—N) bond is broken and a strong (C—C) bond is created.

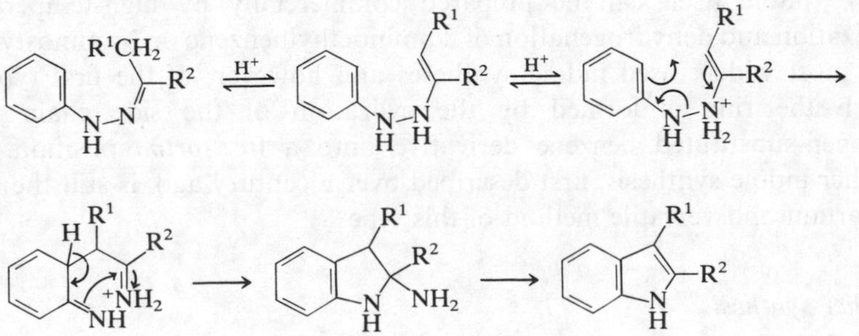

Fig. 6.23 Mechanism of the Fischer indole synthesis.

Synthesis from anilines and β-ketosulphides[80]

Another method based on *ortho* cyclization, which is of much more recent origin than the Fischer method, is shown schematically in Fig. 6.24. The starting material is an aniline, which is N-chlorinated and then converted to a sulphonium salt by reaction with an α-ketosulphide, which provides the

79 M. Nakazaki and K. Yamamoto, *J. Org. Chem.*, 1976, **41**, 1877.
80 P. G. Gassman, T. J. van Bergen, D. P. Gilbert, and B. W. Cue, *J. Am. Chem. Soc.*, 1974, **96**, 5495.

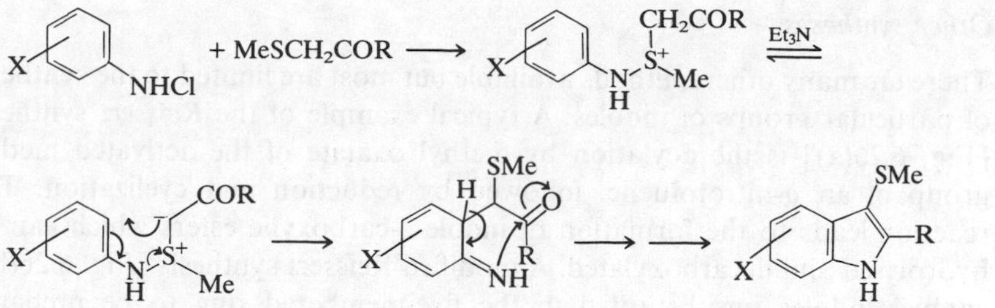

Fig. 6.24 Indoles by sulphonium ylide cyclization.

additional carbon atoms needed to construct the indole system. The cyclization step bears some relationship to that of the Fischer synthesis in that it is a sigmatropic rearrangement in which a weak (N—S) bond is broken and a strong (C—C) bond is made. This type of sulphonium ylide rearrangement is classified as a [3,2] sigmatropic shift. The methylthio group can be removed from the indoles by reaction with Raney nickel. 2,3-disubstituted indoles can also be made by a variant of the method.

Bischler synthesis

This reaction is essentially an acid-catalysed cyclization of an α-(arylamino)-ketone. The reaction is, however, complicated by the fact that in the presence of an excess of the amine and an acid catalyst the arylaminoketone can isomerize before cyclization. This can lead to mixtures of indoles unless the substituents R^1 and R^2 are identical (Fig. 6.25) or unless the equilibrium lies strongly to one side. An example of the latter is the cyclization of 2-(phenylamino)acetophenone ($R^1 = Ph$, $R^2 = H$) which, when heated in the presence of aniline and HBr, gives 2-phenylindole rather than 3-phenylindole.

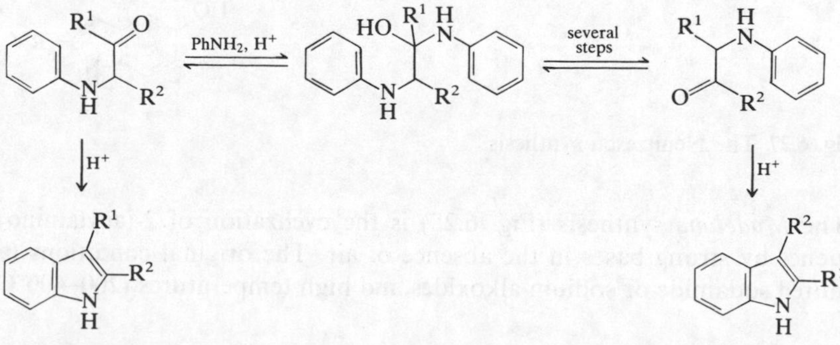

Fig. 6.25 Isomeric indoles from the Bischler synthesis.

Other syntheses

There are many other methods available but most are limited to the synthesis of particular groups of indoles. A typical example of the *Reissert* synthesis [Fig. 6.26(a)] is the acylation by diethyl oxalate of the activated methyl group of an *o*-nitrotoluene, followed by reduction and cyclization. The reaction leads to the formation of indole-2-carboxylic esters which can be hydrolysed and decarboxylated. A modified Reissert synthesis [Fig. 6.26(b)] enables indoles unsubstituted in the five-membered ring to be prepared directly.[81] The reaction shown in Table 4.4 (iii) is another variant of the Reissert synthesis.

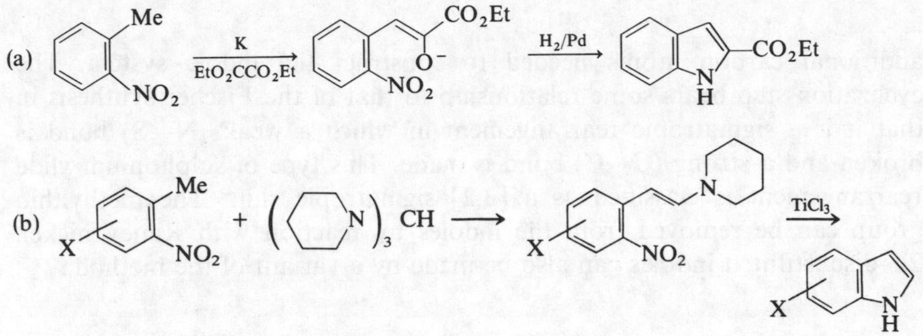

Fig. 6.26 The Reissert synthesis.

The *Nenitzescu* synthesis is useful specifically for the preparation of 5-hydroxyindoles, some of which have biological activity (see structures **67** and **68**). The reaction is a conjugate addition of a vinylogous primary or secondary amide to benzoquinone, followed by cyclization (Fig. 6.27).

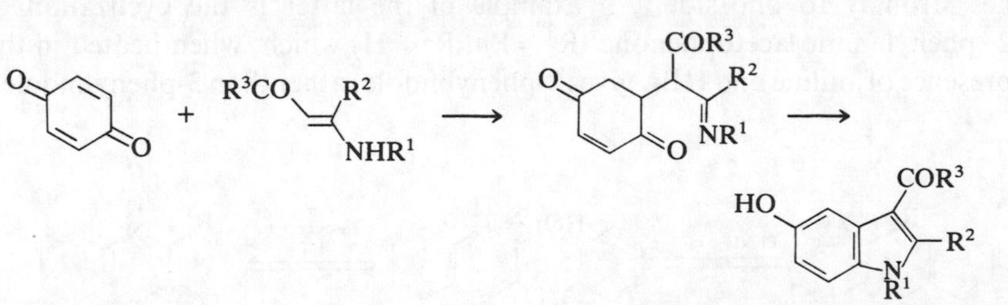

Fig. 6.27 The Nenitzescu synthesis.

The *Madelung* synthesis (Fig. 6.28) is the cyclization of 2-(acylamino)-toluenes by strong bases in the absence of air. The original conditions (a) required sodamide or sodium alkoxides and high temperatures (200–400°C)

81 D. H. Lloyd and D. E. Nichols, *Tetrahedron Lett.*, 1983, **24**, 4561; W. Haefliger and H. Knecht, *Tetrahedron Lett.*, 1984, **25**, 285.

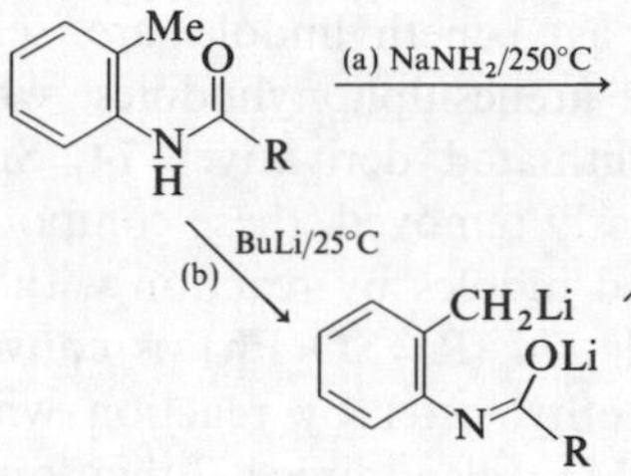

Fig. 6.28 The Madelung synthesis.

so the use of the method was limited to the synthesis of simple indoles such as 2-methylindole. A modern variant (b) is performed at room temperature with butyllithium (2–3 mol) as the base:[82] 2-substituted indoles bearing other groups can be synthesized in good yield by this method. The route to indoles based on isonitrile cyclization which is illustrated in Table 4.8 is related to the Madelung synthesis.

6.5.3   *Acidity: metallated indoles*

Indole is a weak acid ($pK_a = 16.97$) of comparable strength with pyrrole and with aliphatic alcohols. It can be converted into the *N*-sodio derivative by reaction with sodamide in liquid ammonia, or by reaction with sodium hydride in an organic solvent. Salts of other metals can be made by using the appropriate strong bases, such as potassium t-butoxide, Grignard reagents, and butyllithium.

The *N*-metallated indoles are ambident nucleophiles and can react with electrophiles either at nitrogen or at C-3. The more ionic sodium and potassium salts tend to react preferentially at nitrogen, particularly with 'hard' electrophiles. Substitution at nitrogen is also favoured by the use of dipolar aprotic solvents. Thus, the sodium salt of indole can be *N*-methylated with bromomethane in tetrahydrofuran, and it can be *N*-tosylated in dimethylformamide. On the other hand, 1-indolylmagnesium halides are preferentially alkylated and acylated at C-3. A typical example is the preparation of indole-3-ethanol, **73**, from 1-indolylmagnesium bromide and oxirane. Very reactive alkylating agents such as ethyl chloroformate give both 1- and 3-substituted indoles.

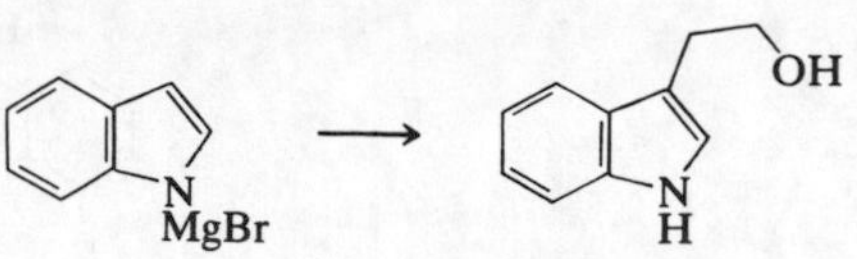

73

82 W. J. Houlihan, V. A. Parrino, and Y. Uike, *J. Org. Chem.*, 1981, **46**, 4511.

Simple *N*-substituted indoles such as 1-methylindole are selectively lithiated at C-2. This also applies to 1-arenesulphonylindoles, which are converted by t-butyllithium into the lithiated derivatives **74**. Since the arenesulphonyl groups can be subsequently removed, these compounds are very useful for preparing C-2 substituted indoles by reaction with electrophiles. For example, the lithiated indole, **74** (R = SO₂Ph), is converted in good yield into the 2-carboxylic acid ethyl ester by reaction with ethyl chloroformate. A 1-substituent CH₂NMe₂ also directs lithiation to the 2-position, and indoles **74** (R = CH₂NMe₂) are readily prepared by reaction with butyllithium.[83] The 1-substituent can subsequently be removed by reaction with sodium borohydride.

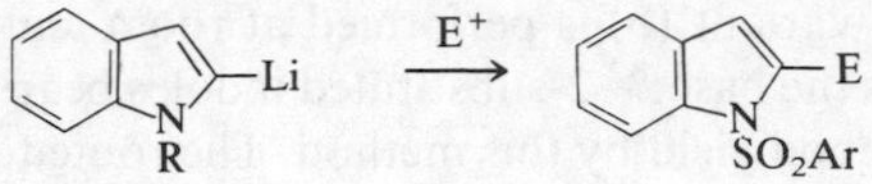

74

6.5.4   *Reaction with electrophiles*

Indole is a nucleophilic heterocycle and it reacts very easily with electrophiles. The preferred site for electrophilic substitution is C-3 rather than C-2, in contrast to pyrrole. The cation formed by attack at C-3 is more stable than that formed by attack at C-2, because the positive charge can be delocalized without involving the benzenoid part of the molecule (Fig. 6.29). If we assume that the transition states for electrophilic substitution resemble these intermediates, the preference for C-3 substitution can be accounted for. The highest occupied π-orbital of indole also has greater electron density at C-3 than at C-2, so that even in cases of electrophilic attack with an 'early' transition state, attack at C-3 is to be expected. The preferred site of reaction of indoles with electrophiles is thus the same as for simple enamines: β to the nitrogen atom.

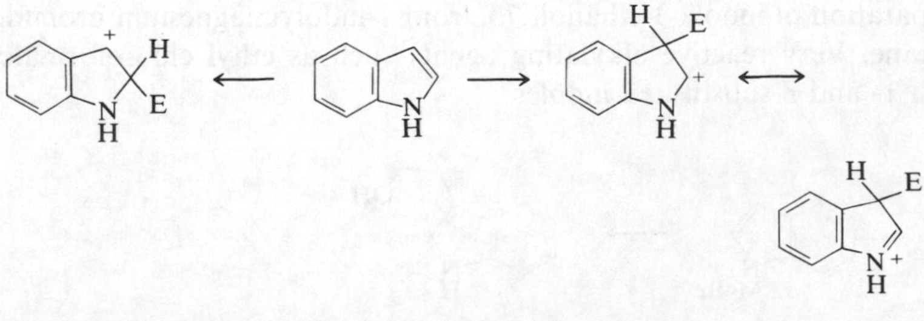

Fig. 6.29 Intermediates formed by electrophilic attack at C-2 and at C-3.

83 A. R. Katritzky, P. Lue, and Y.-X. Chen, *J. Org. Chem.*, 1990, **55**, 3688.

Table 6.4 Electrophilic substitution of indole.[a]

Functional group introduced	Reagents and conditions	Refs. and notes
NO_2	$PhCONO_2$, 0°C	b
Br	NBS, CCl_4, 80°C	c
Cl	NCS, MeOH, 20°C	d
CHO	$POCl_3$, Me_2NCHO, 20–35°C	e
COMe	$(MeCO)_2O$, heat	f
$CH_2CH_2NO_2$	$CH_2{=}CHNO_2$, 0–20°C	g
CH_2CH_2COMe	$CH_2{=}CHCOMe$, $MeCO_2H$, $(MeCO)_2O$, 100°C	h
$CH_2C{=}NOH.CO_2Et$	$BrCH_2C{=}NOH.CO_2Et$, Na_2CO_3, 20°C	i
CH_2NMe_2	CH_2O, Me_2NH, $MeCO_2H$, 20°C	j
$N{=}NPh$	$PhN_2^+Cl^-$, KOH aq., 0°C	k
SO_3H	SO_3–pyridine, heat	l

a All give 3-substituted indoles in good yield, except where indicated. b G. Berti, A. Da Settimo, and E. Nannipieri, *J. Chem. Soc. (C)*, 1968, 2145; 35% yield. c M. Mousseron-Canet and J.-P. Boca, *Bull. Soc. Chim. Fr.*, 1967, 1294. d Reaction carried out on 2-methylindole; J. C. Powers, *J. Org. Chem.*, 1966, **31**, 2627. 3-Chloroindole is conveniently prepared by *N*-chlorination of indole with sodium hypochlorite, followed by base-catalysed thermal rearrangement; M. De Rosa and J. L. T. Alonso, *J. Org. Chem.*, 1978, **43**, 2639. e G. F. Smith, *J. Chem. Soc.*, 1954, 3842. f 3-Acetylindole is obtained in good yield if an antioxidant is added to the reaction mixture; G. Hart, D. R. Liljegren, and K. T. Potts, *J. Chem. Soc.*, 1961, 4267. g D. Ranganathan, C. B. Rao, S. Ranganathan, A. K. Mehrotra, and R. Iyengar, *J. Org. Chem.*, 1980, **45**, 1185. h J. Szmuszkovicz, *J. Am. Chem. Soc.*, 1957, **79**, 2819. i The electrophile is $CH_2{=}C(NO)CO_2Et$; T. L. Gilchrist and T. G. Roberts, *J. Chem. Soc., Perkin Trans. 1*, 1983, 1283. j The Mannich reaction; H. Kuhn and O. Stein, *Chem. Ber.*, 1937, **70**, 567. k Yields of coupled products are low with indole itself but good with 2-alkylindoles; B. C. Challis and H. S. Rzepa, *J. Chem. Soc., Perkin Trans. 2*, 1975, 1209. l G. F. Smith and D. A. Taylor, *Tetrahedron*, 1973, **29**, 669.

Indole is a very weak base ($pK_a = -3.63$) and the most stable cation results from protonation at C-3 rather than at nitrogen. The nitrogen atom is easily protonated, even in neutral water, so hydrogen–deuterium exchange takes place readily, but the reaction is reversible and the equilibrium lies on the side of the unprotonated form. Indoles with a 3-alkyl substituent are also protonated preferentially at the substituted 3-position.

Examples of other reactions with electrophiles (using indole itself as the substrate) which give 3-substituted indoles are shown in Table 6.4. These illustrate the ease with which electrophilic substitution takes place at C-3; thus, indole is halogenated and acetylated without a Lewis acid catalyst, it undergoes the Vilsmeier and Mannich reactions with ease, and it can be coupled to arenediazonium salts.

Not all reactions of indoles with electrophiles lead simply to substitution at the 3-position. Many of the important indoles already carry a substituent at the 3-position, and their reaction with electrophiles is necessarily more complicated. With these indoles the nature of the substitution product is dependent upon the nature of the electrophile and the reaction conditions. Thus, 3-methylindole is formylated at nitrogen, but sulphonation, diazo coupling, and acetylation (in the presence of a Lewis acid) result mainly in 2-substitution. Products of 2-substitution are in many instances formed by initial attack of the electrophile at C-3, followed by a 1,2-shift in the intermediate cation. Two examples of such reactions of 3-methylindole, for which there is good evidence for initial substitution at C-3, are shown in Fig. 6.30.

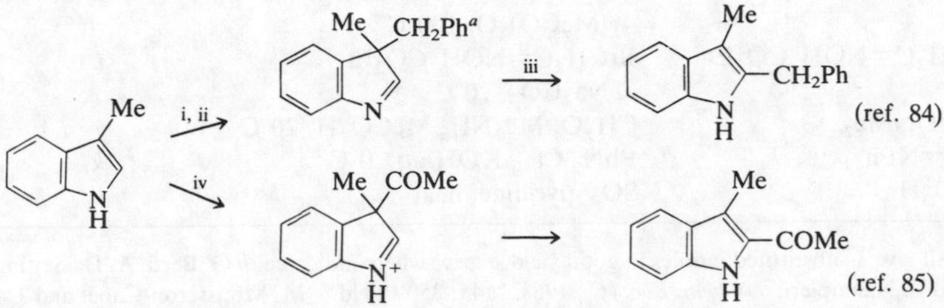

Fig. 6.30 Electrophilic substitution at C-2 by attack at C-3 followed by rearrangement. Reagents: i, EtMgBr; ii, PhCH$_2$Br; iii, HCl, iv, (MeCO)$_2$O, BF$_3$-ether. ^{a}Isolable intermediate.

Electrophilic substitution of indoles in strongly acidic media can also lead to products other than those resulting from 3-substitution. This is because the indole is protonated at C-3 in the strong acid and the heterocyclic ring is thus deactivated to further attack. For example, 2-methylindole is nitrated in the six-membered ring (at C-5) rather than at C-3 by concentrated nitric acid in sulphuric acid.[86]

6.5.5 *Oxidation and reduction*

Indoles are easily oxidized by air and by a wide variety of other oxidants. The reactions are often complex and in many cases lead to the cleavage of the five-membered ring. The primary site of attack by the oxidant is usually C-3. Indole itself is first oxidized by air and light to indoxyl, **75**, which then reacts further, by radical coupling, to give the dimer **76** and then the dyestuff indigo (see Section 6.5.6) among other products. The radical formed by

84 A. H. Jackson and P. Smith, *Tetrahedron*, 1968, **24**, 2227.
85 A. H. Jackson, B. Naidoo, A. E. Smith, A. S. Bailey, and M. H. Vandrevala, *J. Chem. Soc., Chem. Commun.*, 1978, 779.
86 W. E. Noland, L. R. Smith, and D. C. Johnson, *J. Org. Chem.*, 1963, **28**, 2262.

hydrogen abstraction from C-2 of indoxyl, **75**, is a stabilized one[87] and can dimerize or can add to further molecules of indole. 3-Substituted indoles react with oxygen at the 3-position to give peroxides, **77**, which can then react further in a number of ways. Singlet oxygen (produced by dye-sensitized irradiation) forms the same type of peroxide but probably by electrophilic attack at C-3, since the zwitterionic intermediate **78** to be expected for such an attack on 1,3-dimethylindole has been detected.[88]

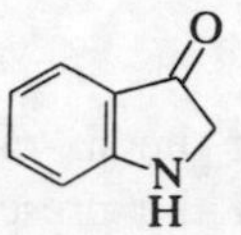

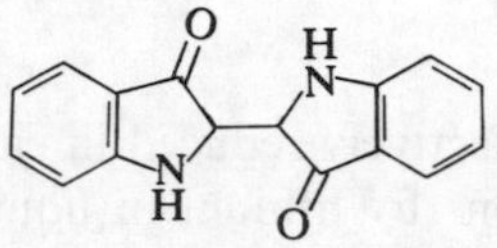

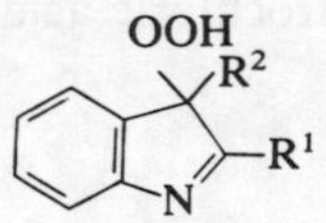

As an illustration of the further reactions that can take place after photooxygenation, the degradation of tryptophan to formylkynurenine, **79**, is shown in Fig. 6.31.[89] Tryptophan is degraded to formylkynurenine *in vivo*, probably by a similar route.

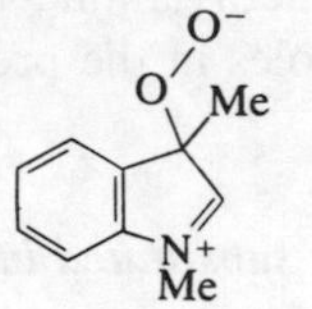

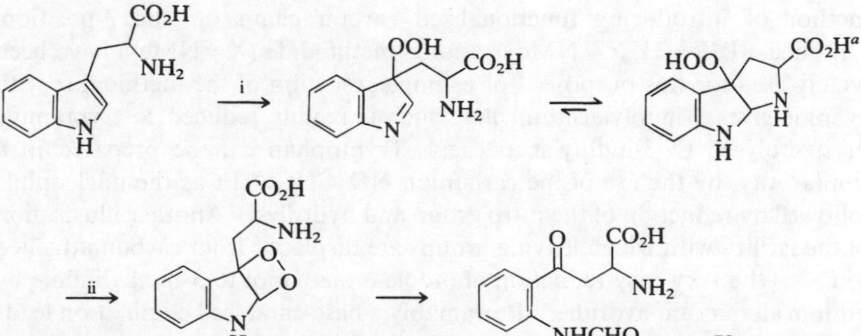

Fig. 6.31 Oxidation of tryptophan. Reagents: i, O_2, sens. *hv*; ii, Na_2CO_3 aq. [a] Isolable intermediate.

87 H. G. Viehe, R. Merényi, L. Stella, and Z. Janousek, *Angew. Chem. Int. Edn Engl.*, 1979, **18**, 917.
88 I. Saito, S. Matsugo, and T. Matsuura, *J. Am. Chem. Soc.*, 1979, **101**, 7332.
89 M. Nakagawa, S. Kato, S. Kataoka, and T. Hino, *J. Am. Chem. Soc.*, 1979, **101**, 3136.

A good method for the oxidation of tryptophans and other 3-substituted indoles at the 2-position involves the use of a mixture of concentrated hydrochloric acid and dimethyl sulphoxide. Oxindoles, **80**, are formed.[90]

80

Indoles can be selectively reduced in either the five- or the six-membered ring. Birch reduction (by lithium in liquid ammonia in the presence of a source of protons) gives 4,7-dihydroindoles. The five-membered ring is reduced in acidic media (probably by way of the cation) by a number of reagents. For example, trimethylamine–borane complex can be used to reduce indoles to 2,3-dihydroindoles in the presence of hydrochloric acid.[91]

6.5.6 Properties of some substituted indoles

The electron-releasing character of the indole nucleus affects the properties of substituents on the five-membered ring, particularly those at the 3-position, in much the same way as for the other five-membered heterocycles described earlier. For example, the reactivity of the carbonyl group in indole-3-carboxaldehyde is reduced by electron release from the ring system, but the departure of leaving groups on carbon α- to the ring (as in structure **81**) is accelerated. The latter type of interaction is important since it provides a method of introducing functionalized carbon chains at the 3-position. Gramine, **81** ($R = H$, $X = NMe_2$), and its methiodide ($X = \overset{+}{N}Me_3$) have been widely used for this purpose. For example, reaction of the methiodide with cyanide gives 3-indolylacetonitrile, which is readily reduced to tryptamine or hydrolysed to 3-indolylacetic acid. Tryptophan can be prepared in a similar way, by the use of the carbanion $NO_2\bar{C}HCO_2Et$ as the nucleophile followed by reduction of the nitro group and hydrolysis. Another illustration of the facility with which leaving groups are displaced from carbon attached to C-3 is the very easy reduction of indole-3-methanol to 3-methylindole by lithium aluminium hydride.[92] Presumably a base-catalysed elimination leads to the formation of an intermediate, **82**, which is then intercepted by hydride. This is consistent with the fact that when the nitrogen atom bears a methyl group, and the intermediate, **82**, cannot be formed, the alcohol is not reduced.[92]

90 W. E. Savige and E. Fontana, *J. Chem. Soc., Chem. Commun.*, 1976, 599.
91 J. G. Berger, *Synthesis*, 1974, 508.
92 E. Leete and L. Marion, *Can. J. Chem.*, 1953, **31**, 775.

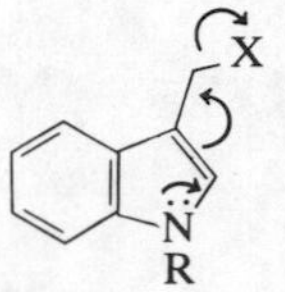

81

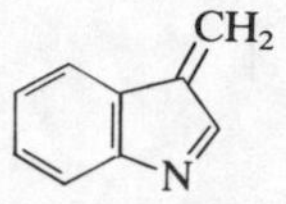

82

A different type of activation is shown in the increased acidity of the protons of 2-alkylindoles. The 2-methyl group in 2,3-dimethylindole selectively undergoes acid-catalysed deuterium exchange; this and other reactions are consistent with the existence of a tautomeric equilibrium 83⇌84 for 2-methylindoles. A remarkable example of the lability of the protons in the 2-methyl group is provided by the acylation of the imine 85. A transient diene 86 is generated, which then undergoes intramolecular Diels–Alder cycloaddition.[93] This reaction has provided an important route to complex indole alkaloids.

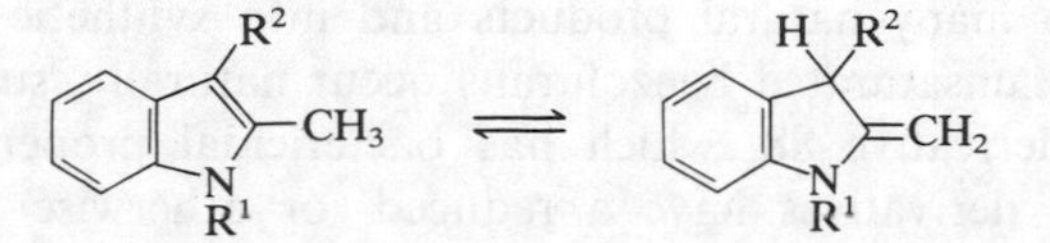

83

84

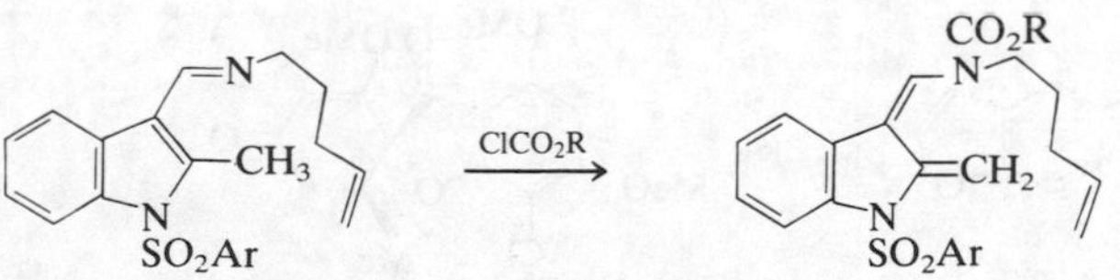

85

86

Indole 2- and 3-carboxylic acids are both decarboxylated by heating in acid solution; again, protonation at C-3 assists the reaction.

2- and 3-hydroxyindoles exist predominantly in the keto forms 80 and 75. The protons on the carbon atoms adjacent to the carbonyl groups, are, of course, very acidic. The instability of indoxyl 75 has been referred to earlier because of its tendency to form a radical by hydrogen abstraction. Oxidation of the radical dimer 76 is also very easy and it results in the formation of the insoluble blue dyestuff, indigo, 87. This is applied to the fabric in the form of the water-soluble sodium salt of the reduced species 76, which is then oxidized to indigo by exposure to the air. The colour of indigo

93 C. Exon, T. Gallagher, and P. Magnus, *J. Am. Chem. Soc.*, 1983, **105**, 4739.

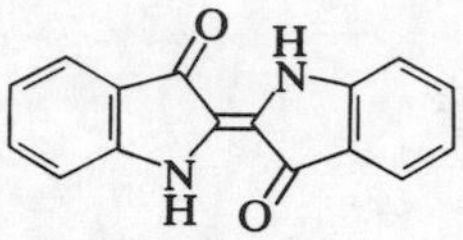

87

is due to the presence of the double vinylogous amide function, and is not much dependent upon the benzene rings or their substituents.[94] Derivatives of the indigo system occur naturally and have been used since ancient times as blue and purple dyes.

6.6   Other benzo[*b*]-fused heterocycles

6.6.1   Benzo[b]furans[95]

The benzofuran ring system, although by no means as common as indole, is incorporated into many natural products and into synthetic pharmaceuticals. Some fully unsaturated benzofurans occur naturally, such as the simple 5-methoxy derivative **88**, which has bactericidal properties. The majority of natural derivatives have a reduced, or otherwise modified, five-membered ring; an important example is the antifungal antibiotic, griseofulvin, **89**.[96]

88

89

Benzofuran occurs in coal tar but is prepared on a semicommercial scale by vapour-phase dehydrogenation and cyclization of 2-ethylphenol. Simple alkylbenzofurans can be made in an analogous way. Cyclization reactions of various types have been used to produce substituted benzofurans: one such reaction is outlined in Table 4.4. These reactions are

94 J. Griffiths, *Colour and Constitution of Organic Molecules*, Academic Press, London, 1976, p. 195.
95 Reviews: A. Mustafa, *Benzofurans*, Wiley-Interscience, New York, 1974; P. Cagniant and D. Cagniant, *Adv. Heterocycl. Chem.*, 1975, **18**, 337; R. Livingstone, in *Rodd's Chemistry of Carbon Compounds*, Vol. IVA, ed. S. Coffey, Elsevier, Amsterdam, 1973, p. 141.
96 Reviews: J. F. Grove, in *Progress in the Chemistry of Organic Natural Products*, ed. L. Zechmeister, Springer-Verlag, Vienna, 1964, p. 203; J. W. Corcoran and F. E. Hahn, in *Antibiotics* Vol. 3, Springer-Verlag, New York, 1975, p. 606.

often used to produce functionalized benzofurans because ring synthesis provides an easier route than electrophilic substitution of the parent heterocycle; for example, 2-acetylbenzofuran, **90**, is conveniently prepared from 2-hydroxybenzaldehyde and chloroacetone.[97]

90

The aromatic character of the five-membered ring is weak and the ring is readily cleaved by oxidizing and reducing agents; it is polymerized by concentrated mineral acids and by Lewis acids. When electrophilic substitution does take place, there is a preference for 2- over 3-substitution, in contrast to indole: evidently the oxygen atom provides insufficient stabilization to the $3H$-cation. Benzofuran undergoes the Vilsmeier reaction to give the 2-carboxaldehyde in moderate yield, and it can be nitrated at the 2-position by nitric acid in acetic acid. Benzofuran reacts with halogens to give 2,3-addition products which can then give mixtures of 2- and 3-halobenzofurans by loss of hydrogen halide. Benzofuran, like furan, can be specifically lithiated at the 2-position[10] and this provides an alternative, mild method for the introduction of electrophiles at the 2-position. 3-Lithiobenzofuran, **91**, can also be prepared, by halogen–metal exchange from 3-halobenzofurans, but it is thermally unstable and is cleaved to 2-hydroxyphenylacetylene. This type of cleavage is typical of nonaromatic β-lithiated enol ethers.

91

The 2,3-bond is susceptible to cycloaddition reactions. Photochemical [2+2] additions are the most important of these. An example of such a reaction occurs with the benzofuran **92**. This is used in ultraviolet treatment of psoriasis, a disease which is associated with overproduction of DNA. It has been shown to add to thymine, one of the nucleic acid bases, on irradiation to form a [2+2] adduct across the double bond of the five-membered ring. This provides an indication of its probable mechanism of action *in vivo*.[98]

97 E. D. Elliott, *J. Am. Chem. Soc.*, 1951, **73**, 754.
98 E. J. Land, F. A. P. Rushton, R. L. Beddoes, J. M. Bruce, R. J. Cernik, S. C. Dawson, and O. S. Mills, *J. Chem. Soc., Chem. Commun.*, 1982, 22.

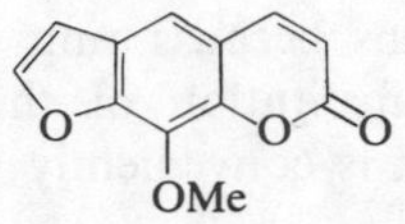

92

6.6.2   Benzo[b]thiophenes[99]

The benzothiophene ring system occurs naturally but its more important derivatives are prepared by laboratory synthesis. Derivatives are used as dyestuffs, pharmaceuticals, pesticides, and for many other purposes. Benzothiophene is a thermally stable, low-melting solid with a naphthalene-like odour which is present in coal tar. The most important methods of ring synthesis are cyclization reactions which use *ortho*-disubstituted benzenes as starting materials.[100] For example, 2-acetylbenzothiophene, **93**, can be

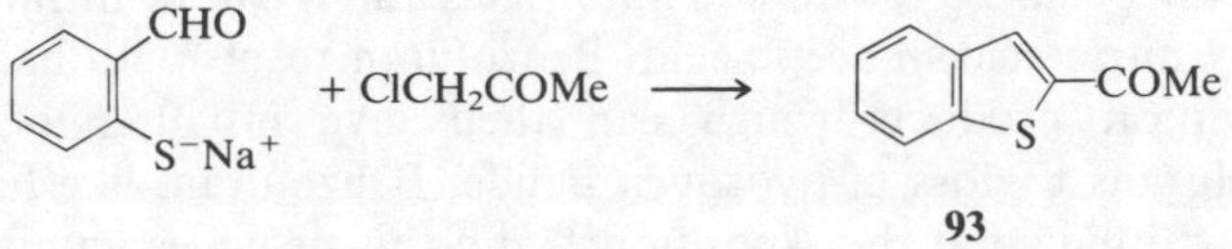

93

prepared in a manner analogous to that for the corresponding benzofuran **90**, described in the preceding section. Benzothiophenes can also be constructed from appropriate monosubstituted benzenes by intramolecular cyclization of the Friedel–Crafts type, as illustrated for the preparation of 3-methylbenzothiophene, **94**.

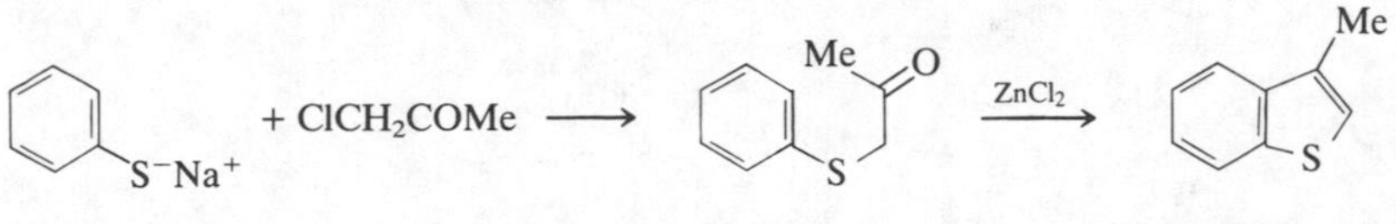

94

The five-membered ring system is more stable than that of benzofuran. Oxidation by hydrogen peroxide gives benzothiophene-*SS*-dioxide. The 2,3-bond can be reduced by sodium in alcohol. The 3-position is more reactive than the 2-position in electrophilic substitution, although both isomers are usually formed, so that in this respect benzothiophene resembles indole more

99 Reviews: B. Iddon and R. M. Scrowston, *Adv. Heterocycl. Chem.*, 1970, **11**, 177; R. M. Scrowston, *Adv. Heterocycl. Chem.*, 1981, **29**, 171.
100 B. Iddon, in *New Trends in Heterocyclic Chemistry*, ed. R. B. Mitra, Elsevier, Amsterdam, 1979, p. 250.

than benzofuran. Benzothiophene is, however, less reactive towards electrophiles than thiophene.[57,101] Nitration can be carried out using nitric acid in acetic anhydride; 3-nitrobenzothiophene is the major product. Friedel–Crafts acylations require the use of a Lewis acid catalyst and give 3-substituted derivatives. 3-Chloro and 3-bromo derivatives can be formed without the use of a catalyst. The presence of an electron-withdrawing substituent at the 3-position deactivates the five-membered ring towards electrophilic substitution, and the six-membered ring is then substituted preferentially; for example, 3-acetylbenzothiophene is nitrated to give a mixture of 4-, 5-, 6-, and 7-nitro derivatives but there is no substitution at the 2-position.[102]

Many of the other reactions of benzothiophenes are similar to those of thiophenes. The ring system is readily lithiated at the 2-position and the lithiated benzothiophenes are useful intermediates for the formation of 2-substituted derivatives. Magnesium derivatives can be formed from 3-halobenzothiophenes but the 3-lithio derivatives are thermally unstable and undergo a ring-opening reaction analogous to that of 3-lithiobenzofuran, **91**. If kept at $-80°C$, however, 3-lithiobenzothiophene can be intercepted by a range of electrophiles to give the 3-substituted derivatives in good yield.

Benzothiophene undergoes cycloaddition reactions of various types across the 2,3-bond; examples of such reactions are shown in Fig. 6.32.

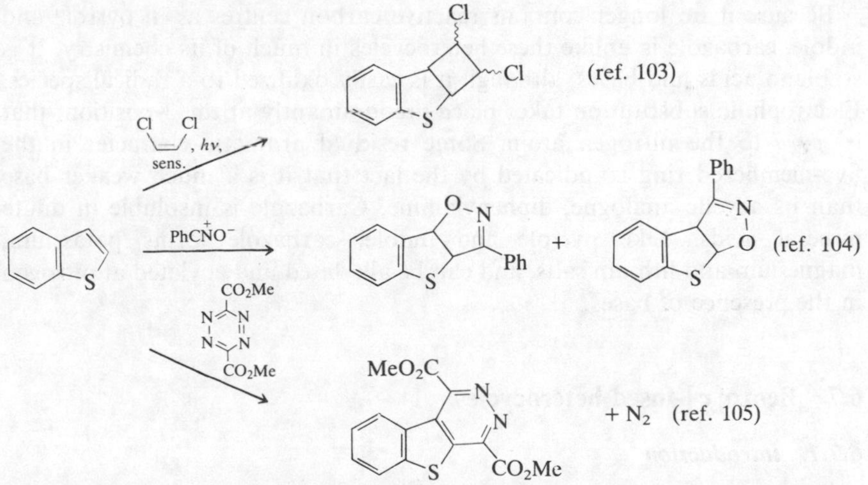

Fig. 6.32 Cycloaddition reactions of benzothiophene.

101 S. Clementi, P. Linda, and G. Marino, *J. Chem. Soc. (B)*, 1971, 79.

102 G. C. Brophy, S. Sternhell, N. M. D. Brown, I. Brown, K. J. Armstrong, and M. Martin-Smith, *J. Chem. Soc. (C)*, 1970, 733.

103 D. C. Neckers, J. H. Dopper, and H. Wynberg, *J. Org. Chem.*, 1970, **35**, 1582.

104 P. Caramella, G. Cellerino, P. Grünanger, F. M. Albini, and M. R. ReCellerino, *Tetrahedron*, 1978, **34**, 3545.

105 G. Seitz and T. Kämpchen, *Arch. Pharm.*, 1976, **309**, 679.

6.6.3   *Carbazoles*[106]

The dibenzo-fused heterocycle carbazole occurs in coal tar, and a few derivatives, including 1-methoxycarbazole-3-carboxaldehyde (murrayanine), **95**, occur as plant alkaloids.[76] Some carbazole derivatives are used as dyestuffs, and *N*-vinylcarbazole is manufactured as a monomer for plastics.

95

There are several general ring syntheses available, including the Fischer indole synthesis (Section 6.5.2) of cyclohexanone derivatives, which leads to the formation of tetrahydrocarbazoles; these are easily aromatized by heating with sulphur or palladium. The decomposition of 2-azidobiphenyls (Table 4.10) provides another method, as does the decomposition of 1-arylbenzotriazoles (the Graebe–Ullmann synthesis), which is referred to in Section 8.4.

Because it no longer contains reactive carbon centres as in pyrrole and indole, carbazole is unlike these heterocycles in much of its chemistry. It is stable to acids and bases, although it is easily oxidized to a radical species. Electrophilic substitution takes place predominantly at the 3-position; that is, *para* to the nitrogen atom. Some residual aromatic character in the five-membered ring is indicated by the fact that it is a much weaker base than its acyclic analogue, diphenylamine. Carbazole is insoluble in dilute mineral acids. Like pyrrole and indole, carbazole forms potassium, magnesium and lithium salts, and can be alkylated and acylated at nitrogen in the presence of bases.

6.7   Benzo[*c*]-fused heterocycles

6.7.1   *Introduction*

The benzo[*c*]-fused heterocycles, **96**, and their derivatives do not occur naturally but they have been prepared and studied extensively in order to investigate their structure and properties, and to determine their relationship to the other five-membered aromatic heterocycles. The parent compounds, isoindole, **96** (X = NH),[107] isobenzofuran (X = O),[108] and benzo[*c*]thiophene

106 Review: J. A. Joule, *Adv. Heterocycl. Chem.*, 1984, **35**, 83.
107 Review: R. Bonnett and S. A. North, *Adv. Heterocycl. Chem.*, 1981, **29**, 341.
108 Reviews: W. Friedrichsen, *Adv. Heterocycl. Chem.*, 1980, **26**, 135; B. Rickborn, *Adv. Theor. Interesting Mol.*, 1989, **1**, 1.

$(X=S)$,[109] are all reactive substances which can be isolated and studied only at low temperatures. They were first isolated in the period 1962–72. As Table 2.4 shows, all three have smaller resonance energies than their benzo[b]-fused analogues but they vary considerably in aromatic character. Isobenzofuran, like furan itself, is much the least aromatic of the three and is essentially a nonaromatic *ortho*-quinonoid species. Isoindole and benzo[c]thiophene retain appreciable aromatic character. This is illustrated for isoindole by the fact that it exists in solution as such rather than as a tautomer, **97**, which is normally preferred for imines, even though **97** is a conjugated benzenoid system. Isoindole is, however, kinetically unstable because it is highly electron-rich and can couple and react with oxidizing agents, in a manner typical of pyrroles unsubstituted at the 2-position, very easily. Electron-withdrawing substituents increase the kinetic stability of isoindoles by reducing this tendency to undergo oxidation and polymerization. All three heterocycles **96** are also stabilized by steric protection, particularly by the presence of substituents at the 1- and 3-positions.

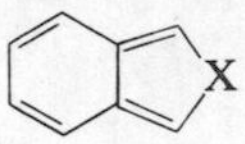

96 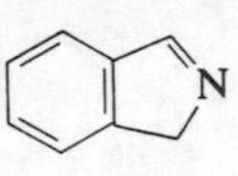   **97**

6.7.2   Synthesis

Cyclization reactions have been used to prepare substituted derivatives of these heterocycles but they are unsuitable for preparing the parent systems. Several such cyclization routes have been used; two examples are illustrated in Fig. 6.33. 1,2-Dibenzoylbenzene is a suitable precursor for the preparation

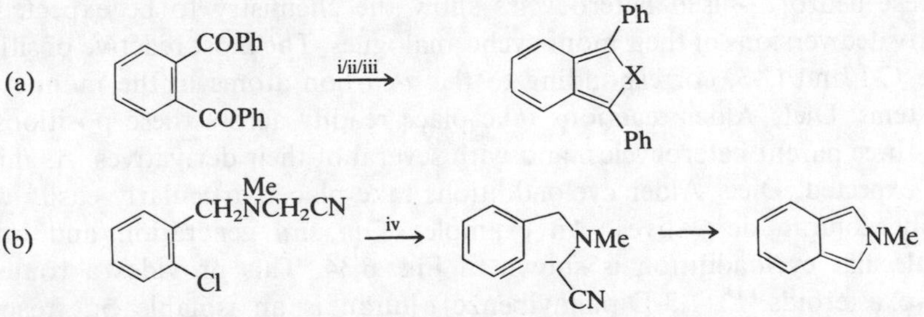

Fig. 6.33 Ring synthesis by cyclization. Reagents: i, (for $X=NMe$), $MeNH_2$, $NaBH_4$ (ref. 110); ii (for $X=O$), KBH_4 (ref. 111); iii (for $X=S$), PCl_5, KSCSEt (ref. 112); iv, KNH_2, NH_3 (ref. 113).

109 Review: B. Iddon, *Adv. Heterocycl. Chem.*, 1972, **14**, 331.
110 M. J. Haddadin and N. C. Chelhot, *Tetrahedron Lett.*, 1973, 5185.
111 M. P. Cava, M. J. Mitchell, and A. A. Deana, *J. Org. Chem.*, 1960, **25**, 1481.
112 A. Schönberg and E. Frese, *Chem. Ber.*, 1968, **101**, 701.
113 B. Jaques and R. G. Wallace, *Tetrahedron*, 1977, **33**, 581.

of 1,3-diphenyl derivatives of all three heterocycles, and the reactions are analogous to the synthesis of the monocyclic compounds from 1,4-diketones. *N*-methylisoindole can be prepared in high yield by an aryne cyclization, as shown (compare the indole synthesis in Table 4.6).

For the preparation of the unstable parent heterocycles it is necessary to use precursors in which the required ring systems are already present in a modified or protected form. The heterocycles are best generated and isolated by using the technique of vapour-phase pyrolysis at reduced pressure, the products being condensed out from the vapour on a cooled surface. Benzo[*c*]thiophene can be conveniently prepared by the thermal dehydration of the *S*-oxide **98** over alumina,[114] and isoindole by pyrolysis of the ester **99**.[115] The retro-Diels–Alder reaction provides a good method for the generation of these species in the vapour phase; for example, isobenzofuran[116] and isoindole[117] have been isolated by the pyrolysis of the compounds **100** (X = O and NH) at 600–650°C and at low pressure.

98   **99**

100

6.7.3 Chemical properties

These benzo[*c*]-fused heterocycles show the chemistry to be expected of activated versions of their monocyclic analogues. The most reactive positions are C-1 and C-3, corresponding to the α-carbon atoms in the monocyclic systems. Diels–Alder reactions take place readily across these positions in all three parent heterocycles, and with several of their derivatives. As might be expected, Diels–Alder cycloadditions take place particularly easily with isobenzofuran derivatives. An example of *in situ* generation and intra-molecular cycloaddition is shown in Fig. 6.34. This provides a route to 11-oxasteroids.[118] 1,3-Diphenylbenzo[*c*]furan is an isolable but reactive diene, which is useful for detecting the presence of transient dienophiles because it is coloured but its adducts are colourless.[108] A related reaction

114 M. P. Cava, N. M. Pollack, O. A. Mamer, and M. J. Mitchell, *J. Org. Chem.*, 1971, **36**, 3932.
115 R. Bonnett, R. F. C. Brown, and R. G. Smith, *J. Chem. Soc., Perkin Trans. 1*, 1973, 1432.
116 U. E. Wiersum and W. J. Mijs, *Chem. Commun.*, 1972, 347.
117 J. Bornstein, D. E. Remy, and J. E. Shields, *Chem. Commun.*, 1972, 1149; E. Chacko, J. Bornstein, and D. J. Sardella, *Tetrahedron*, 1979, **35**, 1055.
118 K. Hildebrandt, T. Debaerdemaeker, and W. Friedrichsen, *Tetrahedron Lett.*, 1988, **29**, 2045.

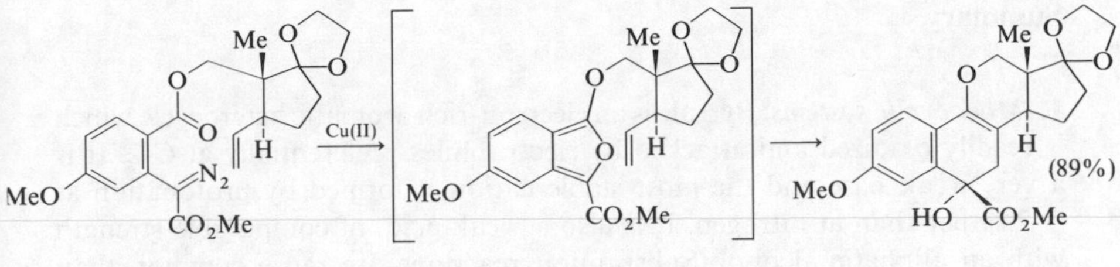

Fig. 6.34 Intramolecular cycloaddition of an isobenzofuran.

is the addition of singlet oxygen across the 1- and 3-positions to give unstable peroxides such as **101**. This compound is converted to 1,3-dibenzoylbenzene in solution; it is therefore important to exclude oxygen from reaction mixtures containing isobenzofurans. The more aromatic isoindoles can undergo substitution rather than addition reactions: for example, 1-phenylisoindole can be acylated at the 3-position and also undergoes the Vilsmeier reaction and coupling to diazonium salts at this position.

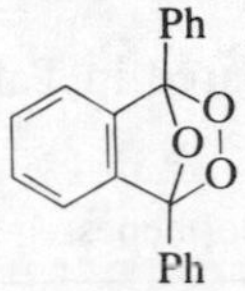

101

6.7.4 Phthalocyanines[119]

Phthalocyanine, **102**, is a stable, greenish-blue crystalline solid which is formally a derivative of isoindole. It is formed by the reductive cyclization of 2-cyanobenzamide with magnesium or antimony. It resembles the porphyrins (Section 6.2.8) in having four nitrogen atoms surrounding a central cavity into which a metal cation can be incorporated. The copper(II) complex is an important pigment in the dyestuff industry. Many related dyestuffs have been prepared in which substituents are attached to the peripheral carbon atoms.

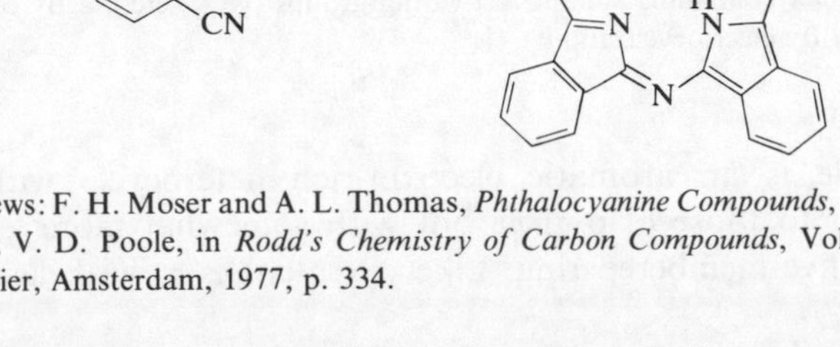

102

119 Reviews: F. H. Moser and A. L. Thomas, *Phthalocyanine Compounds*, Reinhold, New York, 1963; V. D. Poole, in *Rodd's Chemistry of Carbon Compounds*, Vol. IVB, ed. S. Coffey, Elsevier, Amsterdam, 1977, p. 334.

Summary

1. *Monocyclic systems.* Pyrrole is an electron-rich aromatic heterocycle which is readily oxidized and attacked by electrophiles preferentially at C-2. It is a very weak base and the most stable cation is formed by protonation at C-2 rather than at nitrogen. It is also a weak acid, of comparable strength with an aliphatic alcohol. Substitution reactions are more common than addition reactions.

Furan is less aromatic and less nucleophilic than pyrrole. It reacts with electrophiles by addition as well as by substitution and is a moderately reactive diene in cycloaddition reactions.

Thiophene is an aromatic heterocycle which is more reactive towards electrophiles than benzene but less reactive than pyrrole or furan. Nitro-thiophenes and similar species are susceptible to attack by nucleophiles. Substitution reactions are common but addition and cycloaddition reactions occur more readily than in benzene. Alkylation and oxidation can take place at sulphur, and the sulphur can be reductively eliminated by reaction with Raney nickel.

Some of the properties of these heterocycles are compared in Table 6.5.

Table 6.5 Comparison of properties of pyrroles, furans and thiophenes.

Property	Pyrrole	Furan	Thiophene
Resonance energy (REPE) (β)	0.039	0.007	0.032
Position of attack by electrophiles[a]	2- (3-)	2-	2-
Relative rate of electrophilic substitution at C-2[b]	10^6	1	10^{-2}
Stability to mineral acids	Poor	Poor	Fair
Lithiation at C-2	Yes[c]	Yes	Yes
Nucleophilic substitution[d]	Very rare	Not common	Fairly common
Electrophilic addition	Uncommon	Common	Uncommon
Cycloaddition (as diene)	Uncommon	Common	Uncommon
[2+2] Photoaddition	Yes (uncommon)	Yes	Yes
Photorearrangement	Yes (uncommon)	Yes[e]	Yes[e,f]
Formation of 2-Het—$\overset{+}{C}H_2$	Very easy	Easy	Fairly easy

a Preferred position of attack in parent system; attack at C-3 may occur if C-2 and C-5 are substituted. *b* Approximate values for bromination (ref. 101). *c* Pyrrole N must be substituted. *d* Attack on heterocycles containing suitable activating groups (NO_2, etc.). *e* By contraction to 3-membered ring. *f* By electrocyclization.

2. *Indoles.* Indole is an aromatic electron-rich heterocycle with many properties similar to those of pyrrole but with somewhat more 'enamine' character in the five-membered ring. Like pyrrole it is a weak acid and a

very weak base. Attack of electrophiles is strongly favoured at C-3, in contrast to pyrrole. Electrophilic substitution at C-3 takes place in mild conditions and attack may still occur at C-3 even if the position is substituted. It is also easily oxidized at C-3. Indoles can be oxidatively cleaved across the 2,3-bond.

3. *Other benzo-fused heterocycles.* Benzofuran shows a low degree of aromaticity in the five-membered ring and reacts with electrophiles by addition and by substitution. Attack takes place preferentially at C-2. Benzo[*b*]thiophene is more aromatic; it is attacked by electrophiles at C-3 and to a lesser extent than C-2. The benzo[*c*]-fused heterocycles are all reactive species which react readily with dienophiles across the positions adjacent to the heteroatoms.

Problems

1. Suggest reasonable mechanisms for each of the pyrrole syntheses shown in Fig. 6.3.

2. Write down a mechanistic scheme for the Vilsmeier formylation of pyrrole (Table 6.1), showing the intermediates in the reaction.

3. Formulate the steps involved in each of the following:

(a) 3,4-Dimethylpyrrole reacts with formaldehyde in the presence of acetic acid to give the tetramer **A** in high yield.

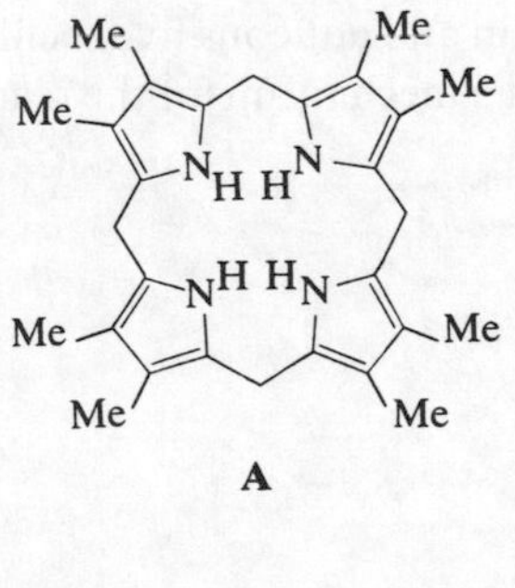

A

(b) 2-Methylindole, when heated with methyl vinyl ketone, gives 2-methylcarbazole.

(c) 3,4-Dinitro-1-methylpyrrole reacts with piperidine in acetonitrile under reflux to give the pyrrole **B**.

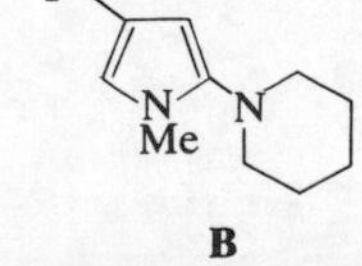

B

4. Predict the most likely positions of attack by electrophiles on each of the following: (a) 1-methylpyrrole; (b) 2,4-dimethylpyrrole; (c) 5-methyl-pyrrole-2-carboxylic acid; (d) furan-2-carboxaldehyde; (e) 2,3-dimethyl-indole; (f) benzothiophene-3-carboxylic acid; (g) 1-phenylisoindole.

5. The following reactions all proceed by way of detectable or isolable intermediates. Suggest structures for the intermediates and write mechanisms for the reactions.

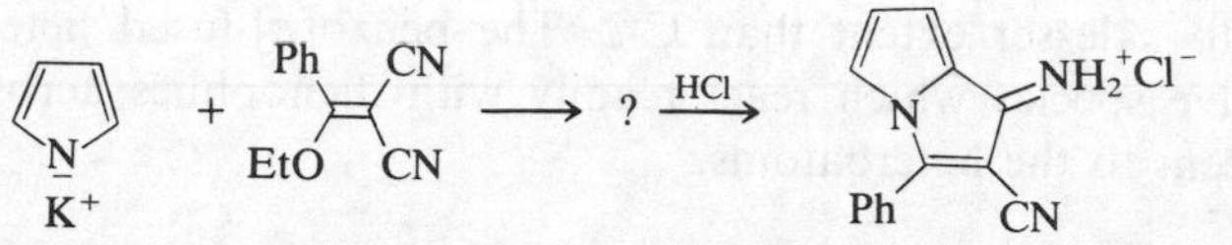

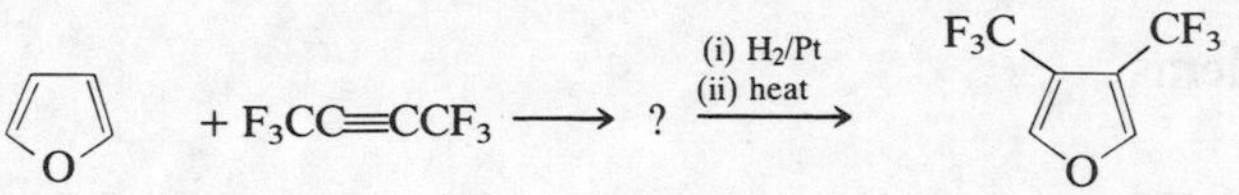

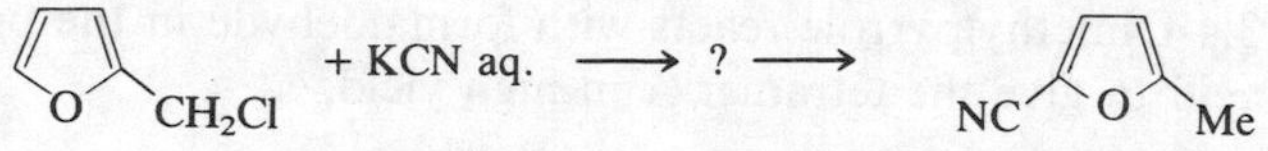

6. 2,5-Dimethylpyrrole, when heated with tin and aqueous hydrochloric acid, gives the dihydroisoindole **C** (57%). Suggest a mechanism for the reaction.

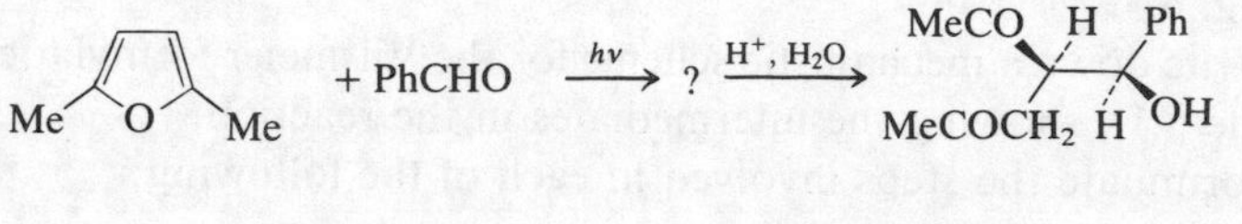

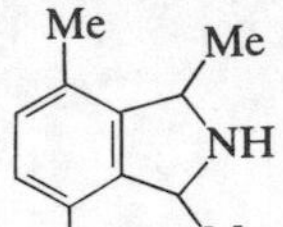

C

7. The indole **D** was heated with potassium carbonate in acetonitrile and gave a low-melting crystalline product $C_{10}H_9N$ which showed no NH absorption in the infrared spectrum. Suggest a structure for the product and explain its formation.

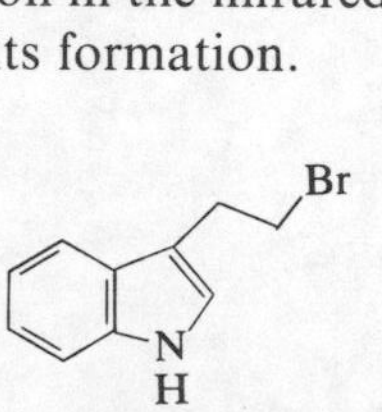

D

8. Optically active indole **E** reacted with the sodium salt of diethyl malonate in toluene to give compound **F**, which was racemic. Suggest a mechanism consistent with the loss of optical activity.

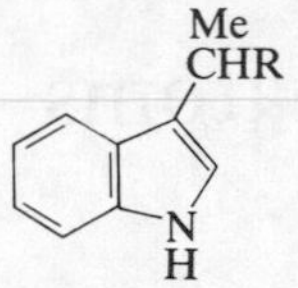

E; R = NHCHMe$_2$
F; R = CH(CO$_2$Et)$_2$

9. The pyrrole **G** was formed as follows. Acetophenone oxime was *O*-alkylated by reaction with dimethyl acetylenedicarboxylate in the presence of sodium methoxide, and the alkylated oxime was then heated at 170–180°C. Using the mechanism of the Fischer indole synthesis as an analogy, formulate a mechanism for the reaction.

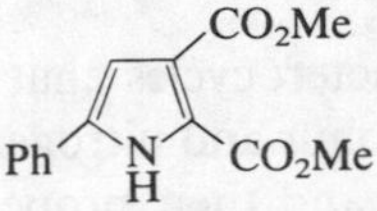

G

10. A method of synthesis of unsymmetrical tetra-arylporphyrins from pyrrole is outlined below. Write mechanisms for each step.

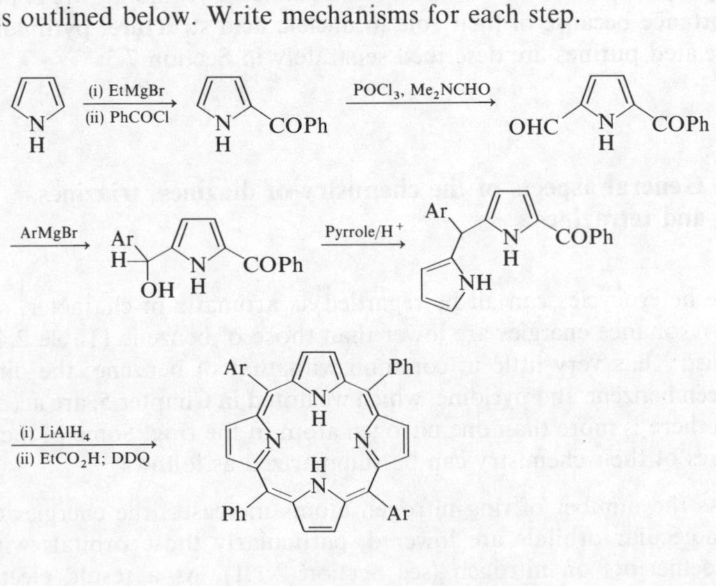

7 Six-membered ring compounds with two or more heteroatoms

7.1 Introduction

Aromatic six-membered heterocycles that contain two, three, and four ring nitrogen atoms are named systematically as diazines, triazines, and tetrazines. The known parent ring systems are illustrated in Chapter 2, Fig. 2.3. The three diazines have the recognized trivial names pyridazine (for 1,2-diazine), pyrimidine (1,3-), and pyrazine(1,4-). The benzodiazines also have recognized trivial names, which are given in Fig. 2.10, but the triazines and tetrazines are named systematically.

There are also many different six-membered heterocycles that contain combinations of oxygen and nitrogen atoms, or sulphur and nitrogen atoms. These ring systems are not aromatic in character and their properties are therefore related to open-chain compounds containing analogous functional groups. Some of the more common ring systems of this type are described briefly in Section 7.6. Most of the chapter is, however, concerned with the chemistry of the aromatic diazines and triazines. Pyrimidines are of particular importance because of their role in nucleic acid structure; pyrimidines and the related purines are described separately in Section 7.3.

7.2 General aspects of the chemistry of diazines, triazines, and tetrazines

These heterocycles can all be regarded as aromatic in character, although their resonance energies are lower than those of benzene (Table 2.4). Their chemistry has very little in common with that of benzene; the differences between benzene and pyridine, which we noted in Chapter 5, are accentuated when there is more than one nitrogen atom in the ring. Some of the general features of their chemistry can be summarized as follows:

(a) As the number of ring nitrogen atoms increases, the energies of the π molecular orbitals are lowered, particularly those orbitals with large coefficients on nitrogen (see Section 2.1.1). As a result, electrophilic attack on the ring carbon atoms becomes increasingly difficult and nucleophilic attack becomes easier.

(b) With the sole exception of C-5 of pyrimidine, all the ring carbon atoms in these heterocycles lie *ortho* or *para* to at least one ring nitrogen atom. Intermediates formed by nucleophilic attack at these positions, or by deprotonation of alkyl substituents at these positions, are particularly well stabilized. This provides selective activation of specific positions in the ring systems. For example, all the carbon atoms in pyridazine and in pyrazine are *ortho* or *para* to one of the ring nitrogen atoms. In pyrimidine, however, C-2, C-4, and C-6 are each activated by both ring nitrogen atoms whereas C-5 is *meta* to both. Activation of the ring carbon atoms in 1,3,5-triazine is even greater because there are more ring nitrogen atoms and because all the carbon atoms are activated by all three nitrogen atoms (Fig. 7.1).

Fig. 7.1 Selective activation of ring carbon atoms. The asterisks indicate the number of *ortho* or *para* nitrogen atoms.

(c) Cations derived from these heterocycles by electrophilic attack on nitrogen are less well stabilized than the corresponding pyridinium cations. The heterocycles are therefore generally more difficult to *N*-alkylate or *N*-oxidize than the corresponding pyridines; they are also weaker bases than pyridine, as illustrated in Table 7.1.

Table 7.1 pK_a values (aqueous solution, 20°C) of some cations derived from diazines.[a]

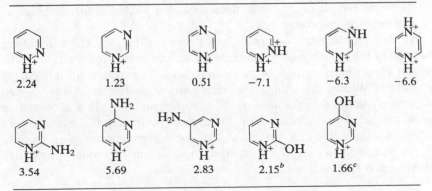

---

[a] Values are from D. D. Perrin, *Dissociation Constants of Organic Bases in Aqueous Solution*, IUPAC, Butterworths, London, 1965 and 1972. [b] Second pK_a (for the loss of a proton from 2-pyrimidinone) = 9.20. [c] Second pK_a = 8.63.

Because of points (a) and (b), electrophilic substitution in these heterocycles is very uncommon unless there are powerfully electron-releasing substituents, such as hydroxy or amino, present to counteract the effect of the ring nitrogen atoms. The tendency towards attack by nucleophiles increases with the increasing number of nitrogen atoms in the ring, however, and within a particular ring system this attack takes place preferentially at activated positions. Because of the lower resonance energies of the ring systems, nucleophilic attack may lead to addition or ring cleavage rather than to substitution. An extreme example is provided by the higher activated 1,3,5-triazine which, although thermally quite stable, is rapidly hydrolysed in water at 25°C.[1] The adduct 1 is probably formed as an intermediate. The ring is also opened by reaction with several other nucleophiles, including ammonia and hydrazine. With the other ring systems it is quite common to find that nucleophiles add reversibly to the heterocycles without cleaving the ring. An example is the adduct 2 formed when 1,2,4-triazine is dissolved in liquid ammonia.[2] Note that nucleophilic attack takes place at a doubly activated position, C-5.

1

2

The effect of nitrogen activation on reactivity towards addition–elimination is illustrated by the relative rate data shown in Table 7.2 for nucleophilic displacement in chlorodiazines.

All the chlorodiazines are much more reactive than the chloropyridines because of the presence of the additional nitrogen atom, and the rates all lie within a fairly narrow range, showing that this is the major factor influencing reactivity. Within this range the displacement is fastest in the 2- and 4-chloropyrimidines, and slowest in 5-chloropyrimidine. This is consistent with the selective activation of these ring positions; for example, an addition–elimination reaction at C-4 of 4-chloropyrimidine goes by way of an intermediate in which both nitrogen atoms provide conjugative stabilization (Fig. 7.2).

1 C. Grundmann, *Angew. Chem. Int. Edn Engl.*, 1963, **2**, 309.
2 A. Rykowski, H. C. van der Plas, and A. van Veldhuizen, *Rec. Trav. Chim.*, 1978, **97**, 273.

Table 7.2 Relative rates of nucleophilic displacement of chloride by 4-nitrophenoxide (4-$NO_2C_6H_4OH$, MeOH, 50°C).[a]

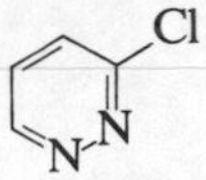

1.2×10^{14} 1.9×10^{14} 4.1×10^{14}

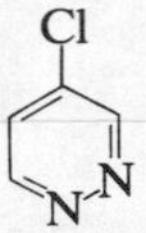

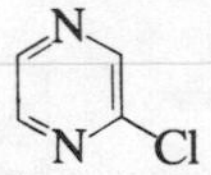

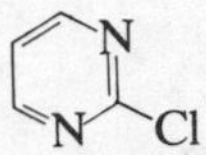

6.3×10^{16} 1.1×10^{15} 1.2×10^{13}

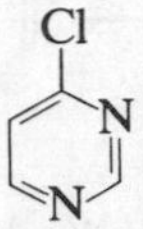

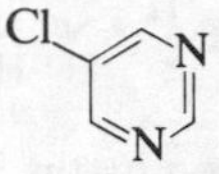

a T. L. Chan and J. Miller, *Aust. J. Chem.*, 1967, **20**, 1595. Relative values for other chloroaromatics are chlorobenzene = 1, 2-chloropyridine = 2.7×10^8, 3-chloropyridine = 9.0×10^4, 4-chloropyridine = 7.4×10^9, and the 2-chloro-*N*-methylpyridinium cation = 1.3×10^{21}.

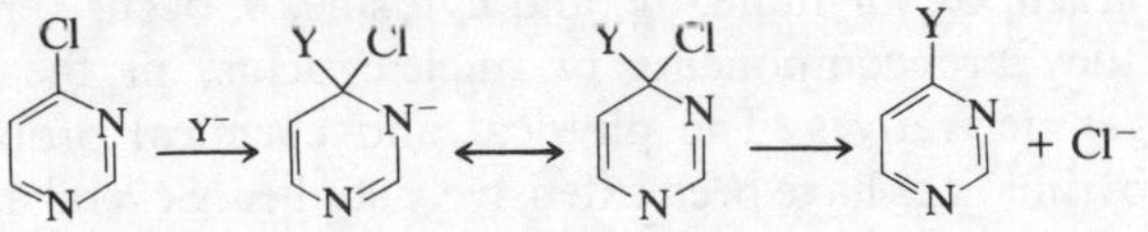

Fig. 7.2 Nucleophilic displacement in 4-chloropyrimidine.

We have so far assumed that all reactions of this type go by way of a conventional addition–elimination pathway, but some such displacements are not as straightforward as they appear to be. A good example is provided by the reaction of 2-chloro-4-phenylpyrimidine with potassamide in liquid ammonia, which gives 2-amino-4-phenylpyrimidine. Although this appears to be an addition–elimination process analogous to that in Fig. 7.2, it is not. When the reaction is carried out with doubly ^{15}N-labelled 2-chloro-4-phenylpyrimidine, only N-3 of the product is labelled, and the other label is found in the exocyclic amino group. Evidently the ring is opened in the reaction. The mechanism shown in Fig. 7.3, involving initial attack of the amide anion at C-6, can explain the location of the labelled nitrogen atoms. This type of mechanism, which has been termed the S_N(ANRORC) mechanism (*A*ddition of *N*ucleophile, *R*ing *O*pening, and *R*ing *C*losure) is quite common in substitution reactions of diazines and triazines which involve potassamide as the nucleophile.[3]

3 Review: H. C. van der Plas, *Accounts Chem. Res.*, 1978, **11**, 462.

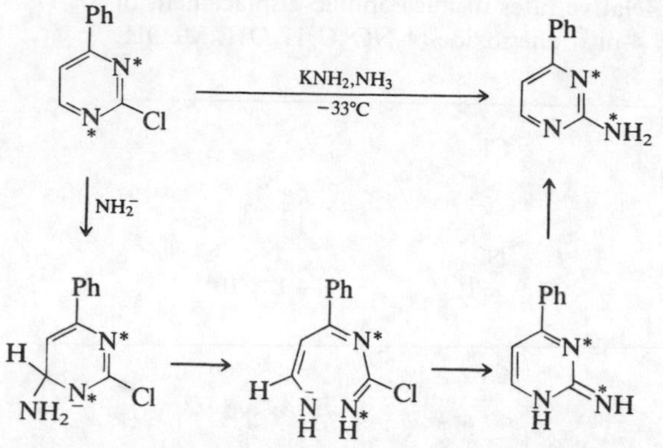

Fig. 7.3 Substitution by the S_N(ANRORC) mechanism.

7.3 Pyrimidines and purines[4]

7.3.1 Introduction

The pyrimidines uracil, **3a**, thymine, **3b**, and cytosine, **4**, occur very widely in nature since they are components of nucleic acids, in the form of N-substituted sugar derivatives. The physical and chemical properties of these and related pyrimidines have been extensively studied. Several analogues have been used as compounds that interfere with the synthesis and functioning of nucleic acids: example are fluorouracil, **3c**, and the anti-AIDS drug AZT (Chapter 1). There are several other important groups of pyrimidines with medicinal uses. The barbiturates, which have the general structure **5** ($R^1 = H$ or Me, R^2, R^3 = alkyl), were for a long period the major group of sedative–hypnotic drugs, but because of problems with toxicity, dependence, and abuse, they are gradually being replaced by other drugs. Some diaminopyrimidines, including pyrimethamine, **6**, and trimethoprim, **7**, are antimalarial agents; trimethoprim is also an effective antibacterial agent when used in combination with a sulphonamide. Minoxidil, **8**, is a vasodilator which has been used in the treatment of hypertension. Vitamin B₁ (structure **41** in Chapter 8) is also a pyrimidine derivative.

The tautomeric forms **3** and **4** shown for uracil, thymine, and cytosine are those which are strongly favoured in the solid and in solution.[5] For example, there are five other (nonzwitterionic) tautomeric forms which can be written

---

4 Reviews: (a) D. J. Brown, *The Pyrimidines*, Wiley-Interscience, New York, 1962, 1970; (b) J. H. Lister, *Fused Pyrimidines. Part II, The Purines*, Wiley-Interscience, New York, 1971; (c) D. T. Hurst, *An Introduction to the Chemistry and Biochemistry of Pyrimidines, Purines, and Pteridines*, Wiley, Chichester, 1980; (d) J. T. Bojarski, J. L. Mokrosz, H. J. Barton, and M. H. Paluchowska, *Adv. Heterocycl. Chem.*, 1985, **38**, 229 (barbituric acid derivatives).

5 Review: J. S. Kwiatkowski and B. Pullman, *Adv. Heterocycl. Chem.*, 1975, **18**, 199.

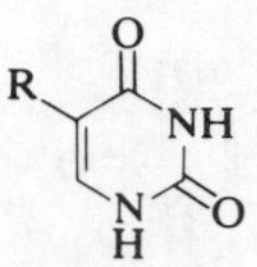

3a; R = H
b; R = Me
c; R = F

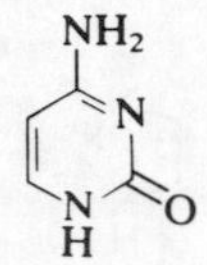

4

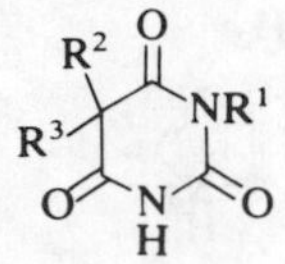

5

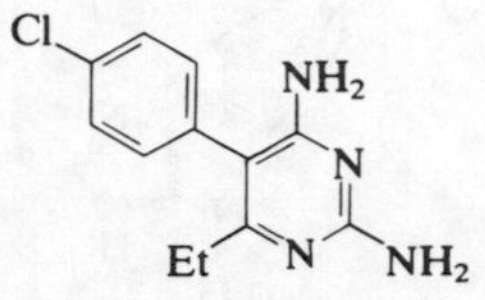

6

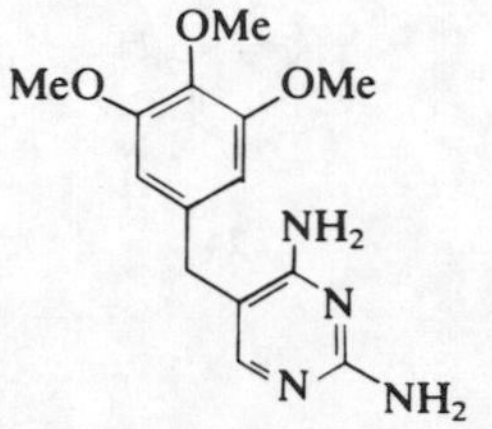

7

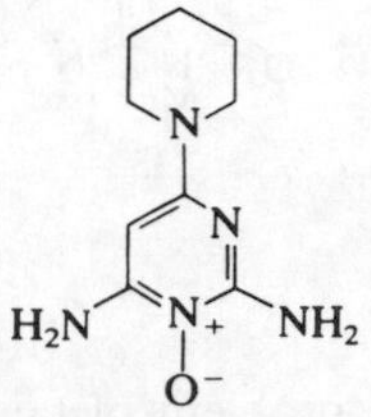

8

for cytosine, but it is estimated to exist as the tautomer **4** to the extent of over 99.8% in aqueous solution. The tautomeric forms are important in determining the nature of the hydrogen bonding in nucleic acids: this is discussed in Section 7.3.6.

Pyrimidine is a water-soluble hygroscopic compound, m.p. 22.5°C and b.p. 124°C. The biological pyrimidines have much higher melting points because they form intermolecular hydrogen bonds. For example, uracil, thymine, and cytosine all crystallize as hydrates with melting points above 300°C. As shown in Table 7.1, pyrimidine is a weak base ($pK_a = 1.23$). Uracil is considerably weaker ($pK_a = -3.38$), but cytosine is a stronger base ($pK_a = 4.61$) because of the stabilizing effect of the 4-amino group on the cation **9**. The acid pK_a of uracil is 9.5.

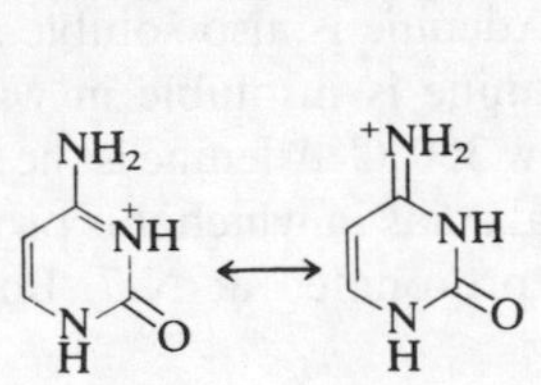

9

The purine ring system forms the skeleton of the nucleic acid bases adenine, **10**, and guanine, **11a**. Note the abnormal numbering system for purines, illustrated in structure **10**, which has become established in the literature. Other naturally occurring purines include uric acid, **12**, and caffeine, **13**. As with the pyrimidines, many analogues of adenine and guanine have been investigated as potential chemotherapeutic agents. Thioguanine, **11b**, and mercaptopurine, **14**, have been successfully used in the treatment of acute leukaemia. Azathioprine, **15**, is used as an immunosuppressive agent.

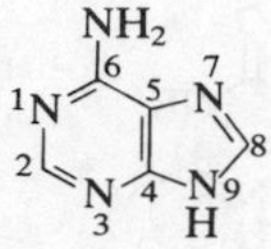

The problem of tautomerism in the biological purines is more complex than in the pyrimidines because, in addition to the usual possibilities of equilibria in the amide and amidine functions, the position of the proton in the five-membered ring has also to be considered, as shown for guanine, **11a**. Guanine exists as a mixture of both these tautomers in solution, although adenine is predominantly in the form of the tautomeric structure **10**.[6]

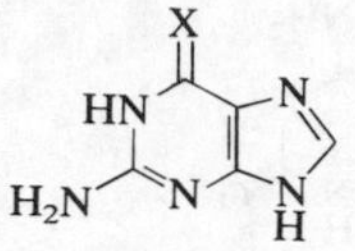

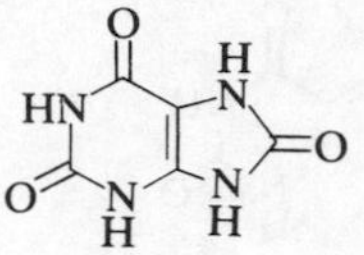

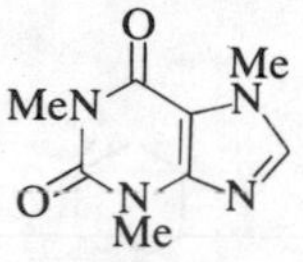

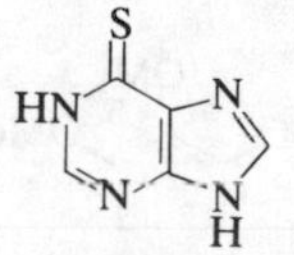

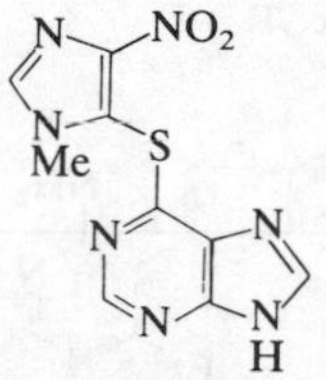

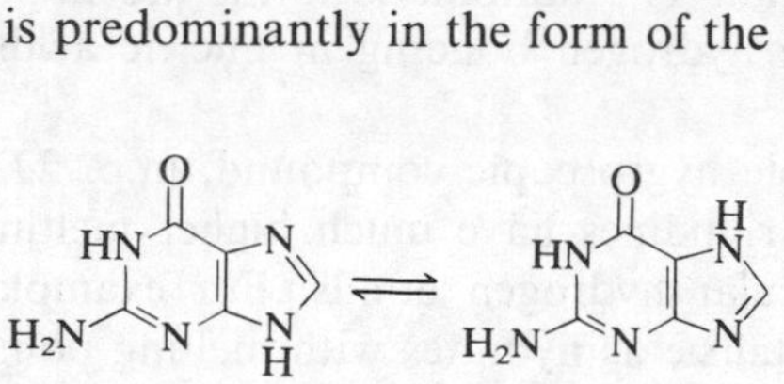

Purine is a water-soluble solid, m.p. 216°C, which is a weak base ($pK_a = 2.30$) and a weak acid ($pK_a = 8.96$). Adenine is also soluble in water and has a high melting point (352°C). Guanine is insoluble in water and most organic solvents, and does not melt below 350°C. Adenine is the stronger base ($pK_a = 4.25$) and forms salts with mineral acids in which it is protonated at N-1. In contrast, guanine ($pK_a = 3.0$) is protonated at N-7. Both form crystalline picrates.

7.3.2 *Synthesis of pyrimidines*

The most general and widely used route to pyrimidines involves the combination of a reagent containing the N—C—N skeleton with one containing a C—C—C unit. These syntheses are typical examples of the 'bis-nucleophile plus bis-electrophile' method of constructing heterocycles,

6 M.-T. Chenon, R. J. Pugmire, D. M. Grant, R. P. Panzica, and L. B. Townsend, *J. Am. Chem. Soc.*, 1975, **97**, 4636.

which was discussed in Section 4.2.1: both the nitrogen atoms of the N—C—N reagent act as nucleophiles and both the terminal carbon atoms of the C—C—C reagent are electrophilic. Urea, thiourea, and guanidine are commonly used as N—C—N reagents. $\alpha\beta$-Unsaturated ketones or acid derivatives can also be used to provide the C—C—C fragment of the ring system. The choice depends upon the substituents required in the final product. An analysis of three pyrimidines (Fig. 7.4) shows how they might be constructed by this type of reaction.

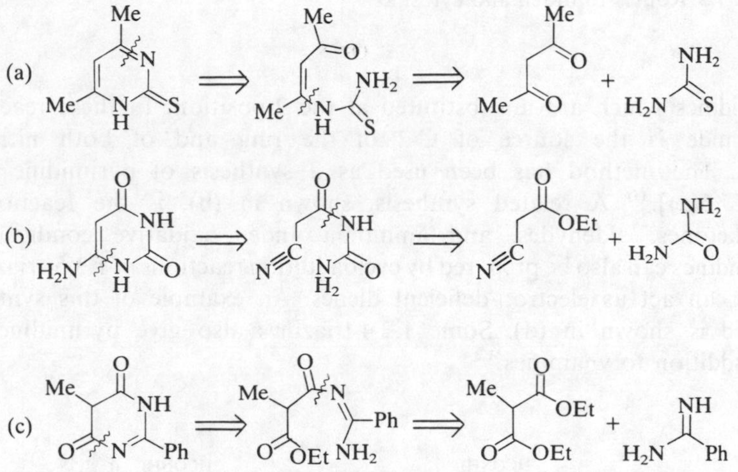

Fig. 7.4 Analysis of routes to pyrimidines.

In practice all these syntheses can readily be achieved by reaction of the components shown: reaction (a) is carried out in concentrated hydrochloric acid under reflux, and (b) and (c) with sodium ethoxide in ethanol under reflux.[7] This synthetic approach is extremely versatile and well established. Two further examples are the syntheses of uracil[8] and of cytosine[9] shown in Fig. 7.5. In both these examples urea is the source of the N—C—N fragment in the pyrimidine ring system. The method of pyrimidine synthesis outlined in Table 4.6 is a reaction of the same general type.

There are many other methods of pyrimidine ring synthesis which are of more limited scope. The reaction of a 1,3-dicarbonyl compound or an equivalent reagent with formamide provides a route to several

7 (a) R. R. Hunt, J. F. W. McOmie, and E. R. Sayer, *J. Chem. Soc.*, 1959, 525; (b) W. R. Sherman and E. C. Taylor, *Org. Synth., Coll. Vol. 4, 1963, 247;* (c) M. Robba and R. Moreau, *Bull. Soc. Chim. Fr.*, 1960, 1648.
8 K. Harada and S. Suzuki, *Tetrahedron Lett.*, 1976, 2321.
9 J. A. Hill and W. J. Le Quesne, *J. Chem. Soc.*, 1965, 1515.

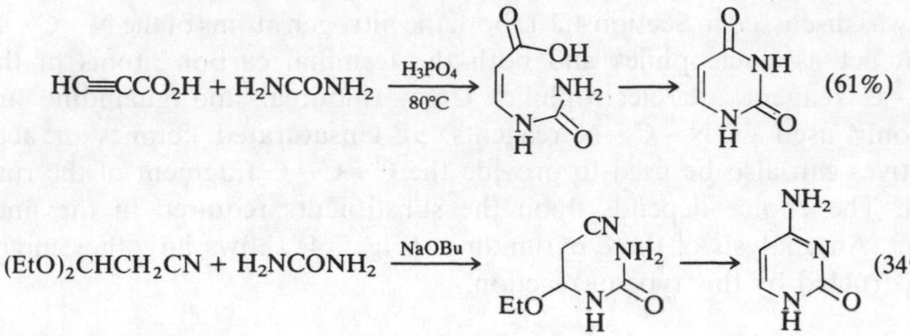

$$HC\!\equiv\!CCO_2H + H_2NCONH_2 \xrightarrow[80°C]{H_3PO_4} \quad \longrightarrow \quad (61\%)$$

$$(EtO)_2CHCH_2CN + H_2NCONH_2 \xrightarrow{\text{NaOBu}} \quad \longrightarrow \quad (34\%)$$

Fig. 7.5 Routes to uracil and cytosine.

pyrimidines which are unsubstituted at the 2-position. In these reactions formamide is the source of C-2 of the ring and of both nitrogen atoms. The method has been used as a synthesis of pyrimidine itself [Fig. 7.6(a)].[10] A related synthesis, shown in (b), is the reaction of 1,3-diketones, aldehydes, and ammonia under oxidative conditions.[11] Pyrimidines can also be prepared by cycloaddition reactions of 1,3,5-triazines, which can act as electron-deficient dienes. An example of this synthetic method is shown in (d). Some 1,2,4-triazines also give pyrimidines by cycloaddition to ynamines.[12]

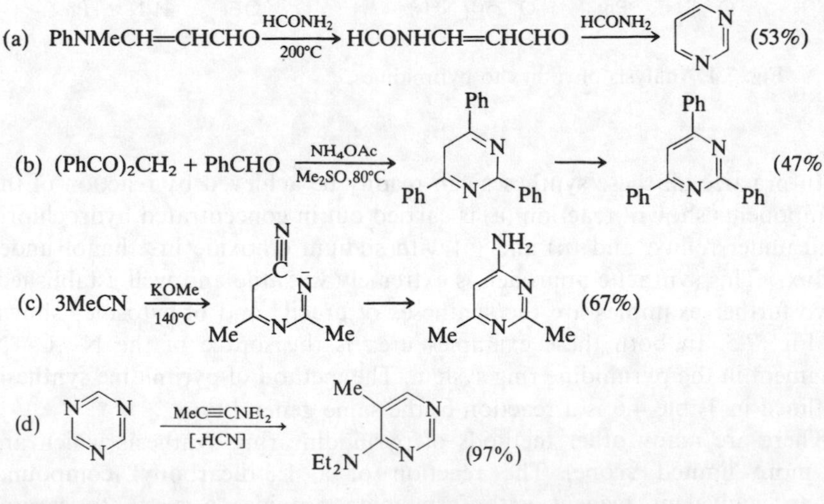

(a) $PhNMeCH\!=\!CHCHO \xrightarrow[200°C]{HCONH_2} HCONHCH\!=\!CHCHO \xrightarrow{HCONH_2}$ (53%)

(b) $(PhCO)_2CH_2 + PhCHO \xrightarrow[Me_2SO,80°C]{NH_4OAc}$ $\longrightarrow$ (47%)

(c) $3MeCN \xrightarrow[140°C]{KOMe}$ $\longrightarrow$ (67%)

(d) $\xrightarrow[\text{[-HCN]}]{MeC\!\equiv\!CNEt_2}$ (97%)

Fig. 7.6 Examples of other methods of pyrimidine synthesis.

10 H. Bredereck, R. Gompper, and H. Herlinger, *Chem. Ber.*, 1958, **91**, 2832.
11 A. L. Weiss and V. Rosenbach, *Tetrahedron Lett.*, 1981, **22**, 1453.
12 Review: D. L. Boger and S. M. Weinreb, *Hetero Diels–Alder Methodology in Organic Synthesis*, Academic Press, San Diego, 1987, Chapter 10.

7.3.3 Synthesis of purines

As the purine ring system is a fusion of two aromatic heterocycles, pyrimidine and imidazole, a logical starting point for ring synthesis is an appropriately substituted pyrimidine or imidazole from which the second ring can be constructed by a cyclization process. This is illustrated in Fig. 7.7.

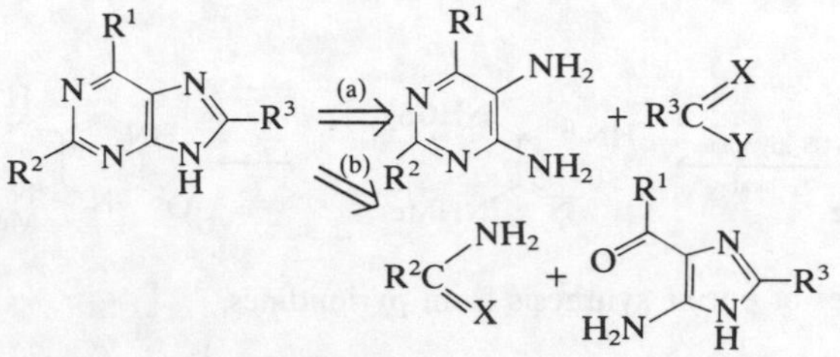

Fig. 7.7 Analysis of two routes to purines.

Both of these methods have been used for the synthesis of purines. Method (a), which is the more common, is the *Traube* synthesis. This is a versatile method because of the variety of substituents possible in both components. Formic acid was originally used as the cyclization reagent; its reaction with 2,5,6-triaminopyrimidin-4-one gave guanine by way of the intermediate 5-*N*-formyl derivative, **16**.

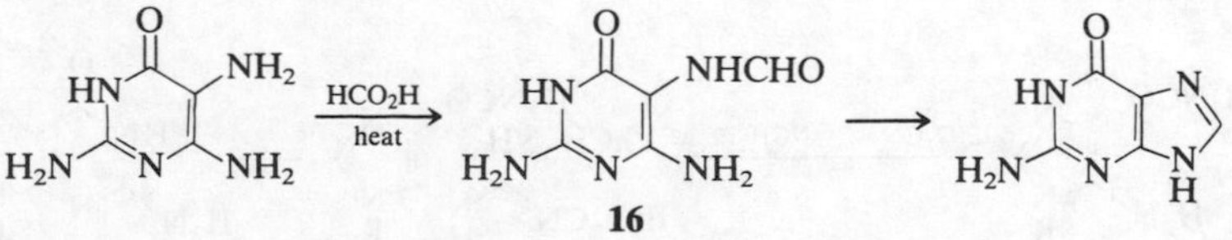

Formamide, formamidine, ethyl orthoformate, and similar electrophilic reagents can be used in place of formic acid in this type of synthesis. 8-Substituted purines can be constructed by using a different type of acylating agent such as acetic anhydride. Fusion of the diaminopyrimidine with urea results in the formation of 8-oxopurines. Examples of such reactions are shown in Fig. 7.8.

The synthesis of purines by route (b) of Fig. 7.7 is a useful alternative to that based on diaminopyrimidines. 5-Aminoimidazole-4-carboxamide and related compounds are commonly used as precursors. For example, the carboxamide can be formylated with formic acid and the formamide is cyclized on heating [Fig. 7.9(a)]. This method of synthesis is closely related to the biological route to purines: 5-aminoimidazole-4-carboxamide occurs naturally (as a ribonucleotide) and it is the precursor of the purine nucleotides.

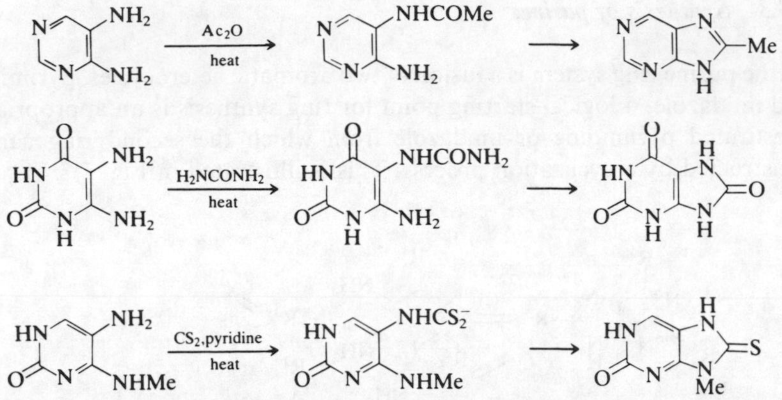

Fig. 7.8 Examples of purine synthesis from pyrimidines.

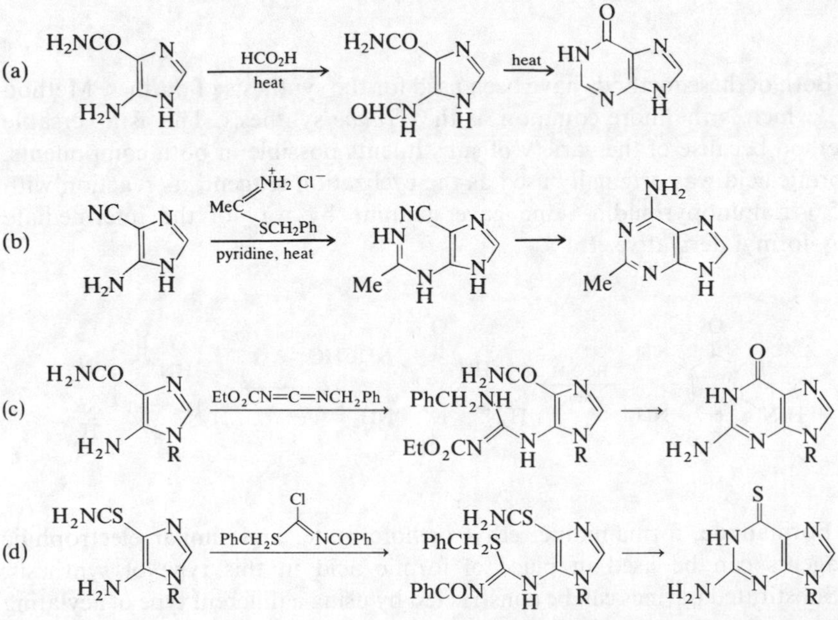

Fig. 7.9 Examples of purine synthesis from imidazoles.

An example of a related reaction, which has been adapted to the chemical synthesis of *C*-nucleotides, is shown in (b).[13] The use of two other reagents for cyclization is illustrated in (c) and (d). The carbodiimide used in route (c) allows the rapid construction of guanosine derivatives.[14] The ring closure

13 T. Huynh-Dinh, A. Kolb, C. Gouyette, and J. Igolen, *J. Heterocycl. Chem.*, 1975, **12**, 111.
14 M. P. Groziak and L. B. Townsend, *J. Org. Chem.*, 1986, **51**, 1277.

is achieved by reductive removal of the benzyl group, which allows the intermediate to cyclize with loss of ammonia. The ethoxycarbonyl group is then removed by basic hydrolysis. A related sequence (d) has been used to prepare nucleosides of the thioguanosine type.[15] The intermediate is cyclized by heating with aqueous ammonia; further heating with base removes the benzoyl protecting group.

Adenine ($C_5H_5N_5$) is a pentamer of hydrogen cyanide, and it has been suggested that it could have been formed on the primitive Earth when hydrogen cyanide was abundant. This suggestion seems reasonable because it has proved possible to produce adenine from hydrogen cyanide under a variety of conditions; for example, from frozen aqueous ammoniacal solutions of hydrogen cyanide[16] and by heating ammonium cyanide absorbed in a zeolite.[17] The likely route to adenine from hydrogen cyanide (Fig. 7.10) is related to those in Fig. 7.9 because it goes by way of an imidazole intermediate.[18]

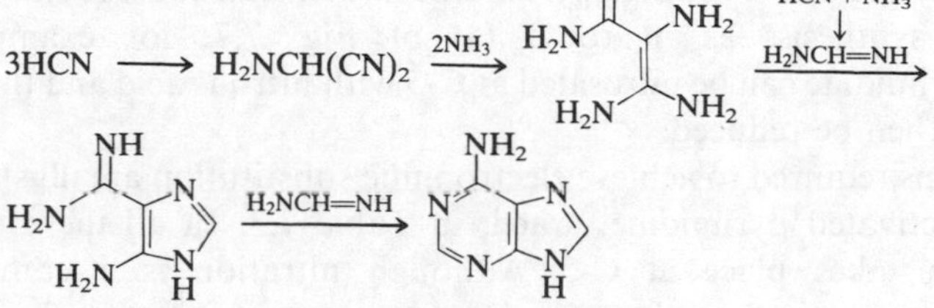

Fig. 7.10 A route to adenine from hydrogen cyanide.

7.3.4 Reactions of pyrimidines

The ring nitrogen atoms of pyrimidines are nucleophilic and can be alkylated, for example, with iodomethane. Pyrimidines with electron-releasing substituents can be *N*-acylated: for example, uracil and thymine are acetylated to give the 1-acetyl derivatives **17** with acetic anhydride, and they give the analogous benzoyl derivatives with benzoyl chloride in acetonitrile and pyridine.[19] Pyrimidines with alkyl or alkoxy substituents can be *N*-oxidized with 3-chloroperbenzoic acid or with hydrogen peroxide in acetic acid, and they are *N*-aminated with *O*-mesitylenesulphonylhydroxylamine.[20]

15 E. M. Acton and K. J. Ryan, *J. Org. Chem.*, 1984, **49**, 528.
16 A. W. Schwartz, H. Joosten, and A. B. Voet, *BioSystems*, 1982, **15**, 191.
17 F. Seel, M. von Blon, and A. Dessauer, *Z. Naturforsch. (B)*, 1982, **37**, 820.
18 C. U. Lowe, M. W. Rees, and R. Markham, *Nature, Lond.* 1983, **199**, 219.
19 K. A. Cruickshank, J. Jiricny, and C. B. Reese, *Tetrahedron Lett.*, 1984, **25**, 681.
20 Review: Y. Tamura, J. Minamikawa, and M. Ikeda, *Synthesis*, 1977, 1.

$(R^1 = H, Me)$ $(R^2 = Me, Ph)$

17

For the reasons outlined in Section 7.2, electrophilic substitution of simple pyrimidines is difficult. Substitution reactions of pyrimidinones and amino-pyrimidines do occur, however. Electrophilic substitution nearly always takes place at the 5-position, unless it is already substituted. This is because C-5 is the least deactivated by the ring nitrogen atoms, being *meta* to both, and, more importantly, it is usually *ortho* or *para* to activating substituents. One activating substituent (such as an amino group) is usually sufficient to allow halogenation or nitration to take place, but at least two such substituents are usually required for nitrosation, diazo coupling, and the Mannich reaction. Nitration and nitrosation reactions at C-5 are particularly useful because they provide a route to pyrimidines which can subsequently be used in purine synthesis, as in route (a) of Fig. 7.7; for example, 4,6-diaminopyrimidine can be nitrosated at C-5 with nitrous acid and the nitroso group can then be reduced.

Conditions required to achieve electrophilic substitution are illustrated for a typical activated pyrimidine, uracil, in Table 7.3. In all these reactions substitution takes place at C-5. Although nitration is a conventional electrophilic substitution, bromination in aqueous solution is not: the reaction goes by way of an intermediate addition product, **18**.[21]

Table 7.3 Electrophilic substitution of uracil.

Electrophile	Reagents and conditions	Yield (%)	Refs.
NO_2^+	HNO_3 (d. 1.5), 75°C	90	a
Br^+	Br_2, H_2O, 100°C	90	b
Cl^+	NCS, AcOH, 50°C	52	c
F^+	F_2, AcOH, 10°C	92	d
SO_2Cl^+	$ClSO_3H$, 40–100°C	—	e
$CH_2{=}\overset{+}{N}Me_2$	$(CH_2O)_n$, Me_2NH, 78°C	76	f
CH_2Cl^+	$(CH_2O)_n$, HCl, 80°C	57	g

a D. J. Brown, *J. Appl. Chem.*, 1952, **2**, 239. b S. Y. Wang, *J. Org. Chem.*, 1959, **24**, 11. c R. A. West and H. W. Barrett, *J. Am. Chem. Soc.*, 1954, **76**, 3146. d D. Cech, H. Meinert, G. Etzold, and P. Langen, *J. Prakt. Chem.*, 1973, **315**, 149. e G. R. Barker, N. G. Luthy, and M. M. Dhar, *J. Chem. Soc.*, 1954, 4206. f T. J. Delia, J. P. Scovill, W. D. Munslow, and J. H. Burckhalter. *J. Med. Chem.*, 1976, **19**, 344. g W. A. Skinner, M. G. M. Schelstraete, and B. R. Baker, *J. Org. Chem.*, 1960, **25**, 149.

21 O. S. Tee and C. G. Berks, *J. Org. Chem.*, 1980, **45**, 830.

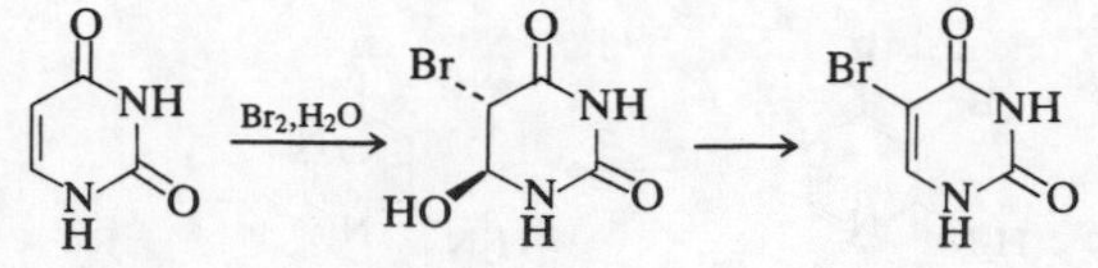

The nucleophilic displacement of good leaving groups at the 2-, 4-, and 6-positions of pyrimidines is, as indicated in Section 7.2, a common reaction. By reaction with phosphorus oxychloride, 2-, 4-, and 6-chloropyrimidines can be made from the corresponding pyrimidinones (compare the analogous conversion of 2-pyridone into 2-chloropyridine in Fig. 5.30), and displacement of chloride by amines, alkoxides, thiols, and hydrazine takes place very readily. In general, chloride is more easily displaced at C-4 than at C-2, and 2,4-dichloropyrimidines may react selectively at C-4 with nucleophiles. Other substituents can act as leaving groups: methylsulphinyl is a particularly good leaving group. Even methoxy groups can be displaced by better nucleophiles; an example is a synthesis of cytosine from 4-methoxy-2-pyrimidinone and ammonia. This and other examples of nucleophilic displacement are shown in Fig. 7.11.

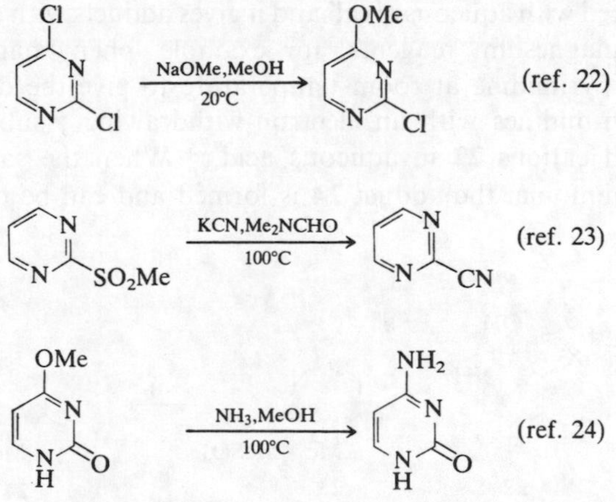

Fig. 7.11 Examples of nucleophilic displacement.

An example of displacement by the S_N(ANRORC) mechanism was given in Fig. 7.3. Another example is provided by the reaction of 4-substituted-5-bromopyrimidines with potassamide in liquid ammonia, which leads to the formation of 6-aminopyridines (Fig. 7.12). This reaction was first interpreted

22 G. W. Kenner, C. B. Reese, and A. R. Todd, *J. Chem. Soc.*, 1955, 855.
23 D. J. Brown and P. W. Ford, *J. Chem. Soc.* (C), 1967, 568.
24 A. Holý, *Coll. Czech. Chem. Commun.*, 1979, **44**, 2846.

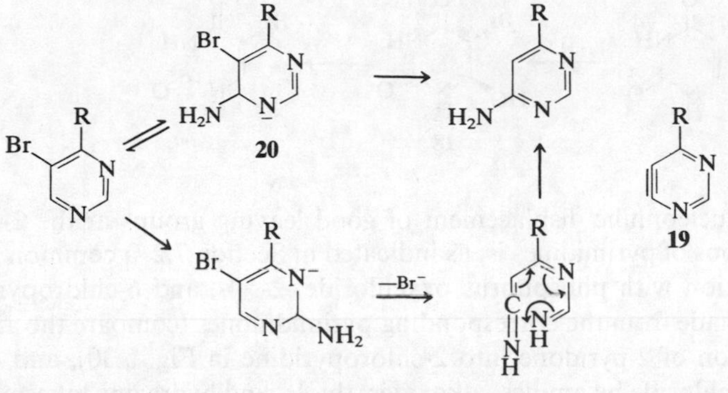

Fig. 7.12 Amination of 5-bromopyrimidines.

as an elimination–addition process in which the aryne intermediates **19** were involved, but more detailed studies have shown that the reaction is instead initiated by addition of amide at C-2 and C-6. The intermediates **20** have been detected by NMR spectroscopy.[3]

Several nucleophilic addition reactions are known. Pyrimidine is hydrolysed when heated with aqueous alkali and it gives adducts with organo-lithium and organomagnesium reagents: for example, phenylmagnesium bromide reacts with pyrimidine at room temperature to give the dihydro-pyrimidine, **21**.[10] Pyrimidines with an electron-withdrawing 5-substituent exist as the hydrated cations **22** in aqueous acid.[25] When the salt **23** is dissolved in liquid ammonia, the adduct **24** is formed and can be detected by NMR.[26]

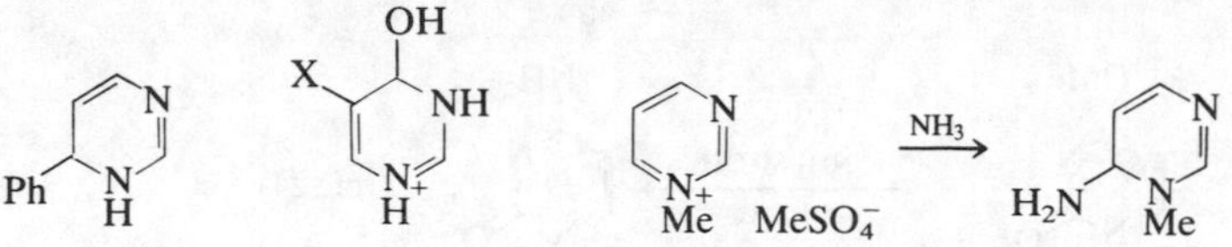

Another example of the relative ease with which some pyrimidines undergo addition and ring cleavage is provided by the Dimroth rearrangement. 1-Alkyl-2-iminopyrimidines, **25**, are rearranged in aqueous alkali to 2-alkyl-aminopyrimidines, **26**. The reaction can be represented by the sequence shown in Fig. 7.13: the intermediate aldehyde (R = Me) has been intercepted as its oxime by carrying out the reaction in the presence of hydroxylamine.

10 H. Bredereck, R. Gompper, and H. Herlinger, *Chem. Ber.*, 1958, **91**, 2832.
25 D. J. Brown, P. W. Ford, and M. N. Paddon-Row, *J. Chem. Soc. (C)*, 1968, 1452.
26 E. A. Oostveen, H. C. van der Plas, and H. Jongejan, *Rec. Trav. Chim.*, 1974, **93**, 114.

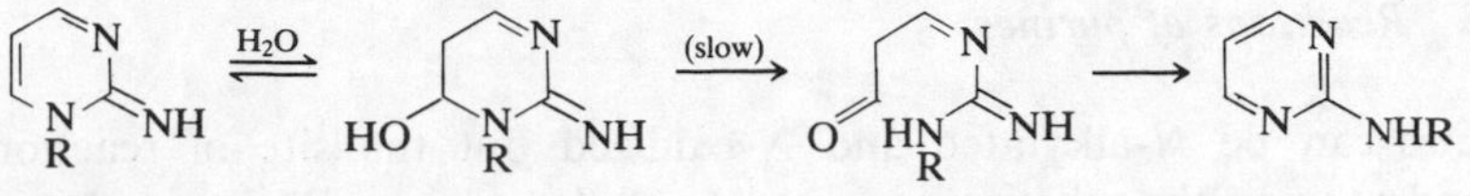

Fig. 7.13 The Dimroth rearrangement.

Addition reactions of uracil, thymine, cytosine, and other biological pyrimidines can take place under the influence of ultraviolet light.[27] For example, uracil is reversibly hydrated across the 5,6-bond to give the adduct **27**; cytosine is hydrated in a similar way.

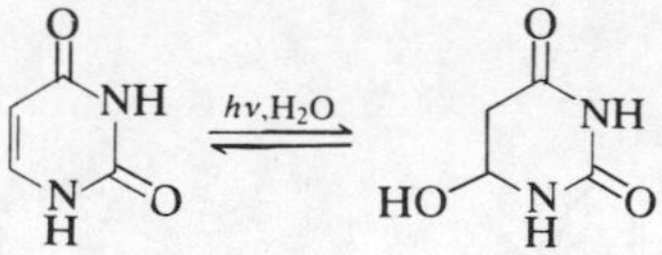

When these pyrimidines are irradiated in frozen aqueous solution, cyclic dimers are formed. The products formed when thymine is irradiated at or near its absorption maximum (260 nm) are the four dimers **28**, **29**, **30**, and **31**. The dimerization can be reversed by irradiation at wavelengths below 240 nm. The biological pyrimidines also undergo addition and [2+2] cycloaddition reactions when irradiated with ethers, olefins, and other organic compounds. Such reactions may be important in the degradation of DNA by ultraviolet light.

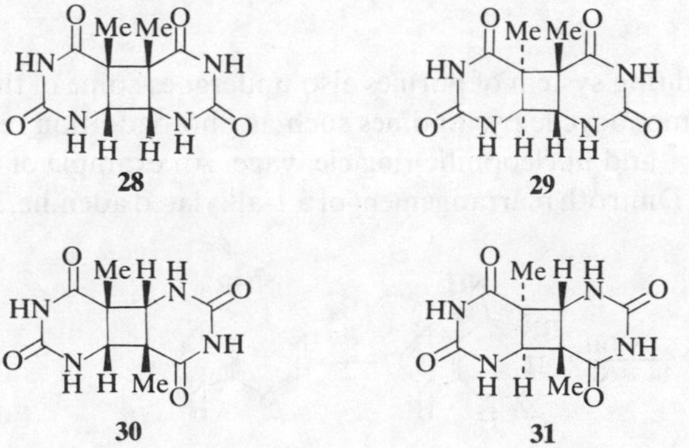

27 Reviews: S. T. Reid, *Adv. Heterocycl. Chem.*, 1970, **11**, 1; 1982, **30**, 239.

7.3.5   Reactions of purines

Purines can be *N*-alkylated and *N*-oxidized but the site of reaction is dependent upon the substituents present and the reagent. Purine is converted into 9-methylpurine by reaction with dimethyl sulphate in aqueous solution, but adenine can be alkylated at N-9, N-3, and N-1 under different conditions. Adenine is oxidized by peracids to give the 1-oxide but guanine is converted into the 3-oxide by pertrifluoroacetic acid. Adenine, guanine, and several other purines can be brominated at C-8, but otherwise electrophilic substitution is of limited value. On the other hand, nucleophilic displacement of halides is a useful procedure. The chloropurines can be made, like the chloropyrimidines, by reaction of oxo derivatives with phosphorus oxychloride. Chloride is more easily displaced at C-6 than at C-2: for example, 2,6-dichloropurine gives 6-amino-2-chloropurine, **32**, with methanolic ammonia at 100°C.[28]

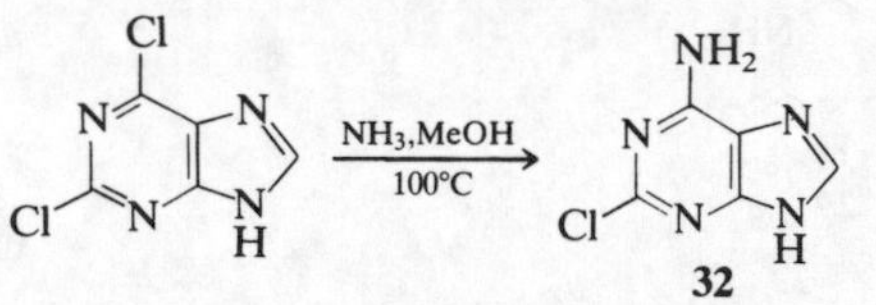

With 2,6,8-trichloropurine, the chloride at C-8 is the least easily displaced in alkaline media because the proton at the adjacent nitrogen atom is removed by the base. If this nitrogen atom is alkylated, however, the order of displacement of halide in the purine **33** is $8 > 6 > 2$.[29]

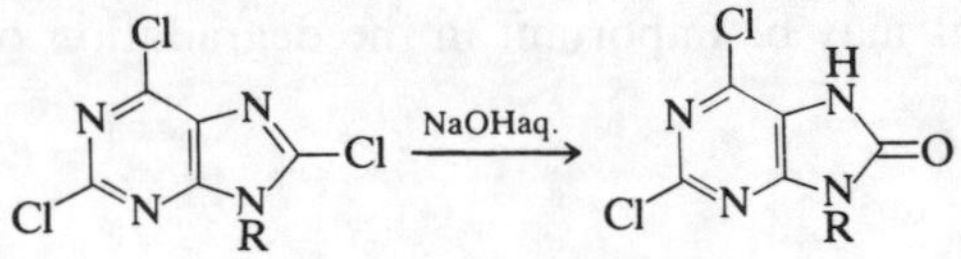

The six-membered ring system of purines also undergoes some of the other reactions typical of monocyclic pyrimidines such as photoaddition reactions across N-1 and C-6[27] and nucleophilic ring cleavage. An example of nucleophilic cleavage is the Dimroth rearrangement of a 1-alkylated adenine, **34**.[30]

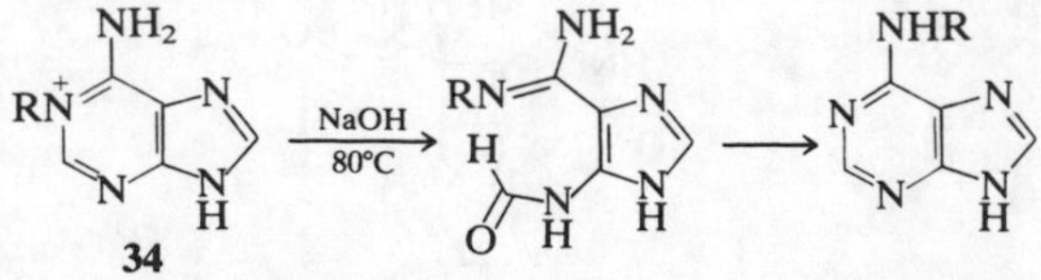

28 G. B. Brown and V. S. Weliky, *J. Org. Chem.*, 1958, **23**, 125.
29 E. Y. Sutcliffe and R. K. Robins, *J. Org. Chem.*, 1963, **28**, 1662.
30 G. Frenner and H.-L. Schmidt, *Chem. Ber.*, 1977, **110**, 373.

7.3.6 Pyrimidines and purines in nucleic acids[31]

Nucleic acids are linear polymers of which the backbone is composed of alternating sugar and phosphate units; pyrimidine and purine bases are attached to C-1 of the sugar residues. Deoxyribonucleic acid (DNA) is present in all living cells and carries the genetic information. In DNA, four heterocyclic bases, thymine, cytosine, adenine, and guanine, usually occur, attached to the deoxyribose units [Fig. 7.14(a)]. The order in which these bases occur on the polymer chain carries the genetic information. Ribonucleic acids (RNAs) are similar polymers in which the sugar units are riboses (with a hydroxyl group at C-2) and which contain uracil in place of thymine. They are responsible for protein synthesis.

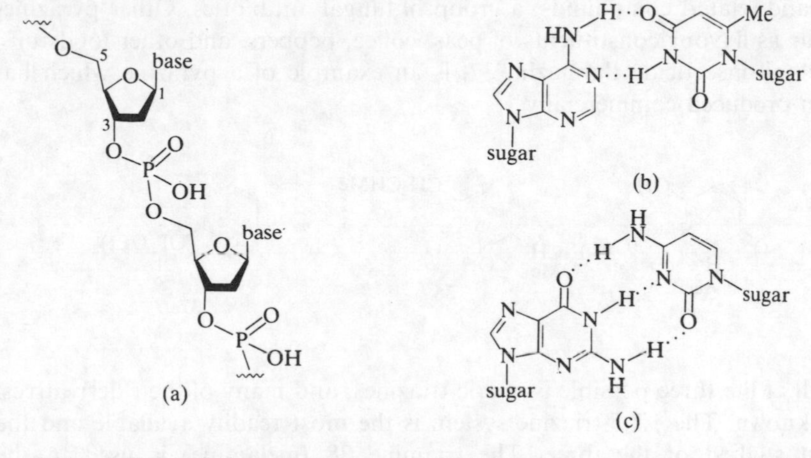

Fig. 7.14 (a) Two of the repeating units in the primary structure of DNA; (b) adenine–thymine pairing; (c) guanine–cytosine pairing.

In DNA the number of thymine units present is equal to the number of adenine units, and the number of cytosine units is equal to the number of guanine units. The 'double helix' structure for DNA first proposed by Watson and Crick accounts for this equivalence by invoking specific hydrogen bonding between pairs of bases as shown in Fig. 7.14, which holds two polymer strands together. The DNA can replicate itself by separation of the strands, the bases on each separate strand then bonding specifically to units containing the complementary bases required for hydrogen bonding: cytosine to guanine, and thymine to adenine.

Although hydrogen bonding is no longer thought to be solely responsible for holding the two strands together, it does provide specificity. For example,

31 For a general description of the structure of nucleic acids see G. M. Blackburn, in *Comprehensive Organic Chemistry* Vol. 5, ed. E. Haslam, Pergamon, Oxford, 1979, p. 23.

in the tautomers shown, guanine cannot efficiently form hydrogen bonds to thymine. It may be that alkylating agents produce their mutagenic effect by altering the tautomeric forms of the pyrimidine and purine bases in the nucleic acids and thus introducing the possibility of error in the replication process.

7.4   Other diazines, triazines, and tetrazines[32]

7.4.1   *Introduction*

There are very few examples of naturally occurring pyridazines (1,2-diazines), but many synthetic derivatives show biological activity. For example, the pyridazinone **35** (maleic hydrazide) is a plant growth inhibitor. There are more naturally occurring pyrazines (1,4-diazines), which include aspergillic acid, **36**, and related compounds, a group of fungal antibiotics. Other pyrazines occur as flavour constituents of peas, coffee, peppers, and other foodstuffs. The soil insecticide thionazin, **37**, is an example of a pyrazine which has been produced commercially.

All of the three possible isomeric triazines, and many of their derivatives, are known. The 1,3,5-triazine system is the most readily available and the most studied of the three. The triamine **38** (melamine) is used in the manufacture of resins. Trichloro-1,3,5-triazine is the starting point for the manufacture of several herbicides such as simazine, **39**, and of dyestuffs. 'Reactive' dyes have proved to be particularly important. These are of the general structure **40**, in which two of the chloro substituents have been replaced by different amines containing the chromophores and the third chloro substituent remains. This enables covalent linkages to be formed with cellulose fibres, in which the OH groups can act as nucleophiles. A wide range of very 'fast' colours can be incorporated into the fibres by the use of these dyes.

32 Reviews: (a) *Pyridazines*, ed. R. N. Castle, Wiley-Interscience, New York, 1973; (b) M. Tišler and B. Stanovnik, *Adv. Heterocycl. Chem.*, 1990, **49**, 385 (pyridazines); (c) G. B. Barlin, *The Pyrazines*, Wiley-Interscience, New York, 1982; (d) H. Neunhoeffer and P. F. Wiley, *Chemistry of 1,2,3-Triazines and 1,2,4-Triazines, Tetrazines, and Pentazines*, Wiley-Interscience, New York, 1978; (e) E. J. Modest, in *Heterocyclic Compounds*, Vol. 7, ed. R. C. Elderfield, Wiley, New York, 1961, p. 627 (1,3,5-triazines); (f) V. N. Charusin, S. G. Alexeev, O. N. Chupahkin, and H. C. Van der Plas, *Adv. Heterocycl. Chem.*, 1989, **46**, 73 (1,2,4-triazines).

| **38** | **39** | **40** |

All the fully unsaturated ring systems are planar and delocalized. There is, however, some bond fixation in pyridazine, for which the localized structure **41** is the best representation; the N—N bond has mainly single-bond character. Similarly, 1,2,4-triazine is best represented by the Kekulé structure **42**. A determination of the empirical resonance energy of pyrazine gave a value of $22.3 \, \text{kcal mol}^{-1}$ ($93.3 \, \text{kJ mol}^{-1}$), which is 62% of the value for benzene. This is in reasonable agreement with the calculated REPE value for pyrazine of 0.049β (Table 2.4). These heterocycles are thus less aromatic than benzene.

| **41** | **42** |

All the parent heterocycles and simple alkyl derivatives are liquids or solids with low melting points. 1,2,4,5-Tetrazines are unusual in being highly coloured (red or violet) because of a weak $n \to \pi^*$ absorption in the visible spectrum at 520–570 nm. 1,2,4,5-Tetrazines have a low lying π^* molecular orbital, which accounts not only for their colour but also their reactivity as electrophilic dienes.[33] In contrast, the $n \to \pi^*$ absorption bands in the other parent heterocycles are at much shorter wavelength, as shown in Table 7.4.

Table 7.4 Ultraviolet spectra of diazines and triazines.

Heterocycle	Solvent	λ_{max} (nm)(ε)		Refs.
		$\pi \to \pi^*$	$n \to \pi^*$	
Pyridazine	cyclohexane	246 (1300)	340 (315)	a
Pyrimidine	cyclohexane	243 (2030)	298 (326)	a
Pyrazine	cyclohexane	260 (5600)	328 (1040)	a
1,2,3-Triazine	ethanol		b	c
1,2,4-Triazine	methanol	248 (3020)	374 (400)	d
1,3,5-Triazine	cyclohexane	222 (150)	272 (890)	a

a S. F. Mason, *J. Chem. Soc.*, 1959, 1247. *b* λ_{max} 232 (shoulder), 288 (860) and 325 (shoulder) nm; bands not assigned. *c* A. Ohsawa, H. Arai, H. Ohnishi, and H. Igeta, *J. Chem. Soc., Chem. Commun.*, 1981, 1174. *d* H. Neunhoeffer and H. Hennig, *Chem. Ber.*, 1968, **101**, 3952.

33 R. Gleiter, V. Schehlmann, J. Spanget-Larsen, H. Fischer, and F. A. Neugebauer, *J. Org. Chem.*, 1988, **53**, 5756.

7.4.2   Methods of ring synthesis

The standard method of synthesis of the pyridazine ring is the action of hydrazine on 1,4-dicarbonyl compounds or their equivalent.[32a] Saturated 1,4-diketones give dihydropyridazines which are easily oxidized to the aromatic compounds [Fig. 7.15(a)], but 2,3-unsaturated 1,4-diketones give the aromatic ring system directly. With γ-oxocarboxylic acids or esters, pyridazinones are formed, as shown in the example in Fig.. 7.15(b).[34] The ring system can also be produced by the cycloaddition of azocarbonyl compounds to dienes (Table 4.21) and by the Diels–Alder reactions of 1,2,4,5-tetrazines (Section 7.4.3).

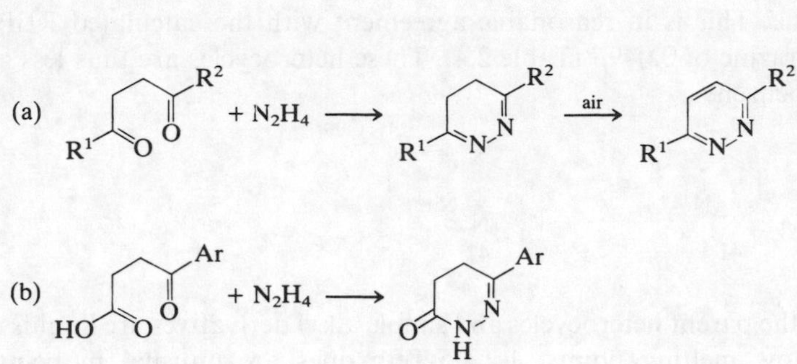

Fig. 7.15 Routes to the pyridazine ring system.

Pyrazines are normally made either by the self-condensation of α-aminocarbonyl compounds, or by the reaction of 1,2-diketones with 1,2-diamines. The two methods lead to pyrazines with different substitution patterns (Fig. 7.16).

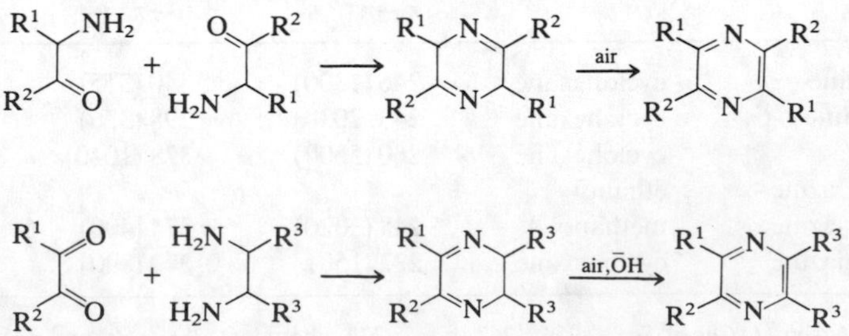

Fig. 7.16 General routes to pyrazines.

34 W. V. Curran and A. Ross, *J. Med. Chem.*, 1974, **17**, 273.

Monocyclie 1,2,3-triazines with aromatic substituents are prepared by an unusual method, the thermal rearrangement of 2-azidocyclopropenes [Fig. 7.17(a)]. The most general route to 1,2,4-triazines is the reaction of 1,2-dicarbonyl compounds with amidrazones [Fig. 7.17(b)]: the method has been used to prepare the parent heterocycle.[32d]

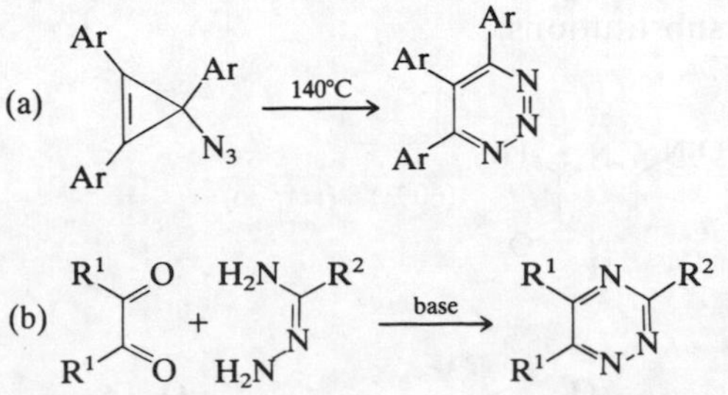

Fig. 7.17 Routes to 1,2,3- and 1,2,4-triazines.

Symmetrical 1,3,5-triazines can be prepared by the trimerization of nitriles under a wide range of conditions. The method has been used as an industrial synthesis of trichlorotriazine (from cyanogen chloride) and of melamine, **38** (from cyanamide). Aliphatic nitriles are not easily trimerized to triazines because they can give pyrimidines instead (Fig. 7.6) but aromatic nitriles are easily trimerized. A convenient method is the use of chlorosulphonic acid at $0°C$.[32e] The parent 1,3,5-triazine can be prepared from formamide by trimerization and dehydration.

There are several routes to 1,2,4,5-tetrazines but the most generally useful one is the oxidation of 1,2-dihydro derivatives. Nitrous acid is the oxidant most commonly used. The dihydrotetrazines can be prepared by the reaction of nitriles with hydrazine hydrate in ethanol under reflux (Fig. 7.18).

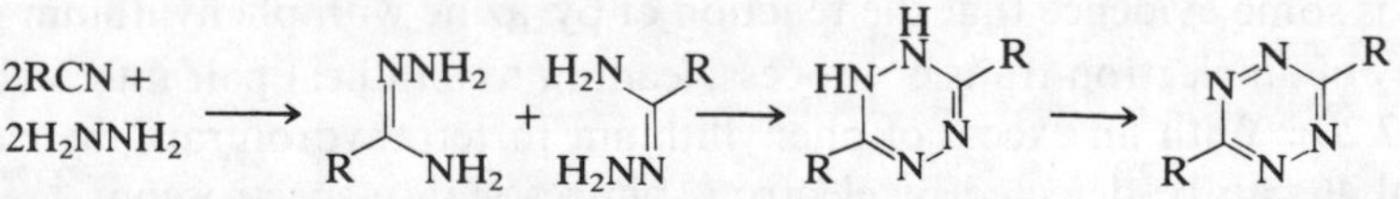

Fig. 7.18 A route to 1,2,4,5-tetrazines.

7.4.3 Chemical properties

The main features of the chemistry of these compounds have been outlined in Section 7.2. Pyridazines and pyrazines can be *N*-alkylated and pyrazine has been dialkylated (using triethyloxonium tetrafluoroborate) although the product is unstable. Pyridazine and alkylpyridazines can be N-oxidized and N-aminated.[35]

Electrophilic substitution at carbon is rare and requires the presence of activating groups. Two examples are shown in Fig. 7.19. As with pyridine-N-oxide, the N-oxide function provides a means of stabilizing the intermediate cation in electrophilic substitution. Pyrazine can be directly chlorinated and brominated at high temperatures, but it is unlikely that these are simple electrophilic substitutions.

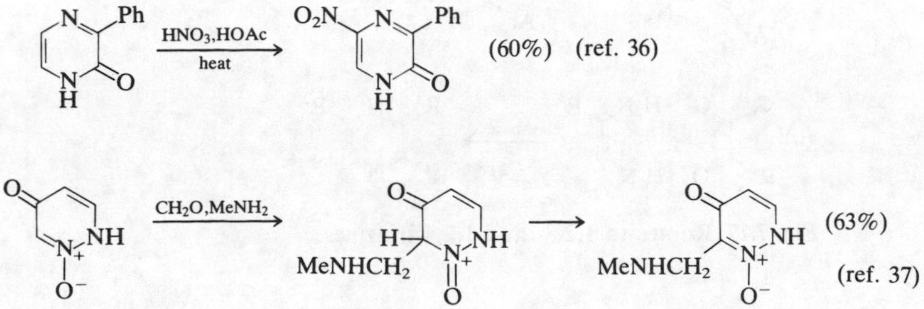

Fig. 7.19 Examples of electrophilic substitution.

The addition of nucleophiles to 1,2,4- and 1,3,5-triazine has been referred to in Section 7.2. The diazines also undergo addition reactions with nucleophiles. For example, pyrazine forms the adduct **43** with potassamide in liquid ammonia.[38]

These 1,2-adducts, such as **43**, are oxidized to the aromatic aminoheterocycles by potassium permanganate. In this way, pyrazine, pyridazine, 1,2,4,5-tetrazines (and pyrimidine) have been aminated in good yield. The method provides an alternative to the Chichibabin reaction, which either fails or is very inefficient with these heterocycles.[39]

Organolithium reagents also form addition compounds; for example, pyridazine and pyrazine give the adducts **44** and **45** with phenyllithium. There is some evidence that the reaction of pyrazine with phenylithium goes by way of an electron transfer process, leading to a radical pair intermediate (Fig. 7.20). With an excess of phenyllithium in tetrahydrofuran, the cation radical **46** can be detected by electron spin resonance spectroscopy.[41]

35 For a review of oxidation, amination, and alkylation of diazines and triazines see M. R. Grimmett and B. R. T. Keene, *Adv. Heterocycl. Chem.*, 1988, **43**, 127.
36 G. Karmas and P. E. Spoerri, *J. Am. Chem. Soc.*, 1956, **78**, 4071.
37 S. Kamiya, G. Okusa, M. Osada, M. Kumagai, A. Nakamura, and M. Koshinuma, *Chem. Pharm. Bull.*, 1968, **16**, 939.
38 J. A. Zoltewicz and L. S. Helmick, *J. Org. Chem.*, 1973, **38**, 658.
39 Review: H. Vorbrüggen, *Adv. Heterocycl. Chem.*, 1990, **49**, 117.
40 R. E. van der Stoel and H. C. Van der Plas, *Rec. Trav. Chim.*, 1978, **97**, 116.
41 Review: W. Kaim, *Angew. Chem. Int. Edn Engl.*, 1983, **22**, 171.

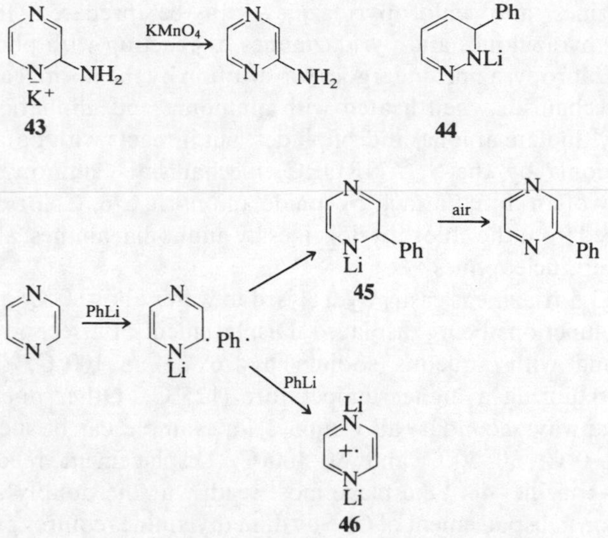

Fig. 7.20 The reaction of pyrazine with phenyllithium.

1,2,4,5-Tetrazines also form addition compounds with nucleophiles such as liquid ammonia and sodium borohydride. Addition takes place across the N-1 to C-6 bond. The adducts show some unusual structural features; for example, in the NMR spectra of the 1,6-dihydrotetrazines **47**, H-6 is abnormally shielded and appears at high field ($\delta\,2.13$ to 2.26).[42] The crystal structure of compound **47** (R = Et) has been recorded; the molecule is boat-shaped and atoms N-1 to N-5 are all sp^2-hybridized, forming a delocalized system (Fig. 7.21);[43] H-6 lies above the plane of this delocalized system.

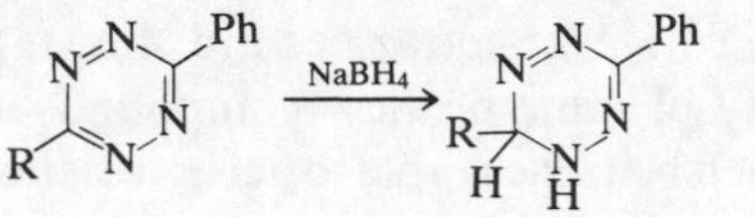

47 (R = H, Me, Et)

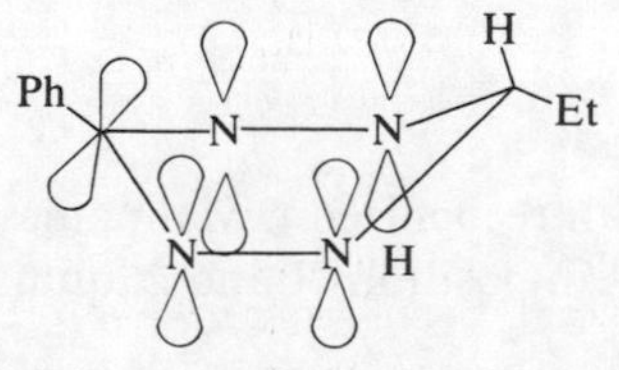

Fig. 7.21 Boat structure of compound **47** (R = Et).

42 A. Counotte-Potman, H. C. van der Plas, and B. van Veldhuizen, *J. Org. Chem.*, 1981, **46**, 2138.

43 For a review of this and related structures see A. Weiss, *Adv. Heterocycl. Chem.*, 1985, **38**, 1.

Chloropyrazines and chloropyridazines can be prepared from the corresponding pyrazinones and pyridazinones by reaction with phosphorus oxychloride. Chloropyrazine undergoes substitution by the normal addition–elimination mechanism when heated with ammonia and aliphatic amines, alkoxides, alkylthiolate anions, and sulphides, but it reacts with potassamide in liquid ammonia by the S_N(ANRORC) mechanism:[3] aminopyrazine is formed by way of an initial attack by amide anions at C-6. Chloride is also readily displaced from the chloropyridazines by ammonia, amines, alkoxides, and many other nucleophiles.

Trichloro-1,3,5-triazine is easily hydrolysed in water above 0°C, one of the three chloride functions being displaced. Displacement of a second chloride requires heating with aqueous sodium hydroxide at 100°C, the third displacement requiring a higher temperature (125°C). Other nucleophiles react in a similar way: secondary alkylamines, for example, can be successively introduced at 0°C, 30–50°C and 80–100°C. Displacement reactions of trichloro-1,2,4-triazine, **48**, take place most readily at the doubly activated 3- and 5-positions; displacement of Cl-6 by dimethylamine requires prolonged heating at 195°C.[44]

The Diels–Alder reactions of 1,2,4- and 1,2,3-triazines have been illustrated earlier (see Table 5.1 and Fig. 7.6). 1,2,4,5-Tetrazines also act as electron-deficient dienes in a similar way, and are useful partners in the Diels–Alder reaction because they react even with simple alkenes. The course of the Diels–Alder reaction with a tetrazine is easily followed because the red or violet colour of the tetrazine is discharged as the adduct is formed. Some examples of Diels–Alder reactions of 1,2,4,5-tetrazines are shown in Table 7.5. The remarkably high reactivity of some of these compounds as dienes is highlighted by example (v), in which benzene is able to act as a dienophile.

7.5 Some fused ring systems

7.5.1 Introduction

There are four benzodiazines, all with recognized trivial names.[45] These are cinnoline, which has nitrogen atoms in positions 1 and 2, quinazoline (1,3-),

44 H. Neunhoeffer and B. Lehmann, *Chem. Ber.*, 1976, **109**, 1113.
45 Reviews: (a) *Condensed Pyridazines Including Cinnolines and Phthalazines*, ed. R. N. Castle, Wiley-Interscience, New York, 1973; (b) *Fused Pyrimidines, Part 1: Quinazolines*, ed. W. L. F. Armarego, Wiley-Interscience, New York, 1967; (c) W. L. F. Armarego, *Adv. Heterocycl. Chem.*, 1979, **24**, 1 (quinazolines); (d) G. W. H. Cheeseman and R. F. Cookson, *Condensed Pyrazines*, Wiley-Interscience, New York, 1979 (quinoxalines).

Table 7.5 Diels–Alder reactions of 1,2,4,5-tetrazines.[a]

R in 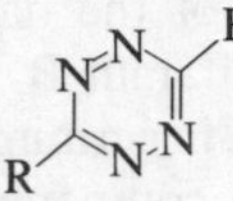 (tetrazine, 3,6-R₂-1,2,4,5-tetrazine)	Dienophile, conditions	Product
(i) CO_2Me	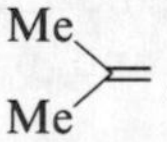 $Me_2C=CMe_2$ (Me₂C=), CH_2Cl_2, 25°C	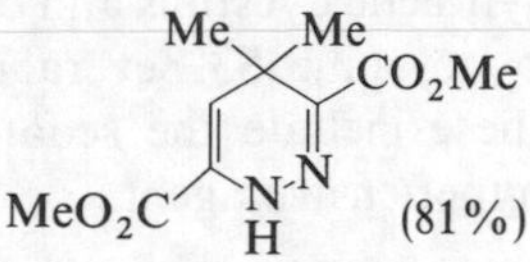 (Me, Me / CO_2Me / MeO_2C, N–N, H) dihydropyridazine (81%)
(ii) Ph	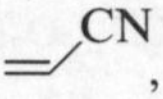 $CH_2=CH$–CN, 100°C/5 days	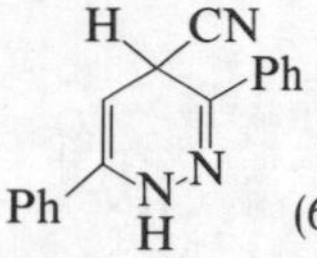 (H, CN, Ph / Ph, N–N, H) (64%)
(iii) SMe	$PhC\equiv CH$, 166°C/12 h	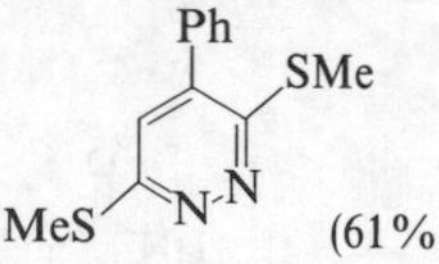 (Ph, SMe / MeS, N–N) (61%)[b]
(iv) CO_2Me	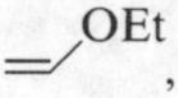 $CH_2=CH$–OEt, dioxane, 25°C	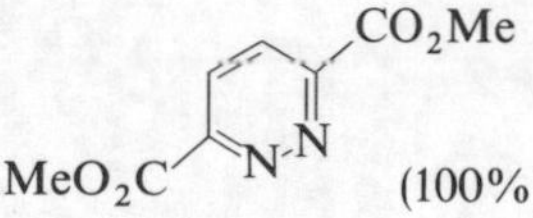 (CO_2Me / MeO_2C, N–N) (100%)
(v) CF_3	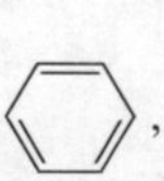 cyclohexene, 140°C/24 h	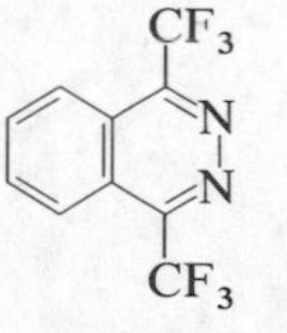 (phthalazine with CF_3 at 1,4-positions) (87%)[c]

[a] See ref. 12. [b] D. L. Boger and S. M. Sakya, *J. Org. Chem.*, 1988, **53**, 1415. [c] The product is formed by *in situ* dehydrogenation: G. Seitz, R. Hoferichter, and R. Mohr, *Angew. Chem. Int. Edn Engl.*, 1987, **26**, 332.

quinoxaline (1,4-), and phthalazine (2,3-). The structures of these heterocycles are shown in Fig. 2.10. Two other fused systems which, because of their relative importance, have recognized trivial names are pteridine, **49**,[4c] and phenazine, **50**.

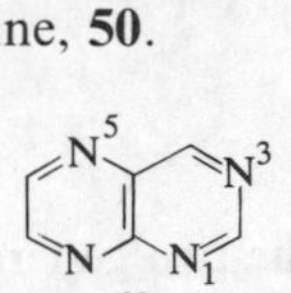

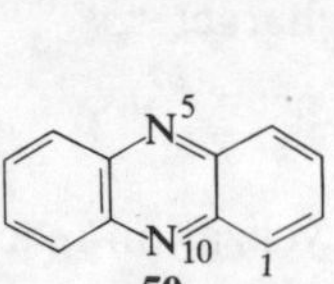

(Structures **49** pteridine and **50** phenazine shown)

The pteridine derivative, folic acid, **51**, is a naturally occurring growth factor which is required by all higher animals. It is involved in the biological conversion of serine to glycine and of homocysteine to methionine. A synthetic folic acid, lacking two of the nitrogen atoms of the natural material, is is a powerful antitumour agent.[46] Riboflavin (vitamin B_2), **52**, is a benzo[g]pteridine which occurs, in a phosphorylated form, in germinating seeds and in milk and eggs. The phenazine ring system is a component of synthetic dyestuffs and of natural pigments such as the blue bacterial pigment pyocyanine, **53**. Several pharmaceuticals contain the quinazoline ring system: these include the sedative methaqualone, **54**, and prazosin, **55**, an antihypertensive agent.

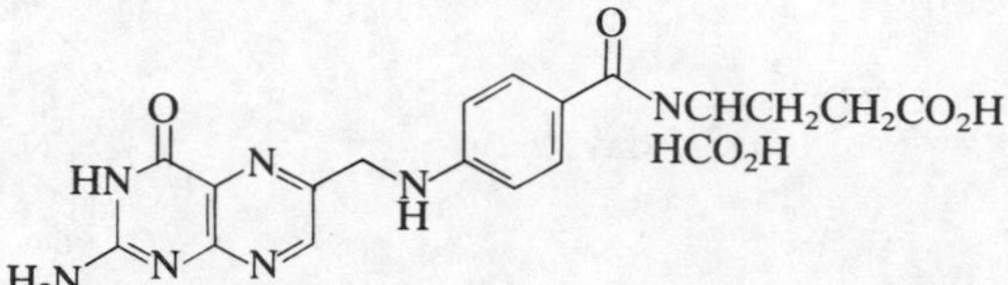

51

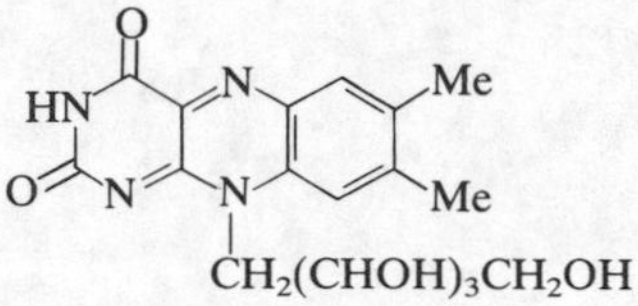

52

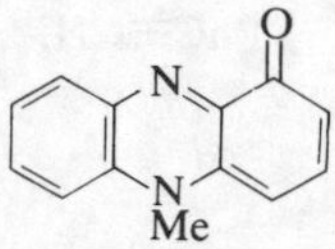

53

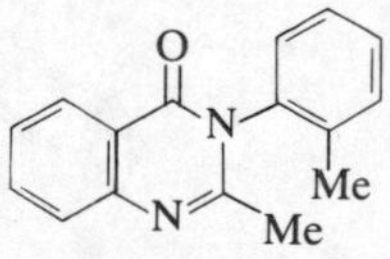

54

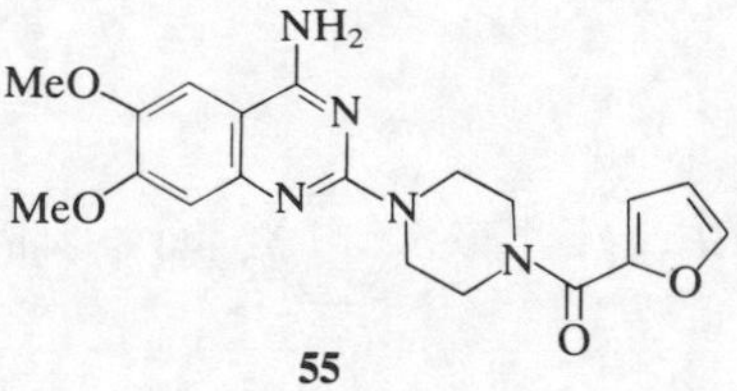

55

All the parent benzodiazines are crystalline solids with low melting points, which are soluble in water and organic solvents. They are all weak bases with pK_a values comparable with those of the corresponding monocyclic diazines: cinnoline, 2.29; quinazoline, 1.95; quinoxaline, 0.56; and phthalazine, 3.47. As in naphthalene there is appreciable bond localization in the ring systems, the 1,2- and 3,4-bonds having double-bond character, and the 2,3-bonds single-bond character.

7.5.2   Ring synthesis

Cinnolines can be prepared by cyclization reactions of the two general types described in Section 4.2.1, in which either monosubstituted benzenes or 1,2-disubstituted benzenes are the precursors.[45] The second type of synthesis

46 E. C. Taylor and G. S. K. Wong, *J. Org. Chem.*, 1989, **54**, 3618.

is the more general; it is exemplified by the preparation of 3,4-disubstituted cinnolines (R^1 = aryl or alkyl) from diazotized 2-vinylaniline derivatives [Fig. 7.22(a)] The other type of synthesis is exemplified by the acid-catalysed cyclization of benzile monophenylhydrazone (b).

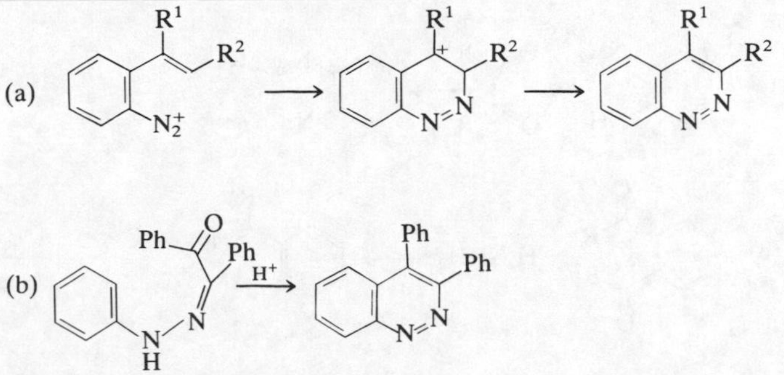

Fig. 7.22 Routes to cinnolines.

Many quinazolines can be prepared from 2-aminobenzaldehydes, 2-aminophenyl ketones, or anthranilic acids.[45b] Thus, a simple route to quinazolines which have alkyl or aryl substituents in the 2- and 4-positions is the cyclization of 2-acylaminobenzaldehydes or 2-acylaminophenyl ketones with alcoholic ammonia [Fig. 7.23(a)]. 4-Quinazolinones can be prepared in a similar way from anthranilic acid and its derivatives. The reaction can be carried out by heating anthranilic acid with an amide to about 180°C, but there are many possible variations: for example, benzoxazinones react with primary amines at or above room temperature to give 4-quinazolinones [Fig. 7.23(b)].[47]

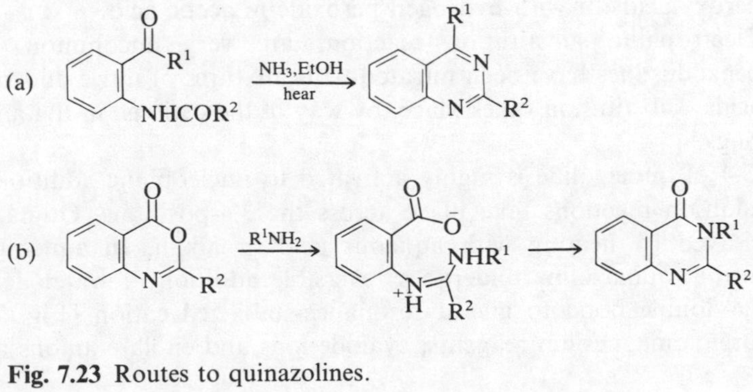

Fig. 7.23 Routes to quinazolines.

47 L. A. Errede, J. J. McBrady, and H. T. Oien, *J. Org. Chem.*, 1977, **42**, 656.

The majority of quinoxalines have been prepared by the reaction of 1,2-diaminobenzene, or an equivalent reagent, with a 1,2-dicarbonyl compound [Fig. 7.24(a)].[45d,48] 1,4-Disubstituted phthalazines can be made by reaction of 1,2-diacylbenzenes with hydrazine (b). 2-Acylbenzoic acids react with hydrazine to give 1-phthalazinones in a similar way.[45a]

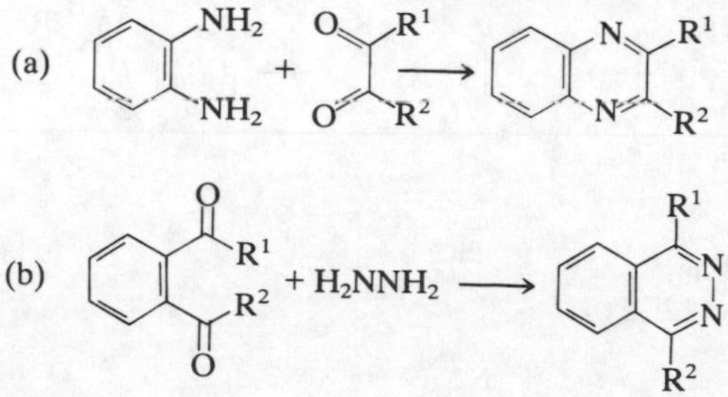

Fig. 7.24 Routes to quinoxalines and phthalazines.

7.5.3   Chemical properties

The chemistry of the benzodiazines is broadly similar to that of the monocyclic analogues; in particular, substituents attached to carbon in the heterocyclic rings are activated by the nitrogen atoms. Some of the main features are as follows.

(a) The nitrogen atoms are basic and nucleophilic. Salts are formed with strong acids. All the benzodiazines can be N-alkylated: cinnoline is methylated at N-2 and quinazoline at N-3 and N-1 (the ratio of products being 5:1 with iodomethane). All can be N-oxidized by reaction with peroxy acids or with hydrogen peroxide in acetic acid.
(b) Electrophilic substitution reactions are very uncommon. All the benzodiazines have been nitrated with mixtures of nitric and sulphuric acids; substitution takes place, by way of the cations, in the all-carbon rings.
(c) C-4 of quinazoline is highly activated to nucleophilic addition. Many addition reactions take place across the 3,4-positions. Quinazoline is cleaved by heating with aqueous acid or alkali. In aqueous acidic media, quinazoline undergoes reversible addition of water across the 3,4-double bond to give a resonance-stabilized cation [Fig. 7.25(a)]. Organomagnesium reagents, cyanide ions and enolate anions also add

48 Examples are syntheses of quinoxaline (R. G. Jones and K. C. McLaughlin, *Org. Synth., Coll. Vol.* 4, 1963, 824) and of pyocyanine 53 (A. R. Surrey, *Org. Synth., Coll. Vol.* 3, 1955, 753).

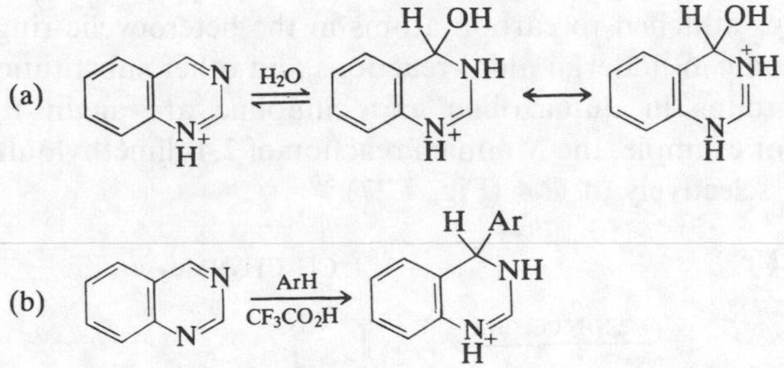

(a)

(b)

Fig. 7.25 Addition reactions of 1-quinazolinium cations.

to the 4-position. It has even proved possible to add activated aromatic compounds such as phenol, pyrrole, and indole to the 4-position by reaction in acid media [Fig. 7.25(b)].[49]

Phthalazine, although less activated than quinazoline, forms addition products by reaction with organomagnesium and organolithium reagents at the 1-position. Pseudobases **56** are produced from the quaternary salts in alkaline media, and Reissert compounds **57**, analogous to those from isoquinolines, can be formed by reaction with acid chlorides and potassium cyanide.

56

57

(d) Nucleophilic displacement of chloride takes place very readily in the heterocyclic rings. The chloro compounds are most easily prepared by reaction of the oxo compounds with phosphorus oxychloride. A typical sequence with 4-cinnolinone is shown in Fig. 7.26. In both cinnoline and quinazoline, chloride at C-4 is the most easily displaced. 4-Chloroquinazoline reacts spontaneously with methanol, and the reaction is autocatalytic; the hydrochloric acid formed in the reaction protonates the heterocycle at N-1 and so promotes further nucleophilic attack.

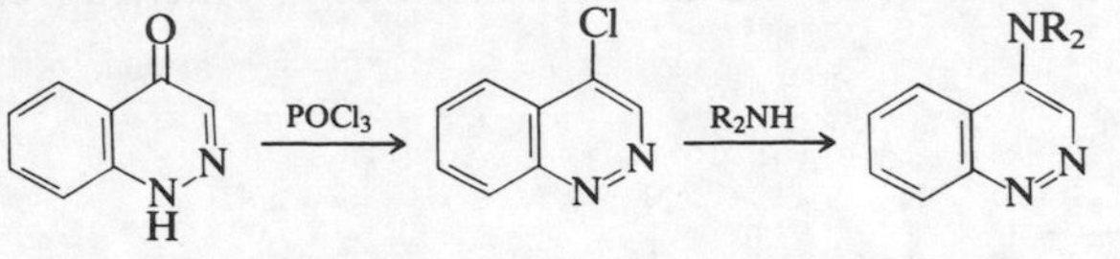

Fig. 7.26 Formation and reaction of 4-chlorocinnoline.

49 W. P. K. Girke, *Chem. Ber.*, 1979, **112**, 1348.

(e) Alkyl groups attached to carbon atoms in the heterocyclic ring can be alkylated and can undergo aldol reactions and other substitutions. The 4-methyl groups in quinazoline and cinnoline are again the most activated; for example, the Mannich reaction of 2,4-dimethylquinazoline takes place selectively at C-4 (Fig. 7.27).[50]

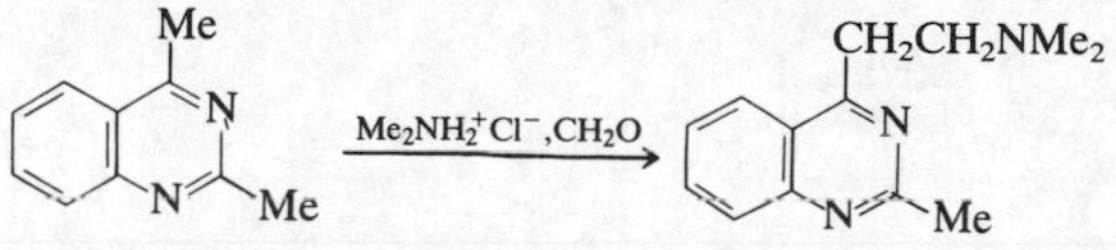

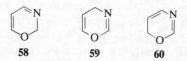

Fig. 7.27 Mannich reaction of 2,4-dimethylquinazoline.

7.6 Oxazines and thiazines

7.6.1 *Introduction*

Six-membered heterocycles with one nitrogen and one oxygen atom are named systematically as oxazines. There are three positional isomers (1,2-, 1,3-, and 1,4-) and each has double-bond isomers. Thus, there are three isomers, **58** to **60**, of the 1,3-oxazine system, which are named as 2*H*-, 4*H*-, and 6*H*-1,3-oxazine, respectively.

None of these ring systems is aromatic because they all contain an sp³-hybridized carbon atom. It is, however, possible to derive an aromatic cation **61** from them, by removal of a hydride ion. Substituted cations of this type have been synthesized. On the other hand, removal of a proton from the heterocycles should produce the anion **62** with eight π-electrons. For the reasons discussed in Section 2.3.2, an anion of this type might be expected to be very unstable, and this is borne out in practice (see Section 7.6.2).

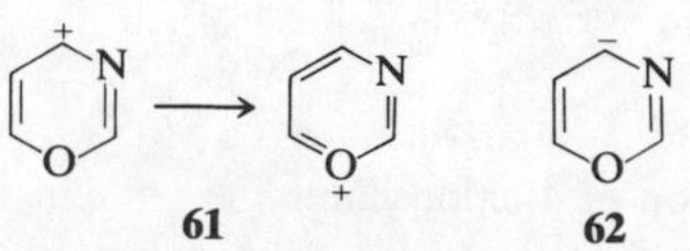

50 J. Siegle and B. E. Christensen, *J. Am. Chem. Soc.*, 1951, **73**, 5777.

A similar series of heterocycles exists which contain one nitrogen and one sulphur atom; these are called thiazines. Thus, **63** is 4*H*-1,3-thiazine and **64** is 2*H*-1,4-thiazine. The dibenzo analogue of 4*H*-1,4-thiazine, **65**, has the recognized trivial name of phenothiazine. Drugs based on this ring system are widely used in medicine. They are important in the management of psychiatric illness and have other useful properties as sedatives, anti-histamines, and in the treatment of nausea. Two phenothiazine drugs are chlorpromazine, **66a**, and triflupromazine, **66b**.

63 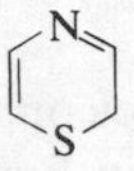   64 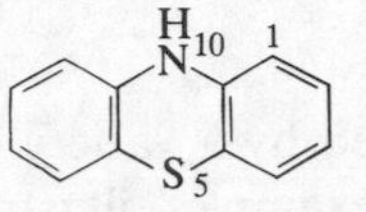   65 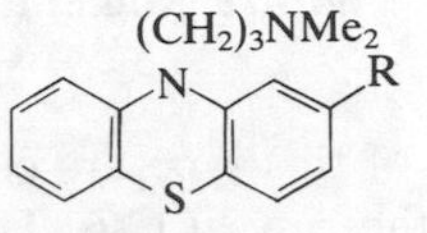   66a; R = Cl
                                     b; R = CF$_3$

The 1,3-oxazines and the phenothiazines are the heterocycles in this group which have been the most widely studied, and we shall consider their chemistry briefly.

7.6.2   *1,3-Oxazines*[51]

The 1,3-oxazine ring system occurs in some natural antibiotics, but most of the work on 1,3-oxazines has been concerned with their use as reagents in organic synthesis.[51,52] The derivatives which have been used most are substituted 1,3-oxazinium cations, **67**, which, like pyrylium cations, are subject to nucleophilic attack, and tetrahydro-1,3-oxazines, **68**, which can be regarded as protected aldehydes.

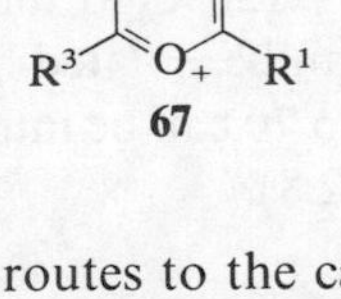

67 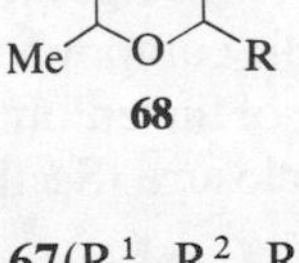   68

Two routes to the cation **67**(R^1, R^2, R^3 = Ph) are illustrated in Fig. 7.28. Note that the triphenylmethyl cation is capable of producing the oxazinium cation by hydride abstraction, showing that the cation **67** is a relatively stable species.

51 Reviews: (a) R. R. Schmidt, *Synthesis*, 1972, 333; (b) Z. Eckstein and T. Urbanski, *Adv. Heterocycl. Chem.*, 1978, **23**, 1.
52 Review: A. I. Meyers, *Heterocycles In Organic Synthesis*, Wiley-Interscience, New York, 1974, p. 201.

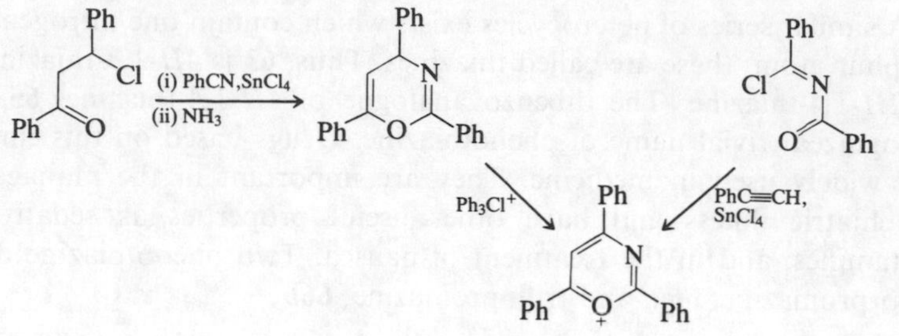

Fig. 7.28 Routes to 2,4,6-triphenyloxazinium salts.

Salts of this type are cleaved by a variety of nucleophiles, which usually attack the ring at C-6. Two examples, illustrated in Fig. 7.29, are the ring opening with ammonia and with benzylmagnesium bromide, followed by cyclization.[51a]

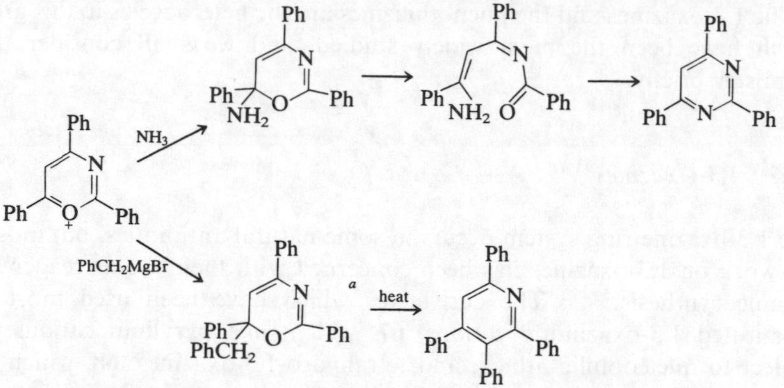

Fig. 7.29 Reactions with nucleophiles. [a] Isolable intermediate.

In contrast to the cations **67**, the corresponding anions are extremely unstable. The deep blue anion **69** rearranges above $-120°C$ to the yellow valence tautomer **70**, and the bicyclic compound **71** can be isolated from the reaction mixture.[53] The conversion of the anion **69** into **70** can be interpreted as a 6π-electron electrocyclic ring closure (Section 4.2.8).

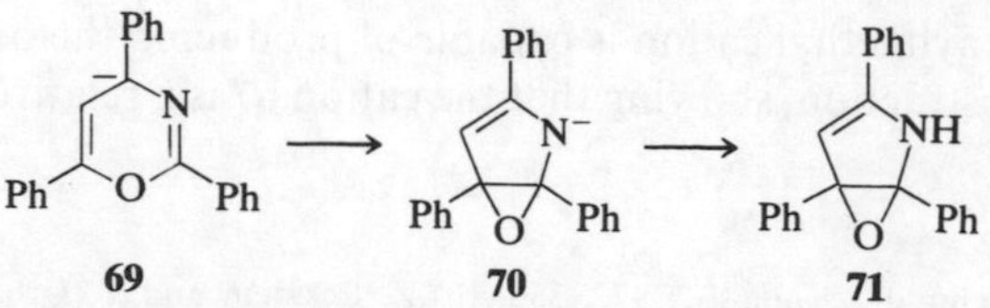

53 Review: R. R. Schmidt, *Angew. Chem. Int. Edn Engl.*, 1975, **14**, 581.

An example of the generation and use of tetrahydro-1,3-oxazines, **68**, is shown in Fig. 7.30. The dihydrooxazines **72** are prepared from the readily available diol 2-methylpentane-2,4-diol and nitriles in acidic media. They can be lithiated and alkylated efficiently at the exocyclic methylene group. The functionalized oxazines are then converted into tetrahydrooxazines by reduction and the aldehyde carbonyl group is released by acidic hydrolysis.[52]

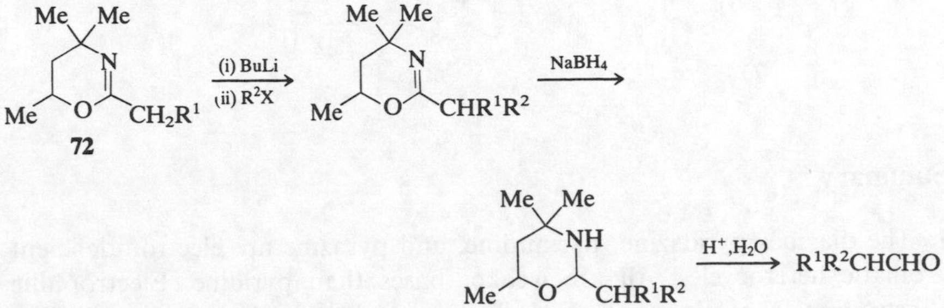

Fig. 7.30 Formation of aldehydes by way of tetrahydrooxazines.

7.6.3 Phenothiazines[54]

Phenothiazine exists as the 10*H*-tautomer **65**, and the phenothiazine drugs are 10-alkylated compounds. One of the oldest and most widely used methods of synthesis is the reaction of diarylamines with sulphur, iodine being used as a catalyst. The reaction is illustrated as part of a synthesis of the drug triflupromazine in Fig. 7.31.[55] As in this example, a major disadvantage of the method is that two isomeric phenothiazines can be formed from *meta*-substituted amines.

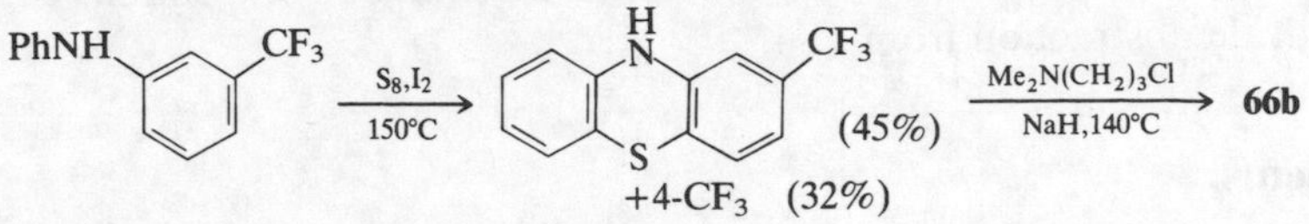

Fig. 7.31 A route to triflupromazine.

Phenothiazine is a yellow solid which is easily oxidized. One electron oxidation gives the cation radical **73**, which may lose a proton or which may be further oxidized to the cation **74**. This cation is attacked by nucleophiles at C-3; for example, 3-nitrophenothiazine can be formed by oxidizing phenothiazine with iron(III) chloride in the presence of sodium nitrite.[56]

54 Review: C. Bodea and I. Silberg, *Adv. Heterocycl. Chem.*, 1968, **9**, 321.
55 H. L. Yale, F. Sowinski, and J. Bernstein, *J. Am. Chem. Soc.*, 1957, **79**, 4375.
56 J. Daneke and H.-W. Wanzlick, *Liebigs Ann. Chem.*, 1970, **740**, 52.

Phenothiazine can be converted into its *S*-oxide by reaction with hydrogen peroxide in the presence of sodium ethoxide. Electrophilic substitution reactions often lead to mixtures of products and are complicated by oxidation of the phenothiazines.

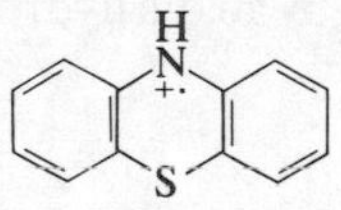

73 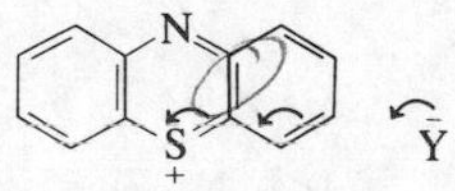   74

Summary

1. The diazines pyridazine, pyrimidine, and pyrazine are electron-deficient aromatic heterocycles. All are weaker bases than pyridine. Electrophilic substitution takes place only in derivatives with strongly electron-releasing groups, the 5-position of pyrimidine being the least deactivated.

2. Nucleophiles attack the diazines, triazines, and tetrazines readily. The rate of displacement of leaving groups increases with the number of nitrogen atoms in the ring. In a particular ring system the rate is greatest at the carbon atoms which are best activated by ring nitrogen atoms.

3. Nucleophilic substitution often occurs by the normal addition–elimination mechanism but there are also many examples of nucleophilic addition and cleavage reactions. Substitution reactions involving the use of potassimide in liquid ammonia often involve ring cleavage (the S_NANRORC mechanism).

4. 1,2,4-Triazines and 1,2,4,5-tetrazines can act as electron-deficient dienes in Diels–Alder reactions.

5. Quinazoline is the most activated of the benzodiazines toward nucleophilic attack and it undergoes nucleophilic addition readily at C-4.

6. 4*H*-1,3-Oxazines can be converted into aromatic 1,3-oxazinium cations by hydride abstraction from C-4.

Problems

1. Rationalize the following routes to pyrimidines:

(a) 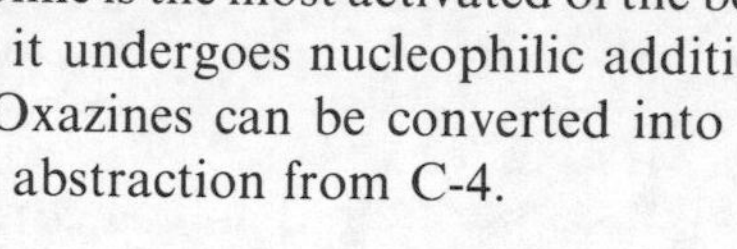 PrCN + MeO⁻ $\xrightarrow{180°C}$ (Pr = MeCH₂CH₂)

(b) 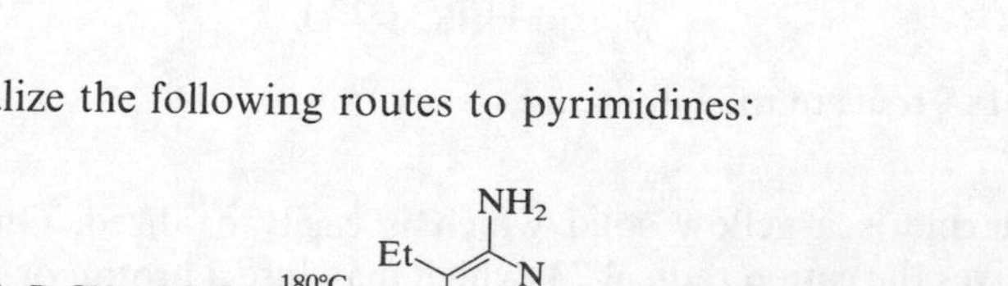 PhCN + MeLi $\longrightarrow$ $\xrightarrow{200-300°C}$

2. Suggest synthetic routes to (a) minoxidil, **8**, starting from ethyl cyanoacetate and guanidine, (b) methaqualone, **54**, starting from anthranilic acid.

3. An alternative synthesis of minoxidil, **8**, has the following steps:

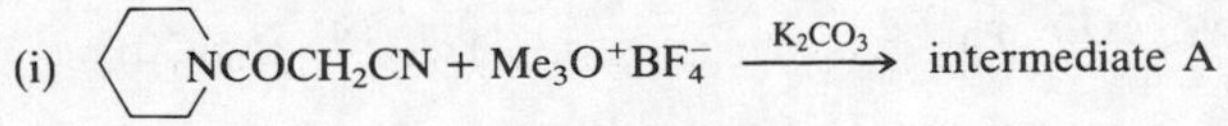

(i) $\langle \text{NCOCH}_2\text{CN} + \text{Me}_3\text{O}^+\text{BF}_4^- \xrightarrow{\text{K}_2\text{CO}_3}$ intermediate **A**

(ii) **A** + H$_2$NCN $\longrightarrow$ intermediate **B**

(iii) **B** + H$_2$NOH $\longrightarrow$ minoxidil

By analysing the route retrosynthetically, or otherwise, deduce structures for A and B and suggest how the final product is formed.

4. The 5-oxide of riboflavin (**52**) has been synthesized as follows:

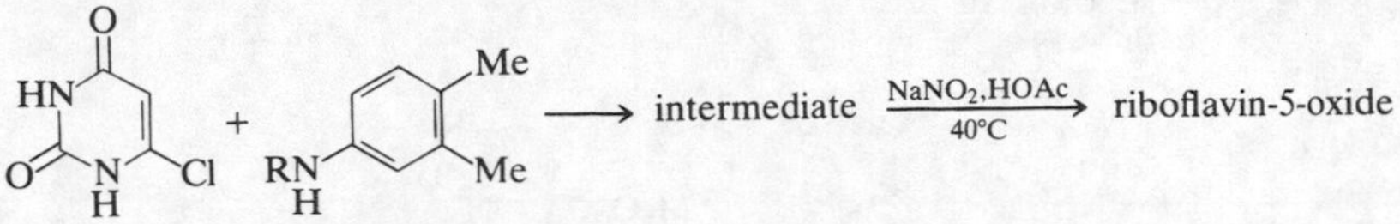

$\longrightarrow$ intermediate $\xrightarrow[\text{40°C}]{\text{NaNO}_2,\text{HOAc}}$ riboflavin-5-oxide

(R = ribityl)

Identify the intermediate and explain the reaction steps.

5. Suggest the explanations for the following:

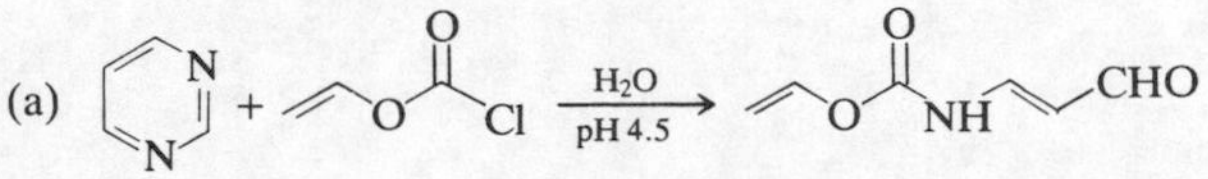

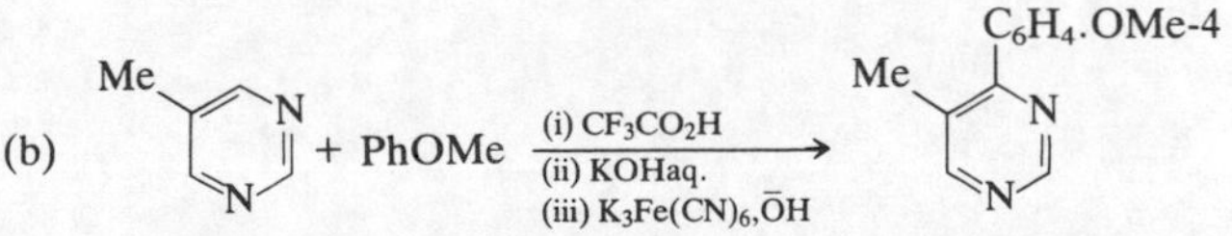

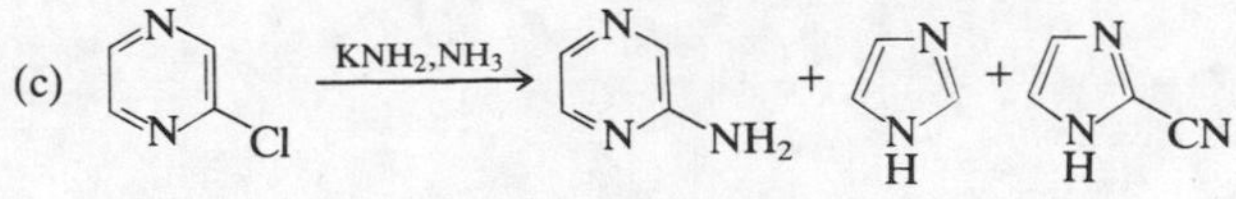

(a)

(b)

(c)

(13%)   (13%)   (30%)

(d)

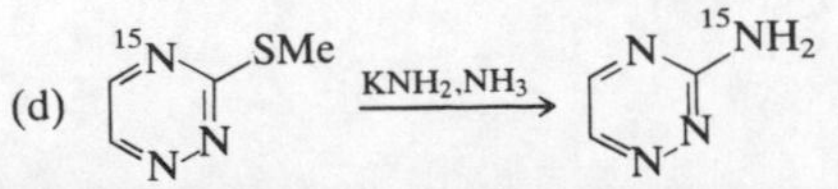

(e)

(f)

(g)

(h)

8 Five-membered ring compounds with two or more heteroatoms

8.1 Introduction

From the discussion of aromatic heterocycles in Chapter 2 we can see that the largest and most diverse group of such compounds is that represented by five-membered rings containing more than one heteroatom. Most of these ring systems are formally derived from furan, pyrrole, or thiophene by replacement of one or more of the CH groups by sp^2-hybridized nitrogen. It is the variation in the number and positions of these nitrogen atoms which leads to the structural diversity of this group of heterocycles. The preferred nomenclature for the ring systems is made up of a mixture of trivial and systematic (Hantzsch–Widman) names; however, nearly all the preferred names of the nitrogen-containing rings end in -azole. There are two monocyclic diazoles; pyrazole and imidazole. Triazoles and tetrazoles, with respectively three and four nitrogen atoms in the rings, are named systematically, as are most of the heterocycles containing nitrogen in combination with oxygen or sulphur.[1]

The presence of additional nitrogen atoms in the rings has important effects on the properties of the ring system, in comparison with those discussed in Chapter 6. The additional nitrogen atoms bear lone pairs of electrons in the plane of the rings. These lone pairs are not involved in the π-electron system and thus provide a favourable point of attack for protons and other electrophiles. The additional nitrogen atoms also cause a lowering in the energy levels of the π-orbitals (compare benzene and pyridine, Fig. 2.2) so that the heterocycles are less 'π-electron rich'. The overall result is to make electrophilic attack at carbon less easy than in pyrrole, furan or thiophene. On the other hand, the additional nitrogen atoms have an inductive electron-withdrawing effect and can provide stabilization to negatively charged reaction intermediates. Two examples of reactions which are rarely observed in pyrrole, furan or thiophene, but which are much more common in the azoles, are nucleophilic addition–elimination [Fig. 8.1(a)] and the

---

1 For comparative reviews of the diazoles, triazoles, and tetrazoles, see M. R. Grimmett, in *Comprehensive Organic Chemistry*, Vol. 4, ed. P. G. Sammes. Pergamon Press, Oxford, 1979, p. 357; K. Schofield, M. R. Grimmett, and B. R. T. Keene, *The Azoles*, Cambridge University Press, Cambridge, 1976.

(a)

(b)

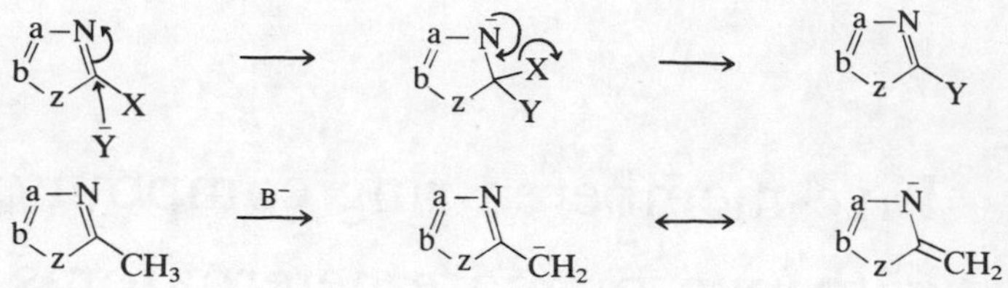

Fig 8.1 Examples of azole reactions involving negatively charged intermediates.

deprotonation of substituent methyl groups [Fig. 8.1(b)]. Reactions of this type become increasingly favoured as the stabilization of the intermediate anions increases.

The effects of the additional nitrogen atoms can also be seen in the acidity and basicity of these heterocycles. The lone pairs on nitrogen provide sites for protonation, and most azoles are stronger bases than pyrrole. On the other hand the stability of azolyl anions is greater than that of the pyrryl anion, so that azoles containing NH groups are stronger acids than pyrrole. The pK_a values listed in Table 8.1 illustrate these points. Imidazole is a

Table 8.1 pK_a values of some azoles and azolium cations.[a]

| 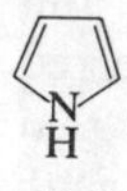 16.5 | 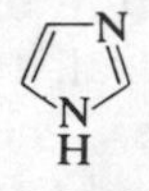 14.52 | 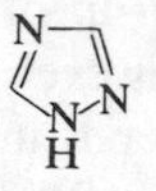 10 | 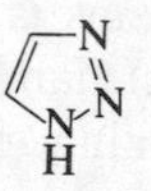 9.4 | 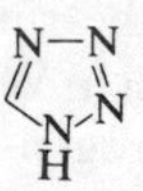 4.8 | 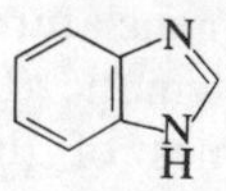 13.2 |

| 7.00 | 2.53[b] | 0.8[b] | 5.5 |

| 2.5 | −0.51[b] | −2.97[b] | −4.7[b] | −2.20[b] |

| 2.2 | 1.2 | −3.0[c] |

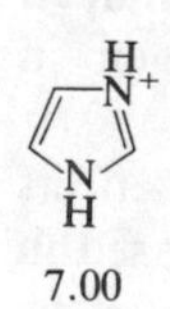

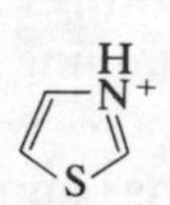

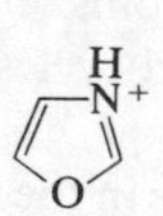

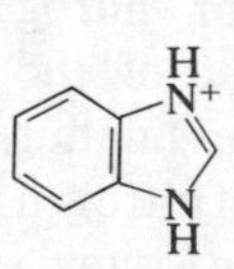

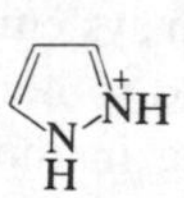

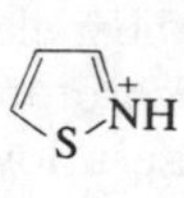

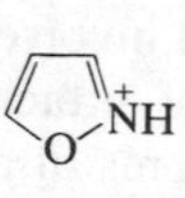

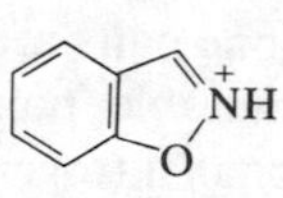

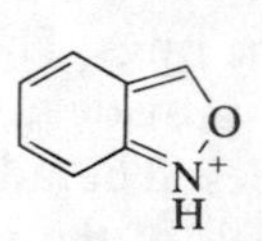

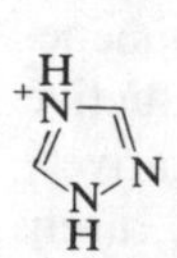

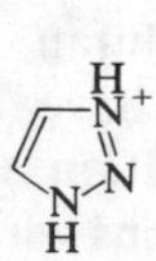

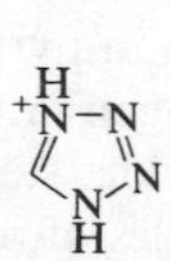

a Values are from ref. 1, except where indicated. *b* Values from J. A. Zoltewicz and L. W. Deady, *Adv. Heterocycl. Chem.*, 1978, **22**, 71. *c* M. M. Sokolova, V. A. Ostrovskii, G. I. Koldobskii, V. V. Mel'nikov, and B. V. Gidaspov, *Zh. Org. Khim.*, 1974, **10**, 1085. See also J. Catalan, J. L. M. Abboud, and J. Elguero, *Adv. Heterocycl. Chem.*, 1987, **41**, 187.

moderately strong base but the other azoles are weak bases, the base strength generally decreasing as the number of nitrogen atoms increases because of the inductive electron-withdrawing effect of the additional nitrogen atoms. The oxygen-containing heterocycles are the least basic, as might be expected because of the inductive effect of oxygen, and the base-weakening effect is greater when the heteroatom is adjacent to the site of protonation. Conversely the acidity of the azoles increases with the number of nitrogen atoms: the triazoles are comparable in acid strength with phenol and 1*H*-tetrazole is about as strong an acid as acetic acid.

One consequence of the easy mobility of protons in the *N*-unsubstituted azoles is that they are extensively hydrogen bonded in the liquid and solid phases. They are also much more soluble in water than is pyrrole. Hydrogen bonding is no longer possible when the nitrogen is substituted, nor in oxazoles and thiazoles. The effect of hydrogen bonding is most marked on the properties of 1-unsubstituted imidazoles. Imidazole is a solid, m.p. 90°C, b.p. 256°C, whereas 1-methylimidazole is a liquid, m.p. −6°C, b.p. 198°C. The boiling points of thiazole (117°C) and oxazole (69°C) are also much lower than that of imidazole.

Another important effect of proton mobility is that in the *N*-unsaturated diazoles, triazoles, and tetrazoles, tautomeric forms exist which are interrelated by intermolecular proton transfer (Section 2.5). This proton transfer is usually much too rapid to permit the isolation of separate structures. The tautomers of asymmetrically substituted pyrazoles and imidazoles are shown in Fig. 8.2; triazoles and tetrazoles, having more nitrogen atoms, have a greater number of tautomeric forms. Note from Fig. 8.2 that, because of this rapid proton transfer, a 3-substituted pyrazole and a 5-substituted pyrazole do not normally have separate existence; similarly, 4- and 5-substituted imidazoles interconvert too rapidly to allow their separate isolation. The tautomers may, however, exist in solution predominantly in one form. For example, imidazoles with a nitro group at the 4(5)-position show a strong preference for the 4-substituted tautomer (Fig. 8.3), as do other imidazoles with electron-withdrawing substituents. If N-1 is substituted in these heterocycles, the possibility of prototropic tautomerism, is, of course, removed.

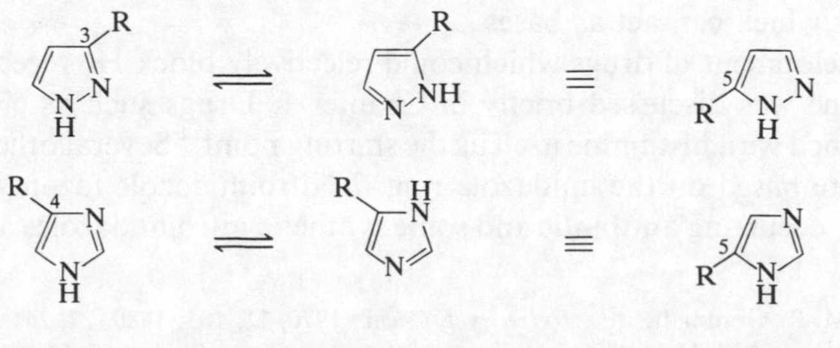

Fig. 8.2 Tautomerism in pyrazoles and imidazoles.

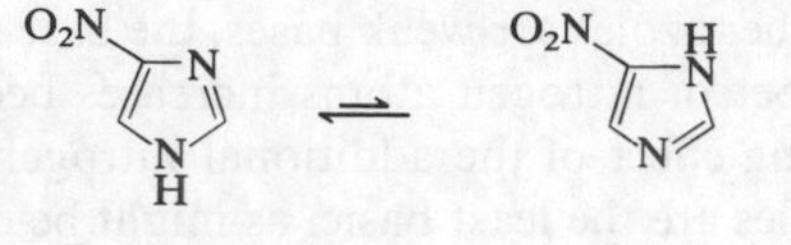

Fig. 8.3 Preference for the 4-nitro structure in 4(5)-nitroimidazole.

8.2 Imidazoles[2]

8.2.1 *Introduction*

Imidazole is a planar heterocycle which has appreciable resonance energy, somewhat higher than that of pyrrole.[3] The ring system is of particular importance because it is present in the essential amino acid, histidine, **1**, and its decarboxylation product, histamine.

Histidine residues are found at the active sites of ribonuclease and of several other enzymes. In some enzymes their function appears to be the catalysis of proton transfer (see also Section 8.2.3). The imidazole system can act both as a base and as an acid. Free imidazole is a moderately strong organic base (pK_a 7.0) and it can also act as a weak acid (pK_a 14.5) (Table 8.1). Both the cation and the anion are symmetrical delocalized structures (Fig. 8.4).

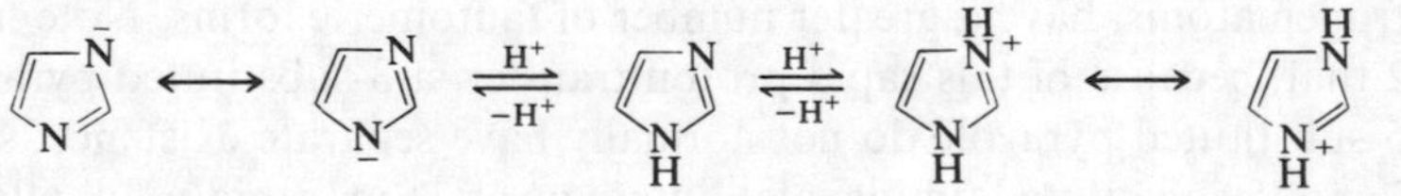

$$pK_a = 14.5 \qquad pK_a = 7.0$$

Fig. 8.4 Proton transfer equilibria.

At the physiological pH of about 7.4, appreciable concentrations of both protonated and neutral imidazole units are likely to be present. Thus, there are species present at the active sites of the enzymes which can act as acids and others which can act as bases.

The development of drugs which could selectively block H_2 receptor sites of histamine was discussed briefly in Chapter 1. Drugs such as cimetidine were designed with histamine itself as the starting point.[4] Several other classes of drugs are based on the imidazole ring. 2-Nitroimidazole (azomycin) **2** is a naturally occurring antibiotic and some synthetic nitroimidazoles are active

2 Reviews: M. R. Grimmett, *Adv. Heterocycl. Chem.*, 1970, **12**, 103; 1980, **27**, 241.
3 M. J. S. Dewar, A. J. Harget, and N. Trinajstić, *J. Am. Chem. Soc.*, 1969, **91**, 6321.
4 See ref. 5 in Chapter 1.

against intestinal infections.[5] Metronidazole **3** is used for this purpose and also as a radiosensitizer in X-ray therapy. Other imidazoles are useful antifungal agents: these include bifonazole **4** and clotrimazole **5**.

8.2.2 Ring synthesis

There is no general, widely applicable method for the synthesis of the imidazole ring, but a variety of cyclization reactions are used to produce specifically substituted imidazoles. One of these, based on isonitrile cyclization, is useful for imidazoles unsubstituted at C-2 and is outlined in Chapter 4, Table 4.8. Another good method for 2-unsubstituted imidazoles is the Bredereck reaction[6] in which an α-hydroxyketone or an α-haloketone is heated with formamide [Fig. 8.5(a)]. α-Aminoketones are intermediates in the synthesis of several types of imidazole; reaction with thiocyanates or isothiocyanates gives imidazole-2-thiols (the *Marckwald* synthesis), and with cyanamide, 2-aminoimidazoles are formed [Fig. 8.5(b)]. A cyclization route to 4-aminoimidazoles is shown in (c).[7] A different type of cyclization process, illustrated by example (d), is the photolysis of 1,5-disubstituted tetrazoles, which probably goes by way of an imidoylnitrene intermediate.[8] This method has been used to prepare a range of 2,4-disubstituted imidazoles. 1,3-Dipolar cycloaddition reactions of nitrile ylides and of azomethine ylides have also been used to prepare imidazoles; an example is given in Table 4.18.

5 Reviews: *Nitroimidazoles; Chemistry, Pharmacology and Clinical Applications*, eds A. Breccia, B. Cavalleri, and G. E. Adams, Plenum Press, New York, 1982; J. H. Boyer, *Nitrazoles*, VCH, Deerfield Beach, Florida, 1986.

6 Review: H. Bredereck, R. Gompper, H. G. v. Shuh, and G. Theilig, in *Newer Methods of Preparative Organic Chemistry*, Vol. III, ed. W. Foerst, Academic Press, New York, 1964, p. 241. See also B. H. Lipschutz and M. C. Morey, *J. Org. Chem.*, 1983, **48**, 3745.

7 A. Edenhofer, *Helv. Chim. Acta*, 1975, **58**, 2192. For a review of imidazole syntheses based on HCN oligomers see D. S. Donald and O. W. Webster, *Adv. Heterocycl. Chem.*, 1987, **41**, 1.

8 M. Casey, C. J. Moody, C. W. Rees, and R. G. Young, *J. Chem. Soc., Perkin Trans. 1*, 1985, 741.

(a)

R¹—C(=O)—CH(R²)—OH + HCONH₂ ⟶ R¹—C(OH)(NHCHO)—CH(R²)—OH ⟶ R¹—CH(NHCHO)—C(=O)R²

$\xrightarrow{\text{HCONH}_2}$ R¹—CH(NHCHO)—C(R²)(NHCHO)—OH ⟶ imidazole (R¹, R², N, NH)

(b)

R¹—C(=O)—CH(R²)—NH₂

$\xrightarrow{\text{KCNS}}$ imidazole with SH at 2-position

$\xrightarrow{\text{H}_2\text{NCN}}$ imidazole with NH₂ at 2-position

(c)

Me—C(OMe)=N—CN + H₃N⁺CH₂XCl⁻ ⟶ Me—C(NHCH₂X)=N—CN $\xrightarrow{\text{NaOEt}}$

H₂N-substituted imidazole with X and Me   (X = CN or CO₂Et)

(d)

Me-tetrazole with N=N–N, CH=CH–CO₂Me $\xrightarrow[-\text{N}_2]{hv}$ [Me—C(=N:)—N=CH—CH=... CO₂Me] ⟶ imidazole with CO₂Me, Me, N, NH (63%)

Fig. 8.5 Some cyclization routes to imidazoles. (See also Tables 4.6 and 4.8)

8.2.3 Reactions

Imidazole is an excellent nucleophile and it can react readily at nitrogen with alkylating and acylating agents. Imidazoles can be *N*-acylated by reaction (in a 2:1 ratio) with the acyl halide in an aprotic solvent.

N-Acylimidazoles are very useful 'acyl transfer' reagents, having reactivity comparable with acyl halides or anhydrides. For example, 1-acetylimidazole has a half-life of only 41 min in water (pH 7.0) at 25°C, whereas normal amides are not hydrolysed at all in these conditions. The usual amide stabilization through the lone pair on nitrogen (Fig. 3.6) is much less effective in *N*-acylimidazoles because the nitrogen lone pair is part of the aromatic sextet.

This is shown by the carbonyl stretching frequency of $1747\,cm^{-1}$ in the infrared spectrum, which is at a much higher value than for NN-disubstituted amides. In addition, N-3 provides a site for protonation. The heterocycle is then an excellent leaving group. The acylimidazoles thus undergo typical nucleophilic addition–elimination reactions, for example with alcohols (to give esters), amines (to give amides), and Grignard reagents (to give ketones).

Imidazole can also catalyse the hydrolysis of esters and other acyl derivatives. With acyl derivatives RCOX, in which X is a good leaving group, imidazole acts as a nucleophile to attack the carbonyl group. Because of its minimal steric requirement, imidazole is an excellent nucleophile, but, unless it is deprotonated, it is also an excellent leaving group and the acylated intermediate is easily attacked by water or another external nucleophile (Fig. 8.6).

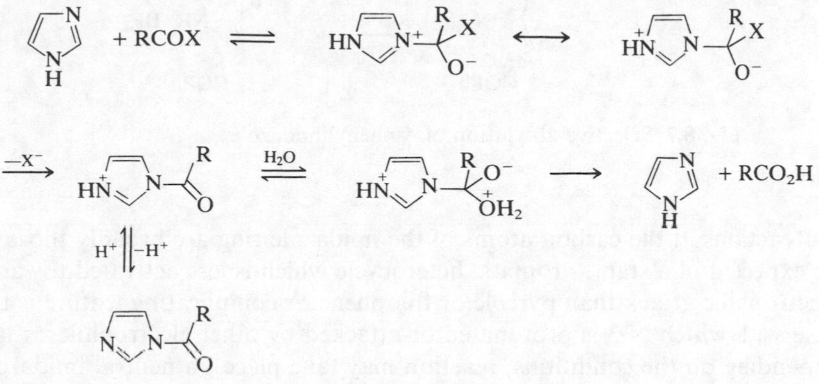

Fig. 8.6 Hydrolysis of N-acylimidazole, **6**, and nucleophilic catalysis by imidazole of hydrolysis of RCOX.

When X in RCOX is a poor leaving group, imidazole can catalyse hydrolysis in a different way, simply by acting as a base to remove a proton from water and thus to generate hydroxide ions. This is probably the role of histidine in some hydrolytic enzymes; a hydroxide ion is produced in the vicinity of the acyl group of the substrate.

The N-alkylation of imidazole is most often carried out in the presence of base, so that the anion is the species alkylated. Alkyl halides and similar compounds can be used. Imidazole can also be alkylated by conjugate addition to acrylonitrile and related species in neutral media. The reaction of imidazole with alkyl halides in neutral conditions can lead to further alkylation and the formation of quaternary salts.[9] With 4-substituted

9 For a survey of conditions for alkylation of 1-methylimidazole and other azoles see M. Begtrup and P. Larsen, *Acta Chem. Scand.*, 1990, **44**, 1050.

imidazoles there are two possible products of *N*-alkylation. The nature of the product depends upon the choice of alkylating conditions, and on the steric and electronic effects of the substituent. Steric effects seem to be important in that it is usually the 1,4-rather than the 1,5-isomer which is formed predominantly. This is illustrated for 4-phenylimidazole in Fig. 8.7. The 1,5-isomer can, however, be obtained by acylating the imidazole, quaternizing it, and removing the acyl group as shown.

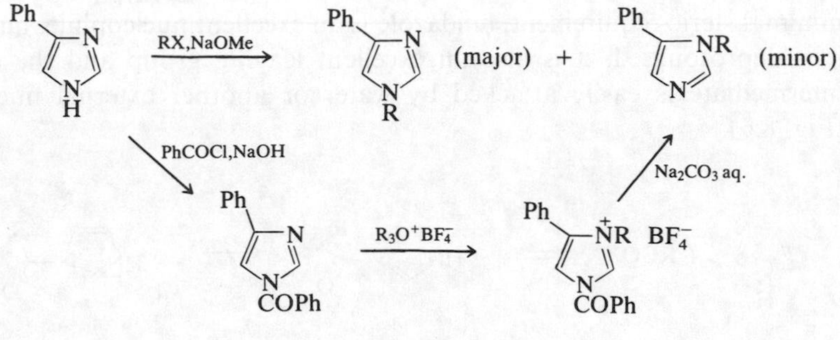

Fig. 8.7 Selective alkylation of 4-phenylimidazole.

Reactions at the carbon atoms of the imidazole ring are broadly those to be expected of a stable aromatic heterocycle which is less activated towards electrophilic attack than pyrrole or thiophene. A complicating feature is the ease with which N-3 is protonated or attacked by other electrophiles: thus, depending on the conditions, reaction may take place on neutral imidazole or on an imidazolium cation. Reactions carried out in strongly acid conditions involve the cation, which is deactivated to electrophilic attack. Friedel–Crafts alkylation or acylation generally fails for this reason, and nitration in sulphuric acid is difficult, giving 4-nitroimidazole as the major product. Sulphonation (with hot 50% oleum) also takes place at the 4-position. It might be expected that electrophilic substitution at the 2-position would be unfavourable, relative to attack at C-5, because the charge in the intermediate cation (Fig. 8.8) has to be delocalized through the imine nitrogen.

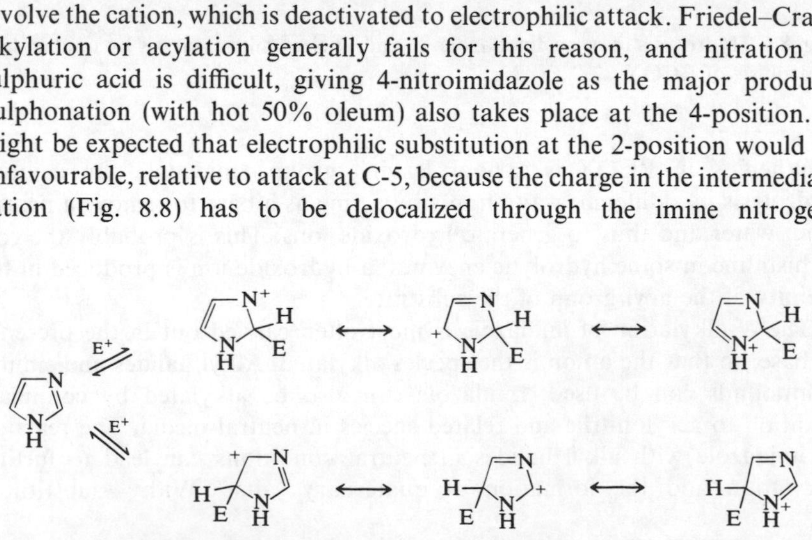

Fig. 8.8 Electrophilic attack at C-2 and C-5 of imidazole.

Nevertheless, imidazole is easily converted into 2,4,5-tribromoimidazole by reaction with bromine in the absence of a catalyst. It also undergoes coupling, mainly at the 2-position, with arenediazonium salts in basic solution; in this case, it is the imidazolyl anion which is attacked by the electrophile.

One unusual feature of the chemistry of imidazoles is the ease with which the proton can be removed from C-2 in neutral or basic conditions. This process seems to involve the slow formation of an imidazolium ylide 7 (R = H). The ylide 7 (R = Me) can be generated from the 1,3-dimethylimidazolium cation by reaction with sodium hydride and can then be substituted at C-2 by alkyl halides and other electrophiles.[10] A reaction which may involve a similar type of ylide intermediate 8 is the useful conversion of 1-benzoylimidazole into 2-benzoylimidazole in the presence of triethylamine and pyridine.[11] 1-Substituted imidazoles can be selectively lithiated at C-2 and this provides an alternative method of C-substitution.

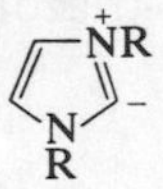

7 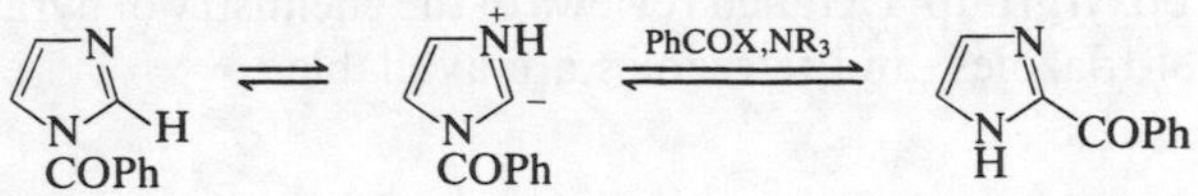   **8**

Nucleophilic displacement reactions of imidazoles take place most readily at C-2 [see Fig. 8.1(a)] but displacement reactions are not easy at any position unless additional activating groups are present. The imidazole ring is cleaved in strongly basic conditions and this precludes some displacement reactions with basic nucleophiles. Some examples[2] of nucleophilic displacement reactions are given in Fig. 8.9.

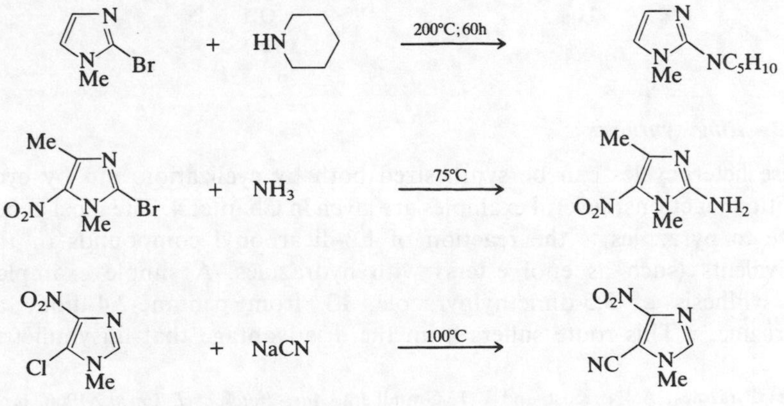

Fig. 8.9 Examples of nucleophilic displacement in imidazoles.

10 M. Begtrup, *J. Chem. Soc., Chem. Commun.*, 1975, 334.
11 L. A. M. Bastiaansen and E. F. Godefroi, *Synthesis*, 1978, 675.

8.3   Pyrazoles, triazoles, and tetrazoles[1]

8.3.1   Introduction

Derivatives of these heterocycles are stable aromatic compounds. Many have been used commercially as pharmaceuticals, pesticides, and dyestuffs. Some examples are the pyrazolium salt **9** (difenzoquat), a herbicide, Raxil **10**, a fungicide used for plant protection, phenylbutazone **11**, an antiinflamatory drug, and fluconazole **12**, a drug used to treat fungal infections. 1,2,3-Triazole derivatives are used as optical brightening agents. Tetrazoles are important in medicinal chemical research because N-unsubstituted tetrazoles can be regarded as analogues of carboxylic acids: their pK_a values are comparable (Section 8.1) and they are planar, delocalized systems with about the same spatial requirement. Analogues of amino acids and many other natural carboxylic acids have been synthesized with a tetrazole ring in place of the carboxyl group. Detailed reviews of the chemistry of pyrazoles, 1,2,3-triazoles, 1,2,4-triazoles, and tetrazoles are available.[12]

8.3.2   Ring synthesis

These heterocycles can be synthesized both by cyclization and by cyclo-addition reactions; several examples are given in Chapter 4. The most general route to pyrazoles is the reaction of 1,3-dicarbonyl compounds or their equivalents (such as enol esters) with hydrazines. A simple example is the synthesis of 3,5-dimethylpyrazole, **13**, from pentane-2,4-dione and hydrazine.[13] This route suffers from the disadvantage that unsymmetrical

12 (a) Pyrazoles: A. N. Kost and I. I. Grandberg, *Adv. Heterocycl. Chem.*, 1966, **6**, 347;
   (b) 1,2,3-triazoles: T. L. Gilchrist and G. E. Gymer, *Adv. Heterocycl. Chem.*, 1974, **16**, 33;
   K. T. Finley, *Triazoles:1,2,3*, ed. J. A. Montgomery, Wiley-Interscience, New York, 1980;
   (c) 1,2,4-triazoles: C. Temple, *Triazoles:1,2,4*, ed. J. A. Montgomery, Wiley-Interscience, New York, 1981; (d) tetrazoles: R. N. Butler, *Adv. Heterocycl. Chem.*, 1977, **21**, 323.
13 R. H. Wiley and P. E. Hexner, *Org. Synth., Coll. Vol. 4*, 1963, 351.

dicarbonyl compounds or their derivatives sometimes give mixtures of isomeric pyrazoles. A good route to several *N*-unsubstituted pyrazoles is provided by the reaction of hydrazides $XCSNHNH_2$ ($X = SR$ or NR_2) with α-haloketones. This reaction gives the thiadiazines **14** as intermediates, which, either spontaneously or on treatment with acids, extrude sulphur to give pyrazoles in good yield.[14]

$$MeCOCH_2COMe + H_2NNH_2 \xrightarrow{\bar{O}H} \underset{\underset{NNH_2}{\|}}{MeCOCH_2CMe} \longrightarrow \mathbf{13}$$

$$R^1COCHR^2Cl + \underset{\underset{S}{\|}}{XCNHNH_2} \longrightarrow \mathbf{14} \xrightarrow{H^+}$$

Pyrazoles are also made by cyclization of acetylenic hydrazones (Table 4.7), by electrocyclization of unsaturated diazo compounds (Table 4.11), and by 1,3-dipolar addition reactions of diazo compounds (Table 4.16) and nitrile imides (Table 4.17).

There are many methods available for the synthesis of 1,2,4-triazoles, the most important of which are based on the construction and cyclization of structures of the types N—C—N—N—C and C—N—C—N—N. An example of the first type is the thermal condensation of an acylhydrazide with an amide or (better) a thioamide (the *Pellizzari* reaction); thus benzoylhydrazide and thiobenzamide react at 140°C to give 3,5-diphenyl-1,2,4-triazole, **15**.[15] The *Einhorn–Brunner* reaction is an example of the second type of cyclization; in this, hydrazine or a monosubstituted hydrazine is condensed with a diacylamine in the presence of a weak acid. Phenylhydrazine and *N*-formalbenzamide combine in this way to give 1,5-diphenyl-1,2,4-triazole, **16**, in good yield.[16]

$$PhCONHNH_2 + PhCSNH_2 \longrightarrow \mathbf{15} \longrightarrow$$

$$PhNHNH_2 + PhCONHCHO \longrightarrow \mathbf{16} \longrightarrow$$

14 H. Beyer, H. Honeck, and L. Reichelt, *Liebigs Ann. Chem.*, 1970, **741**, 45.
15 M. Pesson, S. Dupin, and M. Antoine, *Bull. Soc. Chim. Fr.*, 1962, 1364.
16 M. R. Atkinson and J. B. Polya, *J. Chem. Soc.*, 1952, 3418.

1*H*-1,2,3-Triazoles can be prepared by the 1,3-dipolar cycloaddition of a wide variety of organic azides XN_3 (X=alkyl, vinyl, aryl, acyl, arenesulphonyl, etc.) to acetylenes. An example of this type of reaction is given in Table 4.17. Several other compounds react with azides to give 1,2,3-triazoles; these include enolate anions, enol ethers, enamines, and α-acylphosphorus ylides. For example, the ylide **17** reacts with azidobenzene in solution at 80°C to give 1,5-diphenyl-1,2,3-triazole in good yield.[17] These reactions, in contrast to the additions to acetylenes, are highly regioselective. 2*H*-1,2,3-Triazoles are occasionally formed in azide additions, for example with trimethylsilyl azide and acyl azides; apparently the substituent migrates to *N*-2 after formation of the ring system. A more general approach to 2*H*-1,2,3-triazoles is from 1,2-diketones. A common example of this type is the oxidative cyclization by copper(II) salts of the bis-arylhydrazones of 1,2-diketones. A likely mechanism for the cyclization of *bis*-phenylhydrazones of α-ketoaldehydes, **18**, is shown;[18] note that it is another example of the heterotriene cyclizations illustrated in Fig. 4.19.

Several of the important syntheses of tetrazoles involve the addition of hydrazoic acid or azide ions to compounds containing carbon–nitrogen multiple bonds. Examples are (a) the synthesis of 5-phenyltetrazole by heating benzonitrile and sodium azide in *NN*-dimethylformamide,[19] (b) the formation of 1-benzyltetrazole from benzyl isocyanide and hydrazoic acid in the presence of a trace of sulphuric acid,[20] and (c) the conversion of imidoyl chlorides **19** into 1,5-disubstituted tetrazoles by reaction with hydrazoic acid or sodium azide.[21] 2,5-Disubstituted tetrazoles can be obtained in a different way, by the reaction of amidrazones, **20**, with nitrous acid.

17 P. Ykman, G. L'Abbé, and G. Smets, *Tetrahedron*, 1971, **27**, 845.
18 See ref. 19 in Chapter 4.
19 W. G. Finnegan, R. A. Henry, and R. Lofquist, *J. Am. Chem. Soc.*, 1958, **80**, 3908.
20 D. M. Zimmerman and R. A. Olofson, *Tetrahedron Lett.*, 1969, 5081.
21 E. K. Harvill, R. M. Herbst, E. C. Schreiner, and C. W. Roberts, *J. Org. Chem.*, 1950, **15**, 662.

$$PhCN + N_3^- \longrightarrow$$ [structure: Ph-tetrazole with N-N-N-N ring, NH]

$$PhCH_2\overset{+}{N}\equiv\bar{C} + HN_3 \longrightarrow$$ [structure: triazole ring with N-N-N, CH_2Ph]

[structure 19: R^1, Cl, NR^2] $+ N_3^- \longrightarrow$ [structure: R^1, N_3, NR^2] $\longrightarrow$ [structure: R^1, tetrazole N-N-N-N, R^2]

19

[structure 20: R^1, NH_2, $NNHR^2$] $+ HONO \longrightarrow$ [structure: R^1, N_2^+, $NNHR^2$] $\longrightarrow$ [structure: R^1, tetrazole N=N-N, NR^2]

20

8.3.3 Substitution reactions

The chemistry of these azoles becomes increasingly different from that of pyrrole as the number of nitrogen atoms increases. Electrophilic substitution reactions at carbon are relatively uncommon in these ring systems, because the number of carbon atoms is fewer and because electrophiles preferentially attack nitrogen. Pyrazole can be chlorinated and brominated at the 4-position in mild conditions. 4-Nitropyrazole can also be prepared in an indirect way from pyrazole: reaction with nitronium acetate ($NO_2^+OAc^-$) gives 1-nitropyrazole, **21**, which is converted to 4-nitropyrazole when treated with sulphuric acid. This reaction probably involves transfer of the NO_2^+ group by way of the cation **22**, because 1-nitropyrazole has been shown to be a convenient and effective nitrating agent for aromatic hydrocarbons in the presence of acids.[22] 1,2,3-Triazole can be converted into the 4,5-dibromo derivative by reaction with sodium hypobromite in acetic acid[23] and 1,2,4-triazole is chlorinated at C-3 by way of an isolable 1-chloro derivative.[24] Thus at least some of the C-substitutions by electrophiles in these ring systems go through intermediates in which the initial attack is on nitrogen.

[structure 21: pyrazole with NO_2 on N] $\longrightarrow$ [structure 22: pyrazolium with NH^+ and NO_2] $\longrightarrow$ [structure: 4-nitropyrazole with O_2N, NH]

21   **22**

22 G. A. Olah, S. C. Narang, and A. P. Fung, *J. Org. Chem.*, 1981, **46**, 2706.

23 R. Hüttel and G. Welzel, *Liebigs Ann. Chem.*, 1955, **593**, 207.

24 V. Grinsteins and A. Strazdina, *Khim. Geterotsikl. Soedin.*, 1969, 1114 (*Chem. Abstr.*, 1970, **72**, 121456).

The azoles are *N*-alkylated by a wide variety of alkylating agents but it is often difficult to predict which of the possible isomers will be produced. The isomer ratio is usually very sensitive to the nature of the alkylating agent and the reaction conditions. For example, methyl 5-methylpyrazole-3-carboxylate, **23**, is methylated at N-1 by diazomethane, but at N-2 by iodomethane. Sometimes steric effects determine the products: for example, 4-phenyl-1,2,3-triazole is methylated at N-1 and N-2, but not at N-3, by dimethyl sulphate.[25] Some *N*-trimethylsilylazoles can be selectively alkylated; for example, 1-trimethylsilyl-1,2,4-triazole, **24**, which is easily prepared from 1,2,4-triazole, is alkylated at N-2. The trimethylsilyl group is lost in the process.[26]

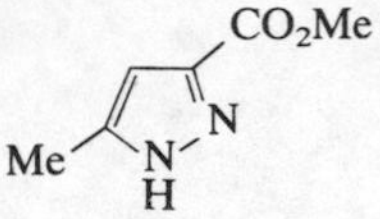

23

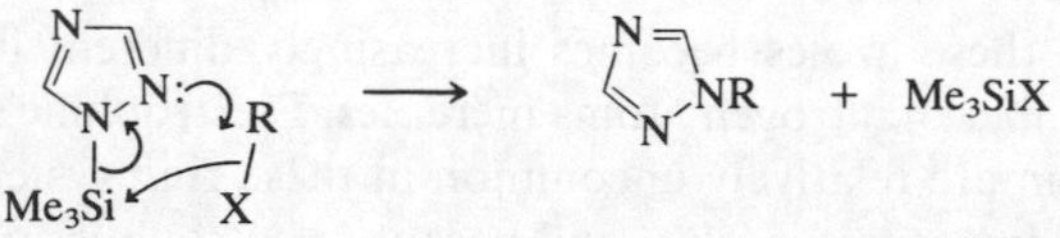

24

There are many examples of the formation of quaternary salts from *N*-substituted azoles and alkylating agents. It is even possible to form diquaternary salts: for example, 1-methyl-1,2,4-triazole is successively methylated at N-4 and N-2 by an excess of trimethyloxonium tetrafluoroborate, and the salt **25** can be isolated as a crystalline solid.[27]

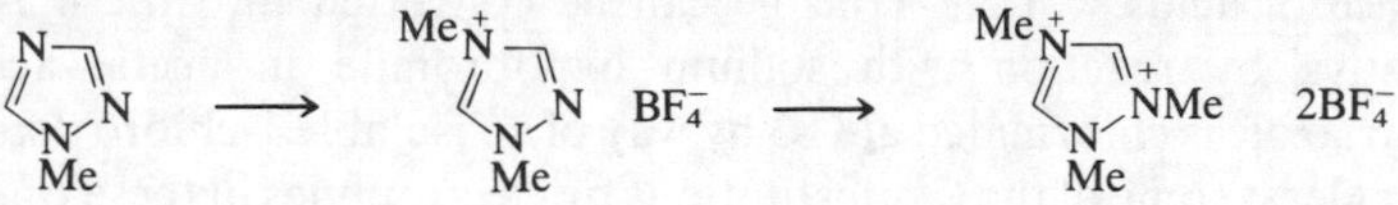

25

Nucleophilic displacement reactions of carbon substituents which are good leaving groups takes place in several of these azoles. 5-Bromo- or 5-chloro-1-substituted tetrazoles can be converted into other 5-substituted tetrazoles by nucleophilic displacement. The 1,5-disubstituted tetrazoles react more readily than their 2,5-isomers, probably because the intermediates are better stabilized (Fig. 8.10).[28]

25 H. Hoberg, *Liebigs Ann. Chem.*, 1967, **707,** 147.
26 J. P. Gasparini, R. Gassend, J. C. Maire, and J. Elguero, *J. Organometal. Chem.*, 1980, **188,** 141.
27 T. J. Curphey and K. S. Prasad, *J. Org. Chem.*, 1972, **37,** 2259.
28 G. B. Barlin, *J. Chem. Soc. (B)*, 1967, 641.

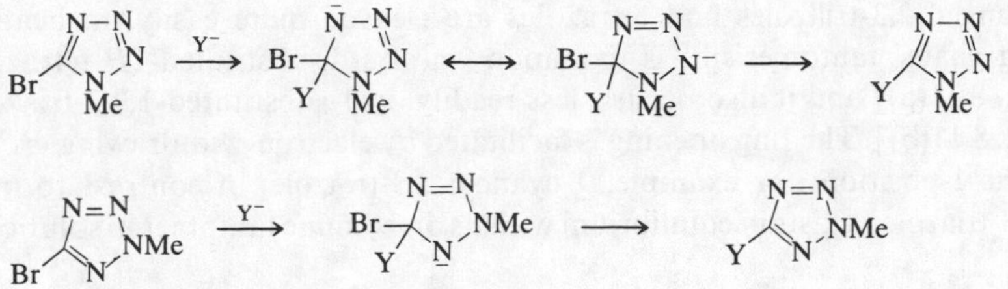

Fig. 8.10 Nucleophilic displacement reactions of 5-bromo-1-methyltetrazole and 5-bromo-2-methyltetrazole.

The base-catalysed reaction of 5-chloro-1-phenyltetrazole with phenols leads to the formation of the ethers **26**, which can be reductively cleaved by reaction with hydrogen over a palladium catalyst.[29] This is a useful procedure for the deoxygenation of phenols.

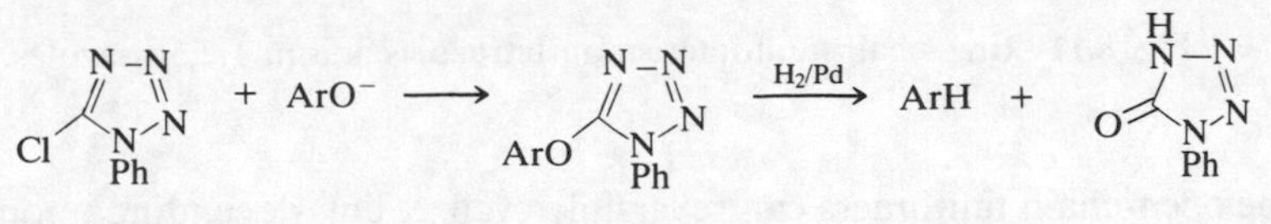

26

The ease of displacement of halide from these azoles is, of course, increased by the presence of additional electron-withdrawing groups on the ring carbon atoms, or by quaternization at nitrogen; for example, in the triazolium cation **27**, the chloride is displaced very easily by azide ions.[30]

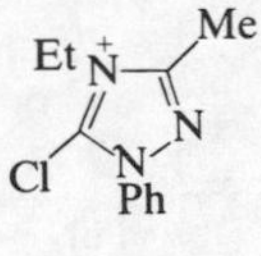

27

8.3.4 Ring cleavage

These ring systems are, in general, remarkably resistant to cleavage reactions. All are resistant to oxidative cleavage and they survive reaction with most reducing agents. Lithium aluminium hydride does cleave 1,5-disubstituted tetrazoles to amines **28**.[31]

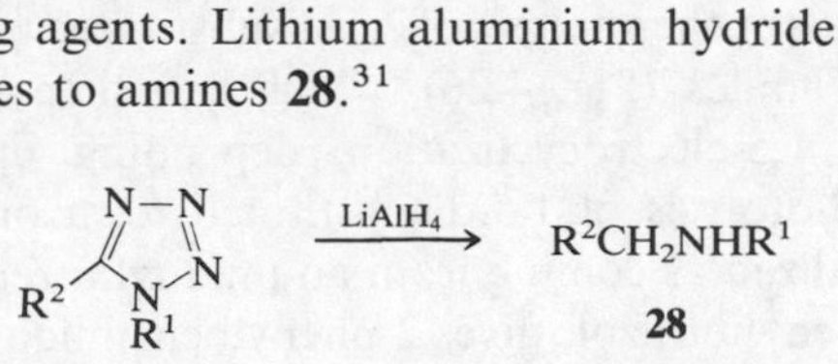

28

29 W. J. Musliner and J. W. Gates, *Org. Synth.*, 1971, **51**, 82.
30 H. Balli and F. Kersting, *Liebigs Ann. Chem.*, 1961, **647**, 1.
31 R. LaForge, C. E. Cosgrove, and A. D'Adamo, *J. Org. Chem.*, 1956, **21**, 988.

Some 1,2,3-triazoles and tetrazoles are cleaved quite easily on heating. Ring–chain tautomerism is common in 1,5-disubstituted-1*H*-tetrazoles [Fig. 8.11(a)] and it also occurs, less readily, in 1-substituted-1,2,3-triazoles [Fig. 8.11(b)]. The ring opening is facilitated by electron-withdrawing groups at the 1-position; for example, 1-cyano-1,2,3-triazole, in contrast to most 1,2,3-triazoles, exists in equilibrium with its diazoimine tautomer in solution.

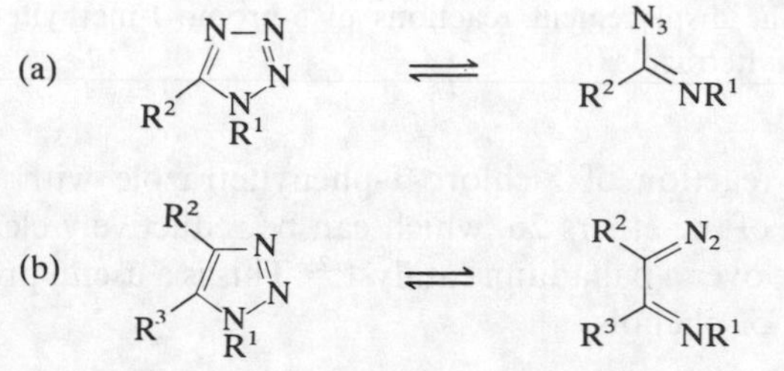

Fig. 8.11 Ring–chain tautomerism in tetrazoles and in 1,2,3-triazoles.

The open-chain tautomers can reversibly cyclize, but, depending upon the nature of the substituents and the conditions, they may also rearrange or undergo other reactions in which nitrogen is extruded. *The Dimroth rearrangement*[32] occurs when tetrazoles and 1,2,3-triazoles having a 5-amino substituent are heated. It is illustrated for 5-amino-1-phenyl-1,2,3-triazoles in Fig. 8.12. The position of equilibrium in such rearrangements is influenced both by the nature of the substituents and by the pH of the solvent.

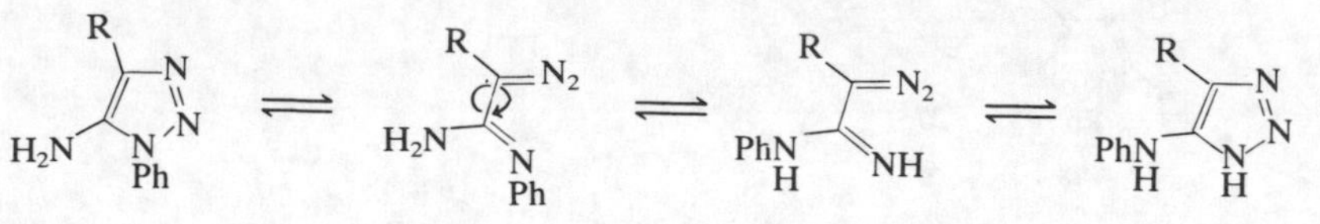

Fig. 8.12 The Dimroth rearrangement.

Nitrogen can be extruded from tetrazoles and from 1,2,3-triazoles by the action of heat or light. Photolysis or thermolysis of 2,5-disubstituted tetrazoles is used as a route to nitrile imides (Fig. 4.26), which can then undergo 1,3-dipolar cycloaddition or 1,5-electrocyclization, depending upon the nature of the substituents. Photolysis of 1,5-disubstituted tetrazoles gives products which can be rationalized as coming from an imidoylnitrene intermediate; for example, 1,5-diphenyltetrazole gives 2-phenylbenzimidazole, **29**. The imidazole synthesis in Fig. 8.5 provides another example.

32 Review: D. J. Brown, in *Mechanisms of Molecular Migrations*, Vol. 1, ed. B. S. Thyagarajan, Wiley-Interscience, New York, 1968, p. 209.

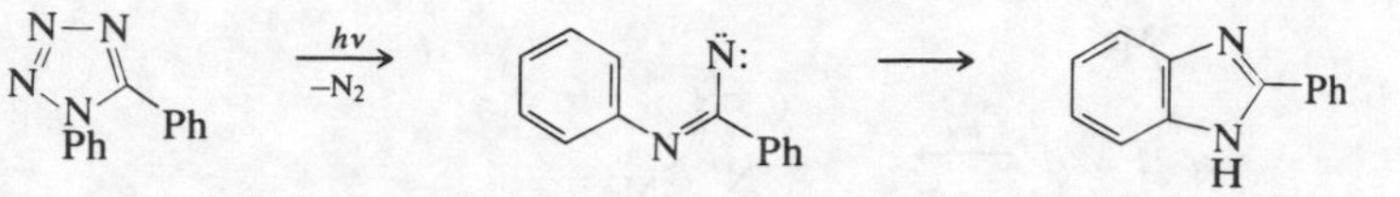

29

1-Substituted-1,2,3-triazoles similarly give products on thermolysis or photolysis which can be derived from intermediate imidoylcarbenes. The products depend very much upon the nature of the substituents and the reaction conditions, but there is good evidence that some of the products isolated from the vapour-phase pyrolysis of 1,2,3-triazoles are formed by way of transient 1*H*-azirine intermediates **30**.[33]

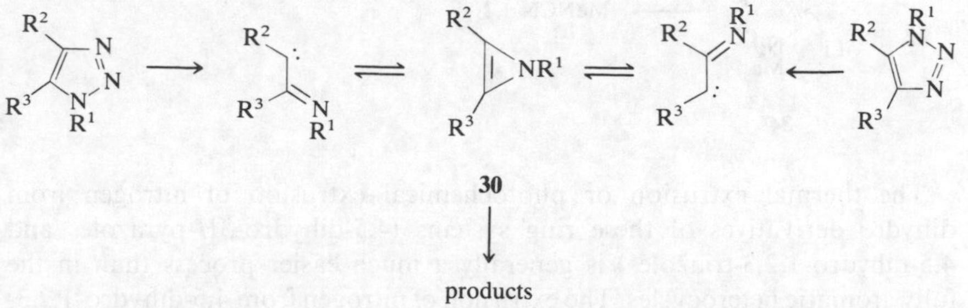

30

↓

products

Perhaps the most remarkable application of the nitrogen extrusion reaction is the use of tetrazole-5-diazonium chloride **31** as a source of carbon atoms. The salt **31** is carefully decomposed thermally in the presence of reagents which can intercept the monatomic carbon produced.[34]

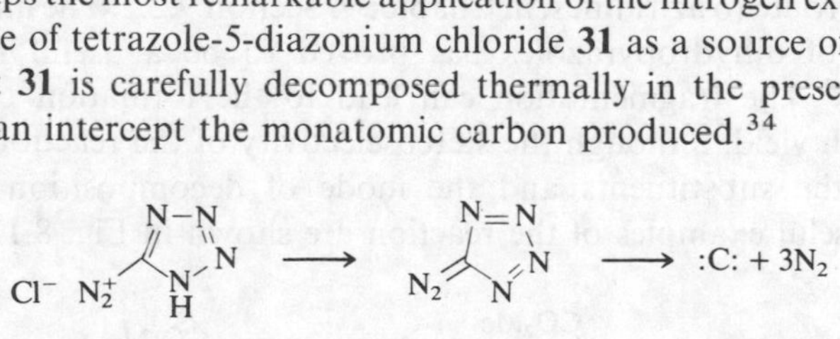

31

The action of heat and light on 1*H*-pyrazoles and 1,2,4-triazoles, which do not have the —N=N— grouping, rarely leads to ring cleavage. Some examples of the cleavage of 1,2,4-triazoles at high temperatures in the vapour phase are recorded.[35] Several pyrazoles undergo photorearrangement to imidazoles[36] (in a manner analogous to the photochemical rearrangements of furans and thiophenes discussed in Chapter 6); for example, 1,3-dimethylpyrazole, **32**, is converted by irradiation into 1,2-dimethylimidazole, **33**.

33 T. L. Gilchrist, G. E. Gymer, and C. W. Rees, *J. Chem. Soc., Perkin Trans, 1*, 1975, 1.
34 Review: P. B. Shevlin, in *Reactive Intermediates*, Vol. 1, ed. R. A. Abramovitch, Plenum, New York, 1980, p. 1.
35 T. L. Gilchrist, C. W. Rees, and C. Thomas, *J. Chem. Soc., Perkin Trans. 1*, 1975, 12.
36 Review: see ref. 73 in Chapter 6.

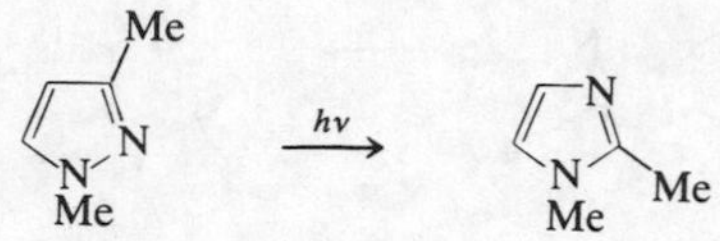

Attempts to carry out substitution reactions by way of *C*-metallated intermediates sometimes fail, however, because they are easily cleaved; for example, both 5-lithio-1-methyltetrazole, **34**, and 5-lithio-1,4-diphenyl-1,2,3-triazole are thermally unstable and extrude nitrogen above about $-50°C$.[37]

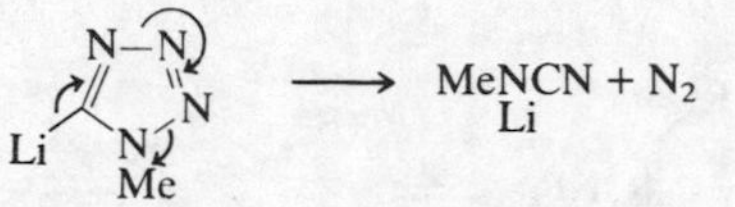

The thermal extrusion or photochemical extrusion of nitrogen from dihydro derivatives of these ring systems (4,5-dihydro-3*H*-pyrazoles and 4,5-dihydro-1,2,3-triazoles) is generally a much easier process than in the fully aromatic heterocycles. The extrusion of nitrogen from 4,5-dihydro-1,2,3-triazoles usually leads to the formation of aziridines: this reaction is described, as a preparative route to aziridines, in Chapter 9, Section 9.2.2. The analogous decomposition of dihydropyrazoles has proved to be a useful route to cyclopropanes.[38] The fragmentation can lead to the formation of cyclopropanes in high yield, although the stereoselectivity of the reaction is very dependent on the substituents and the mode of decomposition.[38] Two preparatively useful examples of the reaction are shown in Fig. 8.13.

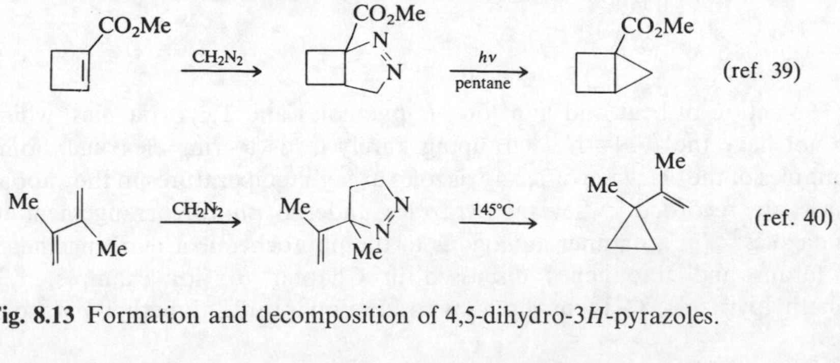

Fig. 8.13 Formation and decomposition of 4,5-dihydro-3*H*-pyrazoles.

37 Review: T. L. Gilchrist, *Adv. Heterocycl. Chem.*, 1987, **41**, 41.
38 Reviews: R. M. Kellogg, in *Photochemistry of Heterocyclic Compounds*, ed. O. Buchardt, Wiley-Interscience, New York, 1976, p. 367; D. Wendisch, in *Methoden der Organischen Chemie* (Houben-Weyl), **4/3**, 42.
39 P. G. Gassman and K. T. Mansfield, *J. Org. Chem.*, 1967, **32**, 915.
40 R. J. Crawford and D. M. Cameron, *Can. J. Chem.*, 1967, **45**, 691.

8.4 Benzodiazoles and benzotriazoles

Three ring systems exist which are benzo-fused diazoles or triazoles: indazole,[41] benzimidazole,[42] and benzotriazole.[43] The benzimidazole ring system is part of the vitamin B_{12} structure (Chapter 6, structure **34**) and several benzimidazoles are commercially available as pharmaceuticals, veterinary products, and fungicides. The benzimidazole omeprazole **35** is a drug used for the treatment of ulcers. Several 2-substituted benzotriazoles and indazoles are used as optical brightening agents and as light stabilizers in plastics. Some simple 1-substituted benzotriazoles are useful reagents in synthetic chemistry. 1-Chlorobenzotriazole (**36**, X = Cl) is an excellent oxidant and chlorinating agent,[44] 1-hydroxybenzotriazole (X = OH) is used as a co-reagent in peptide coupling,[45] and 1-aminobenzotriazole (X = NH_2) is a useful reagent for generating benzyne[46] (see Fig. 8.14). Benzotriazole itself has several commercial uses: it is a good corrosion inhibitor, especially for copper, with which it forms a copper(I) salt on the surface of the metal. As with the monocyclic systems, benzotriazole is the strongest acid of the three heterocycles ($pK_a = 8.2$) whereas benzimidazole is the strongest base (the pK_a of the benzimidazolium cation is 5.5). The stability of the benzotriazolyl anion accounts for the ability of *N*-chlorobenzotriazole to act as a chlorinating agent.

MeO

35

36

Indazoles and benzotriazoles which are substituted on nitrogen can be grouped as 1*H*- and as 2*H*-derivatives. The 2*H*-derivatives, in the 'fixed-bond' structures **37** and **38** shown, are *ortho*-quinonoid, and the question arises as to how much aromatic character these compounds retain. The ultraviolet

41 Review: L. C. Behr, in *Pyrazolines, Pyrazolidines, Indazoles and Condensed Rings*, ed. R. H. Wiley, Wiley-Interscience, New York, 1967, p. 289.

42 Reviews: P. N. Preston, *Chem. Rev.*, 1974, **74**, 279; *Benzimidazoles and Congeneric Tricyclic Compounds*, eds P. N. Preston, D. M. Smith, and G. Tennant, Wiley-Interscience, New York, 1981.

43 Review: J. H. Boyer, in *Heterocyclic Compounds*, Vol. 7, ed. R. C. Elderfield, Wiley, New York, 1961, p. 384.

44 C. W. Rees and R. C. Storr, *J. Chem. Soc.* (*C*), 1969, 1474; M. Fieser and L. F. Fieser, *Reagents for Organic Synthesis*, Vol. 4, Wiley-Interscience, New York, 1974, p. 78.

45 M. Fieser and L. F. Fieser, *Reagents for Organic Synthesis*, Vol. 3, Wiley-Interscience, New York, 1972, p. 156.

46 C. D. Campbell and C. W. Rees, *J. Chem. Soc.* (*C*), 1969, 742.

spectra of simple isomeric alkyl-substituted 1H-and 2H-indazoles and benzotriazoles shown few significant differences at the long-wavelength end of the spectra[47] and the ^{1}H NMR spectra are also not substantially different. The NMR chemical shifts and coupling constants (in $CDCl_3$) are shown for 2-methylbenzotriazole, **39**.[48] Note that the ratio of the coupling constants J_{56} to J_{45} is 0.78 so that, on the basis of this criterion of aromatic character (discussed in Chapter 2, Section 2.2.1), 2-methylbenzotriazole is substantially aromatic. These 2-substituted compounds are best regarded as delocalized aromatic systems rather than as dienes with localized bonding.

Standard methods of synthesis of these ring systems involve cyclization reactions, for which 1,2-disubstituted benzenes of the appropriate type are the most common starting materials. For example, benzene-1,2-diamine reacts with a wide range of carboxylic acid derivatives to give 2-substituted benzimidazoles, **40**. Benzotriazoles can be obtained from the same type of starting material by diazotization and cyclization of the diazonium salt. Indazoles can be made in an analogous way, by diazotization of an aniline with an *ortho* carbon substituent.

These syntheses can usually be adapted to incorporate additional substituents in the benzenoid ring system, if required. For this reason electrophilic substitution reactions have not been widely used, although it is possible to carry out nitration, halogenation and similar reactions. Benzimidazole and indazole are attacked preferentially at the 5-position, and benzotriazole at the 4-position.

The 2-position of benzimidazoles is susceptible to nucleophilic attack: chloride is displaced from the 2-position of benzimidazole and *N*-alkylbenzimidazoles by alkoxides, amines, thiols, and other nucleophiles. Their reactivity towards nucleophiles is lower than for the corresponding benzoxazole and benzothiazole (Section 8.5.3). 1-Substituted benzimidazoles can be directly aminated at the 2-position by heating with sodamide, in a reaction analogous to the Chichibabin reaction of pyridines.

47 J. Derkosch, O. E. Polansky, E. Rieger, and G. Derflinger, *Monatsh.*, 1961, **92**, 1131.

The alkylation of indazole and benzotriazole normally leads to mixtures of the 1- and 2-substituted derivatives, in a ratio dependent upon the alkylating agent.[48] The amination of benzotriazole is a useful reaction because it provides a simple route to 1-aminobenzotriazole; some of the 2-amino compound is also formed. These compounds are both rapidly oxidized by lead(IV) acetate and in both cases the ring system is cleaved.[46] The mode of ring cleavage in the two isomers is quite different, however, as shown in Fig. 8.14. Benzyne generated from 1-aminobenzotriazole can be intercepted by cycloaddition reactions even at −78°C.

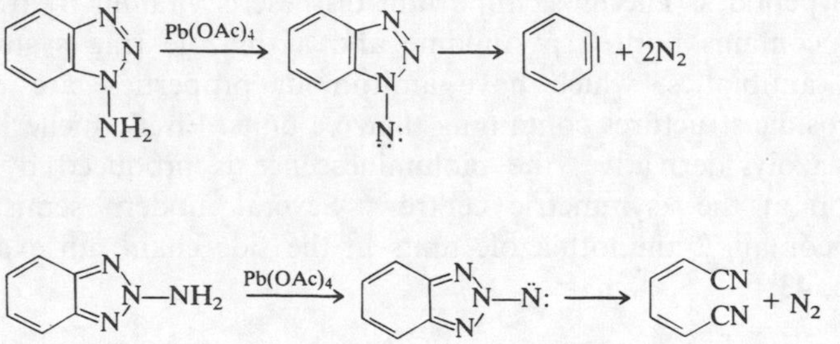

Fig. 8.14 Oxidative fragmentation of 1- and 2-aminobenzotriazole.

1-Substituted benzotriazoles, like the monocyclic 1,2,3-triazoles, can extrude nitrogen on thermolysis or photolysis. In the case of 1-arylbenzo-triazoles the intermediates generated by the elimination of nitrogen can recyclize at the *ortho* position of the aryl substituent. This leads to the formation of a carbazole derivative (the *Graebe–Ullmann* reaction, Fig. 8.15). The method has been used in the synthesis of alkaloids containing this type of ring system.[49]

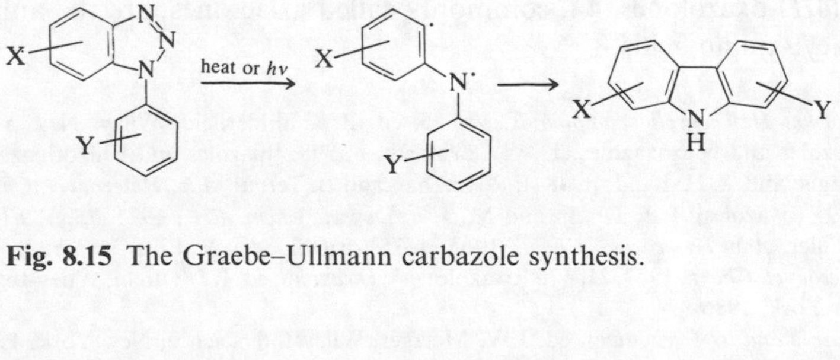

Fig. 8.15 The Graebe–Ullmann carbazole synthesis.

48 M. H. Palmer, R. H. Findlay, S. M. F. Kennedy, and P. S. McIntyre, *J. Chem. Soc., Perkin Trans. 2*, 1975, 1695.
49 R. B. Miller and J. G. Stowell, *J. Org. Chem.*, 1983, **48**, 886.

8.5   Oxazoles[50] and thiazoles,[51] and their benzo derivatives[50]

8.5.1   Introduction

Oxazole and thiazole are formally related to furan and thiophene respectively by the replacement of the CH group at position 3 by nitrogen. The oxazole ring system is poorly represented in nature and there are no important naturally occurring oxazole derivatives. A great deal of work on oxazoles was started in the 1940s when, for a time, penicillin was believed to be an oxazole derivative. In contrast, the thiazole ring system is quite common in natural products, since it can be produced by the cyclization of cysteine residues in peptides. The most important of these is vitamin B_1 (thiamin), 41, which contains both a pyrimidine and a thiazole ring system. The bleomycin antibiotics, which have antitumour properties, are complex aminoglycosidic structures containing thiazole units. Firefly luciferin, 42, is a benzothiazolyl derivative: the bioluminescence is produced by photo-oxygenation at the asymmetric centre.[52] Several modern semisynthetic β-lactams contain 2-aminothiazole units in the side chain; an example is cefotaxime, 43.

2,5-Diaryloxazoles are highly fluorescent and have found commercial application as solutes in liquid scintillators and as optical brightening agents. The 5(4H)-oxazolones, 44, commonly called azlactones, are the anhydrides of N-acyl-amino acids.

50 Reviews: *Heterocyclic compounds*, Vol. 5, ed. R. C. Elderfield, Wiley, New York, 1957 (oxazoles and benzoxazoles, J. W. Cornforth, p. 298; thiazoles and benzothiazoles, J. M. Sprague and A. H. Land, p. 484); R. Lakhan and B. Ternai, *Adv. Heterocycl. Chem.*, 1974, **17**, 99 (oxazoles); I. J. Turchi and M. J. S. Dewar, *Chem. Rev.*, 1975, **75**, 389 (oxazoles); R. Filler, *Adv. Heterocycl. Chem.*, 1965, **4**, 75 (oxazolones); R. Filler and Y. S. Rao, *Adv. Heterocycl. Chem.*, 1977, **21**, 175 (oxazolones); *Oxazoles*, ed. I. J. Turchi, Wiley-Interscience, New York, 1986.
51 *Thiazole and its Derivatives*, ed. J. V. Metzger, Wiley-Interscience, New York, 1979.
52 Review: M. J. Cormier, J. E. Wampler, and K. Hori, in *Progress in the Chemistry of Organic Natural Products*, eds W. Herz, H. Grisebach, and G. W. Kirby, Springer-Verlag, Vienna, 1973, Vol. 30, p. 1.

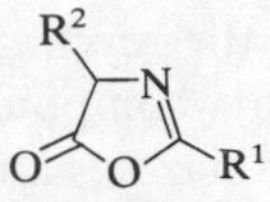

44

These ring systems can be regarded as aromatic, although the bond-length data in Fig. 2.12 indicate that the bonding in oxazole is more localized than in thiazole. There is much the same trend in the imidazole–oxazole–thiazole series as in the pyrrole–furan–thiophene series, the oxygen heterocycle being the least aromatic of the three and showing greatest dienic character. Thus, activated (electron-rich) oxazoles participate in the Diels–Alder reaction with electron-deficient dienophiles whereas the corresponding thiazoles do so only rarely. Nevertheless, the lower aromaticity of the oxazole ring system is not to be equated with instability; some diaryloxazoles are extremely stable to heat and have been recovered unchanged after heating to 400°C.

8.5.2 Ring synthesis

Some of the basic methods of ring synthesis are analogous to those used for furans and thiophenes. For example, an important method of synthesis of oxazoles is the cyclodehydration of α-acylaminoketones (the *Robinson–Gabriel* synthesis). This is analogous to the synthesis of furans by dehydration of 1,4-diketones. There are many variants of the Robinson–Gabriel synthesis, both in the method of construction of the acylaminocarbonyl intermediate and in the choice of dehydrating agent.[50] The example shown in Fig. 8.16(a) provides a route to 5-alkoxy-4-methyloxazoles, which have been used in vitamin B_6 synthesis (as described in Section 8.5.3). Several thiazoles (including thiazole itself) have been made in an analogous way by the reaction of α-acylaminoketones with phosphorus pentasulphide; an example is shown in Fig. 8.16(b).

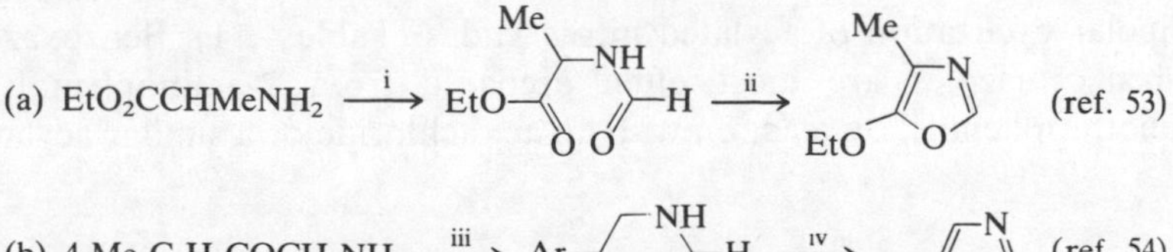

Fig. 8.16 Examples of oxazole and thiazole synthesis from α-acylaminocarbonyl compounds. Reagents; i, HCO_2COMe; ii, P_2O_5; iii, HCO_2H; iv, P_2S_5.

53 E. E. Harris, R. A. Firestone, K. Pfister, R. R. Boettcher, F. J. Cross, R. B. Currie, M. Monaco, E. R. Peterson, and W. Reuter, *J. Org. Chem.*, 1962, **27**, 2705.
54 H. Erlenmeyer, J. Eckenstein, E. Sorkin, and H. Meyer, *Helv. Chim. Acta*, 1950, **33**, 1271.

Another method of ring synthesis is based on the cyclization of intermediates formed from the reaction of α-halocarbonyl compounds and amides or thioamides. This is the most general route to thiazoles and is due to Hantzsch. The reaction has been carried out not only with thioamides but also with thioureas, thiosemicarbazides, and other compounds containing the —N—C=S structural unit. A related thiazole synthesis is shown in Table 4.6. The reaction goes by way of a nucleophilic attack by sulphur on the carbon atom bearing the halogen; the acyclic intermediate formed in this way has been isolated in a few cases. The thiazoles are normally obtained by treating the components together in ethanol or a similar solvent. A preparation of 2,4-diphenylthiazole by the Hantzsch method, showing the likely reaction intermediate, is given in Fig. 8.17(a). 2,4-Disubstituted oxazoles have been made from α-haloketones and amides in an analogous way, although a higher temperature is required and yields are moderate. The reaction of α-haloketones with formamide (the Bredereck reaction) has already been mentioned in Section 8.2 as a route to imidazoles, but in the presence of acid the intermediates can cyclize to oxazoles instead, as shown in Fig. 8.17(b).

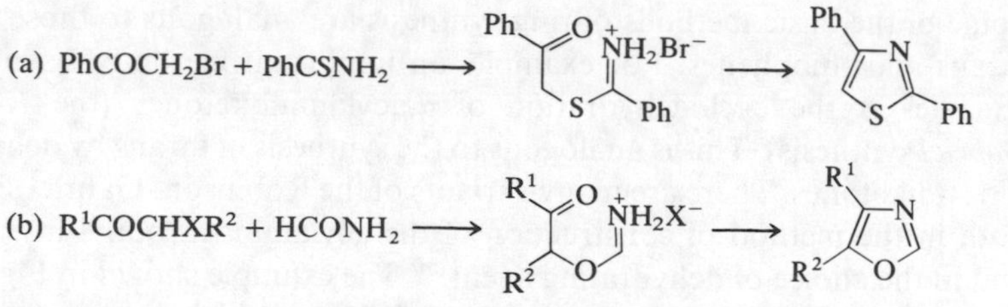

(a) $PhCOCH_2Br + PhCSNH_2 \longrightarrow$

(b) $R^1COCHXR^2 + HCONH_2 \longrightarrow$

Fig. 8.17 Synthesis of thiazoles and oxazoles from α-haloketones.

There are several other good cyclization routes to both oxazoles and thiazoles. Oxazoles and thiazoles unsubstituted at C-2 are conveniently prepared by isonitrile cyclization (Table 4.8). Oxazoles are also formed by 1,5-dipolar cyclization of acylated nitrile ylides (Table 4.11). Benzoxazoles and benzothiazoles are most often prepared from 2-aminophenols or 2-aminothiophenols, by reaction with an acid chloride or a similar acylating agent.

8.5.3   Chemical properties

Oxazole is a very weak base (Table 8.1) and its salts are unstable (for example, the picrate dissociates rapidly in air). *N*-Alkyloxazolium salts have been formed from several substituted oxazoles. Thiazoles can also be quaternized at nitrogen by reaction with a range of alkylating agents. These salts can undergo easy hydrogen–deuterium exchange at C-2 and the ylides **45** are generated as intermediates in this process. A similar reaction of imidazoles was noted in Section 8.2.

Although the rate of exchange is faster in the oxazolium salts, the thiazolium ylides are more stable, presumably because of the ability of sulphur to stabilize an adjacent carbanion. The reaction is of considerable importance because it is likely that an ylide of this type is generated from thiamin, **41**, and that the ylide is involved in the biochemical reactions of the vitamin.[55] Indeed salts of the general structure **46**, derived from a commercially available thiazole, have been extensively used as catalysts in reactions such as the benzoin condensation[56,57] and the addition of aldehydes to activated double bonds.[58] The thiazolium ylides play much the same role as the more familiar cyanide ion catalysts in these processes: they initiate the reaction by nucleophilic attack on the carbonyl group of the aldehyde.

Neutral oxazoles and thiazoles are also deprotonated preferentially at C-2 by strong bases; for example, thiazole forms a magnesium bromide by reaction with ethylmagnesium bromide and it is lithiated at C-2 by phenyllithium at $-60°C$. The metallated thiazoles react with aldehydes, carbon dioxide, and other electrophiles in the normal manner.[51] 4,5-Diphenyloxazole is lithiated at C-2 but this derivative is thermally unstable and undergoes ring opening to the isonitrile **47**.[59] 2-Methyl substituents in oxazoles and thiazoles are activated by the C=N bonds of the ring systems, and these groups can be deprotonated by alkyllithium reagents and other strong bases [see Fig. 8.1(b)].

Electrophiles attack oxazoles and thiazoles preferentially at C-5, then C-4. The oxazole ring system does not undergo electrophilic attack readily unless electron-releasing substituents are present. Bromination and mercuration

55 R. Breslow, *J. Am. Chem. Soc.*, 1958, **80**, 3719.
56 T. Matsumoto, M. Ohishi, and S. Inoue, *J. Org. Chem.*, 1985, **50**, 603.
57 H. Stetter and G. Dämbkes, *Synthesis*, 1977, 403.
58 Review: H. Stetter, *Angew. Chem. Int. Edn Engl.*, 1976, **15**, 639.
59 For a review of lithiation, cycloaddition, and other reactions of oxazoles in organic synthesis see B. E. Maryanoff, in *Oxazoles*, ed. I. J. Turchi, Wiley-Interscience, New York, 1986, p. 963.

(with mercuric acetate) are the most general electrophilic substitution reactions of oxazoles. Reactions other than substitution on the oxazole ring can occur; for example, chlorine gives an addition product (across the 4- and 5-positions) with 4,5-dimethyl-2-phenyloxazole, and phenyloxazoles are nitrated in the benzene ring rather than in the heterocycle. Bromine in methanol converts some 2,5-disubstituted oxazoles into 2,5-dimethoxy-2,5-dihydro derivatives, in exactly the same way as for furan (Section 6.3.3); this illustrates the relatively low aromatic character of the ring system. Similarly, thiazoles require electron-releasing groups to allow electrophilic substitution reactions to go readily; the 5-position is again the most favoured one for attack. An example is the Mannich reaction of 2-acetamido-4-methylthiazole to give the alkylated thiazole **48**.[60]

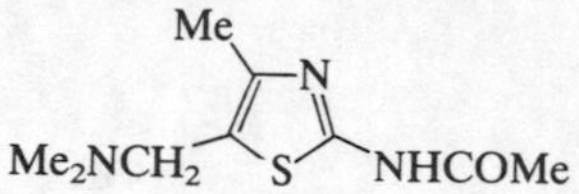

Me

Me$_2$NCH$_2$ S   NHCOMe

48

Oxazoles and thiazoles with a leaving group at the 2-position react readily with nucleophiles to give displacement products by the general process shown in Fig. 8.1(a). Examples showing displacement of halide and nitrite from the 2-position are given in Fig. 8.18. 2-Chlorothiazole is about 100 times more reactive than 2-chloropyridine in nucleophilic displacement processes, and 2-chlorobenzothiazole is still more reactive. More surprising, 4- and 5-halothiazoles are also susceptible to nucleophilic displacement reactions, although the relative rates of reaction of the halothiazoles with different nucleophiles vary widely.[61] An example of nucleophilic amination of a thiazole exists, and is shown in Fig. 8.18. This is apparently analogous to the Chichibabin reaction of pyridines, which is discussed in Section 5.2.7.

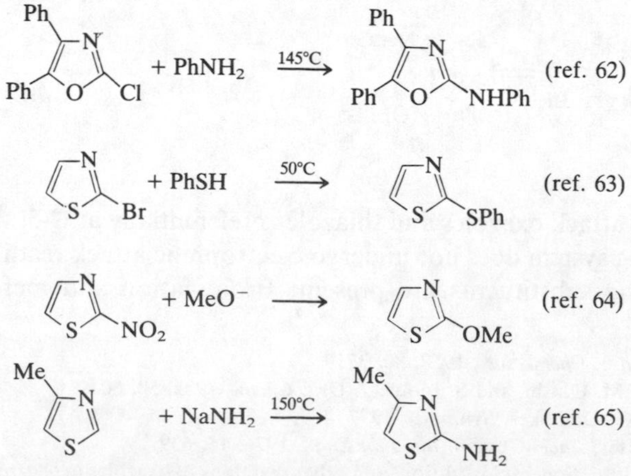

Ph

Ph   O   Cl   + PhNH$_2$ $\xrightarrow{145°C}$ Ph   O   NHPh   (ref. 62)

S   Br   + PhSH $\xrightarrow{50°C}$ S   SPh   (ref. 63)

S   NO$_2$ + MeO$^-$ $\longrightarrow$ S   OMe   (ref. 64)

Me

S   + NaNH$_2$ $\xrightarrow{150°C}$ Me   S   NH$_2$ (ref. 65)

Fig. 8.18 Examples of nucleophilic substitution in oxazoles and thiazoles.

One of the most important reactions of oxazoles is the Diels–Alder reaction.[50,66] The ring system is less nucleophilic at C-2 and C-5 than is furan, and an electron-releasing substituent (usually an ethoxy group) facilitates the reaction with dienophiles. The impetus for the investigation of the reaction was that, in combination with the oxazole synthesis shown in Fig. 8.16(a), it provided a route to pyridoxol (vitamin B_6), **49**. There are several variants of the reaction but their key step is the cycloaddition of an activated olefinic dienophile to an oxazole of this type. An example, in which diethyl maleate is used as the dienophile, is shown in Fig. 8.19.[53] The analogous reaction of 5-ethoxy-4-methylthiazole also gives pyridoxol, although the cycloaddition step requires a much higher temperature (200°C).[67]

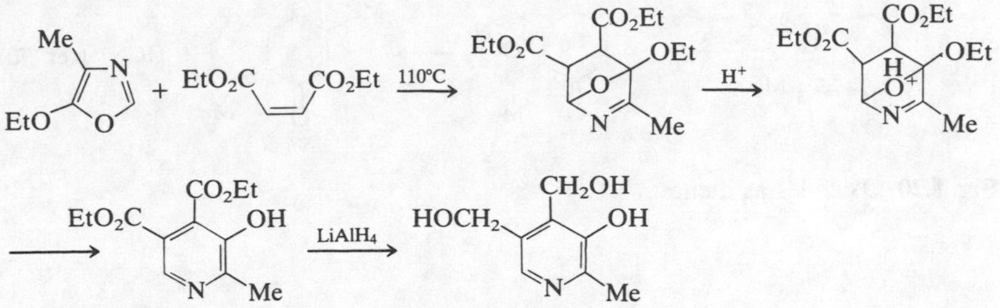

Fig. 8.19 Synthesis of pyridoxol by Diels–Alder addition.

The acid- or base-catalysed cleavage of the ether bridge in the primary cycloadducts provides a good route to pyridines. Oxazole itself reacts with acrylonitrile and similar dienophiles on heating to give the appropriate pyridines in moderate yields [Fig. 8.20(a)]. With acetylenic dienophiles the reaction takes a different course: the addition is normally followed by a retroaddition, with the loss of a nitrile, and the product is a furan. Examples are shown in Figs 8.20(b) and (c). In none of these cases was the primary cycloadduct isolated.

The photooxygenation of oxazoles is an analogous process in that it involves the addition of singlet oxygen across the 2- and 5-positions. The

60 N. F. Albertson, *J. Am. Chem. Soc.*, 1948, **70**, 669.
61 Review: L. Forlani and P. E. Todesco, in ref. 51, Part 1, p. 565.
62 R. Gompper and F. Effenberger, *Chem. Ber.*, 1959, **92**, 1928.
63 M. Bosco, V. Liturri, L. Troisi, L. Forlani, and P. E. Todesco, *J. Chem. Soc., Perkin Trans. 2*, 1974, 508.
64 M. Bosco, L. Forlani, V. Liturri, and P. E. Todesco, *Chim. Ind.*, 1972, **54**, 266.
65 E. Ochiai and F. Nagasawa, *Ber.*, 1939, **72B**, 1470.
66 Review: D. L. Boger, *Tetrahedron*, 1983, **39**, 2869.
67 Takeda Chemical Industries Ltd, Fr. Pat. 1,400,843 (*Chem. Abstr.*, 1965, **63**, 9922). An intramolecular cycloaddition of a thiazole has been observed but also requires a very high temperature: P. A. Jacobi and R. F. Frechette, *Tetrahedron Lett.*, 1987, **28**, 2937.

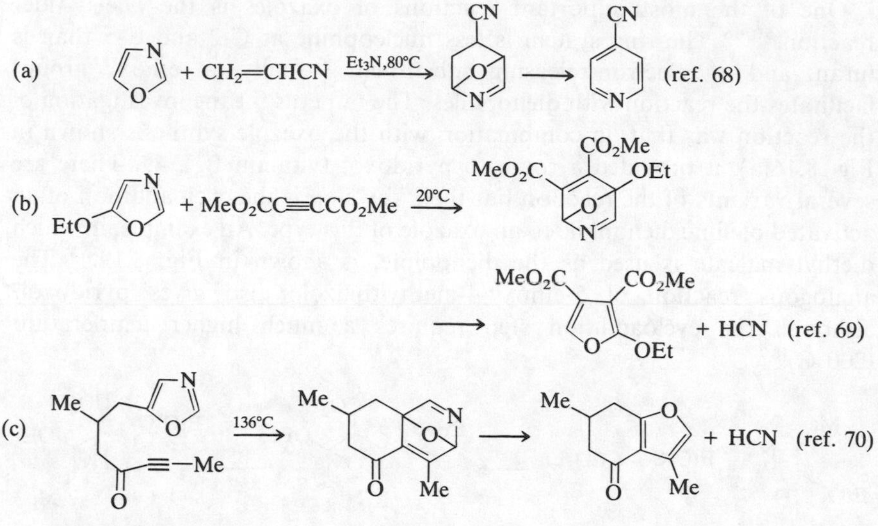

(a) (ref. 68)

(b) + HCN   (ref. 69)

(c) + HCN   (ref. 70)

Fig. 8.20 Oxazoles as dienes.

mode of decomposition of the unstable endoperoxide is solvent dependent, but triamides **50** can be isolated in high yields from trisubstituted oxazoles.[71] This reaction has useful synthetic applications.[72] Trisubstituted thiazoles react with single oxygen in the same way.[71]

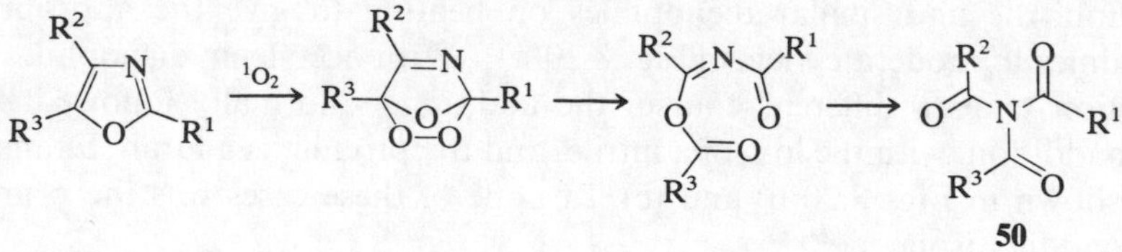

50

Both oxazoles and thiazoles can undergo light-induced rearrangements of the types discussed earlier for thiophene (see Chapter 6, Fig. 6.22).[36] Both the ring contraction–ring expansion mechanism and the electrocyclic mechanism appear to be involved, as shown for the photoisomerization of

68 P. Colin, Fr. Pat. 1,500,352 (*Chem. Abstr.*, 1970, **72**, 31629).
69 R. Grigg and J. L. Jackson, *J. Chem. Soc.* (*C*), 1970, 552.
70 P. A. Jacobi, D. G. Walker, and I. M. A. Odeh, *J. Org. Chem.*, 1981, **46**, 2065.
71 Review: H. H. Wasserman and B. H. Lipschutz, in *Singlet Oxygen*, eds H. H. Wasserman and R. W. Murray, Academic Press, New York, 1979, p. 429.
72 Review: H. H. Wasserman, K. E. McCarthy, and K. S. Prowse, *Chem. Rev.*, 1986, **86**, 845.

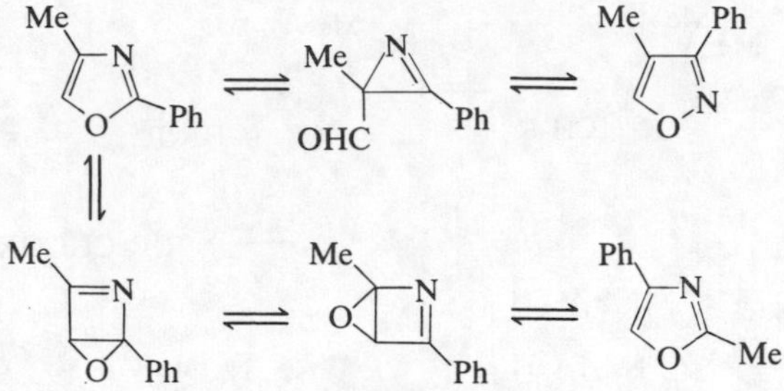

Fig. 8.21 Photorearrangement of 4-methyl-2-phenyloxazole.

4-methyl-2-phenyloxazole in Fig. 8.21. The azirine formed in the first type of rearrangement is an isolable intermediate.[73]

Some 4-acyloxazoles rearrange on heating at 90–120°C to isomeric oxazoles in which the original 4-acyl group becomes part of the ring system. This reaction, which is called the *Cornforth rearrangement* after its discoverer, is proposed to go by way of a 1,5-dipolar intermediate (Fig. 8.22).[50] The ring closure step is thus an example of the 1,5-dipolar cyclization process discussed in Chapter 4, Section 4.2.8.

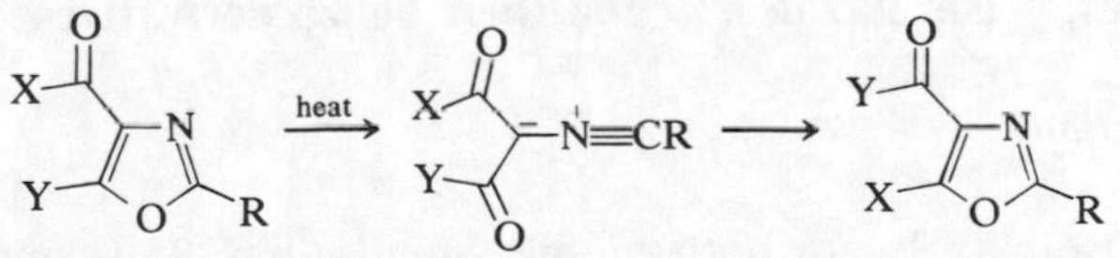

Fig. 8.22 The Cornforth rearrangement ($X = H, Cl, NR_2$; $Y = Cl, OR, O^-$).

The heterocycles in this group (particularly 4,5-dihydrooxazoles[74]) have been used in organic synthesis as reagents in which C-2 of the ring system functions as a protected or latent carbonyl group. For example, carboxylic acids are readily converted into dihydroxazoles by reaction with 2-amino-2-methyl-1-propanol; these compounds can be deprotonated at the carbon atom adjacent to C-2 of the ring, further substituted, and the ring cleaved, as shown in Fig. 8.23(a). A related use of the benzothiazole ring system as a protected aldehyde is shown in Fig. 8.23(b).

73 M. Maeda and M. Kojima, *J. Chem. Soc., Perkin Trans. 1*, 1977, 239.
74 Review: A. I. Meyers and E. D. Mihelich, *Angew. Chem. Int. Edn Engl.*, 1976, **15**, 270.

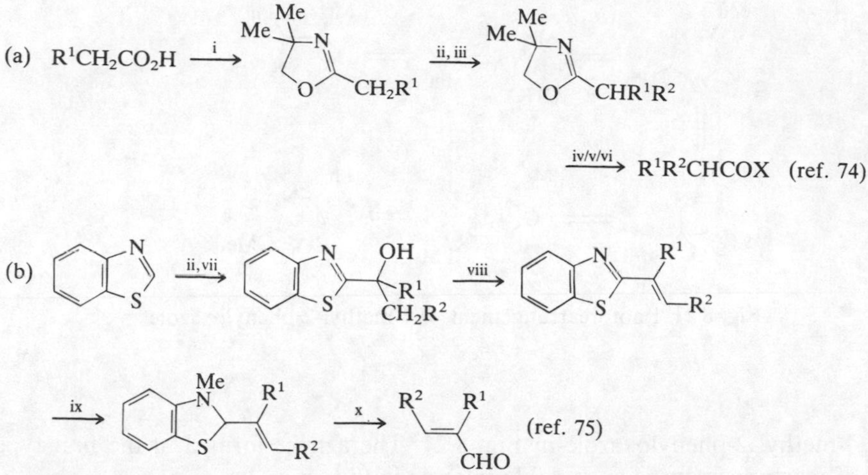

(a) $R^1CH_2CO_2H \xrightarrow{i}$ [structure] $\xrightarrow{ii, iii}$ [structure]

$\xrightarrow{iv/v/vi} R^1R^2CHCOX$ (ref. 74)

(b) [benzothiazole structure] $\xrightarrow{ii, vii}$ [structure] $\xrightarrow{viii}$ [structure]

$\xrightarrow{ix}$ [structure] $\xrightarrow{x}$ [structure] (ref. 75)

Fig. 8.23 Dihydrooxazoles and benzothiazoles as masked carbonyl equivalents. Reagents: i, $H_2NCMe_2CH_2OH$; ii, BuLi; iii, R^2I; iv (for X=OH), H_3O^+; v (for X=OEt), H^+, EtOH; vi (for X=H), MeI (alkylation on N) then $NaBH_4$ followed by hydrolysis; vii, $R^1COCH_2R^2$; viii, P_2O_5, $MeSO_3H$; ix, $MeOSO_2F$ then $NaBH_4$; x, $AgNO_3$, aq. MeCN.

8.6　Isoxazoles,[76] isothiazoles,[77] and their benzo derivatives[78]

8.6.1　Introduction

Isoxazole and isothiazole are both mobile liquids (b.p. 95°C and 113°C, respectively) with pyridine-like odours. The chemistry of isoxazoles was first investigated in the last century whereas that of the monocyclic isothiazoles dates only from the 1950s. During this period, at a time when the isothiazole ring system was virtually unknown, R. B. Woodward devised and carried out a remarkable synthesis of the alkaloid colchicine in which an isothiazole was made in the first step and then used as a template on which to construct the carbon skeleton.[79]

There are several naturally occurring isoxazoles with important

75 E. J. Corey and D. L. Boger, *Tetrahedron Lett.*, 1978, 5.

76 Reviews: A. Quilico, in *Five- and Six-membered Compounds with Nitrogen and Oxygen*, ed. R. H. Wiley, Wiley-Interscience, 1962, p. 1; N. K. Kochetkov and S. D. Sokolov, *Adv. Heterocycl. Chem.*, 1963, **2**, 365; B. J. Wakefield and D. J. Wright, *Adv. Heterocycl. Chem.*, 1979, **25**, 147.

77 Review: K. R. H. Wooldridge, *Adv. Heterocycl. Chem.*, 1972, **14**, 1.

78 Reviews: K. H. Wünsch and A. J. Boulton, *Adv. Heterocycl. Chem.*, 1967, **8**, 277 (benzisoxazoles); M. Davis, *Adv. Heterocycl. Chem.*, 1972, **14**, 43 (benzisothiazoles).

79 For a discussion of the role of isothiazole in this synthesis see I. Fleming, *Selected Organic Syntheses*, Wiley-Interscience, London, 1973, p. 202.

pharmacological activity. Muscimol, **51**, from the mushroom *Amanita muscaria* (fly agaric), has powerful psychotropic effects. Muscimol shows activity in brain nerve cells, which use γ-aminobutyric acid (GABA) as a neutrotransmitter.[80] This structure has been used as the basis for the design of a number of synthetic isoxazoles, such as **52**, as potential analgesics. Cycloserine, **53**, is a naturally occurring antituberculosis antibiotic, and the isoxazoline **54** is a natural antitumour antibiotic.[81] Some commercial semi-synthetic penicillins (oxacillin, cloxacillin and dicloxacillin) have isoxazolyl side chains. 4-Hydroxyisoxazole is a seed-germination inhibitor which also occurs naturally.[82]

There are two types of benzo-fused isoxazole and isothiazole: those having structure **55** are given the systematic names 1,2-benzisoxazole (X=O) and 1,2-benzisothiazole (X=S) by *Chemical Abstracts*. Compounds **56** are 2,1-benzisoxazole and 2,1-benzisothiazole, although the older trivial name 'anthranil' is still often found in the literature for compound **56** (X=O).

The monocyclic systems and their benzo analogues can all be regarded as aromatic but all contain a weak N—X bond which is a potential site of ring cleavage. Isoxazoles, in particular, have been used widely as synthetic reagents because the ring system is sufficiently stable to allow substitution and manipulation of functional groups, yet is easily cleaved when necessary by reduction or other means. The bond lengths of isoxazole are given in Fig. 2.12.

8.6.2 Ring synthesis

The two most important general methods for constructing the isoxazole ring are (a) the reaction of hydroxylamine with a three-carbon atom component, such as 1,3-diketone or an αβ-unsaturated ketone, and (b) the reaction of a

80 Review: P. Krogsgaard-Larsen, L. Brehm, and K. Schaumburg, *Acta Chem. Scand. (B)*, 1981, **35**, 311.
81 D. G. Martin, D. J. Duchamp, and C. G. Chidester, *Tetrahedron Lett.*, 1973, 2549.
82 T. Kusumi, C. C. Chang, M. Wheeler, I. Kubo, K. Nakanishi, and H. Naoki, *Tetrahedron Lett.*, 1981, **22**, 3451.

nitrile oxide with an alkene or an alkyne. The second of these methods has been discussed in Section 4.3.2; it provides a versatile route to isoxazoles since the substituents on both components can be varied widely. The cycloaddition reactions of nitrones also lead to the formation of the isoxazole ring system, but at a lower oxidation level. Examples of these reactions are given in Tables 4.16 and 4.17. The reaction of hydroxylamine with 1,3-diketones, outlined in Table 4.4, is a good method for 3,5-disubstituted isoxazoles. There are many variants of the reaction which enable other isomers to be formed. For example, the enol ether 57 and similar compounds react with hydroxylamine to give 4,5-disubstituted oxazoles [Fig. 8.24(a)]. Acetylenic carbonyl compounds can also be used (Table 4.7). In these reactions, if the carbonyl group is aldehydic [Fig. 8.24(b)] or otherwise reactive, the initial process is formation of the oxime, but in other cases the structure of the isoxazole formed indicates that the hydroxylamine reacts by conjugate addition to the triple bond.

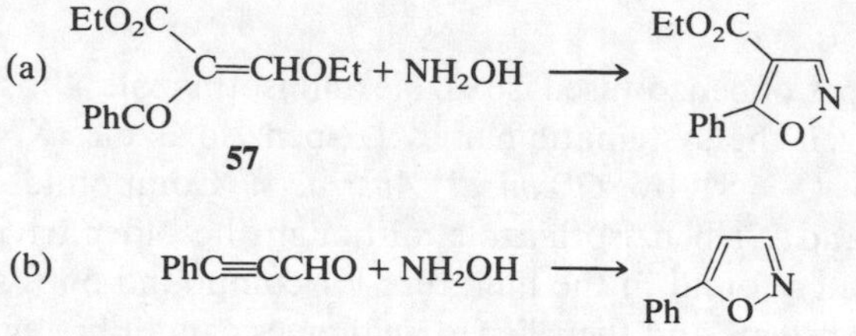

(a) EtO_2C C=CHOEt + NH_2OH $\longrightarrow$ EtO_2C
       PhCO   Ph
       57

(b)   PhC≡CCHO + NH_2OH $\longrightarrow$
       Ph

Fig. 8.24 Isoxazoles by cyclization reactions.

These general approaches to isoxazoles are not readily translated into synthetic routes for isothiazoles. 'Thiohydroxylamine' is not available and the nearest equivalent is potassium thiohydroxylamino-*S*-sulphonate, 58, which is reported to react with propargaldehyde at the carbonyl group to give a protected thiooxime intermediate; this can be cyclized to isothiazole in the presence of sodium hydrogencarbonate.[77]

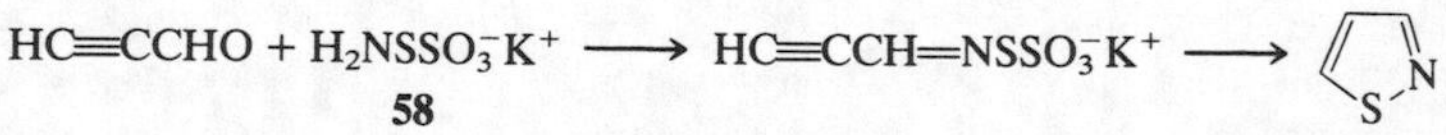

HC≡CCHO + $H_2NSSO_3^-K^+$ $\longrightarrow$ HC≡CCH=$NSSO_3^-K^+$ $\longrightarrow$
58

1,3-Dipolar cycloaddition reactions of nitrile sulphides provide a synthesis of isothiazoles which is equivalent to the nitrile oxide route to isoxazoles, but the reactions are more limited in terms of the range of substituents that can be incorporated.[83]

The most important routes to isothiazoles are based on cyclization reactions in which the N—S bond is formed. Nucleophilic attack can take place on sulphur when a good leaving group is attached. There are several

83 Review: R. M. Paton, *Chem. Soc. Rev.*, 1989, **18**, 33.

variants of the method,[77] of which two are illustrated in Figs 8.25(a) and (b). Isothiazoles can also be obtained from the corresponding isoxazoles by reductive cleavage followed by conversion to the thioamide and oxidative cyclization. This is a cyclization of the same general type [Fig. 8.25(c)].

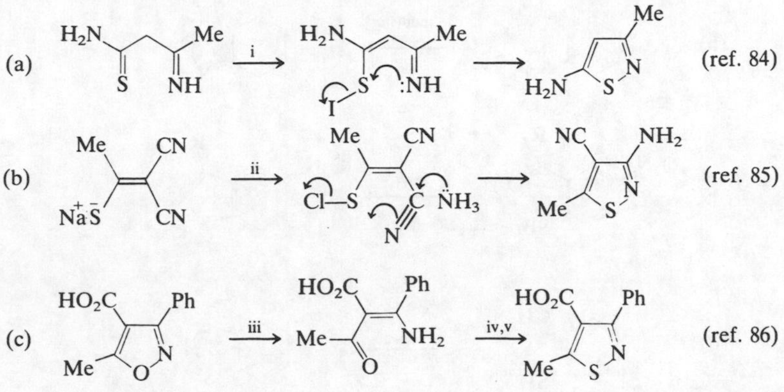

Fig. 8.25 Isothiazole synthesis by formation of the N—S bond. Reagents: i, I_2; ii, NH_2Cl; iii, H_2/Ni; iv, P_2S_5; v, chloranil.

Some of the general methods used for preparing the benzo-fused systems are shown in Fig. 8.26. Those illustrated for benzisothiazoles are related to the methods used for the monocyclic systems, in that the N—S bond is formed oxidatively. In the general method (b) shown for 2,1-benzisoxazoles it has been established that the ring oxygen atom is derived from the nitro group rather than the carbonyl group.

2,1-Benzisoxazoles can also be made by the thermolysis of 2-azidoaryl ketones. This is an example of a more general heterocyclic synthesis in which aryl azides bearing an *ortho* substituent are cyclized with loss of nitrogen:[87] the reaction has also been used to make 2,1-benzisothiazoles, for which 2-azidothioketones are the starting materials.[88] It has been established that the substituent adjacent to the azide accelerates the decomposition of the azide if it is a conjugated π-bonding group, but not otherwise. This is most reasonably accounted for by a transition state in which nucleophilic displacement takes place on nitrogen in the plane of the molecule, while at the same time the aromatic character of the new heterocycle is being developed by overlap of the π-orbitals perpendicular to the plane (Fig. 8.27). The reaction is closely related to the Boulton–Katritzky rearrangement of isoxazoles which is discussed in Section 8.6.3.

84 J. Goerdeler and H. W. Pohland, *Chem. Ber.*, 1961, **94**, 2950.
85 K. Hartke and L. Peshkar, *Angew. Chem. Int. Edn Engl.*, 1967, **6**, 83.
86 D. N. McGregor, U. Corbin, J. E. Swigor, and L. C. Cheney, *Tetrahedron*, 1969, **25**, 389.
87 L. K. Dyall, *Aust. J. Chem.*, 1977, **30**, 2669.
88 J. Ashby and H. Suschitzky, *Tetrahedron Lett.*, 1971, 1315.

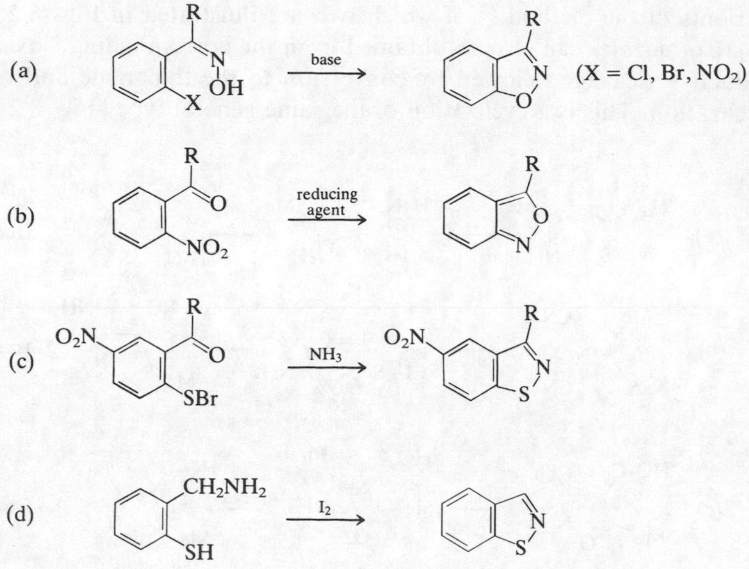

(a) ... base ... (X = Cl, Br, NO₂)

(b) ... reducing agent ...

(c) ... NH₃ ...

(d) ... I₂ ...

Fig. 8.26 Routes to benzisoxazoles and benzisothiazoles.

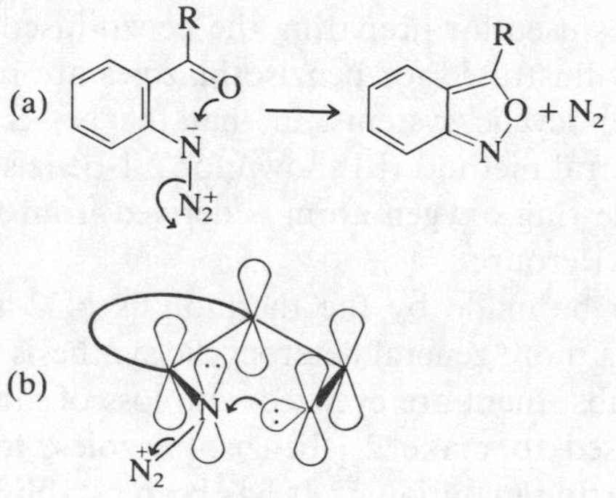

(a) ... ⟶ ... + N₂

(b)

Fig. 8.27 (a) 2,1-Benzisoxazoles from 2-azidoaryl ketones; (b) generalized transition state for these displacements.

8.6.3 Chemical properties

The ease of cleavage of the N—O bond in isoxazoles has been exploited in a wide variety of ways by synthetic chemists. The most general methods of cleavage are (i) catalytic reduction to give vinylogous amides [Fig. 8.28(a)], (ii) metal–ammonia reduction in the presence of an alcohol to give β-aminoketones and enones [Fig. 8.28(b)], and (iii) cleavage of 3-isoxazolyl anions to give α-cyanoketones [Fig. 8.28(c)].

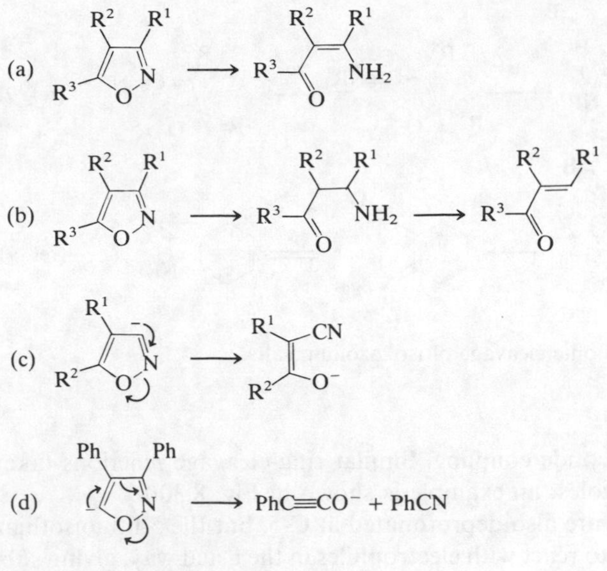

Fig. 8.28 Cleavage of isoxazoles.

A related fragmentation of a 5-isoxazolyl anion is also known [Fig. 8.28(d)].[89] Among many applications, the catalytic reductive cleavage has been used in a strategy for the synthesis of vitamin B_{12} and other corrins:[90] isoxazoles with suitably functionalized side chains were constructed, and on reduction provided the skeleton necessary to construct adjacent pyrrolic units. The relevant steps in the sequence are shown in outline in Fig. 8.29.

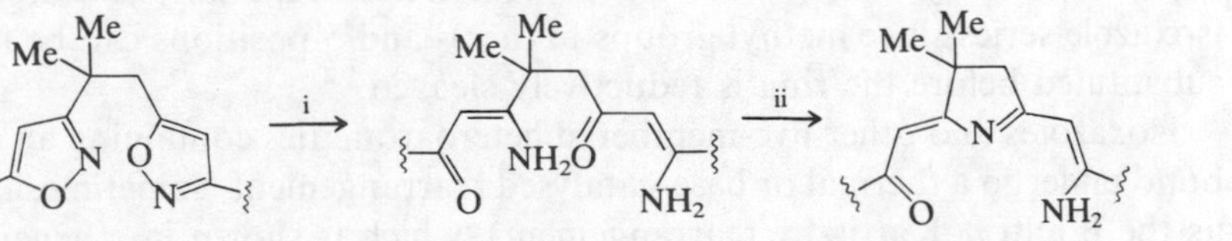

Fig. 8.29 Reductive cleavage of isoxazoles in corrin synthesis. Reagents: i, H_2/Ni; ii, Et_3N.

The anionic cleavage shown in Fig. 8.28(c) can be achieved in several ways. 3-Unsubstituted isoxazoles are easily deprotonated at C-3, even by hydroxide ions. Isoxazole-3-carboxylic acids are cleaved on heating by the same type of mechanism, and 3-acylisoxazoles are deacylated and cleaved by nucleophiles. The cleavage of 3-unsubstituted isoxazolium salts by bases is particularly easy [Fig. 8.30(a)]. Isoxazolium salts have been used as

89 U. Schölkopf and I. Hoppe, *Angew. Chem. Int. Edn Engl.*, 1975, **14**, 765.
90 R. V. Stevens, J. M. Fitzpatrick, P. B. Germeraad, B. L. Harrison, and R. Lapalme, *J. Am. Chem. Soc.*, 1976, **98**, 6313.

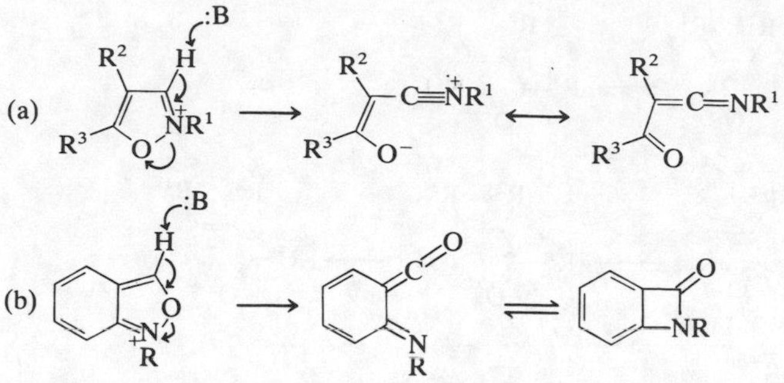

(a) ⟶ ⟷   (ref. 91a)

(b) ⟶ ⇌   (ref. 91b)

Fig. 8.30 Anionic cleavage of isoxazolium salts.

reagents for peptide coupling. Similar ring-cleavage reactions take place in the benzisoxazoles: an example is shown in Fig. 8.30(b).

Isothiazoles are also deprotonated at C-5, but the 5-lithioisothiazoles are stable enough to react with electrophiles in the usual way, giving substitution products. When the 5-position is blocked, the 3-isothiazolyl anion can be formed and it undergoes the same type of cleavage as for the isoxazole. The sulphur atom of isothiazoles can be reductively removed by hydrogenolysis using Raney nickel. Both isoxazoles and isothiazoles can undergo electrophilic substitution reactions and both are substituted preferentially at the 4-position. Both types of ring system can undergo photoisomerization and photocleavage reactions by way of the general routes described earlier[36] (see Fig. 8.21).

Alkyl groups at the 3- and 5-positions are slightly activated and can be deprotonated by strong bases. This reaction is especially useful in the isoxazole series since methyl groups in the 3- and 5-positions can be further substituted before the ring is reductively cleaved.[92]

Isoxazoles and other five-membered heteroaromatics containing an N—O bond undergo a thermal or base-catalysed rearrangement (sometimes known as the Boulton–Katritzky rearrangement), which is shown in a generalized form in Fig. 8.31.[93] The reaction is reversible only when atom W is oxygen.

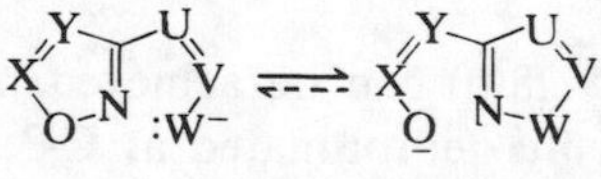

Fig. 8.31 Rearrangement of isoxazoles and related heterocycles.

91 (a) R. B. Woodward and R. A. Olofson, *Tetrahedron Suppl.*, 1966, **7**, 415; (b) R. A. Olofson, R. K. Van derMeer, and S. Stournas, *J. Am. Chem. Soc.*, 1971, **93**, 1543.
92 Review: C. Kashima, Y. Yamamoto and Y. Tsuda, *Heterocycles*, 1977, **6**, 805.
93 Review: M. Ruccia, N. Vivona, and D. Spinelli, *Adv. Heterocycl. Chem.*, 1981, **29**, 141.

It is probably mechanistically related to the assisted azide displacement shown in Fig. 8.27: in both cases, nucleophilic displacement at nitrogen in the plane of the ring is assisted by the development of an aromatic π-bonding system orthogonal to the plane. The reaction has provided a useful method of preparing other five-membered heterocycles, starting either from isoxazoles or from 2,1-benzisoxazoles. An example is the preparation of the 1,2,4-thiadiazole derivative **59** from the isoxazole **60**: the reaction can be brought about by the action of heat or by a base at room temperature.[94]

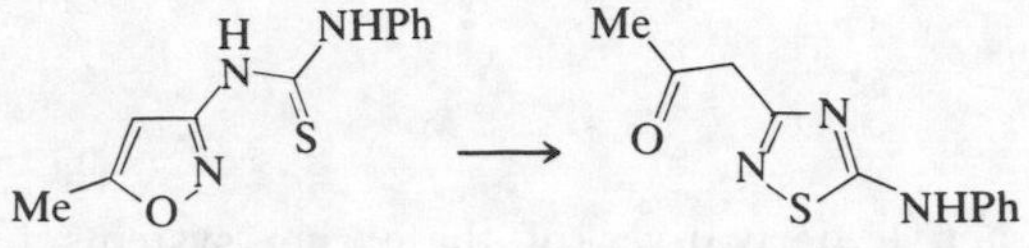

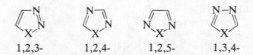

8.7 Oxadiazoles, thiadiazoles, and related systems

8.7.1 Introduction

There are four isomeric types of oxadiazole and thiadiazole, as shown in Fig. 8.32.

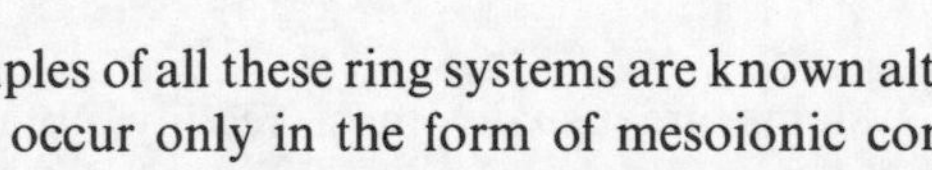

1,2,3-	1,2,4-	1,2,5-	1,3,4-

Fig. 8.32 Isomeric oxadiazoles and thiadiazoles.

Examples of all these ring systems are known although the 1,2,3-oxadiazole systems occur only in the form of mesoionic compounds (see Section 8.8). Potential 1,2,3-oxadiazoles exist entirely in the open diazoketone tautomeric form. The cyclic tautomer is the preferred structure for the 1,2,3-thiadiazoles, however. Benzo-fused analogues of 1,2,3-thiadiazole and of the 1,2,5-isomers also exist. 1,2,5-Oxadiazole is often referred to by the trivial name furazan in the literature; 1,2,5-oxadiazole-2-oxide, a common derivative, has the

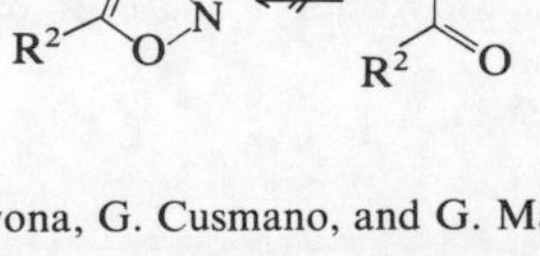

94 N. Vivona, G. Cusmano, and G. Macaluso, *J. Chem. Soc., Perkin Trans. 1*, 1977, 1616.

trivial name furoxan. There are reviews on the chemistry of 1,2,4-[95], 1,2,5-[96], and 1,3,4-[97] oxadiazoles, and on 1,2,3-[98], 1,2,4-[99], 1,2,5-[100] and 1,3,4-[101] thiadiazoles. Several derivatives of the 1,2,3,4-thiatriazole system **61** are also known;[102] the 1,2,3,5-thiatriazole system is known in the form of non-aromatic derivatives **62**.[103]

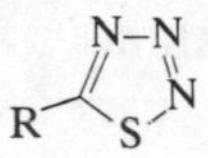

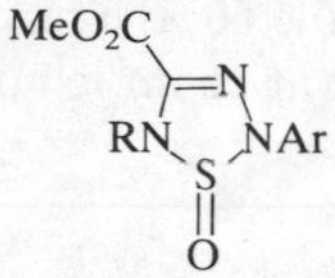

Many biologically active derivatives of these ring systems have been synthesized and several are incorporated into commercial drugs and pesticides. The 1,2,5-thiadiazole, timolol, **63**, is used as its maleate salt in the form of eyedrops to treat glaucoma and is also used as a protective drug for patients who have suffered heart attacks. The sulphonamide derivative, acetazolamide, **64**, is a diuretic. Terrazole, **65**, and other 1,2,4-thiadiazoles are systemic soil fungicides. Several thiadiazolyldiazonium salts are used in coupling reactions to produce azo dyes; the azo dyes derived by coupling the diazonium cation **66** are particularly important.

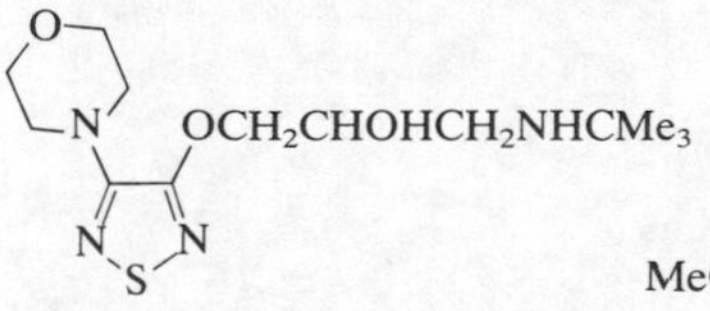

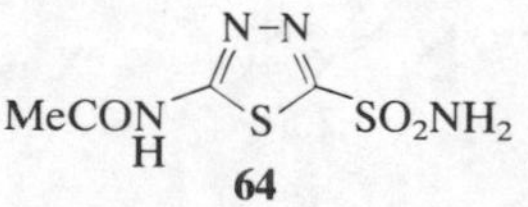

8.7.2   Ring synthesis

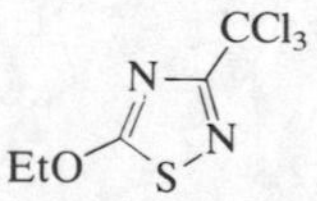

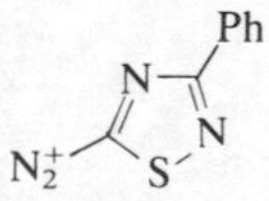

Derivatives of these ring systems are usually most readily obtained by cyclization reactions although some can also be made by 1,3-dipolar

95 L. B. Clapp, *Adv. Heterocycl. Chem.*, 1976, **20**, 65.
96 K. L. Stuart, *Heterocycles*, 1975, **3**, 651; A. Gasco and A. J. Boulton, *Adv. Heterocycl. Chem.*, 1981, **29**, 251.
97 A. Hetzheim and K. Möckel, *Adv. Heterocycl. Chem.*, 1966, **7**, 183.
98 M. Davis, in *Organic Compounds of Sulphur, Selenium, and Tellurium*, Vol. 5, The Chemical Society, London, 1979, p. 431, and earlier vols.
99 F. Kurzer, *Adv. Heterocycl. Chem.*, 1982, **32**, 285.
100 V. G. Pesubm *Russ. Chem. Rev.*, 1970, **39**, 923: L. M. Weinstock and P. I. Pollak, *Adv. Heterocycl. Chem.*, 1968, **9**, 107.
101 J. Sandström, *Adv. Heterocycl. Chem.*, 1968, **9**, 165.
102 A. Holm, *Adv. Heterocycl. Chem.*, 1976, **20**, 145.
103 G. Heubach, *Liebigs Ann. Chem.*, 1980, 1376.

cycloaddition (for example, 1,2,4-oxadiazoles can be made from nitrile oxides and nitriles, and 1,2,3-thiadiazoles from diazoalkanes and isothiocyanates). Some general methods of synthesis of oxadiazoles and thiadiazoles by cyclization reactions are outlined in Fig. 8.33.

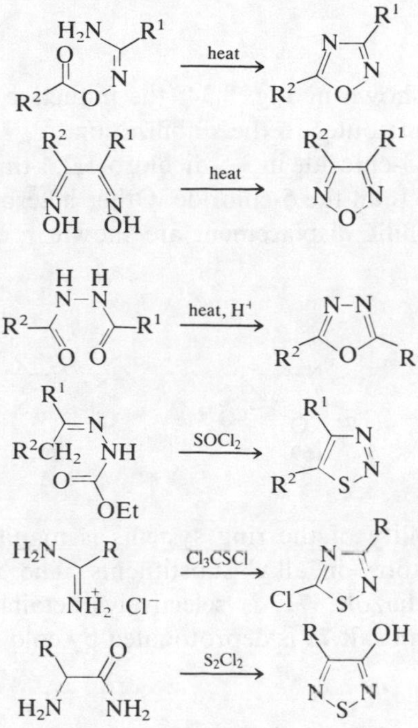

Fig. 8.33 Some general cyclization methods for preparing oxadiazoles and thiadiazoles.

8.7.3 Chemical properties

The increased numbers of nitrogen atoms in these ring systems have a considerable effect on their properties. The typical 'electron-rich' properties of the pyrrole, furan and thiophene ring systems no longer apply to the oxadiazoles and thiadiazoles. Electrophilic substitution reactions on carbon are extremely rare, and nucleophilic substitution reactions are common, particularly in the thiadiazoles. The types of reaction outlined in Fig. 8.1 take place very readily in these ring systems, especially when both nitrogen atoms can assist in stabilizing the intermediates. An example is provided by the high reactivity of 5-chloro-3-phenyl-1,2,4-thiadiazole towards nucleophiles: this compound reacts faster with nucleophiles than many activated six-membered heteroaromatics such as 2-chloropyridine and 2-chloro-4,6-dimethylpyrimidine. This can be ascribed to the stabilization available to

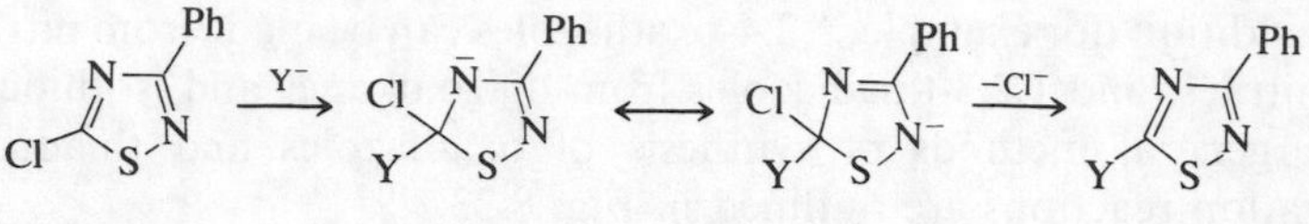

Fig. 8.34 Nucleophilic displacement of chloride in 5-chloro-3-phenyl-1,2,4-thiadiazole (ref. 104).

the tetrahedral intermediate, as shown in Fig. 8.34; the inductive effect of the sulphur undoubtedly also contributes to the stabilization.

This effect is selective, since the 3-chloride in 3,5-dichloro-1,2,4-thiadiazole is much more difficult to displace than the 5-chloride. Other heterocycles in this group that undergo nucleophilic displacement are shown in examples **67** to **70**.

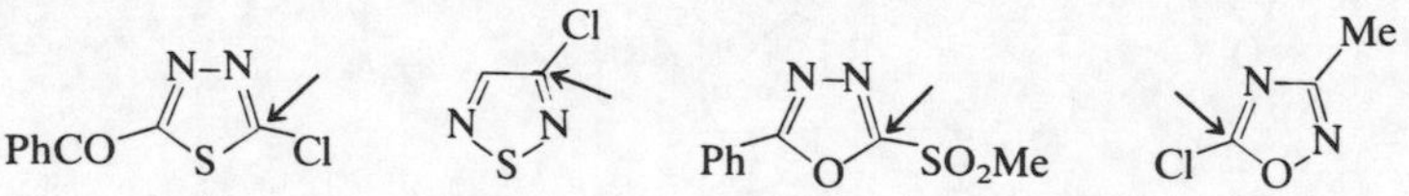

The same anion-stabilizing ability of the ring systems is manifested in the increased acidity of the protons on alkyl substituents. The 5-methyl group in 3,5-dimethyl-1,2,4-thiadiazole, **71**, is selectively metallated and carboxylated.[105] The phosphonium salt **72** is deprotonated by cold aqueous base to give a stabilized ylide.[106]

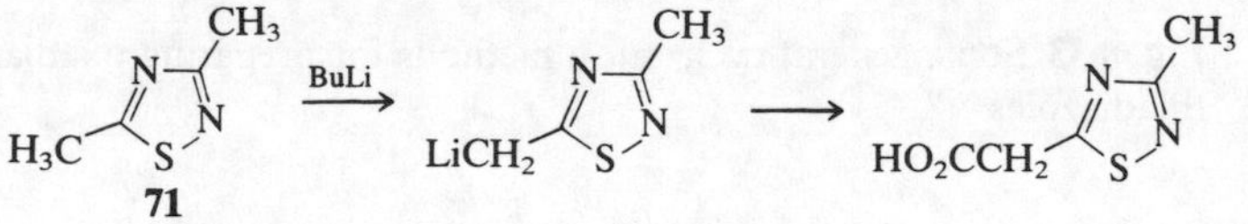

A further illustration of the electron-withdrawing nature of these ring systems is provided by the properties of the diazonium salts of the types **73** and **74**.[99,101] These compounds, which can be obtained by diazotization of the aminothiadiazoles, undergo coupling reactions with great ease: for example, both types are reactive enough to couple with mesitylene. The

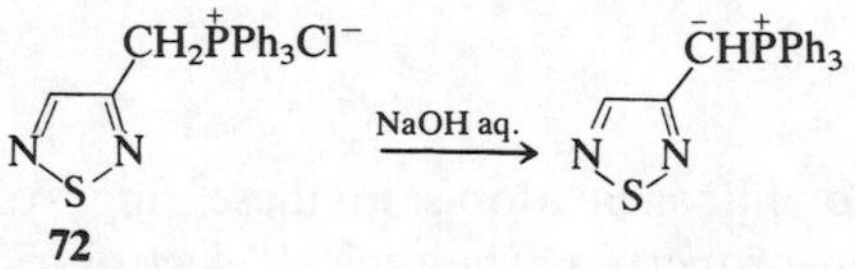

104 H. Grube and H. Suhr, *Chem. Ber.*, 1969, **102**, 1570.
105 R. G. Micetich, *Can. J. Chem.*, 1970, **48**, 2006.
106 D. Mulvey and L. M. Weinstock, *J. Heterocycl. Chem.*, 1967, **4**, 445.

electrophilicity of the diazonium cations is apparently enhanced by the electron-withdrawing character of the heterocycles. As mentioned in Section 8.7.1, azo dyes derived from these cations have commercial importance, particularly for dyeing polyamides and polyesters.

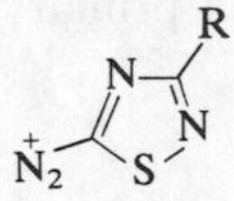

The deprotonation of unsubstituted ring positions of these heterocycles often leads to ring cleavage; examples of such reactions are shown in Fig. 8.35.

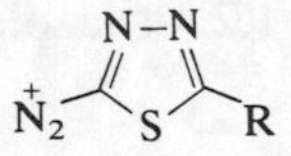

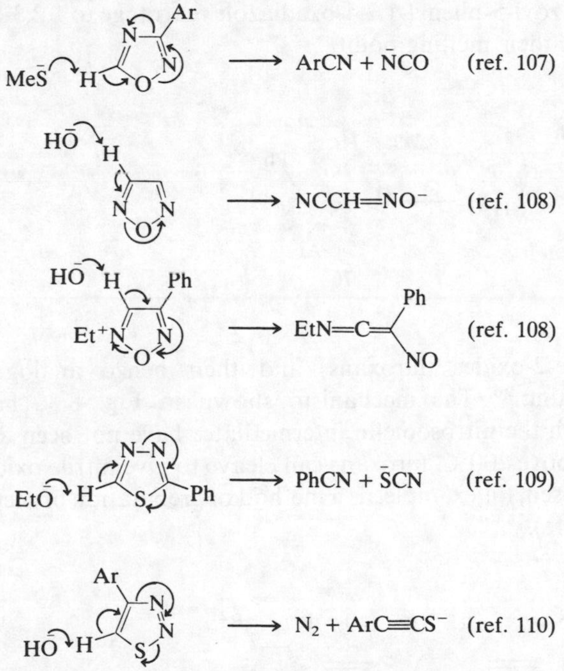

Fig. 8.35 Examples of anionic cleavage.

107 J. A. Claise, M. W. Foxton, G. I. Gregory, A. H. Sheppard, E. P. Tiley, W. K. Warburton, and M. J. Wilson, *J. Chem. Soc., Perkin Trans. 1*, 1973, 2241.
108 R. A. Olofson and J. S. Michelman, *J. Am. Chem. Soc.*, 1964, **86**, 1863.
109 A. Alemagna, T. Bacchetti, and C. Rizzi, *Gazz. Chim. Ital.*, 1972, **102**, 311.
110 A. Shafiee and I. Lalezari, *J. Heterocycl. Chem.*, 1973, **10**, 11.

Other modes of ring cleavage, particularly of oxadiazoles, are common. For example, nucleophiles readily attack 2,5-dialkyl-1,3,4-oxadiazoles and they are cleaved by acids and bases in a reaction which is the reverse of the ring closure shown in Fig. 8.33. Many nucleophilic ring cleavage reactions are followed by recyclization to give different heterocycles. For example, 2,5-dimethyl-1,3,4-oxadiazole, when heated with primary amines, is converted into the 4-substituted 1,2,4-triazoles **75**.[97]

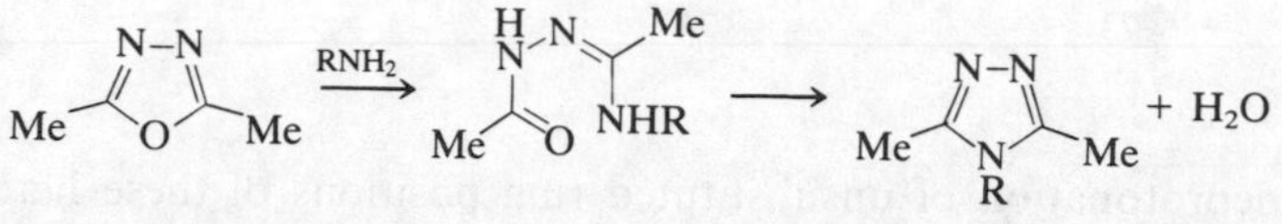

Suitably substituted 1,2,4- and 1,2,5-oxadiazoles undergo rearrangements of the general type illustrated earlier in Fig. 8.31. For example, arylhydrazones derived from 3-benzoyl-5-phenyl-1,2,4-oxadiazole rearrange to 1,2,3-triazoles **76** when heated at their melting point.

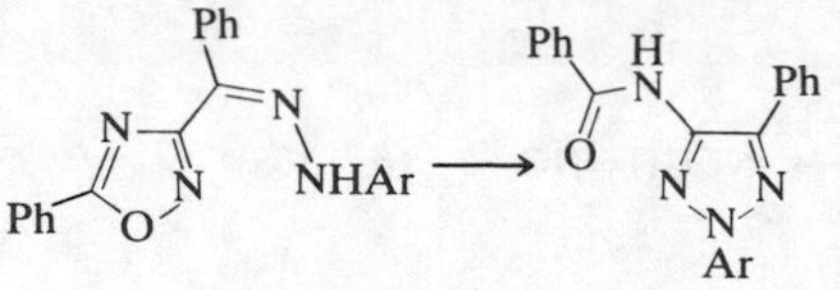

76

1,2,5-Oxadiazole-2-oxides (furoxans) and their benzo analogues can isomerize on heating.[96] The mechanism shown in Fig. 8.36 has been suggested, although the nitrosoolefin intermediates have not been detected. At temperatures above 200°C, furoxans can cleave to give nitrile oxides. The reaction has been used, for example, as a method of preparation of acetonitrile oxide ($R^1 = R^2 = Me$).[111]

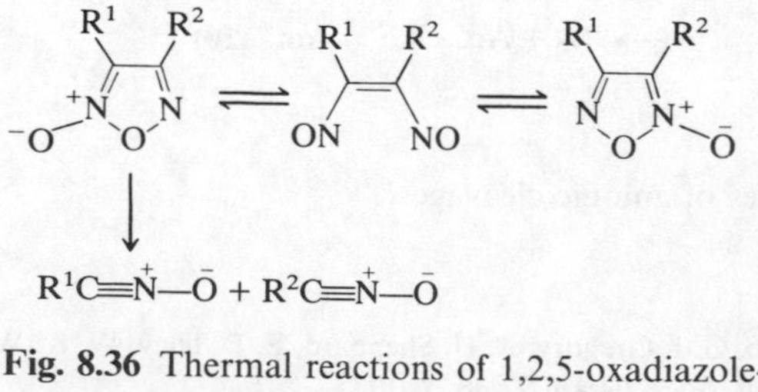

Fig. 8.36 Thermal reactions of 1,2,5-oxadiazole-2-oxides.

111 W. R. Mitchell and R. M. Paton, *Tetrahedron Lett.*, 1979, 2443.

When heated or irradiated, 1,2,3-thiadiazoles extrude nitrogen and give a range of products derived from the intermediates generated in this way. One of the most useful applications of the reaction is to the synthesis of thioketenes; these are formed in good yield by flash pyrolysis of the thiadiazoles **77**.[112] The isolation of thiirenes from the matrix photolysis of 1,2,3-thiadiazoles has been referred to in Chapter 2, Section 2.3.2.

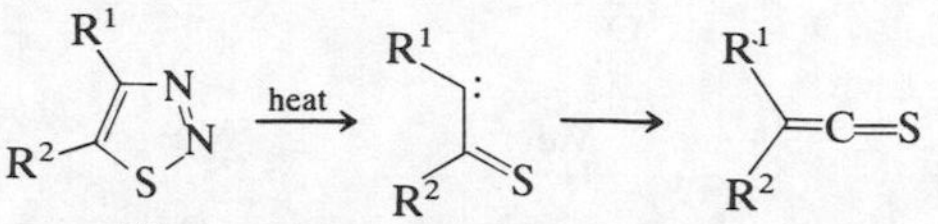

A related, and very useful, reaction is the thermolysis of 1,2,3-selenadiazoles **78**. These compounds, which are easily prepared from ketones, lose both nitrogen and selenium on mild heating and give acetylenes in good yield.[113]

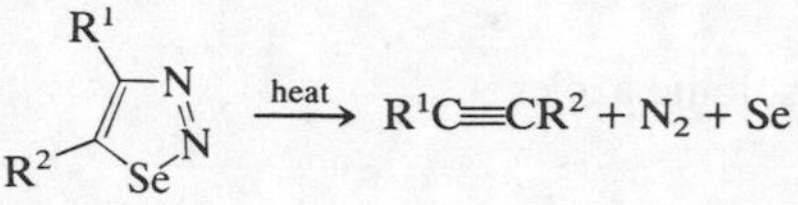

8.8 Betaines and mesoionic compounds

Some fully unsaturated five-membered heterocycles cannot be represented by valence bond structures without using charges. Examples of such compounds are shown in Fig. 8.37. These compounds all contain a cyclic array of five p-orbitals, which can be represented as containing six π-electrons, and a sixth, exocyclic, p-orbital containing two π-electrons. The structures can be divided into two broad groups: (1) those such as compounds (a) and (b) in which the exocyclic atom is joined to the ring system at a heteroatom, and (2) those such as (c) and (d) in which it is attached through carbon. These are all *betaines* (that is, they are represented by dipolar structures) but the first group are best classified as *oxides* or *imides*, whereas the latter are given the general description of *mesoionic compounds*.

8.8.1 N-*Oxides and* N-*imides*

Five-membered heteroaromatic compounds rarely undergo efficient *N*-oxidation and the known *N*-oxides are usually prepared by cyclization

112 G. Seybold and C. Heibl, *Angew. Chem. Int. Edn Engl.*, 1975, **14**, 248.
113 Review: I. Lalezari, A. Shafiee, and M. Yalpani, *Adv. Heterocycl. Chem.*, 1979, **24**, 109.

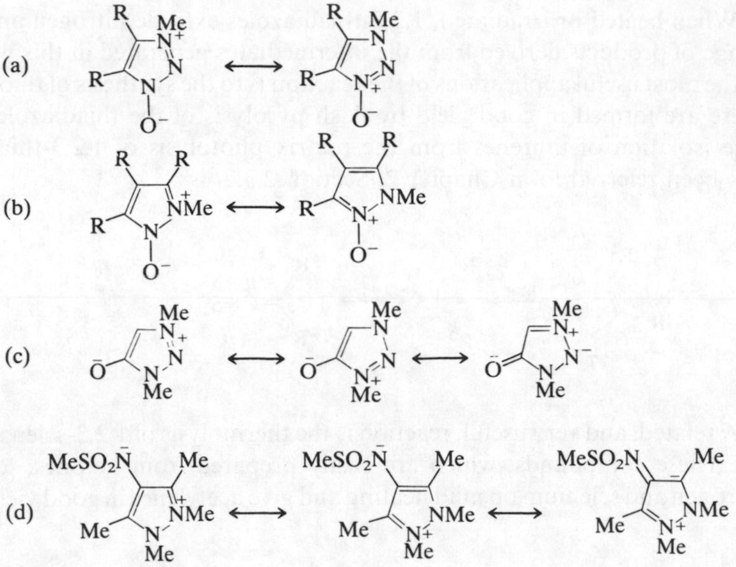

Fig. 8.37 Betaines and mesoionic azoles.

processes. The furoxans (1,2,5-oxadiazole-1-oxides) described in the preceding section provide one example. Another is provided by the benzimidazole-N-oxides, which are formed by cyclization reactions starting from 2-nitro- or 2-nitrosoaniline derivatives. Thus, 1-methylbenzimidazole-3-oxide is formed by the reduction of 2-nitro-N-methylformanilide.[114] This compound can act as a 1,3-dipole: it undergoes cycloaddition, as illustrated in Fig. 8.38, with methyl propionate and other dipolarophiles.

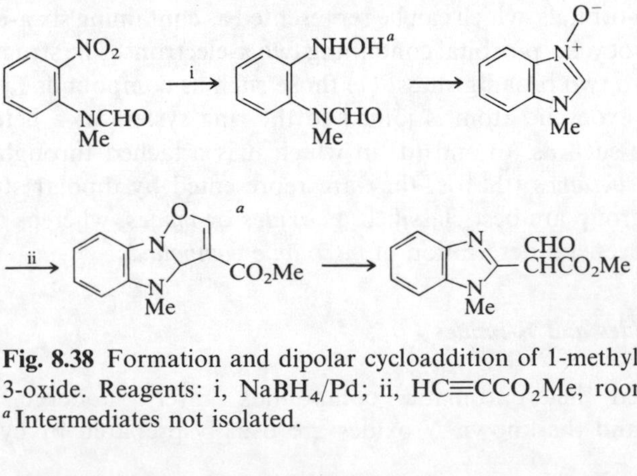

Fig. 8.38 Formation and dipolar cycloaddition of 1-methylbenzimidazole-3-oxide. Reagents: i, NaBH$_4$/Pd; ii, HC≡CCO$_2$Me, room temperature. ^{a}Intermediates not isolated.

114 Review: D. M. Smith, in ref. 42, p. 287.

N-Imides can be formed by similar types of cyclization process, and also, in some cases, by *N*-amination of the parent heterocyclic compound. The highly electrophilic aminating agent *O*-mesitylenesulphonylhydroxylamine, **79**, has been used to aminate several azoles, and *N*-imides can be formed from the *N*-aminoazolium salts by deprotonation.[115] An example, shown in Fig. 8.39, is the *N*-amination of thiazole and its conversion into the imide by acylation and deprotonation. This was intercepted in a 1,3-dipolar addition reaction with dimethyl acetylenedicarboxylate.[116]

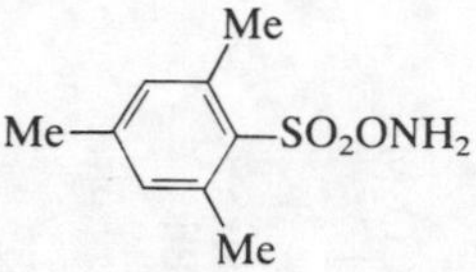

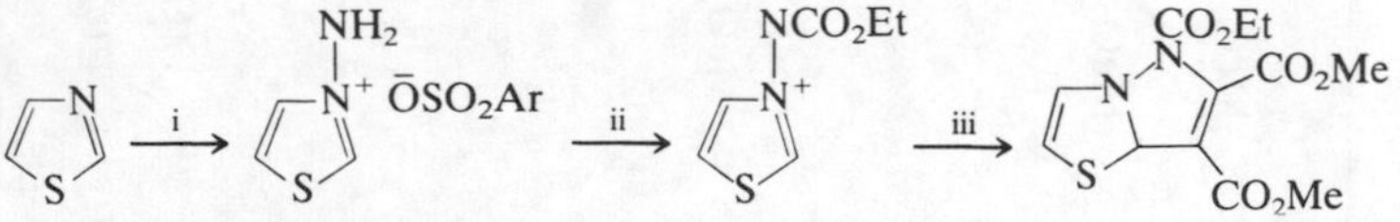

Fig. 8.39 *N*-Amination of thiazole and generation of thiazolium *N*-imide. Reagents: i, **79**; ii, $ClCO_2Et$, K_2CO_3; iii, $MeO_2cc\equiv CCO_2Me$.

8.8.2 Mesoionic compounds[117]

When *N*-nitroso-*N*-phenylglycine is treated with acetic anhydride, a mixed anhydride is produced which, when heated, forms the heterocycle **80** with the loss of acetic acid. This reaction was discovered by Earl and Mackney in 1935, and products of this type were given the trivial name 'sydnones' after the University of Sydney in which they were discovered. Baker and Ollis coined the term 'mesoionic' to describe the structure of *N*-phenylsydnone, which cannot satisfactorily be represented by a single covalent or dipolar structure. They also recognized that the sydnones were one of several groups of compounds which could be represented only as resonance hybrids of dipolar structures. Since the time of their introduction of the term in 1949, many other groups of mesoionic compounds have been synthesized.

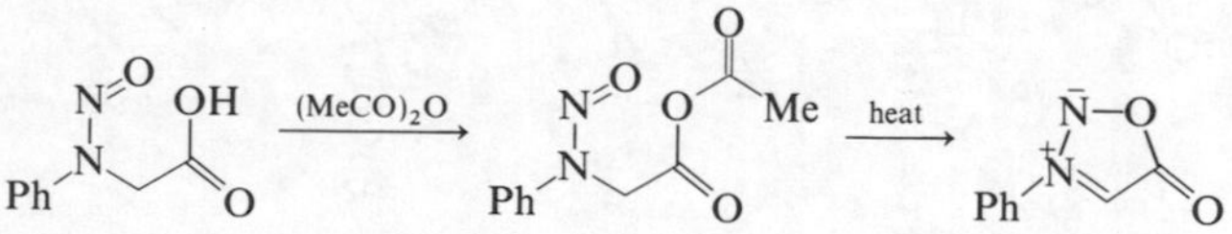

115 Review: Y. Tamura and M. Ikeda, *Adv. Heterocycl. Chem.*, 1981, **29**, 71.
116 H. Koga, M. Hirobe, and T. Okamoto, *Chem. Pharm. Bull.*, 1974, **22**, 482.
117 Reviews: W. D. Ollis and C. A. Ramsden, *Adv. Heterocycl. Chem.*, 1976, **19**, 1; C. G. Newton and C. A. Ramsden, *Tetrahedron*, 1982, **38**, 2965.

Table 8.2 Examples of 1,3-dipolar addition of mesoionic heterocycles.[a]

Mesoionic compound	Dipolarophile	Product[b]	Refs. and notes
(i) 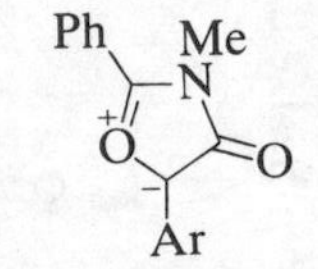	(E)-MeO$_2$CCH=CHCO$_2$Me, 80°C		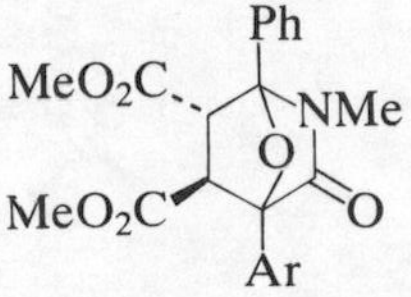 c
(ii)	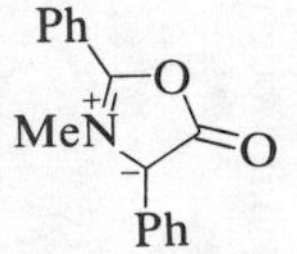 HC≡CH, 120–130°C	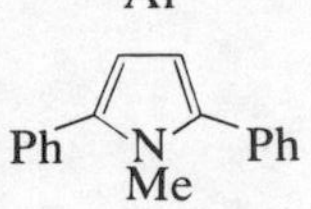	d
(iii) 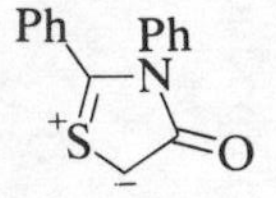	MeO$_2$CC≡CCO$_2$Me, 80°C		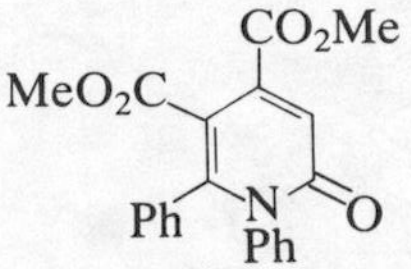 e
(iv)	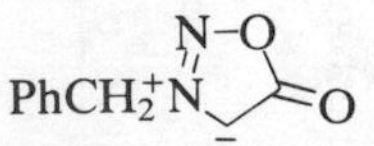 PhCH=CH$_2$, 140°C	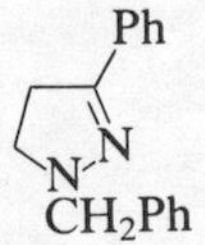	f
(v) 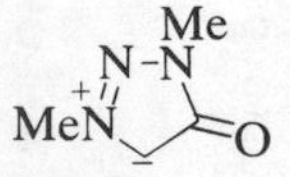	EtO$_2$CN=NCO$_2$Et, 140°C		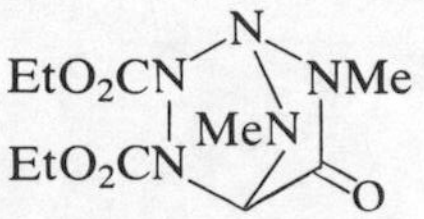 g

a For general reviews and further examples see ref. 117. b Products shown are all isolated in high yield. c Ar = 4-NO$_2$.C$_6$H$_4$; M. Hamaguchi and T. Ibata, *Tetrahedron Lett.*, 1974, 4475. d Product formed from initial adduct by loss of CO$_2$; R. Huisgen, H. Gotthardt, H. O. Bayer, and F. C. Schaefer, *Chem. Ber.*, 1970, **103**, 2611. e Product formed by extrusion of sulphur from the primary cycloadduct; K. T. Potts, E. Houghton, and U. P. Singh, *Chem. Commun.*, 1969, 1129. f Product formed by loss of CO$_2$ from the primary cycloadduct followed by proton shift; R. Huisgen, *Angew Chem. Int. Edn Engl.*, 1963, **2**, 565. g K. T. Potts and S. Husain, *J. Org. Chem.*, 1972, **37**, 2049.

Mesoionic compounds such as *N*-phenylsydnone are best regarded as cyclic 1,3-dipoles: thus, both *N*-phenylsydnone and the triazolium betaine shown in Fig. 8.37(c) can be represented as cyclic azomethine imides. They are, however, significantly more stable than their acyclic counterparts. There is the possibility that sydnones and other mesoionic compounds should be regarded as aromatic because their structures can be represented as cyclic arrays of p-orbitals containing six π-electrons. In the case of *N*-phenylsydnone represented as in **80**, for example, these six would be made up of four from the C=N—N system and two from the lone pair on oxygen. Evidence based on chemical reactivity is inconclusive: the reactions of sydnones include both substitution and addition. For example, 3-phenylsydnone, **80**, undergoes electrophilic chlorination (with chlorine and acetic anhydride), bromination (with bromine and sodium hydrogencarbonate) and nitration (with nitric and sulphuric acids) at the 4-position. The ring system is, however, cleaved by catalytic reduction and, hydrolytically, by moderately concentrated hydrochloric acid. It can also be cleaved oxidatively by reaction with fuming nitric acid, potassium permanganate, and other reagents. In the infrared spectra the carbonyl stretching frequency of most sydnones is in the range 1770–1750 cm^{-1}, which is close to that of other γ-lactones, and X-ray structural measurements show that the exocyclic C—O bond is close in length to that of a normal carbonyl group. It does not, therefore, seem useful or appropriate to classify sydnones and other mesoionic compounds as aromatic, despite the existence of the cyclic array of p-orbitals.

The most important general reaction of these mesoionic heterocycles is 1,3-dipolar cycloaddition. Some specific examples of 1,3-dipolar addition reactions of mesoionic heterocycles are given in Table 8.2.

A useful extension of this type of reaction is provided by the discovery that some heterocycles can apparently tautomerize to mesoionic structures which are then trapped by dienophiles. For example, the oxazolone **81** gives the furan **83** when heated with dimethyl acetylenedicarboxylate, by way of the mesoionic tautomer **82**.[118]

Summary

1. In the diazoles, triazoles and tetrazoles, two types of nitrogen atom occur, one with a lone pair in a p-orbital orthogonal to the plane of the ring, and the other with a lone pair in an sp^2-orbital in the plane of the ring. The second type is the more basic and the more nucleophilic centre. All the azoles have at least one nitrogen atom of this second type.

118 K. T. Potts and J. L. Marshall, *J. Org. Chem.*, 1979, **44**, 626.

2. Unsaturated five-membered heterocycles tend to be increasingly electron deficient in character as the number of heteroatoms increases. This is revealed, for example, by the following trends:

(a) decreasing basicity;

(b) a decreasing tendency to undergo electrophilic substitution;

(c) an increasing tendency to undergo nucleophilic substitution, particularly at carbon atoms bearing leaving groups and activated by imine nitrogen at the α- or γ-positions in the ring.

(d) increasing acidity of alkyl groups attached to carbon and activated by imine nitrogen at the α- or γ-position.

3. Imidazole is amphoteric in character; both the imidazolium cation and the imidazolyl anion are symmetrical delocalized structures. It is an excellent nucleophile, easily alkylated and acylated, which can act as a catalyst in acylation reactions. Electrophilic substitution in strongly acidic media is inhibited by protonation at nitrogen, but takes place preferentially at C-4. Nucleophilic substitution, although diffiucult, occurs mainly at C-2. Substitution of imidazoles can also occur by deprotonation at C-2 to give an intermediate imidazolium ylide, followed by rearrangement.

4. In pyrazole, electrophilic substitution can take place at C-4 but in at least some cases the electrophiles attack first at nitrogen. In the triazoles and tetrazoles, electrophilic substitution at carbon is uncommon. Nucleophilic substitution at carbon is easier than in the diazoles, so that, for example, 5-chlorotetrazoles undergo nucleophilic displacement quite easily. 1,2,3-Triazoles and tetrazoles which contain the —N=N— grouping can be cleaved by heat or light with the loss of nitrogen.

5. Oxazoles have low aromatic character and undergo addition and cycloaddition reactions. The cycloaddition process, in which oxazoles act as dienes, is easier when electron-releasing groups are present on the oxazoles. In oxazoles and thiazoles, and their benzo analogues, nucleophilic displacement reactions take place readily at C-2. The deprotonation of C-2 in thiazolium salts, giving thiazolium ylides, also occurs readily.

6. In isoxazoles, the weak N—O bond provides a point at which the ring can be cleaved; this is useful in synthesis.

7. In several oxadiazoles and thiadiazoles, nucleophilic substitution takes place readily at positions activated by one or more nitrogens. The most activated position is C-5 of the 1,2,4-thiadiazole system. A leaving group is readily displaced at this position and an alkyl group is selectively activated to deprotonation.

8. Mesoionic compounds are unsaturated five-membered betaines with an exocyclic anionic group joined to the ring through a carbon atom. Sydnones (1,2,3-oxadiazolium-5-oxides) are the best-known examples of such compounds. Sydnones and some other mesoionic compounds act as 1,3-dipoles in cycloaddition reactions.

Problems

1. 1,1-Diacetylcyclopropane **A** forms a monooxime which has the following spectra: infrared v_{max} 3420 and 1610 cm^{-1}; ^{1}H NMR, δ1.03 (4 H), 1.33 (3 H), 1.66 (3 H), and 3.90 (1 H). When one drop of concentrated hydrochloric acid was added to the NMR solution, the isoxazole **B** was formed. Account for these observations.

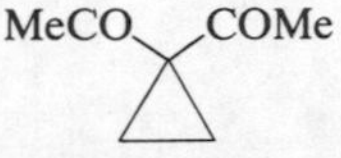

MeCO COMe

A

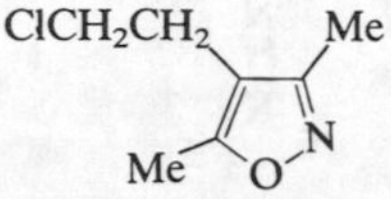

ClCH$_2$CH$_2$ Me

Me   O   N

B

2. The aminoketone **C** was benzoylated with benzoyl chloride enriched with oxygen-18. The product was then cyclized with concentrated sulphuric acid to give 4-methyl-2,5-diphenyloxazole, **D**, in which the oxygen-18 label was completely retained. Formulate a reaction sequence for the preparation of the oxazole which is consistent with this result.

PhCOCHMe
NH$_2$

C

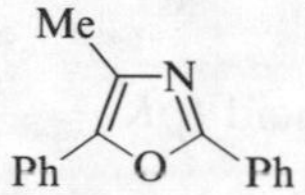

Me
N

Ph   O   Ph

D

3. A method for the exchange of functionality in conjugated ketones **E** and **F** depends upon the formation and cleavage of an isoxazole. Suggest suitable reagents for each of the steps, and draw the structure of the intermediate isoxazole.

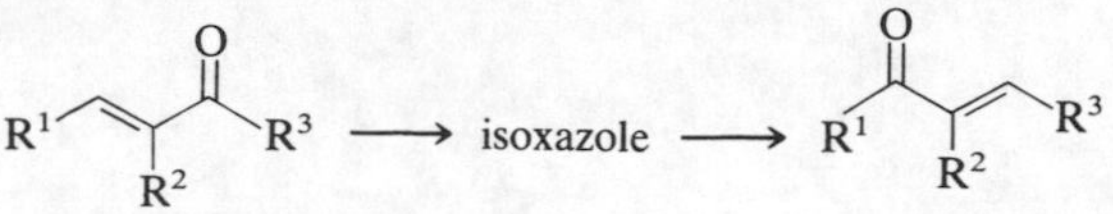

R^1 O   R^3 ⟶ isoxazole ⟶   O   R^3
    R^2 R^1 R^2

E   **F**

4. A general method for the conversion of a carboxylic acid into an aldehyde is the reaction of the acid RCO$_2$H with NN'-carbonyldiimidazole **G** in equimolar proportions followed by the reduction of the product, the acylimidazole **H**, with lithium aluminium hydride. Explain the formation of **H** from **G** and its selective reduction to an aldehyde.

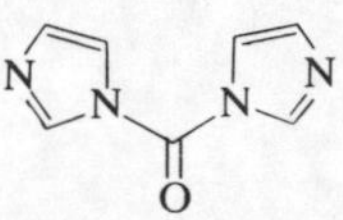

N   N   N   N
    O

G

N
N
COR

H

5. Account for the following:

(a) 1,4-Dinitropyrazole **J** reacts with diethylamine to give the substitution product **K**; 1,3-dinitropyrazole, in contrast, is converted into 3-nitropyrazole by secondary amines.

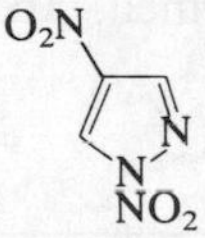

J 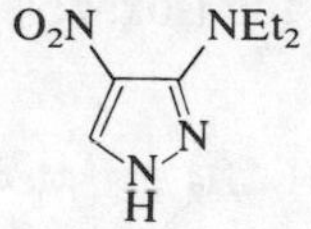   **K**

(b) Base-catalysed H/D exchange occurs selectively at the 5-methyl group of dimethyl-1,2,4-oxadiazole **L**. Under comparable conditions no exchange is observed at the methyl groups of dimethyl-1,3,4-oxadiazole **M**.

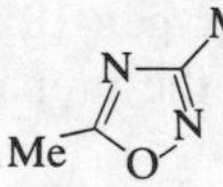

L 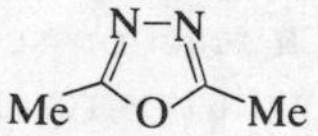   **M**

(c) Compound **N** is a strong acid ($pK_1 = -1.36$, $pK_2 = 7.53$).

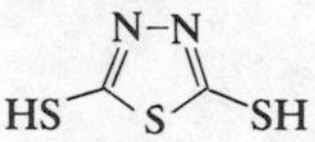

N

(d) The furoxan **P** is converted by sodium ethoxide into an isomer which has the following spectra: infrared v_{max} 3600–2800, 1665, and 1590 cm^{-1}; ^{1}H NMR δ 5.16 (2 H), 7.30–8.08 (5 H, m), and 11.4 (1 H, br).

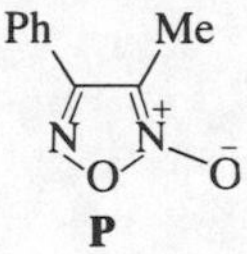

P

(c) The ^{1}H NMR spectrum of the benzisoxazole **R** shows two signals (δ 2.78 and 2.92 in Me$_2$SO-d_6) for the two methyl groups. When the solution is heated at 160–180°C the two signals coalesce. The original spectrum reappears when the solution is cooled.

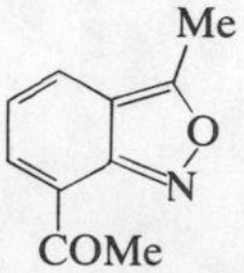

R

6. Explain the following reaction sequences:

(a)

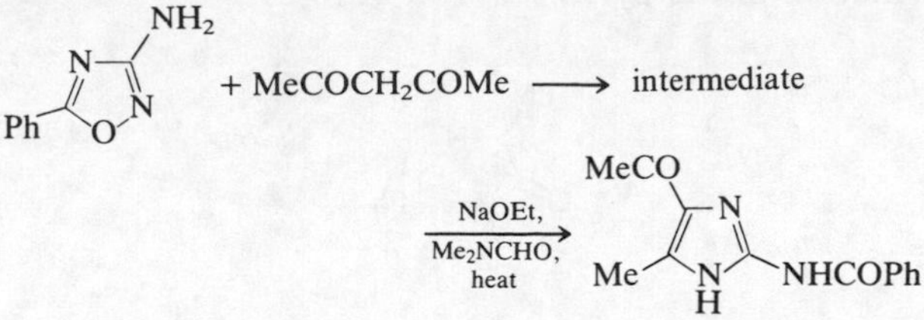

[structure: 3-amino-5-phenyl-1,2,4-oxadiazole with NH$_2$, Ph, N, O, N] + MeCOCH$_2$COMe $\longrightarrow$ intermediate

$\xrightarrow[\text{heat}]{\text{NaOEt,} \atop \text{Me}_2\text{NCHO,}}$ [structure: imidazole ring with MeCO, N, Me, N-H, NHCOPh]

(b) 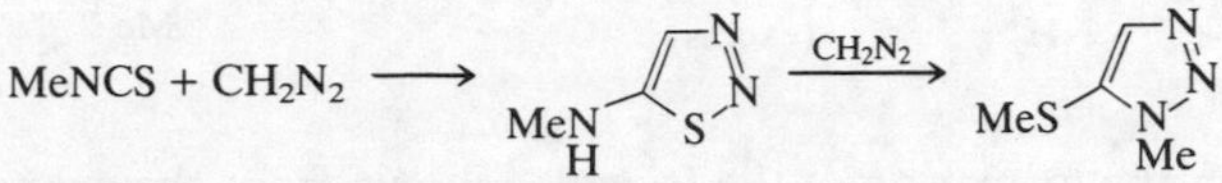 MeNCS + CH$_2$N$_2$ $\longrightarrow$ [structure: MeN(H)—thiadiazole ring S, N] $\xrightarrow{\text{CH}_2\text{N}_2}$ [structure: MeS—triazole ring N, N, N—Me]

(c) 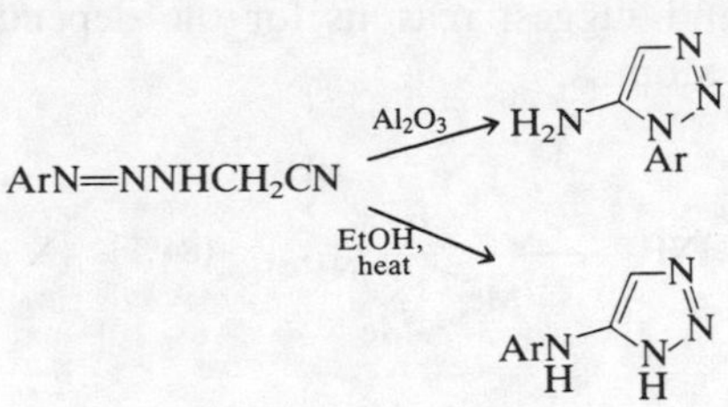 ArN=NNHCH$_2$CN

$\xrightarrow{\text{Al}_2\text{O}_3}$ H$_2$N—[triazole ring N, N, N—Ar]

$\xrightarrow[\text{heat}]{\text{EtOH,}}$ [structure: ArN(H)—triazoline ring N—H, N, N]

(d) 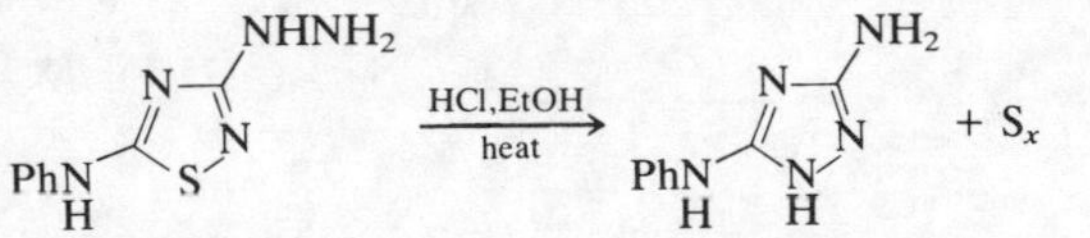 [structure: PhN(H)—thiadiazole with NHNH$_2$, N, S, N] $\xrightarrow[\text{heat}]{\text{HCl,EtOH}}$ [structure: PhN(H)—triazole with NH$_2$, N, N, N—H] + S$_x$

(e) 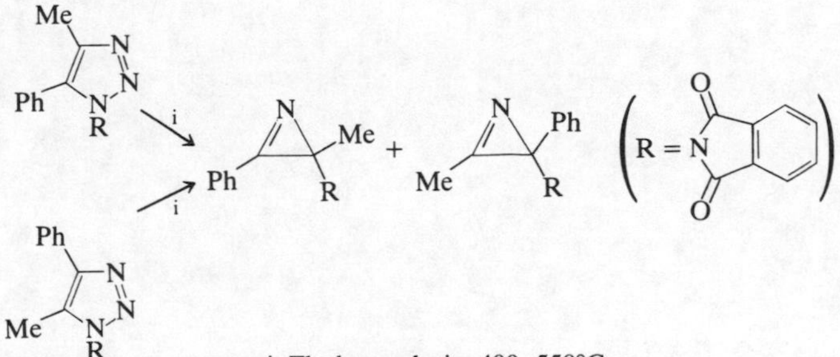 [structure: triazole with Me, Ph, N, N, N—R] $\xrightarrow{i}$ [azirine: Ph, Me, R with N] + [azirine: Me, Ph, R with N] $\left(R = \text{[phthalimido group with two C=O, N]} \right)$

[structure: triazole with Ph, Me, N, N, N—R] $\xrightarrow{i}$

i, Flash pyrolysis, 400–550°C

(f) 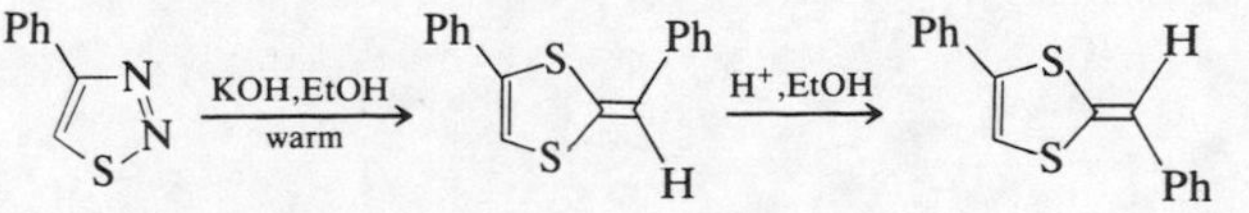 [structure: thiadiazole with Ph, N, N, S] $\xrightarrow[\text{warm}]{\text{KOH,EtOH}}$ [structure: dithiole with Ph, S, Ph, S, H] $\xrightarrow{\text{H}^+\text{,EtOH}}$ [structure: dithiole with Ph, S, H, S, Ph]

7. (a) The oxazolone **S** is a colourless crystalline solid which produces yellow solutions. The extinction coefficient of the band associated with the colour is very solvent dependent (CHCl$_3$:λ_{max} 436, ε 1.4; Me$_2$NCHO:λ_{max} 434, ε 9600). The solid **S** has a carbonyl absorption at 1820 cm^{-1} but in dimethyl sulphoxide there are two carbonyl absorptions at 1820 and 1704 cm^{-1}. Suggest an explanation for these observations and hence suggest a mechanism for the formation of the pyrrole shown in (b).

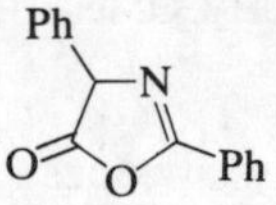

(b) MeCHCO$_2$H + (MeCO)$_2$O + H$_2$C=CClCN $\xrightarrow{\text{100°C}}$
     |
     NH$_2$ (excess)

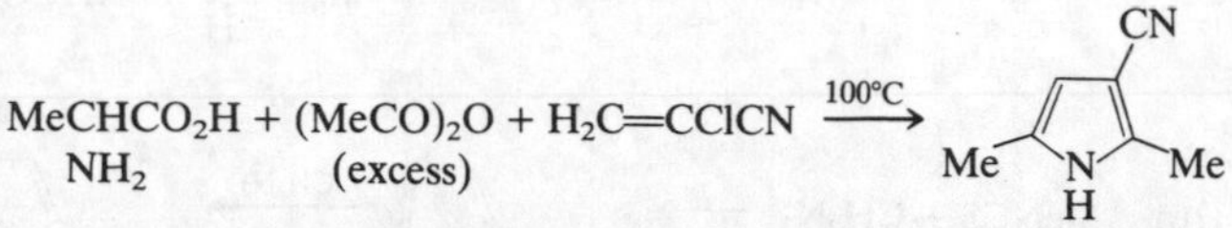

8. Explain the formation of the triazoles shown from the cyano compounds **T** and methylhydrazine, and suggest reasons for the dependence of their structure on the nature of atom X.

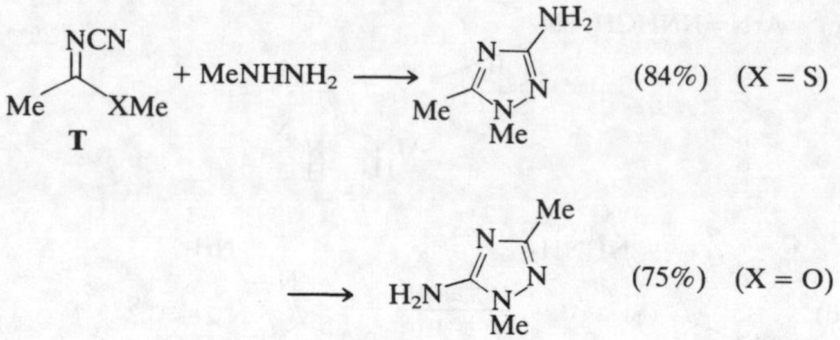

(84%) (X = S)

(75%) (X = O)

9 Three- and four-membered ring compounds

9.1 Introduction

The chemistry of three-membered heterocycles is dominated by ring strain. This leads to enhanced reactivity in processes in which the strain is relieved. Thus, coordination of an electrophile at a ring heteroatom, attack of a nucleophile at a ring carbon atom, heating, or irradiation can all result in opening of the ring. This means that three-membered heterocycles are unusually reactive species and a wide variety of reagents can attack the ring systems. When these reactions are selective (as, for example, are most of the nucleophilic ring cleavages) the heterocycles are useful synthetic intermediates. Oxiranes and aziridines in particular have found wide use both as intermediates in laboratory synthesis and in industry. The four-membered rings show much less evidence of ring strain and they have found correspondingly less use as synthetic intermediates. The β-lactams are exceptional; because of the antibacterial properties of some β-lactams an enormous amount of work has been concerned specifically with the synthesis and chemistry of this ring system.

In this chapter the syntheses and properties of some of the more common three- and four-membered ring compounds are outlined. The physical properties of these ring systems have been discussed earlier, in Chapter 3, as illustrations of the effects of ring strain. Some of the potentially 'antiaromatic' three- and four-membered heterocycles have been mentioned in Chapter 2, Section 2.3.2, and their properties will not be discussed further here.

9.2 Aziridines[1]

9.2.1 Introduction

Aziridines are good alkylating agents because of their tendency to undergo ring-opening reactions with nucleophiles. Many aziridines are thus actively

1 Reviews: J. A. Deyrup, in *Small Ring Heterocycles* Part 1, ed. A. Hassner, Wiley-Interscience, New York, 1983, p. 1; O. C. Dermer and G. E. Ham, *Ethyleneimine and Other Aziridines*, Academic Press, New York, 1969.

mutagenic and toxic. The naturally occurring mitomycin C, **1**, shows antibiotic and antitumour activity which is associated with the presence of an aziridine ring. The alkylating properties of aziridines have been investigated for potential industrial application; for example, as monomers for polymerization and as components of reactive dyes for cellulose fibres.

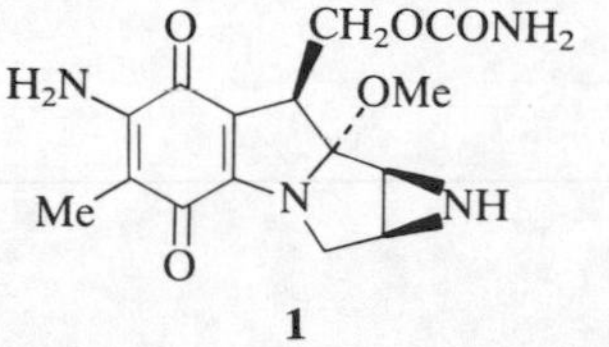

1

9.2.2   Ring synthesis

Many of the preparative routes to aziridines are intramolecular nucleophilic displacement reactions of the type discussed in Chapter 4 and illustrated in Fig. 4.6. The starting materials are often aminoalcohols which can be obtained by the ring opening of oxiranes with amines or by the reduction of α-aminoesters. The OH group is then converted into a good leaving group by reaction with a suitable activating agent and the aziridines are obtained by intramolecular displacement. Alternative procedures have been developed which use olefins as starting materials. Iodine isocyanate, bromine azide, and iodine azide add stereoselectively to olefins. The adducts are then modified by addition of an alcohol to the isocyanato group or by reduction of the azido group to give intermediates which can be cyclized to aziridines. Some of the more general cyclization routes to aziridines are summarized in Table 9.1.

Two other routes to aziridines are based on cycloaddition reactions with olefins. The first of these, and the more general, is shown in Fig. 9.1. 1,3-Dipolar addition of azides to olefins gives 4,5-dihydro-1,2,3-triazoles with retention of olefin stereochemistry. These compounds give aziridines when heated or by direct or sensitized photolysis. The ring-contraction step is not completely stereoselective and the reaction can therefore give mixtures of aziridines from appropriately substituted olefins.

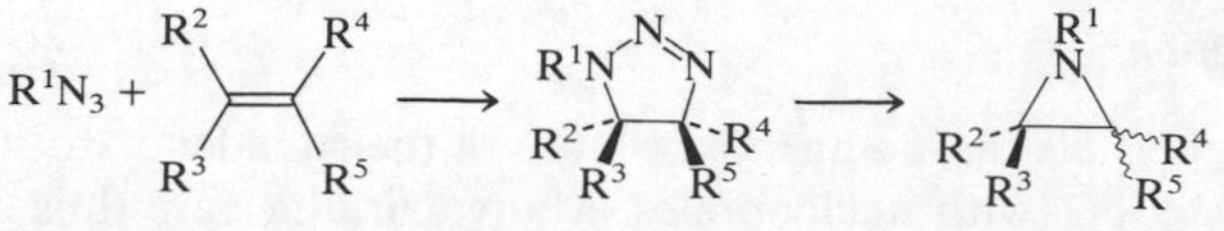

Fig. 9.1 Aziridines from azides and olefins.

Table 9.1 Cyclization routes to aziridines.[a]

$$R^2 \overset{R^1NH}{\underset{R^3}{\overset{|}{-}}}\overset{R^4}{\underset{X}{\overset{|}{-}}}R^5 \longrightarrow R^2 \overset{\overset{\displaystyle R^1}{\underset{\displaystyle N}{}}}{\underset{R^3 \quad R^5}{\triangle}} R^4$$

X	Method	Notes
OSO$_3$H	aminoalcohol + H$_2$SO$_4$ or ClSO$_3$H	b
OSO$_2$R	aminoalcohol + RSO$_2$Cl	c
OPPh$_3^+$Br$^-$	aminoalcohol + Ph$_3$PBr$_2$	d
Cl	Chloroamine + NaH in Me$_2$SO	e
I	Olefin, INCO, ROH	f
I	Olefin, IN$_3$, then LiAlH$_4$ or PPh$_3$	g

a See ref. 1 for primary literature references. *b* The Wenker synthesis; useful for a wide range of substituted aziridines. *c* Used for bulky or deactivated amines which can be attacked selectively at the OH group. *d* Milder alternative to the Wenker method; several other activated phosphorus reagents can be used. *e* The Gabriel synthesis. *f* The NCO group in the adduct is converted by reaction with the alcohol into NHCO$_2$R before cyclization. *g* A mild route to *N*-unsubstituted aziridines.

Many types of azide will participate in this reaction sequence but it has been used mainly for the preparation of aziridines with aryl, vinyl, and arenesulphonyl groups on nitrogen. The other route based on the cycloaddition approach is the addition of nitrenes to olefins, which has been described earlier (Table 4.24). This is particularly useful for preparing aziridines with substituents on nitrogen which are not easily incorporated by the cyclization routes.

9.2.3 Functionalization at nitrogen

Some functionalized aziridines are conveniently prepared from other aziridines. The nitrogen atom of *N*-unsubstituted aziridines or *N*-alkylaziridines is nucleophilic and these compounds behave as secondary or tertiary alkylamines. They can be alkylated by haloalkanes, epoxides, and similar reagents, and they undergo conjugate addition to $\alpha\beta$-unsaturated nitriles and carbonyl compounds. Unlike most secondary amines, aziridines can also react with some aldehydes to give 1:1 addition compounds. *N*-Chloro- and *N*-acylaziridines can also be prepared from the *N*-unsubstituted compounds. In all these reactions, the intermediate aziridinium salts must be deprotonated in order to avoid nucleophilic ring opening. Some examples, in which substitution products can be isolated in good yield, are shown in Table 9.2 (using aziridine itself as a substrate).

Table 9.2 *N*-Substitution of aziridine.[a]

N-Substituent	Reagent
CH_2CO_2Me	$ClCH_2CO_2Me$, Et_3N
$CH(OH)CCl_3$	Cl_3CCHO
$(CH_2)_3Me$	$Me(CH_2)_3Cl$, $PhCH_2\overset{+}{N}Et_3Cl^{-b}$
CH_2CH_2CN	$H_2C{=}CHCN$
$CH{=}CHCOPh$	$ClCH{=}CHCOPh$
$COMe$	$H_2C{=}C{=}O$
SO_2Me	$MeSO_2Cl$
Cl	$NaOCl$

a See ref. 1 for primary references *b* The 'phase transfer' method of alkylation is superior to conventional procedures for aziridine.

9.2.4 *Ring-opening reactions*

The facility with which aziridines are cleaved by nucleophiles depends upon the electron-accepting properties of the nitrogen substituents, steric effects of the carbon substituents, and the nature of the attacking reagents. The ring opening is progressively easier as the electron-accepting ability of the nitrogen substituent increases, and it is most rapid with aziridinium cations. Nucleophilic ring-cleavage reactions of *N*-alkyl- or *N*-unsubstituted aziridines are therefore often acid catalysed and it is the cation which is attacked by the nucleophile. Attack on the ring carbon atoms goes with inversion of configuration. Nucleophilic attack at the less substituted carbon atom of the ring is usually predominant, and is called 'normal' ring cleavage. 'Abnormal' cleavage can predominate if the aziridine contains a dialkyl-substituted carbon atom, however; this is illustrated for the acid-catalysed hydrolysis of 2,2-dimethylaziridine (Fig. 9.2).[2]

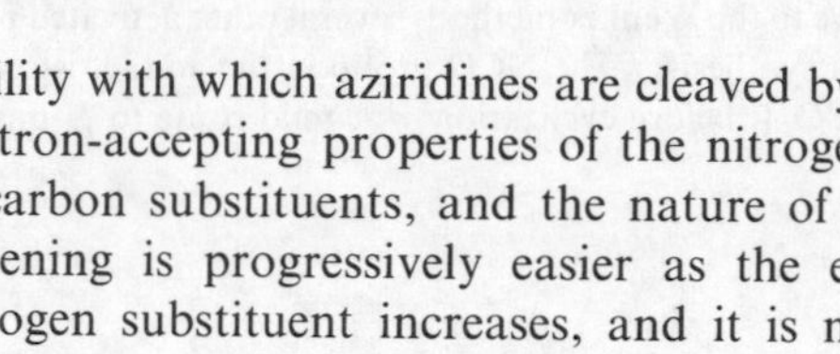

R = H	58	42
R = Me	7	93

Fig. 9.2 Product ratios in the hydrolysis of aziridines ($2\,M\,HClO_4$, 29.5°C). The ring-opening of 2,2-dimethylaziridine is the faster process (J. F. Bunnett, R. L. McDonald, and F. P. Olsen, *J. Am. Chem. Soc.*, 1974, **96**, 2855).

2 For a comparison of 'normal' and 'abnormal' ring opening by different nucleophiles in three-membered heterocycles see M. J. S. Dewar and G. P. Ford, *J. Am. Chem. Soc.*, 1979, **101**, 783. It has been suggested that some 'abnormal' cleavages of aziridines by strong nucleophiles go by way of an electron transfer mechanism: H. Stamm, P. Assithianakis, B. Bucholz, and R. Weiss, *Tetrahedron Lett.*, 1982, **23**, 5021.

N-Substituted aziridines, particularly those with conjugatively electron-withdrawing substituents on carbon, undergo electrocyclic ring opening to azomethine ylides (see Section 4.2.8). The ring opening is normally brought about by heat, and it is then a conrotatory process, but light-induced disrotatory opening can also occur. The azomethine ylides can be intercepted by 1,3-dipolar cycloaddition and this provides a method of preparation of several types of five-membered nitrogen heterocycles, as shown in Fig. 9.3.

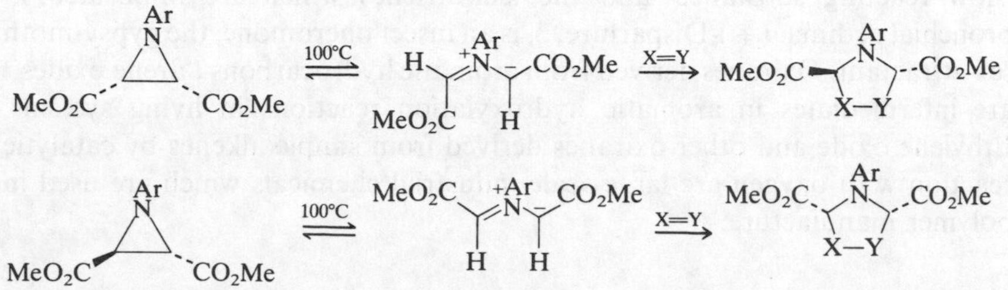

Fig. 9.3 Examples of ring-opening of aziridines to azomethine ylides and subsequent cycloaddition (E. Brunn and R. Huisgen, *Tetrahedron Lett.*, 1971, 473, and references therein). (Ar = C_6H_4.OMe-4; =Y = PhCHO, $MeO_2CCH=CHCO_2Me$, $EtO_2CN=NCO_2Et$, PhCH=NMe, norbornene.)

9.2.5 Fragmentation reactions

N-Unsubstituted aziridines are stereoselectively deaminated by reaction with nitrosyl chloride, nitrous acid, or other nitrosating agents. The *N*-nitrosoaziridines are probably intermediates and the reaction can be formulated as a cheletropic elimination of N_2O. Other related fragmentations include the ready thermal loss of diimide from *N*-aminoaziridines and the fragmentation of *N*-aziridinylhydrazones; examples of these processes are shown in Fig. 9.4.

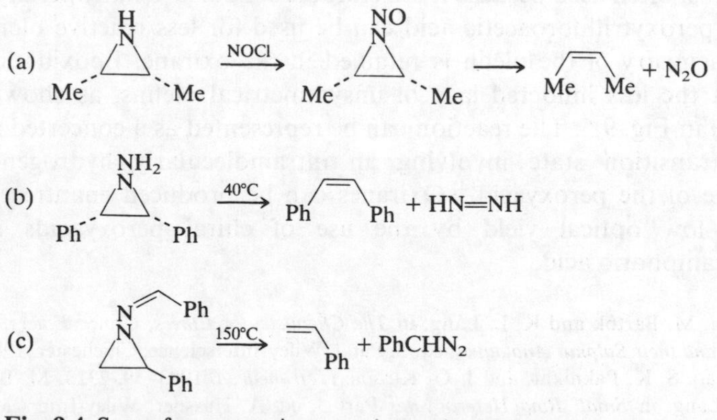

Fig. 9.4 Some fragmentation reactions of aziridines (see ref. 1, and, for (b), see also P. M. Lahti, *Tetrahedron Lett.*, 1983, **24**, 2343).

9.3   Oxiranes[3]

9.3.1   Introduction

Oxiranes are extremely common reagents in synthetic organic chemistry because the ring system can be both created and destroyed easily, and in a highly selective manner. The ring system is also present in many biologically important compounds. Leukotriene A, **2**, is the biosynthetic precursor of the 'slow reacting substance' and the leukotrienes which are implicated in bronchial asthma. (+)-Disparlure, **3**, is an insect pheromone, the gypsy-moth sex attractant. Oxiranes derived from aromatic hydrocarbons ('arene oxides') are intermediates in aromatic hydroxylation reactions in living systems. Ethylene oxide and other oxiranes derived from simple alkenes by catalytic reaction with oxygen are large-scale industrial chemicals which are used in polymer manufacture.

9.3.2   Ring synthesis

The three most important routes to oxiranes in the laboratory are (i) the epoxidation of olefins, (ii) intramolecular cyclization of alcohols bearing a leaving group on the adjacent carbon, and (iii) nucleophilic alkylation of carbonyl compounds. For method (i), peroxycarboxylic acids are the most commonly used laboratory reagents. The process is an electrophilic addition reaction and as such it is facilitated by electron-releasing groups on the double bond and by electron-withdrawing groups on the peroxyacid. Of the many peroxyacids that are suitable, 3-chloroperoxybenzoic acid has been one of the most often used because it is a storable solid. 3,5-Dinitroperoxybenzoic acid or peroxytrifluoroacetic acid can be used for less reactive olefins. The stereochemistry of the olefin is retained in the oxirane. Epoxidation takes place at the less hindered face of unsymmetrical olefins, as shown in the example in Fig. 9.5. The reaction can be represented as a concerted addition via a transition state involving an intramolecularly hydrogen-bonded molecule of the peroxyacid.[4] Oxiranes can be produced enantioselectively but in low optical yield by the use of chiral peroxyacids such as peroxycamphoric acid.

3 Reviews: M. Bartók and K. L. Láng, in *The Chemistry of Ethers, Crown Ethers, Hydroxyl Groups and their Sulphur Analogues*, ed. S. Patai, Wiley-Interscience, Chichester, 1980, p. 609; A. S. Rao, S. K. Paknikar, and J. G. Kirtane, *Tetrahedron*, 1983, **39**, 2323; M. Bartók and K. L. Láng, in *Small Ring Heterocycles* Part 3, ed. A. Hassner, Wiley-Interscience, New York, 1985, p. 1.
4 G. Berti, *Top. Stereochem.*, 1973, **7**, 93; V. G. Dryuk, *Tetrahedron*, 1976, **32**, 2855.

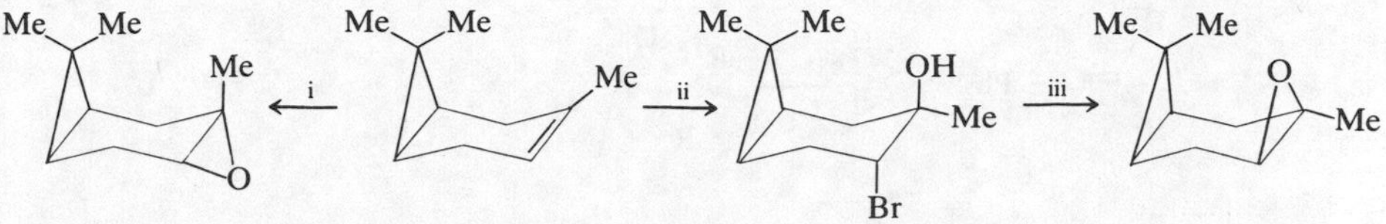

Fig. 9.5 Selectivity of epoxide formation (W. Cocker and D. H. Grayson, *Tetrahedron Lett.*, 1969, 4451). Reagents: i, $MeCO_3H$; ii, NBS, aq. Dioxan; iii, $KOBu^t$.

Olefins can also be epoxidized with organic hydroperoxides in the presence of metal catalysts. An important development of the method is the epoxidation of allylic alcohols, with very high enantioselectivity, using t-butyl hydroperoxide, titanium(IV) isopropoxide, and (+)- or (−)-dialkyl tartrate as a co-reagent. The choice of the enantiomeric tartrate ester determines which enantiomer of the oxirane will be produced. The method has been used in a number of natural product syntheses; for example, it is the key step in a highly efficient asymmetric synthesis of (+)-disparlure (Fig. 9.6).

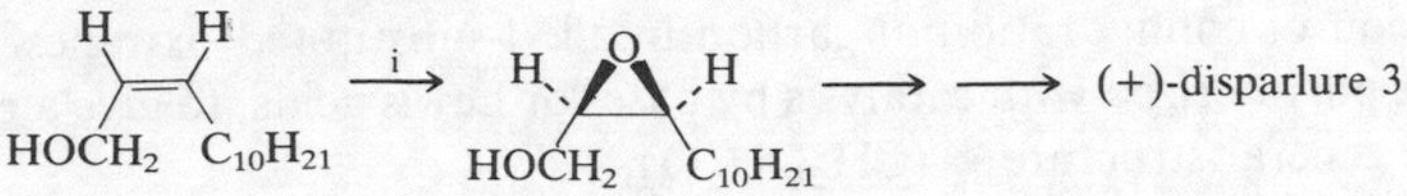

Fig. 9.6 Asymmetric epoxidation of an allylic alcohol (B. E. Rossiter, T. Katsuki, and K. B. Sharpless, *J. Am. Chem. Soc.*, 1981, **103**, 464). Reagents: i, Bu^tOOH, $Ti(OPr^i)_4$, (−)-diethyl tartrate.

Intramolecular S_N2 reactions [method (ii)] provide an alternative method of synthesis of oxiranes. Olefins are again used as the starting materials. HOBr or HOCl is commonly used as the reagent, giving a halohydrin which is converted into the oxirane by base. Halohydrins can also be obtained from α-haloketones by reduction. The HOBr addition–elimination procedure gives oxiranes having a configuration opposite to that obtained by direct epoxidation (Fig. 9.5).

In method (iii) oxiranes can be obtained from carbonyl compounds by reactions with compounds of the general type $R^1R^2\bar{C}X$, in which X is a good leaving group. The Darzens reaction is the classical reaction of this type, but it is generally limited to compounds in which R^1 or R^2 is a carbanion-stabilizing group (CO_2R, CN, etc.). The use of sulphur ylides provides a more versatile alternative because a much wider range of substituents can be introduced. For example, spiro-oxiranes are available from the reaction of the ylide **4** with carbonyl compounds.[5] The general form of the Darzens and sulphur ylide reactions were illustrated earlier (see Fig. 4.8).

5 Review: B. M. Trost and L. S. Melvin, *Sulfur Ylides*, Academic Press, New York, 1975.

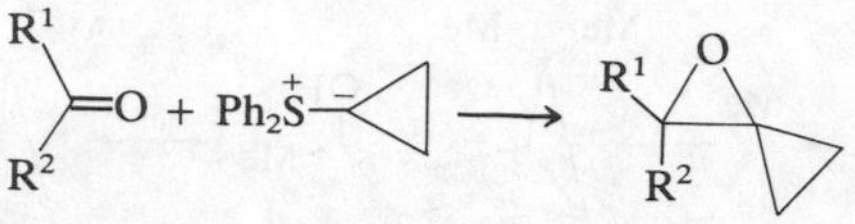

4

9.3.3 Reactions

There is a very wide range of reactions of oxiranes, many of which are markedly dependent upon their substituents and stereochemistry.[3] Only the more important general reactions are outlined here.

Ring-opening reactions of oxiranes are of considerable importance in general synthetic schemes because they occur stereo- and regioselectively in a predictable manner. As with aziridines, ring opening by nucleophiles can be brought about by direct attack, or it can be initiated by coordination of an electrophile at oxygen. Attack by the nucleophile usually occurs predominantly at the less substituted carbon atom unless one of them is benzylic or dialkyl substituted. In these exceptional cases, acid-catalysed ring opening may involve an S_N1-type mechanism. Displacement takes place with inversion of configuration at carbon in alkyl-substituted oxiranes. Oxirane can be polymerized with catalysis by bases or Lewis acids, to give a polyether of the general structure $-(CH_2CH_2O)_n-$.

Allylic alcohols can be obtained from alkyloxiranes by E2 elimination, by using poorly nucleophilic bases such as lithium diisopropylamide (Fig. 9.7). Abstraction of the proton usually takes place at the least substituted carbon atom.[6]

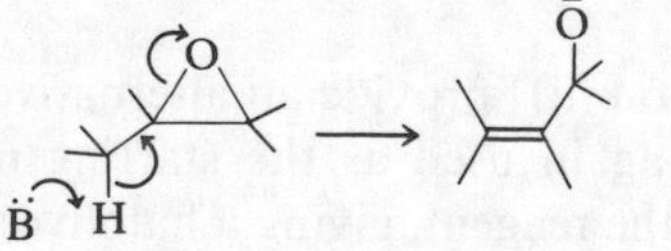

Fig. 9.7 E2 elimination and ring opening of oxiranes.

Oxiranes can be deoxygenated by the use of a variety of low-valent transition metal complexes and other reducing agents. Phosphorus reagents of the type $R_3P=X$ (where $X = S$, Se, or Te)[7] are particularly useful deoxygenating agents because the reaction is stereoselective, the configuration of the oxirane being retained in the olefin.

The rearrangement of vinyloxiranes to dihydrofurans has been illustrated earlier [see example (i) in Table 4.11]. Oxiranes substituted with electron-withdrawing groups also undergo electrocyclic ring-opening reactions to give carbonyl ylides. An example of the reaction, the opening of tetracyanoethylene oxide, is shown in Fig. 4.34.[8]

6 B. Rickborn and R. P. Thummel, *J. Org. Chem.*, 1969, **34**, 3583.
7 F. Mathey and G. Muller, *Compt. Rend (C)*, 1975, **281**, 881.
8 Review: R. Huisgen, *Angew. Chem. Int. Edn Engl.*, 1977, **16**, 572.

9.4 Thiiranes[9]

Thiiranes, **5**, have much in common with the other three-membered heterocycles, but some of their chemistry is associated specifically with the presence of the sulphur atom. For example, two oxidized versions of the ring system exist, the monoxide **6** and the dioxide **7**, because of the ability of sulphur to expand its valence shell.

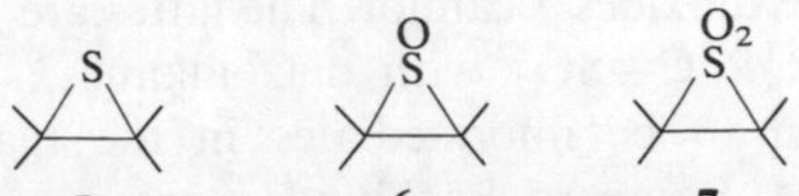

Simple thiiranes are most readily obtained from oxiranes by reaction with thiocyanate ions or with thiourea. The mechanism of the reaction, as shown in Fig. 9.8, is believed to involve a cyclic intermediate although the ring-forming step is a conventional cyclization reaction. Two displacements with inversion of configuration occur in the synthesis.

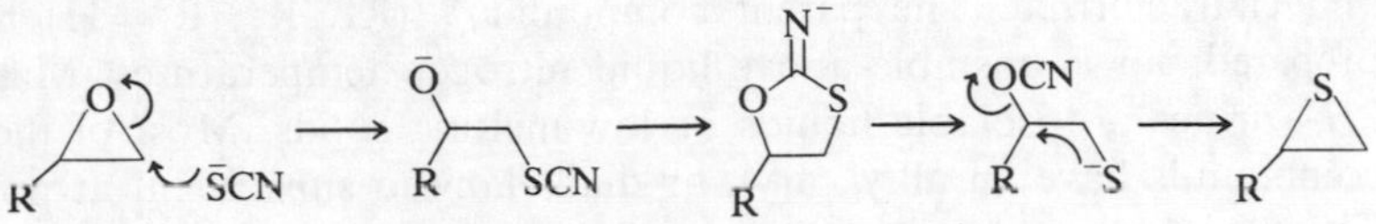

Fig. 9.8 Synthesis of thiiranes from oxiranes.

A useful synthesis of some substituted thiiranes is the electrocyclization of thiocarbonyl ylides (Fig. 4.18).

The thiirane ring is cleaved by good nucleophiles, or by weak nucleophiles in the presence of an acid catalyst, in the same way as the other three-membered ring heterocycles, although thiiranes are generally less reactive than oxiranes. Thiirane itself reacts with primary and secondary amines to provide a useful route to amino-ethanethiols. A complicating factor is that the thiol produced by ring opening may be a better nucleophile than that used in the ring-cleavage, so that dimeric or polymeric products may result. For example, the thiirane **8** forms polymers with hydrazine and similar nucleophiles whereas the corresponding oxirane gives 1:1 adducts.[10]

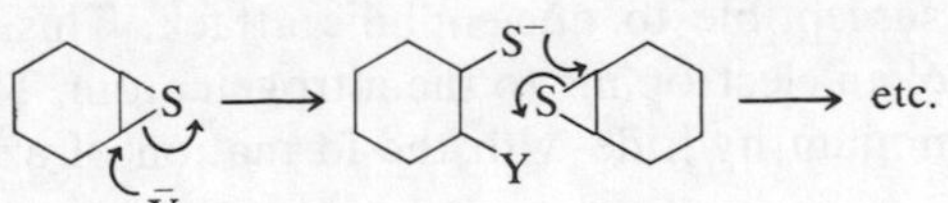

9 Review: U. Zoller, in *Small Ring Heterocycles*, Part 1, ed. A. Hassner, Wiley-Interscience, New York, 1983, p. 333.
10 C. C. J. Culvenor, W. Davies, and N. S. Heath, *J. Chem. Soc.*, 1949, 282.

Nucleophilic attack on thiiranes normally takes place at carbon, but alkyl- and aryllithium reagents may attack at sulphur. The reaction leads to the stereoselective desulphurization of the thiiranes.[11] The desulphurization can be achieved more efficiently by reaction of the thiiranes with trialkyl phosphites or with triphenylphospine. Some thiiranes, particularly those bearing aryl and other conjugative groups, are desulphurized by heat alone, although the process is probably initiated by nucleophilic attack on sulphur.

Thiirane oxides **6** can be prepared by peroxyacid oxidation of the corresponding thiiranes,[9] but the dioxides **7** cannot. The latter are prepared by the reaction of sulphenes $R^1R^2C{=}SO_2$ with diazoalkanes.[9] Thiirane dioxides have also been shown to be intermediates in the reaction of α-halosulphones with base (the Ramberg–Bäcklund rearrangement[9,12]), although this is not a useful synthesis of the heterocycles since they are converted into olefins in the reaction conditions.

9.5   2*H*-Azirines[13]

Of the two possible azirine isomers, only the 2*H*-azirines have been isolated and characterized. The parent compound **9** (R^1, R^2, $R^3 = H$) has been prepared but is unstable above liquid nitrogen temperatures. Many other 2*H*-azirines are isolable liquids or low-melting solids. Most of the known compounds have an alkyl, aryl, or dialkylamino substituent at position 3. There is evidence of ring strain in 2*H*-azirines from the abnormally high $C{=}N$ stretching frequencies in the infrared spectra and from the $^{13}C{-}H$ coupling constants in the NMR (see Section 3.2).

The most general method of synthesis of the ring system is the thermal or photochemical decomposition of vinyl azides, a reaction which may go by way of vinylnitrene intermediates [see entry (iv) in Table 4.10]. Two other methods which have some generality are shown in Fig. 9.9. The reaction of dimethylhydrazone methiodides with a strong base (method B) is a refinement of the Neber reaction, which was the first method of synthesis of 2*H*-azirines. In the Neber reaction, oxime *p*-toluenesulphonates are treated with base and give aminoketones by hydrolysis of the intermediate azirines.

Ring strain plays an important part in determining the chemistry of 2*H*-azirines. Three main types of reaction can be identified: (i) reaction at the $C{=}N$ bond, (ii) thermal cleavage of the $N{-}C2$ bond, and (iii) cleavage of the $C2{-}C3$ bond with ultraviolet light.

(i) The $C{=}N$ bond is susceptible to nucleophilic attack. This may be preceded by coordination of an electrophile to the nitrogen atom. The bond is reduced by lithium aluminium hydride with the formation of aziridines.

11 B. M. Trost and S. D. Ziman, *J. Org. Chem.*, 1973, **38**, 932.
12 Review: L. A. Paquette, in *Mechanisms of Molecular Migrations*, Vol. 1, ed. B. S. Thyagarajan, Wiley-Interscience, New York, 1968, p. 121.
13 Review: V. Nair, in *Small Ring Heterocycles*, Part 1, ed. A. Hassner, Wiley-Interscience, New York, 1983, p. 215.

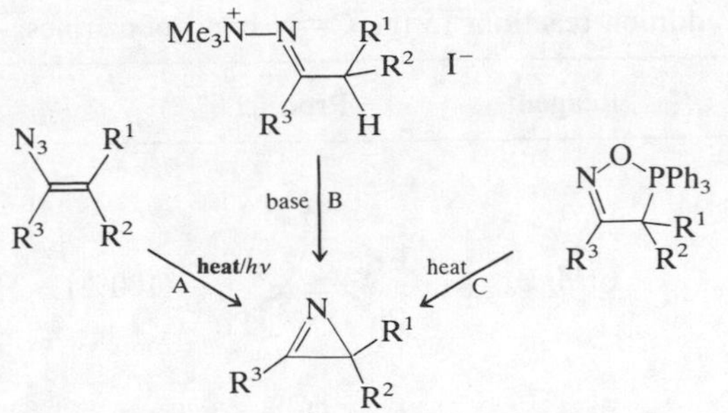

Fig. 9.9 Routes to 2*H*-azirines. Examples using method A: $R^1 = R^2 = H$. $R^3 = Ph$; 110°C; 63%. $R^1 = R^2 = Me$, $R^3 = NEt_2$; 20°C; 94%. $R^1 = H$, $R^2R^3 = (CH_2)_6$; *hv*, pentane; 93%. For method B: $R^1 = H$, $R^2 = Me$, $R^3 = Ph$; NaH, Me$_2$SO; 20°C; 63%. For method C: $R^1 = R^2 = H$, $R^3 = Bu^t$; 120°C; 57%.

The reduction is stereoselective and takes place from the less hindered side; for example, 2,3-diphenylazirine gives *cis*-2,3-diphenylaziridine. A similar reaction occurs with Grignard reagents. Reaction at the C=N bond which is initiated by coordination of an electrophile often results in ring cleavage. Some examples of these reactions are given in Table 9.3. The C=N bond is also more reactive than the corresponding bond of unstrained imines in cycloaddition reactions.

(ii) Thermal cleavage of the N—C2 bond is the reverse of the vinylnitrene cyclization which can be used to prepare azirines. The thermolysis of azirines usually results in cleavage of this bond although products resulting from C—C cleavage are sometimes also found. When azirines bearing a conjugative substituent (phenyl, vinyl, etc.) at C-2 are heated, the major product is often the five-membered ring rearrangement product formed by cleavage of the C=N bond and recyclization. Some examples are shown in Fig. 9.10.

(iii) In contrast to thermolysis, the photolysis of azirines usually involves cleavage of the C—C bond and the formation of nitrile ylides.[14] The reaction

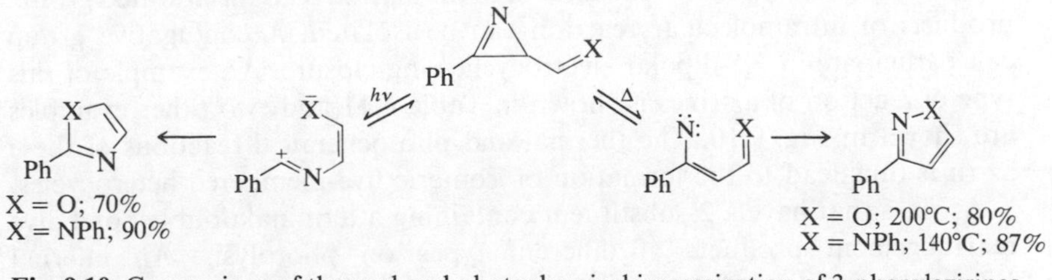

X = O; 70%
X = NPh; 90%

X = O; 200°C; 80%
X = NPh; 140°C; 87%

Fig. 9.10 Comparison of thermal and photochemical isomerization of 3-phenylazirines bearing conjugative substituents (A. Padwa, J. Smolanoff, and A. Tremper, *J. Am. Chem. Soc.*, 1975, **97**, 4682).

14 Review: A. Padwa and P. H. J. Carlsen, in *Reactive Intermediates*, Vol. 2, ed. R. A. Abramovitch, Plenum, New York, 1982, p. 55; see also Fig. 4.28.

Table 9.3 Examples of addition reactions to the C=N bond of azirines.

Azirine	Reagent	Product	Refs.
(i) Ph⟍△⟋Ph (N)	EtMgBr	Et⋯△⋯H (N–H), Ph, Ph (100%)	*a*
(ii) Ph⟍△ with Me, Me (N)	pyridine$^+$–H ClO_4^-	Ph⟍△ (N–H) Me, Py$^+$ Me ClO_4^- (95%)	*b*
(iii) Ph⟍△ (N)	Me$_2$$\overset{+}{S}$$\overset{-}{C}H_2$	⟍△ with N, Ph (68%)	*c*
(iv) Me$_2$N⟍△ with Me, Me (N)	PhNCS	Ph$\overset{+}{}$, S$^-$, N–N, Me$_2$N, Me Me (67%)	*d*

a R. M. Carlson and S. Y. Lee, *Tetrahedron Lett.*, 1969, 4001. *b* N. J. Leonard and B. Zwanenburg, *J. Am. Chem. Soc.*, 1967, **89**, 4456. *c* A. G. Hortmann and D. A. Robertson, *J. Am. Chem. Soc.*, 1972, **94**, 2758. *d* U. Schmid, H. Heimgartner, and H. Schmid, *Helv. Chim. Acta*, 1979, **62**, 160.

has been used, in particular, to produce nitrile ylides from 3-phenylazirines bearing a variety of substituents at the 2-position. The cleavage is probably the result of n→π* excitation; the n→π* band in 3-arylazirines is at about 285 nm. The nitrile ylides can be intercepted in the usual way by dipolarophiles. When the 3-substituent of the azirine is an unsaturated group, products of intramolecular reaction can be isolated. A conjugative group can participate in 1,5-dipolar electrocyclic ring closure. An example of this type of reaction of azirines is shown in Table 4.11 and two other examples are shown in Fig. 9.10. The thermal and photochemical reactions of these azirines thus lead to the formation of isomeric five-membered heterocycles.

Azirines that have a 2-substituent containing a terminal double bond give rearrangement products of different types on photolysis. An internal 1,3-dipolar addition of the nitrile ylide takes place if the alkenyl chain is long enough for the necessary transition state to be achieved [Fig. 9.11(a)]. If it is not, an interesting alternative reaction occurs, in which the 1,3-dipole behaves as a carbene and reacts with the double bond to give a cyclopropane [Fig. 9.11(b)].

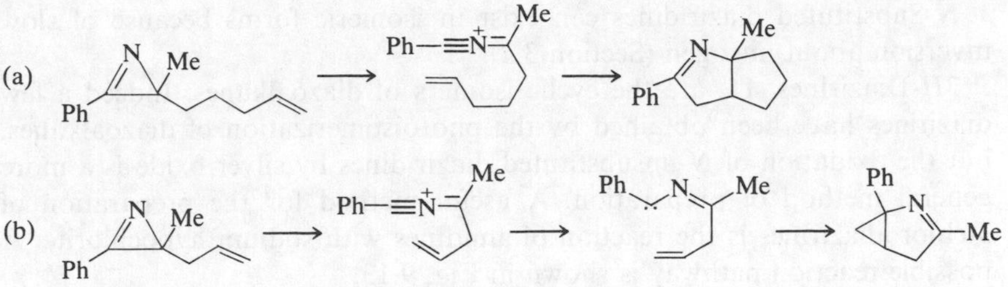

Fig. 9.11 Alternative modes of internal addition of nitrile ylides derived from azirines (see ref. 14).

9.6 Diaziridines and 3*H*-diazirines[15]

Two types of three-membered ring system containing two nitrogen atoms are isolable: the saturated diaziridines, **10**, and 3*H*-diazirines, **11**.

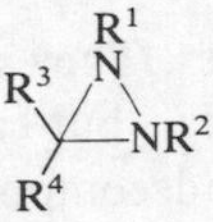

Diaziridines can be prepared from ketones by reaction with ammonia or a primary amine, together with an aminating agent such as chloramine or hydroxylamine-*O*-sulphonic acid. Reactions carried out with ammonia give diaziridines unsubstituted on nitrogen and these can be oxidized to diazirines, **11** [Fig. 9.12(a)]. Diaziridines can also be synthesized by the electrocyclic ring closure of stabilized azomethine imides in the presence of ultraviolet light. An example[16] is shown in Fig. 9.12(b). The reaction is reversed when the diaziridine is heated; other diaziridines with conjugative electron-withdrawing substituents are also cleaved in the same way when they are heated.[15]

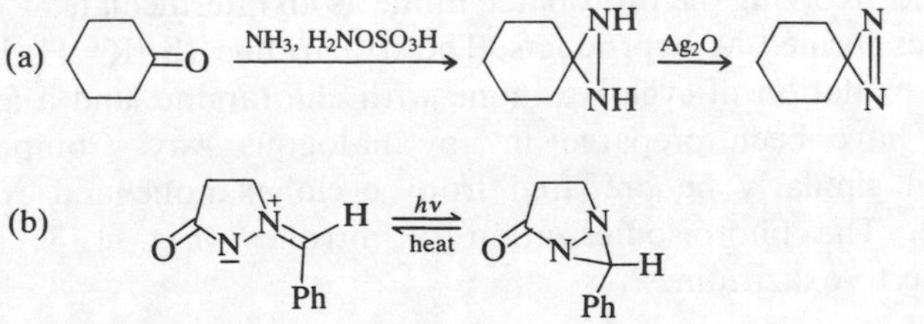

Fig. 9.12 Routes to diaziridines.

15 Review: H. W. Heine, in *Small Ring Heterocycles*, Part 2, ed. A. Hassner, Wiley-Interscience, New York, 1983, p. 547.
16 M. Schulz and G. West, *J. Prakt. Chem.*, 1970, **312**, 161.

N-Substituted diaziridines can exist in isomeric forms because of slow inversion about nitrogen (Section 3.2).

3*H*-Diazirines, **11**, are the cyclic isomers of diazoalkanes. Indeed a few diazirines have been obtained by the photoisomerization of diazoalkanes, but the oxidation of *N*-unsubstituted diaziridines by silver oxide is a more general method of preparation. A useful method for the preparation of 3-chlorodiazirines is the reaction of amidines with sodium hypochlorite. A possible reaction pathway is shown in Fig. 9.13.

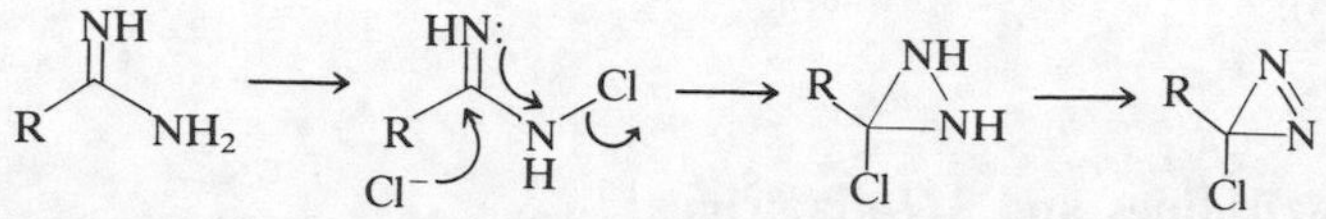

Fig. 9.13 Synthesis of 3-chlorodiazirines from amidines.

3*H*-Diazirines are useful for the generation of carbenes. They are more stable than diazoalkanes, particularly in acidic media, but they can decompose to give carbenes on irradiation. The photodecomposition of halogen-substituted diazirines is a good method for the generation of halocarbenes in neutral media. A diazirine unit has also been incorporated into a phospholipid in order to identify the site of interaction with membrane proteins: on irradiation of the phospholipid–protein complex, a carbene is generated and there is extensive crosslinking because of covalent bond formation by the carbene.[17]

9.7 Oxaziridines[18]

Oxiziridines, **12**, can be prepared by the oxidation of Schiff bases with peroxyacids. In contrast to the epoxidation of olefins the reaction is probably a stepwise one involving the protonated imine as an intermediate.[19] Nitrones are sometimes formed as byproducts. The oxaziridine **13** (R = H) has been obtained by oxidation of cyclohexanone with chloramine, and a few other oxaziridines have been prepared in an analogous way. Compound **13** (R = Me) can similarly be prepared from cyclohexanone and *N*-chloro-methylamine. The photoisomerization of nitrones (Fig. 4.18) provides another route to oxaziridines.

17 B. Erni and H. G. Khorana, *J. Am. Chem. Soc.*, 1980, **102**, 3888.
18 Review: M. J. Haddadin and J. P. Freeman, in *Small Ring Heterocycles*, Part 3, ed. A. Hassner, Wiley-Interscience, New York, 1985, p. 283.
19 D. R. Boyd, D. C. Neill, C. G. Watson, and W. B. Jennings, *J. Chem. Soc., Perkin Trans. 2*, 1975, 1813.

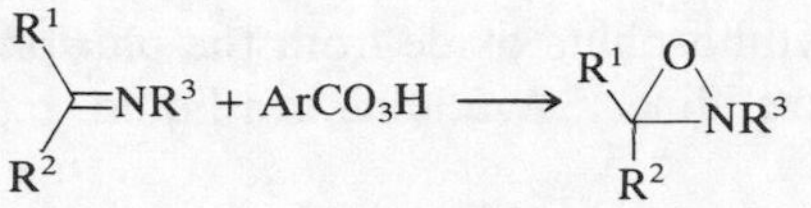

$$\begin{array}{c} R^1 \\ \diagdown \\ R^2 \end{array} =NR^3 + ArCO_3H \longrightarrow$$

12 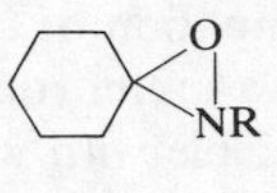   **13**

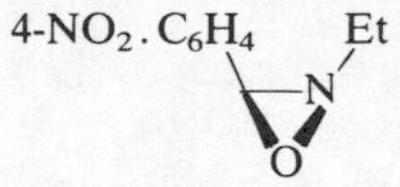

14a 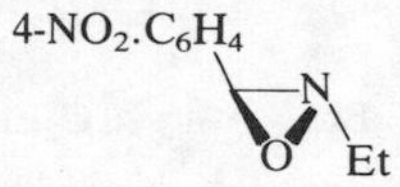   **14b**

Oxaziridines show remarkable configurational stability about nitrogen (Section 3.2) and there are several examples of pairs of isomers which are separable and which differ only in their configuration about nitrogen. The separation of the crystalline isomers **14a** and **14b** was one of the first demonstrations that nitrogen invertomers could be configurationally stable.[20]

The oxaziridine ring can be cleaved by acid either at the C—O bond (giving a nitrone) or at the C—N bond (giving a hydroxylamine). The second type of cleavage has been applied to the synthesis of *N*-hydroxyaminoesters from α-aminoesters (Fig. 9.14).[21]

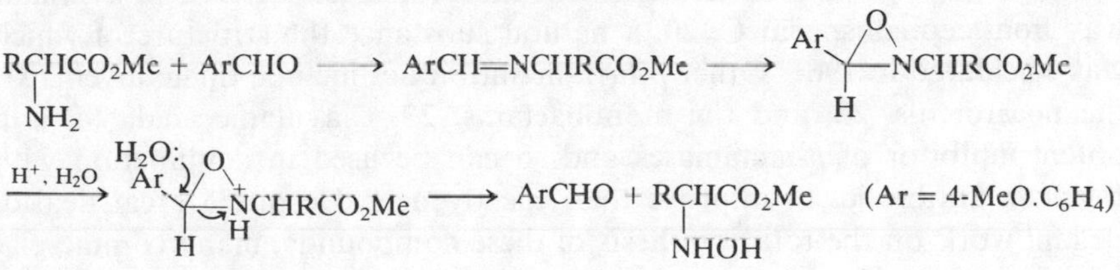

Fig. 9.14 Oxidation of α-aminoesters via oxaziridines.

9.8 Azetidines and azetidinones

The azetidine ring is much less strained than that of aziridine, so its chemistry is much more like that of a normal secondary amine.[22] The ring system can be made, with variable efficiency, by intramolecular nucleophilic displacement reactions. For example, the ester **15** can be prepared in high yield by cyclization of the bromoester **16**, but azetidine itself cannot be prepared efficiently by the analogous route. A modified cyclization, the

20 D. R. Boyd, *Tetrahedron Lett.*, 1968, 4561.
21 T. Poloński and A. Chimiak, *Tetrahedron Lett.*, 1974, 2453.
22 Reviews: N. H. Cromwell and B. Phillips, *Chem. Rev.*, 1979, **79**, 331; J. A. Moore and R. S. Ayers, in *Small Ring Heterocycles*, Part 2, ed. A. Hassner, Wiley-Interscience, New York, 1983, p. 1.

thermal elimination of triphenylphosphine oxide from the phosphine imide **17**, provides a useful route to azetidine.[23] Azetidine is a liquid, b.p. 62.5°C, which is miscible with water.

Azetidinone (β-lactam) chemistry is of great importance because of the use of β-lactam derivatives as antibacterial agents.[24] Since the discovery (in 1945) that the structure of penicillin contains a β-lactam function, a vast amount of effort has been devoted to producing other β-lactam antibiotics with a wider spectrum of activity and a greater resistance to enzymic cleavage by β-lactamases. Some of these are derived by acylation of 6-aminopenicillanic acid, which is an inexpensive and readily available reagent. It can be obtained by cleavage of the side chain from benzylpenicillin, **18**, which is produced on a large scale by fermentation. Ampicillin, **19**, is an example of a 'semisynthetic' penicillin produced by reacylation of 6-aminopenicillanic acid. Many of the more recent semisynthetic antibiotics are derived in a similar way from cephalosporin C, **20**, a natural substance the structure of which was elucidated in 1961. Other β-lactam antibiotics include thienamycin, **21**, the nocardicins, **22**, and the monobactams, **23**. Clavulanic acid, **24**, is a potent inhibitor of β-lactamases and so can be used in conjunction with β-lactam antibiotics to improve their effectiveness. Despite a great deal of elegant work on the total synthesis of these compounds, many commercial β-lactam antibiotics are made by the 'semisynthetic' method. Methods for constructing the β-lactam ring have been illustrated in Chapter 4 (see Tables 4.3, 4.5, and 4.23).

23 J. Szmuskovicz, M. P. Kane, L. G. Laurian, C. G. Chidester, and T. A. Scahill, *J. Org. Chem.*, 1981, **46**, 3562.
24 Reviews: *Recent Advances in the Chemistry of β-lactam Antibiotics*, Royal Society of Chemistry, London, 1981; *Cephalosporins and Penicillins: Chemistry and Biology*, ed. E. H. Flynn, Academic Press, New York, 1972; G. A. Koppel, in *Small Ring Heterocycles*, Part 2, ed. A. Hassner, Wiley-Interscience, New York, 1983, p. 219; C. E. Newall, *Chem. Br.*, 1987, **23**, 976; B. G. Christensen, *Chem. Br.*, 1989, **25**, 371.

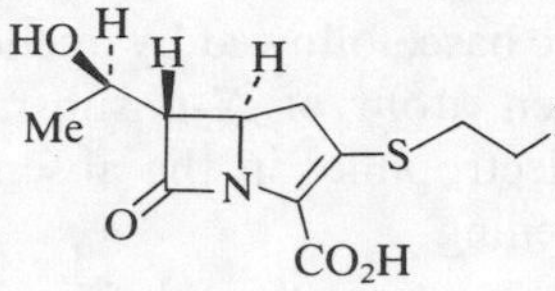

The 4-acetoxyazetin-2-one **25** and 3-substituted derivatives are attractive intermediates for the synthesis of fused β-lactams because **25** is readily available from the addition of chlorosulphonyl isocyanate to vinyl acetate (see Section 4.3.4) and the acetoxy group is easily displaced by nucleophiles. These displacement reactions have been shown to go by an elimination–addition mechanism and intermediate azetinones of this type have been intercepted by dienes in Diels–Alder reactions.[25]

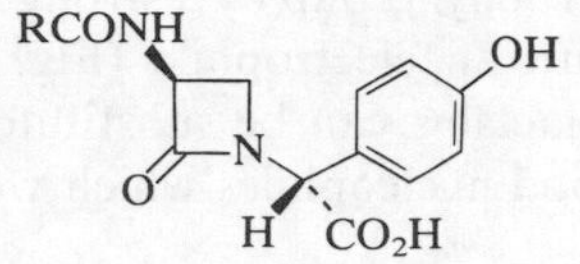

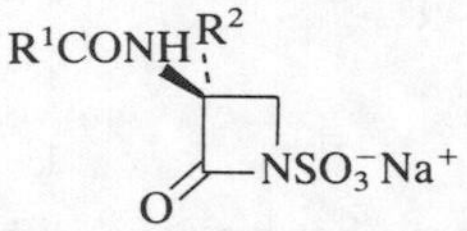

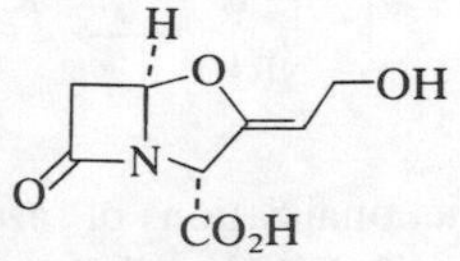

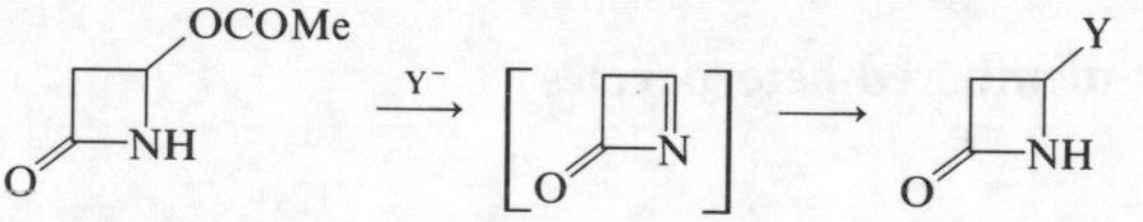

On the basis of the range of β-lactam structures which show biological activity, it is clear that increased ring strain, for example by fusion of a second ring, is not an essential feature. Nor does the inhibition of amide resonance (Section 3.2) seem to be significant in determining biological activity. Several of the biologically active compounds have good leaving groups attached to nitrogen and this can aid breakdown of the tetrahedral intermediate produced by nucleophilic attack on the amide carbonyl group. However, there is not necessarily a direct relationship between chemical reactivity and their antibacterial function of inhibiting cell wall biosynthesis.[26]

Some monocyclic β-lactams can react with nucleophiles without ring cleavage. This was described above for the azetidinone **25**. Similarly, the side-chain acetoxy group in cephalosporins can be displaced by nucleophiles without causing ring opening. Another fairly general method of functionalization of β-lactams is deprotonation at the carbon atom adjacent to the

25 F. Gaviña, A. M. Costero, and M. R. Andreu, *J. Org. Chem.*, 1990, **55**, 434; A. I. Meyers, T. J. Sowin, S. Scholz, and Y. Ueda, *Tetrahedron Lett.*, 1987, **28**, 5103.
26 Review: M. I. Page, *Adv. Phys. Org. Chem.*, 1987, **23**, 165.

carbonyl group by a strong, poorly nucleophilic base, followed by quenching with an electrophile (Fig. 9.15). The nitrogen atom of *N*-unsubstituted β-lactams can be substituted by a range of electrophiles in the absence of good nucleophiles which would cause ring opening.

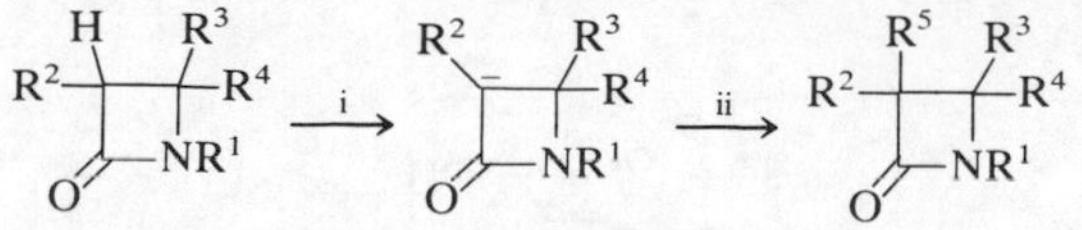

Fig. 9.15 C-3 functionalization of azetidinones. Reagents: i, lithium diisopropylamide; ii, R^5X (R^5 = alkyl, acetyl, trimethylsilyl, etc.)

Benzazetidinones, **26**, can be isolated when they contain bulky substituents (R = t-butyl, 1-adamantyl), and others are implicated as reaction intermediates.[27]

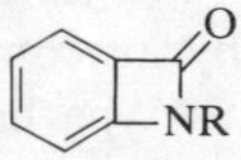

26

9.9   Other four-membered heterocycles

9.9.1   Oxetanes

Derivatives of this cyclic ether system can be prepared by [2 + 2] cycloaddition (Section 4.3.4), particularly by the photochemical Paterno–Büchi reaction. Oxetane itself can be obtained by a cyclization reaction. 3-Chloropropyl acetate, $MeCOO(CH_2)_3Cl$, is heated with a mixture of sodium hydroxide and potassium hydroxide.[28] Oxetane is a water-miscible liquid, b.p. 47°C.

The ring system shows evidence of ring strain in that it is cleaved by nucleophiles, although the conditions required are generally more vigorous than those for ring-opening of oxiranes. Oxetane is cleaved by organolithium and organomagnesium reagents. 2-Methyloxetane is cleaved in the same manner as 2-methyloxirane by reaction with hydrochloric acid (Fig. 9.16),

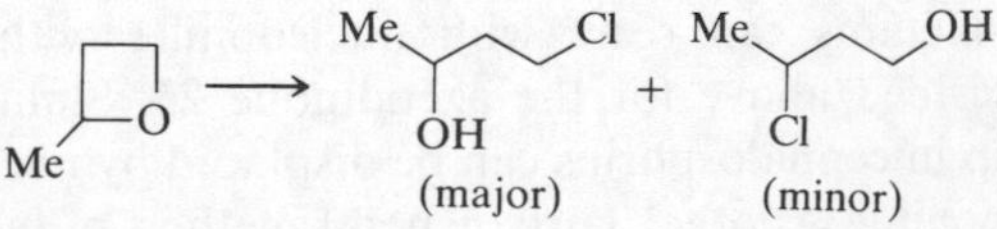

Fig. 9.16 Cleavage of 2-methyloxetane by hydrochloric acid. Reagent: conc. HCl, 5°C.

27 Review: R. J. Kobylecki and A. McKillop, *Adv. Heterocycl. Chem.*, 1976, **19**, 215.
28 S. Searles, *J. Am. Chem. Soc.*, 1951, **73**, 124.

the major product arising from attack on the protonated ether at the less substituted carbon atom.

A derivative of the ring system which has found use as a synthetic intermediate because of its ease of ring opening is 4-methyleneoxetan-2-one, or 'diketene', **27**.[29] This compound, which is formed by dimerization of ketene, is an excellent acylating agent. It reacts with alcohols to give esters of acetoacetic acid, and forms heterocycles with several other nucleophiles; some examples are shown in Fig. 9.17.

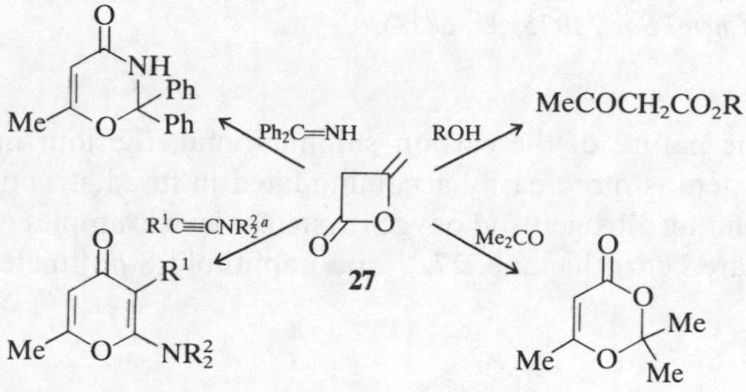

Fig. 9.17 Some synthetic uses of diketene (see ref. 29 and, for *a*, J. Ficini, *Tetrahedron*, 1976, **32**, 1449).

9.9.2 Thietanes[30]

Both cyclization and cycloaddition routes can be used for the preparation of thietanes (Fig. 9.18).

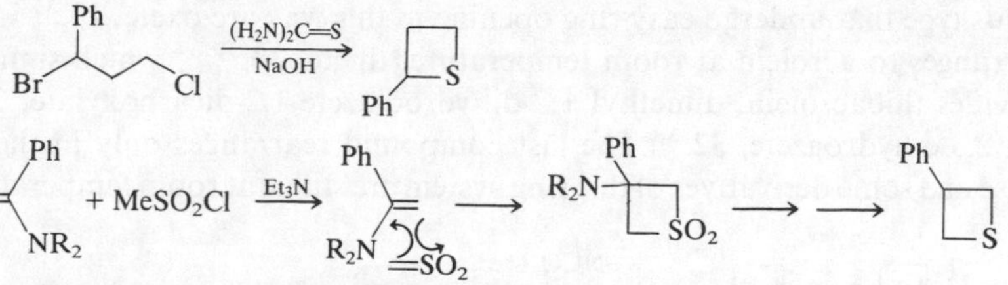

Fig. 9.18 Examples of thietane ring synthesis.

The ring system is, as expected, less susceptible to ring cleavage than the thiirane system. Thietane is cleaved to give ethylene and thioformaldehyde when heated in the gas phase, and the reaction provides a useful method of

29 T. Kato, *Accounts Chem. Res.*, 1974, **7**, 265.
30 Review: D. C. Dittmer and T. C. Segerman, in *Small Ring Heterocycles*, Part 3, ed. A. Hassner, Wiley-Interscience, New York, 1985, p. 431.

generating thioformaldehyde. Thietane is cleaved by alkyllithium reagents, which attack sulphur, as with thiiranes. Several other ring-cleavage reactions are promoted by coordination of an electrophile at sulphur; for example, the ring is cleaved by chlorine, and by reaction with allyl bromide (Fig. 9.19).

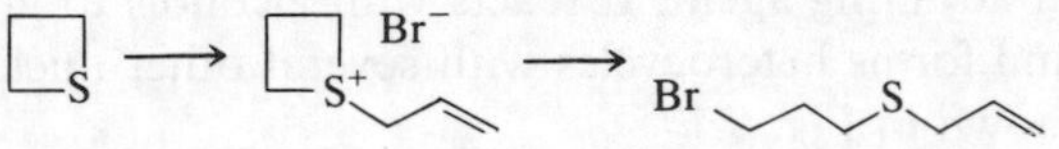

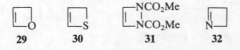

Fig. 9.19 Cleavage of thietane with allyl bromide (E. Vedejs and J. P. Hagen, *J. Am. Chem. Soc.*, 1975, **97**, 6878).

Because of the nature of the carbon–sulphur bond, the four-membered thietane ring system is more easily accommodated in fused structures than are the corresponding nitrogen and oxygen systems. Two examples of isolable fused thietanes are benzothietane, **27**,[31] and naphtho[1,8-*bc*]thiete, **28**.[32]

27   28

9.9.3 Some unsaturated four-membered rings

Monocyclic four-membered heterocycles containing a double bond are, in general, unstable compounds unless they are heavily substituted. This is in part due to their tendency to undergo electrocyclic ring opening to heterodienes (Section 4.2.8). Some examples of four-membered heterocycles of this type that undergo easy ring opening in this way are oxete, **29**,[33] which rearranges to acrolein at room temperature, thiete, **30**,[30,34] which similarly provides thioacrolein, dimethyl 1,2-dihydrodiazete-1,2-dicarboxylate, **31**,[35] and 2,3-dihydroazete, **32**.[36] The last compound rearranges only in the gas phase and some derivatives of this ring system are stable at room temperature.

29   30   31   32

31 R. Schulz and A. Schweig, *Tetrahedron Lett.*, 1980, **21**, 343.

32 J. Meinwald, S. Knapp, S. K. Obendorf, and R. E. Hughes, *J. Am. Chem. Soc.*, 1976, **98**, 6643.

33 P. C. Martino and P. B. Shevlin, *J. Am. Chem. Soc.*, 1980, **102**, 5429; L. E. Friedrich and P. Y. Lam, *J. Org. Chem.*, 1981, **46**, 306.

34 Review: D. C. Dittmer, in *New Trends in Heterocyclic Chemistry*, ed. R. B. Mitra, Elsevier, Amsterdam, 1979, p. 130.

35 E. E. Nunn and R. N. Warrener, *J. Chem. Soc., Chem. Commun.*, 1972, 818.

36 J. C. Guillemin, J. M. Denis, and A. Lablache-Combier, *J. Am. Chem. Soc.*, 1981, **103**, 468.

Summary

1. Aziridines are prepared by cyclization of hydroxyamines, haloamines, and related compounds. The precursors are available from olefins by the stereoselective addition of iodine isocyanate or iodine azide. Aziridines can also be prepared by elimination of nitrogen from triazolines and by nitrene addition to olefins. The ring system is cleaved by nucleophiles which normally attack at the less substituted carbon atom. The displacement goes with inversion of configuration. Conjugated aziridines are cleaved by heat or by irradiation to give azomethine ylides. *N*-Nitroso- and *N*-amino-aziridines are thermally deaminated to give olefins stereoselectively.

2. Oxiranes are available from olefins by epoxidation with peroxyacids. Allylic alcohols can be epoxidized enantioselectively by the use of a titanium(IV) alkoxide, t-butyl hydroperoxide, and a tartrate ester. Oxiranes can also be prepared from carbonyl compounds and from halohydrins. Nucleophilic ring cleavage occurs readily, and is generally of the same type as for aziridines. The ring is opened by E2 elimination and by electrocyclic conversion to carbonyl ylides. Thiirane chemistry is similar to that of oxiranes although nucleophilic attack on sulphur can also occur.

3. 2*H*-Azirines are most readily prepared from vinyl azides by thermal or photochemical elimination of nitrogen. The C=N bond is unusually reactive in addition and cycloaddition reactions. The ring can be cleaved thermally, normally at the N—C bond, and by ultraviolet light, at the C—C bond. When the azirine bears an unsaturated substituent at C-2, these reactions normally result in ring expansion involving the substituent groups.

4. Four-membered rings are generally less strained and so less susceptible to ring cleavage. Azetidinones (*β*-lactams) do show enhanced reactivity at the carbonyl group. The *β*-lactone function of diketene is similarly subject to nucleophilic cleavage. Unsaturated four-membered rings such as oxete and thiete undergo electrocyclic ring opening under very mild conditions and so are generally unstable.

Problems

1. Suggest a mechanism for each of the following reactions which is consistent with the observed stereochemical outcome.

(a)

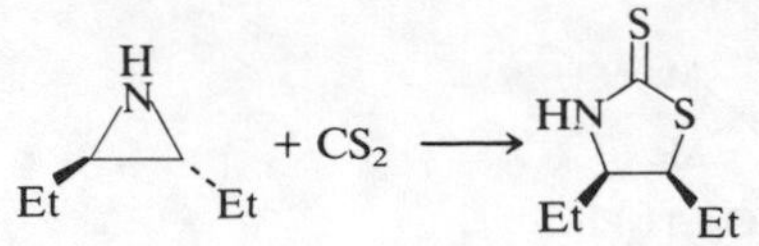

(b)

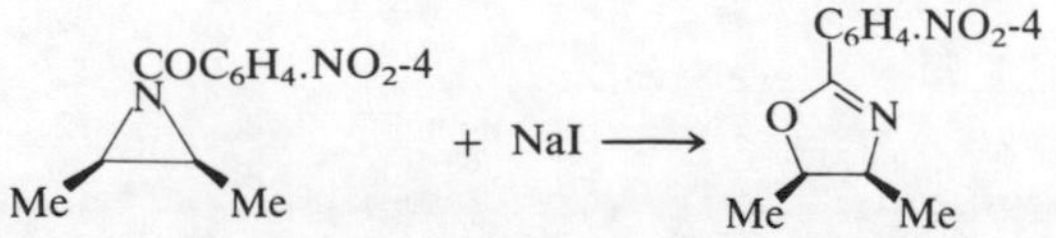

(c)

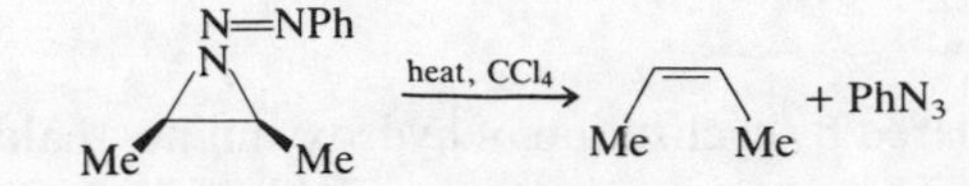

(d)

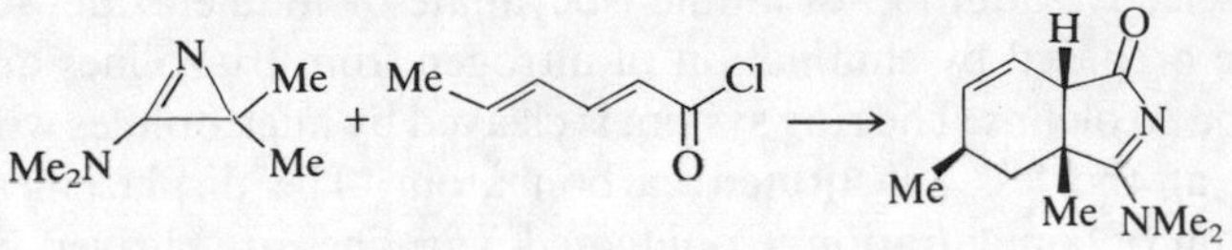

(e)

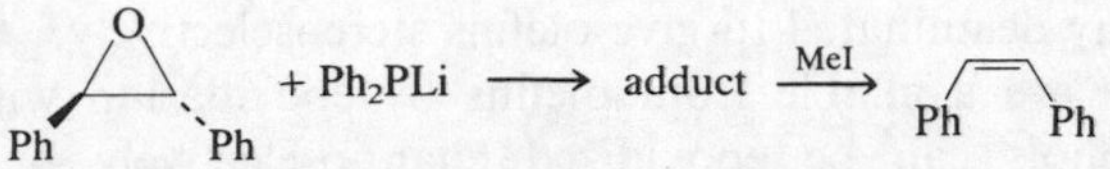

(f)

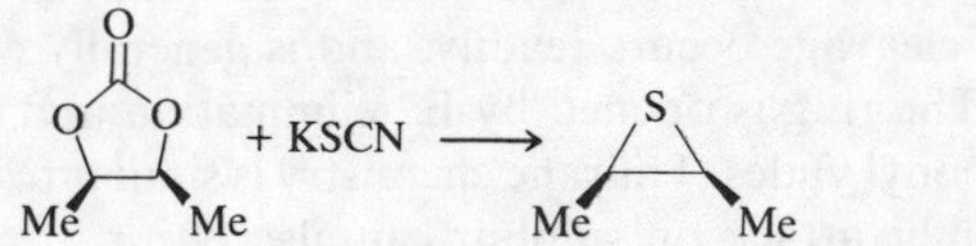

2. Optically active 3-methyl-2-phenyl-2H-azirine was found to racemize when heated in decalin at 120°C. The azirine rearranges to 2-methylindole when heated in decalin about 185°C. How do you interpret these observations?

3. Clavulanic acid, **24**, when heated in methanol, is converted into the pyrrole **A**. Suggest a pathway for the reaction.

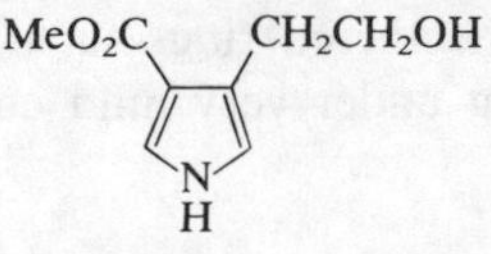

4. The penicillin **B** undergoes Wittig reactions with stabilized ylides such as Ph$_3$P̄C̄HCO$_2$Me (toluene, 110°C) whereas simple amides fail to react in these conditions. Suggest an explanation.

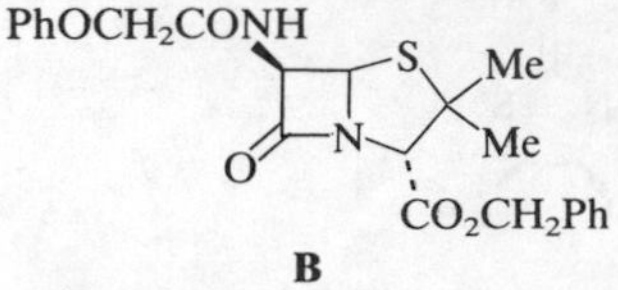

10 Seven-membered ring compounds

10.1 Introduction

The unsaturated seven-membered heterocycles with one nitrogen, oxygen, or sulphur atom are named systematically as azepines, oxepins, and thiepins, respectively. Thiepins are known only with bulky substituents but oxepin and 1H-azepine have both been synthesized.

1H-Azepine and oxepin are nonplanar and there is no evidence that they are delocalized. There is a striking contrast in the properties of pyrrole and 1H-azepine since the azepine is an unstable polyene which rearranges easily to its 3H-tautomer. This can be related to the different numbers of π-electron in the pyrrole and azepine rings: if 1H-azepine were planar, the cyclic electron system would contain eight π-electrons. This planar structure is calculated to have a negative resonance energy in comparison with an acyclic model. The structures which these molecules adopt have therefore been important in developing the concept of aromaticity.

Oxepins 1 exist in equilibrium with bicyclic valence tautomers 2 (the interconversion being a six-electron disrotatory electrocyclic reaction).[1] Oxepin itself exists as an inseparable mixture with benzene oxide at room temperature, since the activation energies for the forward and reverse reactions are so low [9.1 and 7.2 kcal mol^{-1} (38 and 31 kJ mol^{-1}), respectively].[2] Arene oxides are formed as intermediates in the enzymic oxidation of aromatic hydrocarbons, and this has stimulated a great deal of work on oxepin–arene oxide equilibria.

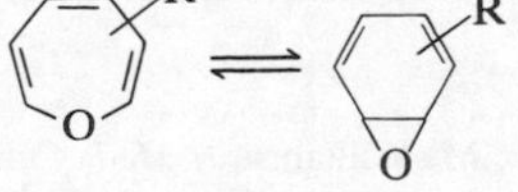

1 Reviews: G. S. Shirwaiker and M. V. Bhatt, *Adv. Heterocycl. Chem.*, 1984, **37**, 67; D. R. Boyd and D. M. Jerina, in *Small Ring Heterocycles*, Part 3, ed. A. Hassner, Wiley-Interscience, New York, 1985, p. 197.
2 E. Vogel and H. Günther, *Angew. Chem. Int. Edn Engl.*, 1967, **6**, 385.

The major stimulus for the investigation of the synthesis and properties of seven-membered heterocycles has, however, come from the discovery of useful biological activity in the benzodiazepines[3] and dibenzazepines. Chlordiazepoxide (Librium), **3**, and diazepam (Valium), **4**, are prototypical of the 1,4-benzodiazepines which are used primarily in the relief of anxiety. All the important active members of the group contain an aryl or a cyclohexenyl substituent at C-5 and an electron-withdrawing group (usually chloro) at C-7. Imipramine, **5**, is one of a group of tricyclic compounds which are in clinical use as antidepressants.

3   **4**   **5**

10.2   Azepines[4]

The most common structural types are derivatives of 3*H*-azepine; 1*H*-azepines unsubstituted at nitrogen readily tautomerize to the 3*H* forms.[4] 1-Substituted azepines **7** have been prepared from the aziridine **6**. This aziridine is derived from 1,4-cyclohexadiene by the addition of iodine isocyanate (see the method in Table 9.1). Alkylation of **6**, followed by addition of bromine to the double bond and dehydrobromination, gives the azepines. Compound **7** (R = CO$_2$Et) can also be formed by the thermolysis or photolysis of ethyl azidoformate in benzene. The reaction goes by way of a nitrene intermediate which adds to the benzene ring. Cyanogen azide and sulphonyl azides have similarly given 1*H*-azepines with benzene.

6   **7**

1*H*-Azepines, unlike oxepins, show little tendency to isomerize to bicyclic structures,[5] but this valence isomerization apparently occurs with protonated 1*H*-azepines since they can be irreversibly converted into benzene derivatives **8** on treatment with acids.

3 Reviews: L. H. Sternbach, *J. Med. Chem.*, 1979, **22,** 1; M. Williams, *J. Med. Chem.*, 1983, **26,** 619.
4 Reviews: (a) J. M. F. Gagan, in *Rodd's Chemistry of Carbon Compounds*, 2nd Edn, Vol. IVK, ed. S. Coffey, Elsevier, Amsterdam, 1979, p. 195; (b) R. K. Smalley, in *Comprehensive Organic Chemistry* Vol. 4, ed. P. G. Sammes, Pergamon, Oxford, 1979, p. 582.
5 L. A. Paquette, *Angew. Chem. Int. Edn Engl.*, 1971, **10,** 11; H. Günther *et al.*, *Chem. Ber.*, 1973, **106,** 984.

$$7 \longrightarrow \text{[structure]} \longrightarrow \text{[structure]}$$

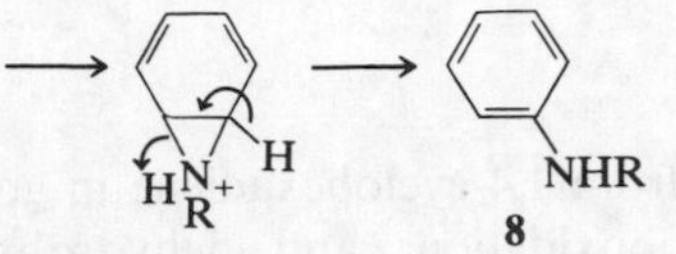

The polyenic character of the π-electron system is shown by the reaction of ethyl 1-azepinecarboxylate with activated dienophiles across C-2 and C-5. Diethyl azodicarboxylate, for example, gives the adduct **9**.[6] When the azepine is irradiated, it is converted into the bicyclic tautomer **10**, by a 4π-electron disrotatory ring closure.[7]

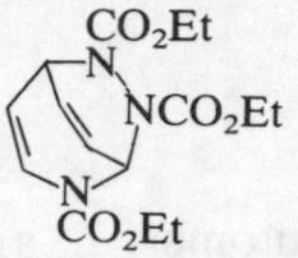

$$\text{[structure 9]} \qquad \text{[structure 10]}$$

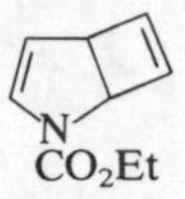

3H-Azepines can be formed by irradiating phenyl azide in primary or secondary aliphatic amines. This reaction goes by way of singlet phenylnitrene, and the azirine **11** has been assumed to be an intermediate in the reaction. The cyclic ketenimine **12** has, however, been detected as a product of low-temperature photolysis of phenyl azide, and it therefore seems probable that azepines are formed by nucleophilic addition to this intermediate (Fig. 10.1).[8]

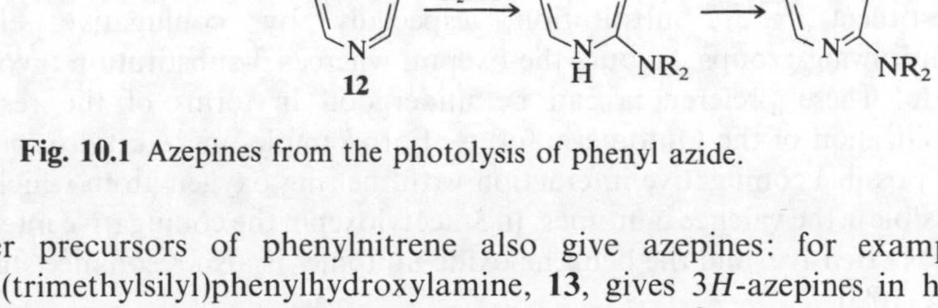

$$\text{PhN}_3 \xrightarrow{h\nu} \text{PhN:}$$

Fig. 10.1 Azepines from the photolysis of phenyl azide.

Other precursors of phenylnitrene also give azepines: for example, NO-bis(trimethylsilyl)phenylhydroxylamine, **13**, gives 3H-azepines in high yield when it is thermolysed in the presence of amines.[9]

6 W. S. Murphy and K. P. Raman, *J. Chem. Soc., Perkin Trans.* 1, 1977, 1824.

7 L. A. Paquette and J. H. Barrett, *J. Am. Chem. Soc.*, 1966, **88**, 1718.

8 Review: E. F. V. Scriven and K. Turnbull, *Chem. Rev.*, 1988, **88**, 297.

9 F. P. Tsui, Y. H. Chang, T. M. Vogel, and G. Zon, *J. Org. Chem.*, 1976, **41**, 3381.

10.3   Oxepins and thiepins[1,10]

Oxepin **1** (R = H) can be prepared from 1,4-cyclohexadiene in good overall yield by addition of bromine, epoxidation, and dehydrobromination (Fig. 10.2).[2,11]

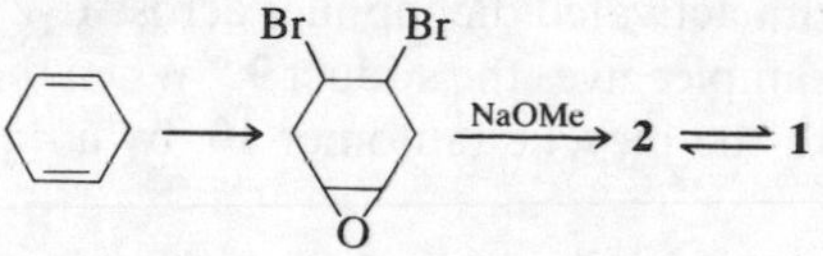

Fig. 10.2 A synthesis of oxepin.

Oxepin exists in equilibrium with its bicyclic tautomer **2**, and both tautomers are present at room temperature. Benzene oxide is the lower energy system $[\Delta H^0 = 1.7 \, \text{kcal mol}^{-1} \, (7.1 \, \text{kJ mol}^{-1})]$, but the entropy gain associated with ring opening displaces the equilibrium slightly in favour of the oxepin at room temperature. The equilibrium shifts toward the benzene oxide at low temperature and in polar solvents. Substituted oxepins can be prepared by related methods; the 4,5-dimethyl ester, for example, is made by debromination of the oxirane **15** with sodium iodide.

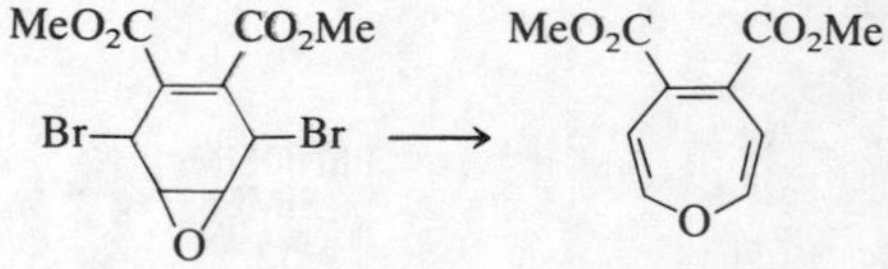

The position of the benzene oxide–oxepin equilibrium in substituted oxepins is very dependent upon the nature and the position of the substituents.[1,12] 2-Substitution, especially by conjugative electron-withdrawing groups, favours the oxepin, whereas 3-substitution favours the oxide. These preferences can be undergood in terms of the resonance stabilization of the tautomeric forms. For example, in 2-acetyloxepin there is a possible conjugative interaction with the ring oxygen atom which is not possible in the valence tautomer. In 3-acetyloxepin the conjugative interaction is less extensive, and the benzene oxide tautomer is also resonance stabilized (Fig. 10.3).

10 Review: A. Rosowsky, in *Seven-membered Heterocyclic Compounds Containing Oxygen and Sulfur*, Wiley-Interscience, New York, 1972.
11 E. G. Lewars and G. Morrison, *Can. J. Chem.*, 1977, **55**, 966.
12 (a) D. M. Hayes, S. D. Nelson, W. A. Garland, and P. A. Kollman, *J. Am. Chem. Soc.*, 1980, **102**, 1255; (b) D. R. Boyd and M. E. Stubbs, *J. Am. Chem. Soc.*, 1983, **105**, 2554.

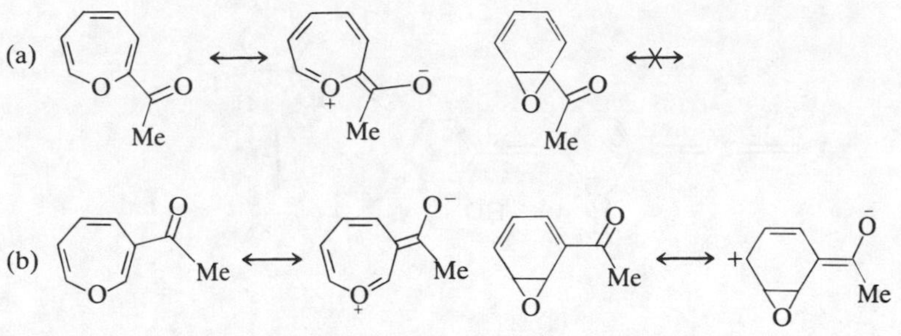

(a)

(b)

Fig. 10.3 Conjugative interactions in (a) 2-acetyloxepin and its valence tautomer, (b) 3-acetyloxepin and its valence tautomer.

The same applies to benzo-fused analogues: For example, naphthalene-1,2-oxide, **16**, exists in the bicyclic form and can be prepared enantiomerically pure, with no tendency to racemize by way of the seven-membered valence tautomer, whereas the 2,3-oxide exists entirely as the benzoxepin, **17**.[12b]

All the monocyclic oxepins can, however, react through their bicyclic tautomers even if this is the less favourable structure. For example, oxepin readily undergoes Diels–Alder reactions through its bicyclic valence tautomer, maleic anhydride giving the adduct **18a**. 2-Acetyloxepin, for which the bicyclic tautomer cannot be detected by NMR, still gives an analogous cycloadduct **18b** with maleic anhydride.

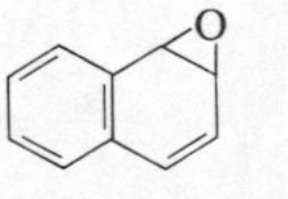

16 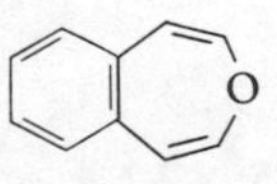   **17**

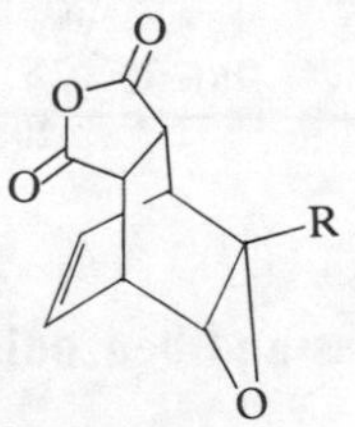

18a; R = H
b; R = COMe

Oxepins are easily rearranged to phenols in acidic media by way of their valence tautomers. The rearrangement can involve a 1,2-hydride shift (Fig. 10.4), which has been established by deuterium labelling experiments. The mechanism of biological hydroxylation of aromatic compounds is analogous in that arene oxides are implicated as intermediates and their rearrangement involves a 1,2-hydride migration. This migration is sometimes called the 'NIH shift' after the US National Institutes of Health where it was first discovered.[13] The oxirane ring of the bicyclic tautomers **2** can also be cleaved by reaction with azide ions and other 'soft' nucleophiles, which attack at C-2.[14]

13 Review: J. W. Daly, D. M. Jerina, and B. Witkop, *Experientia*, 1972, **28**, 1129.
14 A. M. Jeffrey *et al.*, *J. Am. Chem. Soc.*, 1974, **96**, 6929.

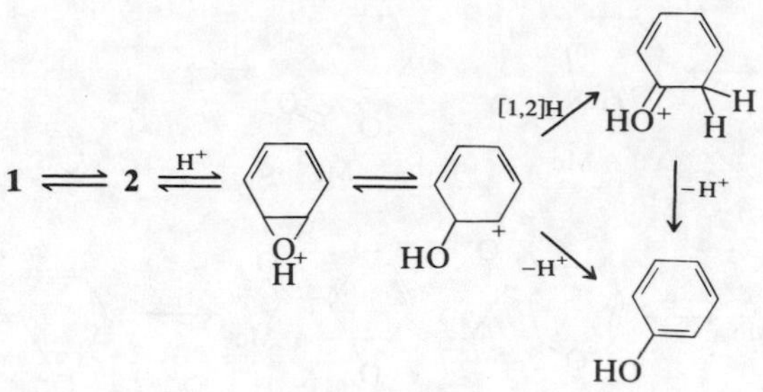

Fig. 10.4 Acid-catalysed rearrangement of oxepin.

Oxepin undergoes a photochemical valence tautomerism to give an isomer **19**, analogous to that formed from 1-carbethoxyazepine, when it is irradiated in a nonpolar solvent at wavelengths above 310 nm. At shorter wavelengths the benzene oxide tautomer is selectively excited and phenol is the major product.

Simple thiepins are thermally unstable because they readily extrude sulphur from the bicyclic valence tautomer. If the tautomerism is sterically inhibited, then the thiepins can be isolated. 2,7-Di-t-butylthiepin, **20**, is thermally stable because the bulky substituents prevent its isomerization to the more crowded benzene sulphide.[15]

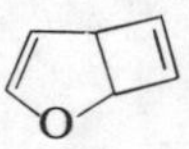

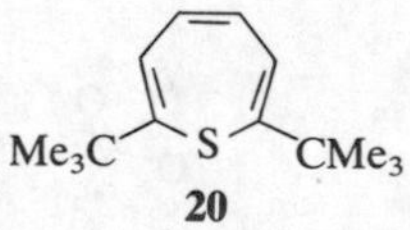

10.4   Diazepines and benzodiazepines[4]

There are three groups of monocyclic diazepines, with nitrogen atoms at the 1,2-, 1,3-, and 1,4-positions. The 1,2-diazepines have been the most studied of the monocyclic systems. The tautomers **21–24** of 1,2-diazepine are shown.

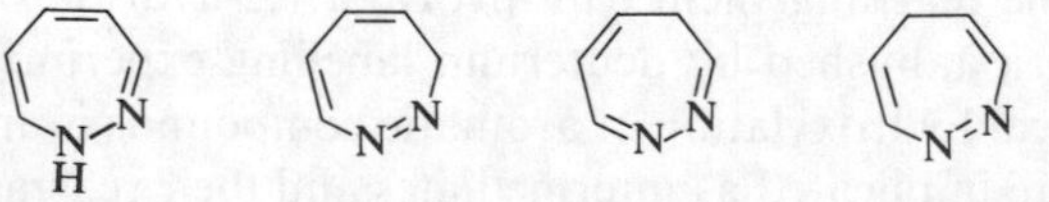

Several 1*H*-1,2-diazepines with electron-withdrawing substituents on N-1 have been prepared by irradiation of pyridinium imides.[16] The ring system is very like that of the 1*H*-azepines, being boat shaped and polyolefinic in

15 K. Yamamoto *et al.*, *Tetrahedron Lett.*, 1982, **23**, 3195.
16 Review: V. Snieckus and J. Streith, *Accounts Chem. Res.*, 1981, **14**, 348.

character. The 1*H*-1,2-diazepines revert to pyridinium *N*-imides above 150°C and are cleaved by bases, which remove the activated proton at C-3. Like the other ring systems of this type they can undergo disrotatory electrocyclization when irradiated (Fig. 10.5).

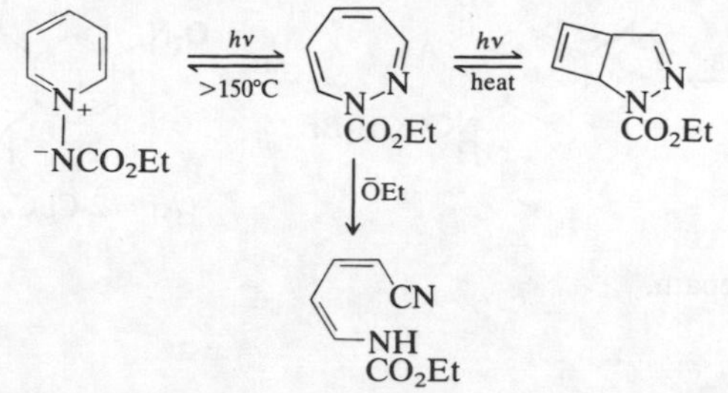

Fig. 10.5 Reactions of ethyl 1,2-diazepine-1-carboxylate.

3*H*-1,2-Diazepines have been prepared by the thermal cyclization of unsaturated diazo compounds (Fig. 10.6).[17]

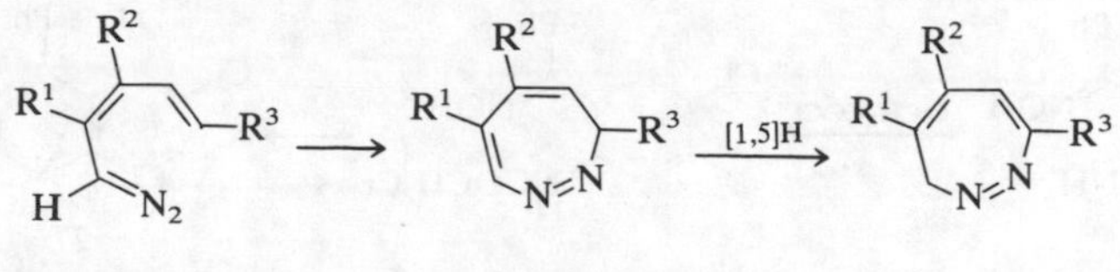

Fig. 10.6 Formation of 3*H*-1,2-diazepines.

5*H*-1,2-Diazepines show a strong preference to exist as bicyclic valence tautomers **25**, because this allows the molecules to exist as conjugated azines rather than as azo compounds. Compound **25** (R = CO_2Me) is hygroscopic and readily forms a covalent hydrate by the addition of water across a carbon–nitrogen double bond.[18]

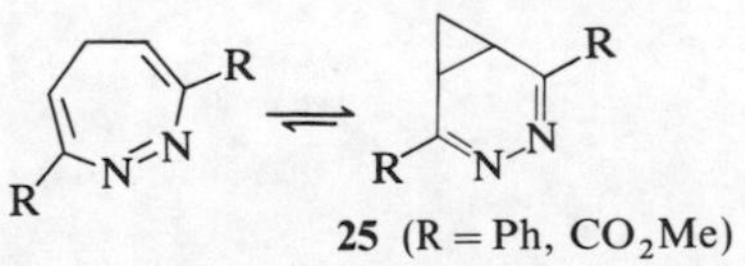

25 (R = Ph, CO_2Me)

Of the benzodiazepines, derivatives of the 1,4-diazepine system are of overwhelming importance; by the late 1970s many thousands of these compounds had been synthesized as potential drugs.[3] the ring system can

17 Review of 1, 7-dipolar cyclizations: G. Zecchi, *Synthesis*, 1991, 181.
18 A. Steigel, J. Sauer, D. A. Kleier, and G. Binsch, *J. Am. Chem. Soc.*, 1972, **94**, 2770.

be prepared by standard methods starting from appropriate 1,2- disubstituted benzenes. A simple general route into the ring system is illustrated by the synthesis of clonazepam, a commercial drug, which is shown in Fig. 10.7.

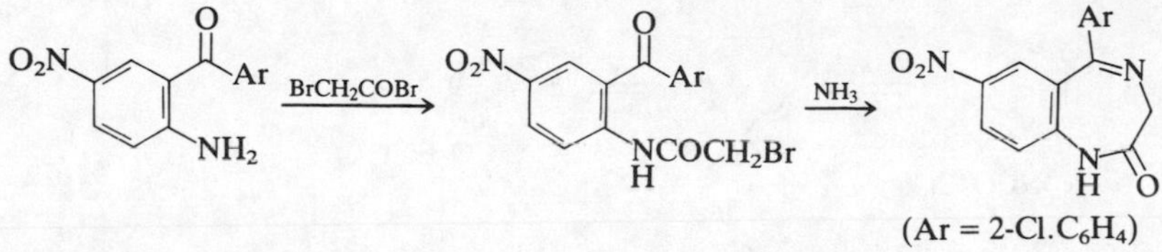

Fig. 10.7 A route to clonazepam.

The original route to chlordiazepoxide (Librium) is a less straightforward one. The oxime **26** reacts with chloroacetyl chloride to give the quinazoline *N*-oxide, **27**, the intermediate amide having cyclized. Attempts to displace the chloride by nucleophiles such as methylamine result in addition to the ring system followed by ring expansion (Fig. 10.8).

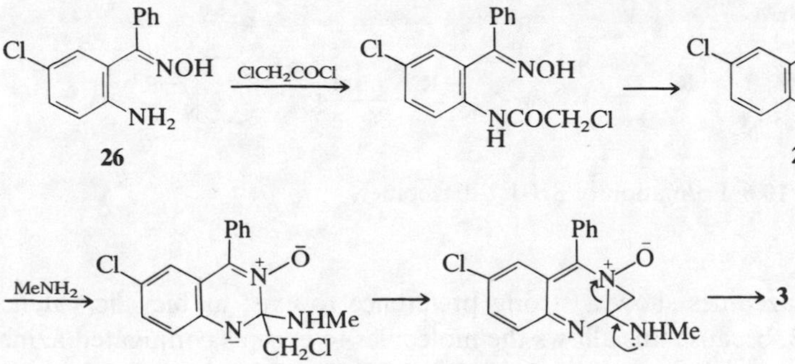

Fig. 10.8 A route to chlordiazepoxide.

Summary

1. Unsaturated seven-membered heterocycles are nonplanar polyenes which show no evidence of cyclic delocalization.
2. Oxepins exist in equilibrium with bicyclic valence tautomers (benzene oxides), the position of the equilibrium being dependent upon the nature and position of substituents. Thiepins are generally unstable because sulphur is easily extruded from the bicyclic tautomers. 1*H*-Azepines show little tendency to tautomerize in this way. 1*H*-Azepines unsubstituted on nitrogen

usually rearrange to the more stable 3*H*-isomers. Azepines and benzene oxides undergo cycloaddition reactions typical of *cis*-dienes.

3. 1,4-Benzodiazepines such as chlordiazepoxide and diazepam are an important class of drugs used in the treatment of anxiety.

Problems

1. Predict the predominant tautomeric forms in the oxepin–benzene oxide equilibria with (a) 2,7-dimethyloxepin; (b) 4,5-dimethyloxepin; (c) 3-chlorooxepin; (d) anthracene-1,2-oxide.

2. The oxirane **A** cannot be made by direct epoxidation of oxepin, but it has been synthesized by an indirect route of which the first step is a Diels–Alder reaction. Suggest an appropriate Diels–Alder reagent and set of reactions which could convert oxepin into compound **A**.

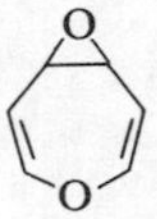

3. Show how the cyclic azo compounds **C** and **D** are formed by thermolysis of the tosylhydrazone salts **B**, and suggest an explanation for the difference in the course of the two cyclizations.

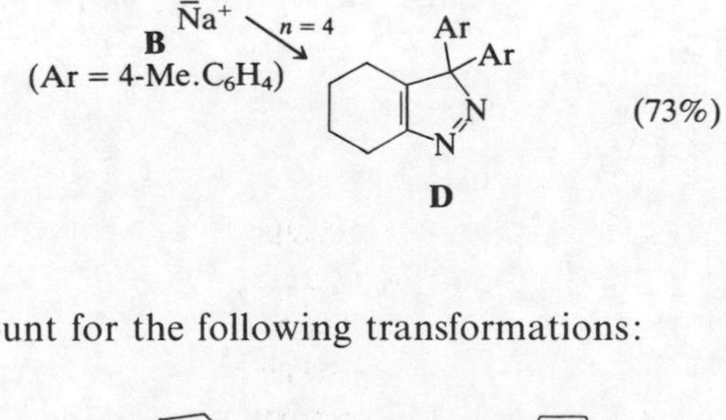

4. Account for the following transformations:

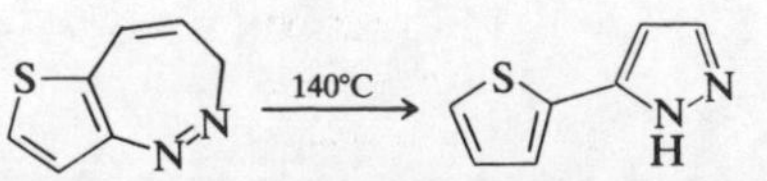

(b)

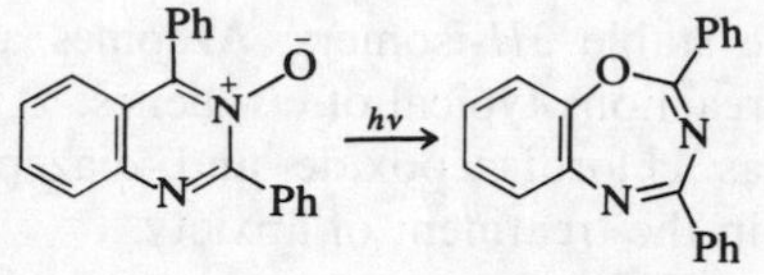

(c)

(R = 2-pyridyl)

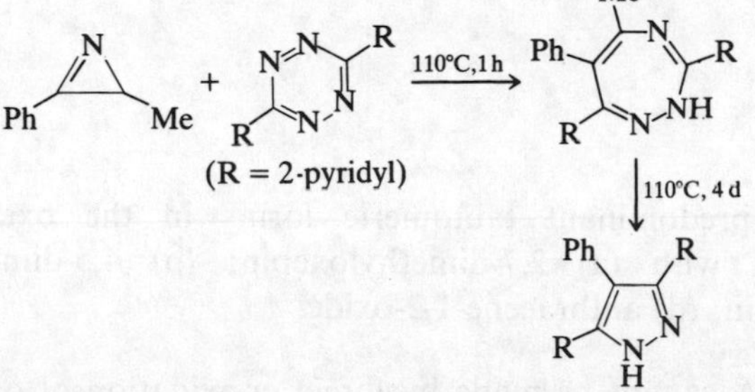

11 Nomenclature

11.1 Introduction

Throughout the earlier chapters, heterocycles have been referred to by systematic names or by 'recognized' trivial names. These names are examples of just two of several different systems of nomenclature which have been devised for heterocyclic compounds. Even in the current literature it is quite common to find three or four different names in use for a particular compound. In such cases it is useful to be able to deduce the structure from the name, if it is a systematic one, and to be able to identify which of the different names is to be preferred.

The most systematic nomenclature so far devised for heterocycles is one based on the names of analogous cyclic hydrocarbons. This has been recommended for heterocyclic compounds[1] but so far it has not been widely adopted. This nomenclature system is described briefly in Section 11.5. The nomenclature system which is in more common use, particularly for aromatic heterocycles, is a hybrid of trivial names and systematic names made up of standard prefixes and suffixes (the Hantzsch–Widman system). These systems are described in Sections 11.2–11.4. The authoritative source of information on the nomenclature of heterocyclic compounds is the set of recommendations published by the International Union of Pure and Applied Chemistry.[2] These recommendations have been summarized and interpreted in a review article.[3]

The material presented here is mainly intended to provide a guide to the way in which the common ring systems are currently named. Section 11.4 includes an introduction to the method of naming more complex fused-ring systems, which is provided mainly for reference purposes. There are some exercises at the end of the chapter if you wish to gain more practice in using the various nomenclature systems. The currently accepted names and numbering system for a heterocycle can be found by making use of *The Ring Index*[4] in conjunction with recent indexes of *Chemical Abstracts*.

1 *Nomenclature of Organic Compounds*, ed. J. H. Fletcher, O. C. Dermer, and R. B. Fox, *Am. Chem. Soc. Advan. In Chem. Ser.*, No. *126*, 1974.
2 *IUPAC Nomenclature of Organic Chemistry*, Definitive rules, sections A to H, Pergamon Press, Oxford, 1979.
3 A. D. McNaught, *Adv. Heterocycl. Chem.*, 1976, **20**, 175.

11.2 Trivial names of common ring systems

In the early days of organic chemistry, names were given to compounds as a means of identifying them, usually before their structures were known. This is still done today, particularly for newly discovered natural products. These names usually have, as their basis, the source of the compound or one of its characteristic properties. For example, the methylpyridines, such as compound **1**, have the trivial name 'picoline', which is based on the Latin *picatus* meaning 'tarry', the compounds being derived from coal tar. The alcohol **2** was called 'furfurol' meaning 'bran oil' since it was isolated by the distillation of bran. The name 'pyrrole' for compound **3** comes from the Greek for 'fiery red' because of the characteristic colour which the compound gives with a pine splint dipped in hydrochloric acid.

These trivial names contain no useful structural information and they are gradually being abandoned. At present, just over sixty trivial and semitrivial names survive as 'recognized' names in the IUPAC system. Of the three names mentioned above, only 'pyrrole' survives as a recognized name These surviving trivial names are, however, important because they are used as the basis for constructing other more systematic names for derivatives or for polycyclic compounds. For the ring systems concerned, the trivial names are currently used in the indexes of *Chemical Abstracts*. These are the names which have been used in earlier chapters. Examples of recognized trivial names are given in Table 11.1; the full list is in refs. 2 and 3. The numbering of these systems usually starts at a heteroatom and follows in sequence round the ring, or, for polycyclic systems, numbering starts at an atom next to a ring junction. The numbering is shown in Table 11.1 for the less straightforward cases.

11.3 Systematic (Hantzsch–Widman) nomenclature for monocyclic compounds

The most widely used systematic nomenclature system for monocyclic heterocycles is based on one introduced in the last century by Hantzsch and by Widman. It has undergone several revisions and extensions since then,

4 A. M. Patterson and L. T. Capell, *The Ring Index*, 2nd Edn, Am. Chem. Soc., 1960.

Table 11.1 Examples of heterocycles with 'recognized' trivial names.

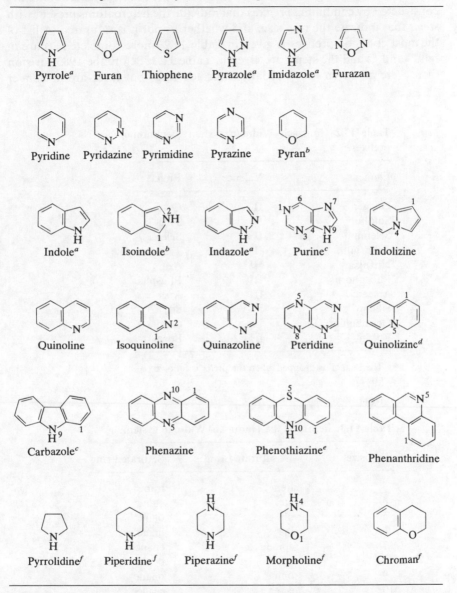

a 1H-tautomer shown. b 2H-tautomer. c 9H-tautomer. d 4H-tautomer. e 10H-tautomer. f Names of these saturated heterocycles are not used in fusion names; for example, compound **X** is 3,4-dihydro-2H-1,4-benzoxazine, not benzomorpholine.

the most recent being in 1983.[5] In this system the name of heterocycles are
constructed by combining *prefixes* that indicate the heteroatoms present with
stems that indicate the ring size, and whether or not it is saturated. A list of
the most common prefixes is given in Table 11.2 (more complete lists are in
refs. 3 and 5) and the stems are given in Table 11.3. Before the 1983 revision
there were also stems for partially saturated rings but these are no longer

Table 11.2 Hantzsch–Widman system: common
prefixes.

Element	Valence	Prefix[a]
Oxygen	II	Oxa
Sulphur	II	Thia
Selenium	II	Selena
Tellurium	II	Tellura
Nitrogen	III	Aza
Phosphorus	III	Phospha
Arsenic	III	Arsa
Silicon	IV	Sila
Germanium	IV	Germa
Boron	III	Bora

a The final 'a' is dropped when the prefix is followed by
a vowel.

Table 11.3 Stems for the Hantzsch–Widman system.

Ring size	Unsaturated ring	Saturated ring
3	irene[a]	irane[b]
4	ete	etane[b]
5	ole	olane[b]
6	ine[c]	inane[d]
7	epine	epane
8	ocine	ocane
9	onine	onane
10	ecine	ecane

a The stem 'irine' is normally used for three-membered rings which
contain nitrogen. *b* The stems 'iridine', 'etidine', and 'olidine' are
preferred for saturated rings containing nitrogen. *c* The stem 'inine'
is used with some phosphorus, arsenic, and boron heterocycles: see
ref. 5. *d* The stem 'ane' is used if immediately preceded by the
prefixes 'ox', 'thi', 'selen', or 'tellur'.

5 IUPAC Commission on Nomenclature of Organic Chemistry, *Pure Appl. Chem.*, 1983, **55**, 409.

recommended for use. Unsaturated stems are used for rings with the maximum number of (noncumulative) double bonds, and saturated stems for rings with no double bonds.

When two or more different heteroatoms are present, prefixes are listed in the order in which they appear in Table 11.2. Two or more heteroatoms of the same type are indicated by 'di-', 'tri-', etc. Numbering starts with the heteroatom highest in Table 11.2. Some examples of systematically named heterocycles are given in Fig. 11.1.

| Thiirene | Oxirane | 1,3-Diazete | 1,2-Oxazetidine |

| 1,3-Dioxolane | 1,2,4-Triazine | Thiepane | Azocine |

Fig. 11.1 Examples of systematically named heterocycles.

One further problem arises when a ring system with the maximum number of double bonds still has a 'saturated' atom in the ring. The position of this atom is numerically indicated, together with the prefix *H*-, as part of the name of the ring system. Where there is a choice of numbering, the indicated position is given the lowest possible number. Some examples are given in Fig. 11.2.

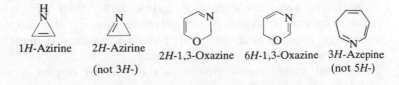

| 1*H*-Azirine | 2*H*-Azirine | 2*H*-1,3-Oxazine | 6*H*-1,3-Oxazine | 3*H*-Azepine |
| | (not 3*H*-) | | | (not 5*H*-) |

Fig. 11.2 Systematic names with 'indicated hydrogen'.

These names do not supersede the recognized trivial names listed in Table 11.1; for example, the trivial name '1*H*-pyrrole' for compound **3** is preferred to the systematic Hantzsch–Widman name of 1*H*-azole.

11.4   The naming of fused ring systems

A very large number of heterocycles contain two or more fused rings. Some of these have recognized trivial names (Table 11.1) but the vast majority have not. The systematic names of fused ring systems are derived by regarding common atoms as belonging to both ring systems. The name is then constructed by combining the names of the individual rings. Two examples are shown in Fig. 11.3.

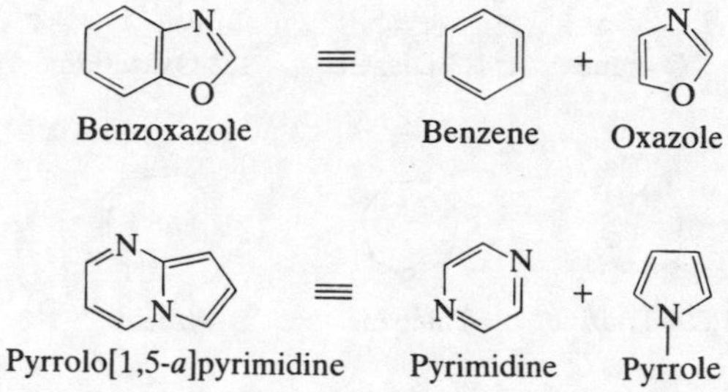

Benzoxazole   =   Benzene   +   Oxazole

Pyrrolo[1,5-*a*]pyrimidine   =   Pyrimidine   +   Pyrrole

Fig. 11.3 Examples of fusion names derived from those of separate ring systems.

The systematic names of known fused ring systems can be obtained from *The Ring Index*,[4] which contains lists of ring structures in order of ring size, arranged according to the numbers and types of atoms present. The following paragraphs give some basic guidance on how the names of fused ring systems are constructed.

(a) Names of the components of the fused system are chosen, where possible, from the list of 'recognized' trivial names.[2] The largest component having a recognized name is chosen (e.g. 'indole' rather than 'pyrrole' if an indole fragment is present in the polycyclic system). If a monocyclic component does not have a recognized trivial name, the systematic name, derived from Tables 11.2 and 11.3, is used.

(b) In a fused system made up of two (or more) separately named components, one is chosen as the *base component* for the composite name. For example, 'pyrimidine' is the base component of the name of the second ring system shown in Fig. 11.3. The Scheme on p. 376 shows a sequence for determining the base components.

(c) The second component is added as a prefix to the name of the base component. This prefix is derived by replacing the final 'e' of the name of the ring system by 'o'. Thus, 'pyrazine' becomes 'pyrazino', etc. There are a few exceptions, which are listed in Table 11.4.

Table 11.4 Non-standard prefixes in fusion names.

Heterocycle	Name as prefix
Furan	Furo
Imidazole	Imidazo
Isoquinoline	Isoquino
Pyridine	Pyrido
Quinoline	Quino
Thiophene	Thieno

(d) Bonds forming the ring system of the *base component* are labelled *a*, *b*, *c*, etc., starting with the side normally numbered 1,2. Atoms forming the ring system of the *second component* are numbered in the normal way. The site of fusion is then indicated by the appropriate letter and numbers, the numbers of the second component being listed in the sequence in which they occur in the base component. Two examples are shown in Fig. 11.4.

(i)

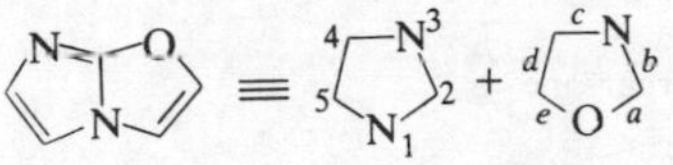

The base component is *oxazole* (step 6 in the Scheme) and fusion is to face *b* of oxazole and face 1,2 of imidazole. Atom 2 of the imidazole occurs before atom 1 on face *b* of the oxazole system.

The system is named imidazo[2,1-*b*]oxazole.

(ii)

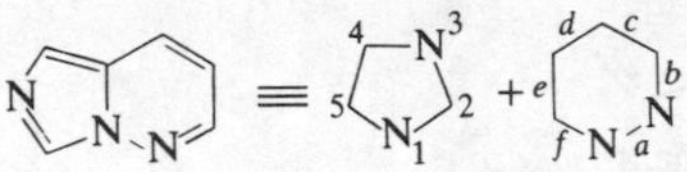

The base component is *pyridazine* (step 4 in the Scheme) and fusion is to face *b* of pyridazine and face 1,5 of imidazole. Atom 1 of the imidazole occurs before atom 5 on face *b* of the pyridazine system.

The system is named imidazo[1,5-*b*]pyridazine.

Fig. 11.4 Examples showing the method indicating the site of fusion.

(e) Numbering of the fused system is carried out independently of the above operations and normally starts at an atom adjacent to a bridgehead position, such that heteroatoms are given the lowest possible numbers. In ambiguous cases, atoms which appear highest in Table 11.2 are given the lowest numbers. 'Indicated hydrogen' is shown as for monocyclic systems. An example of numbering is shown in Fig. 11.5.

Scheme for deriving the base component of a fused ring system

1. Is there only one component which contains nitrogen?

    NO   YES: choose this as base component*

                    e.g.

               base component *pyrrole*

2. Is nitrogen absent from all the rings?

    NO   YES: choose the ring containing a heteroatom
                          which occurs highest in Table 11.2

                    e.g.

               base component *furan*

   (If the rings contain the same heteroatom go to 3.)

3. Are there more than two rings present?

    NO   YES: choose the component containing
                          the greater number of rings

                    e.g.

               base component *quinoline*

4. Are the two rings of different sizes?

    NO   YES: choose the larger one

                    e.g.

               base component *azepine*

5. Do the rings contain different numbers of heteroatoms?

    NO   YES: choose the one with the greater number

                    e.g.

               base component *isoxazole*

6. Are there are differences between the numbers of each kind of heteroatom in the two rings?

    NO   YES: choose the ring with the greater number
                          of atoms listed first in Table 11.2

                    e.g.

               base component *oxazole*

7. Choose as the base component the ring for which the heteroatoms have the lower numbers (before fusion).

                    e.g.

               base component *pyrazole*

*In a polycyclic system, choose the largest nitrogen-containing component with a recognized trivial name (e.g. indole rather than pyrrole).

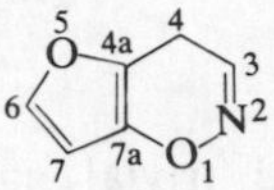

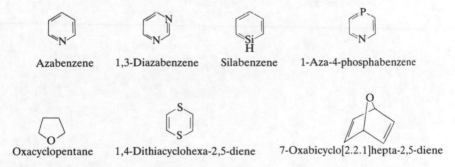

Fig. 11.5 Numbering of 4*H*-furo[2,3-*e*]-1,2-oxazine. Numbering in this direction gives the three heteroatoms the lowest combined numbers (1, 2, and 5).

11.5 Replacement nomenclature

An alternative form of systematic nomenclature is based on the replacement of one or more of the carbon atoms of a carbocycle by heteroatoms. The appropriate carbon ring system is named by the IUPAC rules[2] and the heteroatoms present are indicated as prefixes. The prefixes listed in Table 11.2 are also used in replacement nomenclature. This is the most systematic method of naming heterocyclic compounds since it requires very little elaboration of the rules already developed for carbocycles. Despite this, the nomenclature system is currently used mainly for heterocycles containing unusual heteroatoms and for bridged and spiro ring systems. It can also be used for complex fused ring systems in cases where the Hantzsch–Widman nomenclature breaks down.

Some examples of replacement nomenclature are shown in Fig. 11.6.

Azabenzene   1,3-Diazabenzene   Silabenzene   1-Aza-4-phosphabenzene

Oxacyclopentane   1,4-Dithiacyclohexa-2,5-diene   7-Oxabicyclo[2.2.1]hepta-2,5-diene

Fig. 11.6 Examples of systematic names using replacement nomenclature.

Summary

Three types of name for heterocyclic compounds are currently in general use: (1) recognized trivial names for a limited number of common heterocycles; (2) systematic names based on a combination of standardized prefixes and stems, which are used for monocyclic and fused heterocyclic compounds; and (3) systematic names based on the nomenclature of analogous carbocyclic compounds.

Problems

1. Name each of the following (a) using recognized trivial names, (b) by systematic Hantzsch–Widman nomenclature, and (c) by replacement nomenclature.

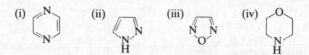

(i) (ii) (iii) (iv)

2. Draw structures for the following: (a) 1,8-diazanaphthalene; (b) 1-oxa-2-azacyclobut-2-ene; (c) selenacyclopentane; (d) 1-thia-3-azacycloheptatriene.

3. Draw structures for the following: (a) 2H-1,2-benzoxazine; (b) thieno[3,4-b]furan; (c) furo[3,2-d]pyrimidine; (d) 4H-[1,3]thiazino [3,4-a]azepine.

4. Name each of the following by the Hantzsch–Widman system.

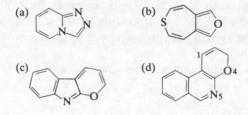

(a) (b) (c) (d)

Answers and References
to Selected Problems

Chapter 2

1. W. L. Jorgensen and L. Salem, *The Organic Chemist's Book of Orbitals*, Academic Press, New York, 1973.
2. **B, D.**
3. W. Lange, W. Haas, H. Schmickler, and E. Vogel, *Heterocycles*, 1989, **28**, 633.
4. J. L. Morris and C. W. Rees, *J. Chem. Soc., Perkin Trans. 1*, 1987, 211.
5. E. Vogel, P. Röhrig, M. Sicken, B. Knipp, A. Herrmann, M. Pohl, H. Schmickler, and J. Lex, *Angew. Chem. Int. Edn Engl.*, 1989, **28**, 1651.

Chapter 3

2. F. Montanari, I. Moretti, and G. Torre, *Gazz. Chim. Ital.*, 1973, **103**, 681.
3. E. Fahr, W. Fischer, A. Jung, L. Sauer, and A. Mannschreck, *Tetrahedron Lett.*, 1967, 161.
4. S. F. Nelsen, *Accounts Chem. Res.*, 1978, **11**, 14.
5. H. Quast and P. Eckert, *Justus Liebigs Ann. Chem.*, 1974, 1727.
7. (a) J. E. Douglass and T. B. Ratliff, *J. Org. Chem.*, 1968, **33**, 355. (b) R. Huisgen and W. Kolbeck, *Tetrahedron Lett.*, 1965, 783. (c) N. J. Leonard and J. A. Klainer, *J. Org. Chem.*, 1968, **33**, 4269.
8. R. Schwesinger, M. Missfeldt, K. Peters, and H. G. von Schnering, *Angew. Chem. Int. Edn Engl.*, 1987, **26**, 1165.

Chapter 4

1. (a) A. Brodrick and D. G. Wibberley, *J. Chem. Soc., Perkin Trans. 1*, 1975, 1910. (b) L. Lalloz and P. Caubere, *J. Chem. Soc., Chem. Commun.*, 1975, 745. (c) A. Padwa and D. Dehm, *J. Org. Chem.*, 1975, **40**, 3139. (d) S. Holand and R. Epsztein, *Bull. Soc. Chim. Fr.*, 1971, 1694. (e) C. D. Anderson, J. T. Sharp, H. R. Sood, and R. S. Strathdee, *J. Chem. Soc., Chem. Commun.*, 1975, 613. (f) C. G. Chavdarian, E. B. Sanders, and R. L. Bassfield, *J. Org. Chem.*, 1982, **47**, 1069.

2. The required reagents are (a) $HN=CHNH_2$, (b) $PhNHNH_2$, (c) $CH_2(CN)_2$, (d) PhCHClCOPh, and (e) $HC\equiv CCO_2Me$.

3. The 1,3-dipolar components are (a) $PhCH=\overset{+}{N}Me\bar{O}$, (b) $Ph\equiv\overset{+}{N}\bar{C}HAr$, (c) $PhC\equiv\overset{+}{N}\bar{S}$, (d) 3-Me-py.-1-oxide, (e) $PhCHN_2$, and (f) the nitrile oxide (see P. N. Confalone, G. Pizzolato, D. L. Confalone, and M. R. Uskokovic, *J. Am. Chem. Soc.*, 1980, **102**, 1954).

4. (a) H. Meyer, *Liebigs Ann. Chem.*, 1981, 1534. (b) G. Simchen and G. Entenmann, *Angew. Chem. Int. Edn Engl.*, 1973, **12**, 119. (c) S. M. Weinreb, N. A. Khatri, and J. Shringarpure, *J. Am. Chem. Soc.*, 1979, **101**, 5073. (d) D. N. Harcourt, N. Taylor, and R. D. Waigh, *J. Chem. Res.* (*S*), 1978, 154. (e) T. Tsuchiya, C. Kaneko, and H. Igeta, *J. Chem. Soc., Chem. Commun.*, 1975, 528. (f) E. Vedejs and G. P. Meier, *Tetrahedron Lett.*, 1979, 4185. (g) E. Rossi, G. Celentano, R. Stradi, and A. Strada, *Tetrahedron Lett.*, 1990, **31**, 903. (h) R. Grigg and H. Q. N. Gunaratne, *Tetrahedron Lett.*, 1983, **24**, 1201. (i) K. A. Parker, D. M. Spero, and J. Van Epp, *J. Org. Chem.*, 1988, **53**, 4628. (j) R. Grigg, M. J. Dorrity, and J. F. Malone, *Tetrahedron Lett.*, 1990, **31**, 1343. (k) U. Rüffer and E. Breitmaier, *Synthesis*, 1989, 623.

Chapter 5

1. A. I. Scott *et al.*, *J. Am. Chem. Soc.*, 1974, **96**, 8054.

3. G. Y. Lesher *et al.*, *J. Med. Pharm. Chem.*, 1962, **5**, 1063.

4. J. Reisch, *Chem. Ber.*, 1964, **97**, 2717.

7. R. M. Anker and A. H. Cook, *J. Chem. Soc.*, 1946, 117.

8. (a) R. A. Abramovitch *et al.*, *J. Am. Chem. Soc.*, 1976, **98**, 5671. (b) M. M. Baradarani and J. A. Joule, *J. Chem. Soc., Perkin Trans. 1*, 1980, 72. (c) G. M. Sanders, M. van Dijk, and H. J. den Hertog, *Rec. Trav. Chim.*, 1974, **93**, 273. (d) J. W. Bunting and W. G. Meathrel, *Org. Prep. Proced. Int.*, 1972, **4**, 9. (e) H. Yamanaka *et al.*, *Chem. Pharm. Bull.*, 1981, **29**, 1056. (f) B. R. Brown and D. L. Hammick, *J. Chem. Soc.*, 1949, 173. (g) T. Tsuchiya *et al.*, *Chem. Pharm. Bull.*, 1980, **28**, 2602. (h) K. Hafner and H. Kaiser, *Org. Synth., Coll. Vol. V*, 1973, 1088. (i) D. E. Minter and M. E. Re, *J. Org. Chem.*, 1988, **53**, 2653. (j) Addition of BuLi: E. W. Thomas, *J. Org. Chem.*, 1986, **51**, 2184. Lithiation at C-3: F. Trécourt, M. Mallet, F. Marsais, and G. Quéguiner, *J. Org. Chem.*, 1988, **53**, 1367. (These conditions are chosen because lithiation by LDA is fast but reversible whereas lithiation by MeLi is slow but irreversible). (k) H. Venkataraman and J. K. Cha, *Tetrahedron Lett.*, 1989, **30**, 3509.

9. J. A. King, P. E. Donohue, and J. E. Smith, *J. Org. Chem.*, 1988, **53**, 6145. (You might be able to come up with two possible structures consistent with the data given in the question).

Chapter 6

3. (a) B. von Maltzan, *Angew. Chem. Int. Edn Engl.*, 1982, **21**, 785. (b) J. Szmuszkovicz, *J. Am. Chem. Soc.*, 1957, **79**, 2819. (c) G. Devincenzis, P. Mencarelli, and F. Stegel, *J. Org. Chem.*, 1983, **48**, 162.

4. (a) 2-; (b) 5-; (c) 4-; (d) 5-; (e) 3-; (f) mixture of 4-, 5-, 6-, and 7-; (g) 3-.
5. (a) R. Neidlein and G. Jeromin, *Chem. Ber.*, 1982, **115**, 714. (b) C. D. Weiss, *J. Org. Chem.*, 1962, **27**, 3693. (c) S. L. Schreiber, A. H. Hoveyda, and H.-J. Wu, *J. Am. Chem. Soc.*, 1983, **105**, 660. (d) S. Divald, M. C. Chun, and M. M. Joullié, *J. Org. Chem.*, 1976, **41**, 2835.
6. R. Bonnett and J. D. White, *Proc. Chem. Soc.*, 1961, 119.
7. J. E. Johansen, B. D. Christie, and H. Rapoport, *J. Org. Chem.*, 1981, **46**, 4914.
8. J. D. Albright and W. R. Snyder, *J. Am. Chem. Soc.*, 1959, **81**, 2239.
9. T. Sheradsky, *Tetrahedron Lett.*, 1970, 25.
10. D. M. Wallace and K. M. Smith, *Tetrahedron Lett.*, 1990, **31**, 7265.

Chapter 7

1. (a) G. W. Miller and F. L. Rose, *J. Chem. Soc.*, 1965, 3357. (b) L. S. Cook and B. J. Wakefield, *Tetrahedron Lett.*, 1979, 1241.
2. (a) See 3. (b) I. R. Ager, D. R. Harrison, P. D. Kennewell, and J. B. Taylor, *J. Med. Chem.*, 1977, **20**, 379.
3. J. M. McCall, R. E. TenBrink, and J. J. Ursprung, *J. Org. Chem.*, 1975, **40**, 3304.
4. F. Yoneda, Y. Sakuma, M. Ichiba, and K. Shinomura, *J. Am. Chem. Soc.*, 1976, **98**, 830.
5. (a) R. F. Pratt and K. K. Kraus, *Tetrahedron Lett.*, 1981, **22**, 2431. (b) W. P. K. Girke, *Chem. Ber.*, 1979, **112**, 1. (c) P. J. Lont and H. C. van der Plas, *Rec. Trav. Chim.*, 1973, **92**, 311; 1971, **90**, 207. (d) A. Rykowski and H. C. van der Plas, *Rec. Trav. Chim.*, 1975, **94**, 204. (e) A. Albert and W. Prendergast, *J. Chem. Soc., Perkin Trans. 1*, 1973, 1794. (f) See ref. 51a in Chapter 7. (g) W. J. Coates and A. McKillop, *Heterocycles*, 1989, **29**, 1077. (h) T. J. Kress, *J. Org. Chem.*, 1985, **50**, 3073. [No mechanisms are established. One possibility: try forming a 'covalent hydrate' followed by N-oxidation (pyrimidone) or expodiation (imidazole)].

Chapter 8

1. N. S. Zefirov, S. I. Kozhushkov, and T. S. Kuznetsova, *Tetrahedron*, 1982, **38**, 1693.
2. H. H. Wasserman and F. J. Vinick, *J. Org. Chem.*, 1973, **38**, 2407.
3. G. Büchi and J. C. Vederas, *J. Am. Chem. Soc.*, 1972, **94**, 9128.
4. H. A. Staab, *Angew. Chem. Int. Edn. Eng.*, 1962, **1**, 351.
5. (a) C. L. Habraken and E. K. Poels, *J. Org. Chem.*, 1977, **42**, 2893. (b) D. J. Brown and P. B. Ghosh, *J. Chem. Soc.* (*B*), 1969, 270. (c) B. Stanovnik and M. Tišler, *Croat. Chim. Acta*, 1965, **37**, 17. (d) A. J. Boulton, D. E. Coe, and P. G. Tsoungas, *Gazz. Chim. Ital.*, 1981, **111**, 167. (e) K. P. Parry and C. W. Rees, *J. Chem. Soc., Chem. Commun.*, 1971, 833.
6. (a) M. Ruccia, N. Vivona, and G. Cusmano, *Tetrahedron*, 1974, **30**, 3859. (b) S. Hoff and A. P. Blok, *Rec. Trav. Chim.*, 1974, **93**, 317. (c) K. M. Baines, T. W. Rourke, K. Vaughan, and D. L. Hooper, *J. Org. Chem.*, 1981, **46**, 856. (d) L. E. A. Godfrey and F. Kurzer, *J. Chem. Soc.*, 1963, 4558. (e) T. L. Gilchrist, G. E. Gymer, and C. W. Rees, *J. Chem. Soc., Perkin Trans. 1*, 1973, 555. (f) See ref. 110 in Chapter 8.

7. H. Gotthardt, R. Huisgen, and H. O. Bayer, *J. Am. Chem. Soc.*, 1970, **92**, 4340.
8. H. Kristinsson and T. Winkler, *Helv. Chim. Acta*, 1983, **66**, 1129.

Chapter 9

1. (a) T. A. Foglia, L. M. Gregory, G. Maerker, and S. F. Osman, *J. Org. Chem.*, 1971, **36**, 1068. (b) H. W. Heine, D. C. King, and L. A. Portland, *J. Org. Chem.*, 1966, **31**, 2662. (c) R. E. Clark and R. D. Clark, *J. Org. Chem.*, 1977, **42**, 1136. (d) R. Prewo, J. H. Bieri, U. Widmer, and H. Heimgartner, *Helv. Chim. Acta.*, 1981, **64**, 1515. (e) E. Vedejs and P. L. Fuchs, *J. Am. Chem. Soc.*, 1973, **95**, 822. (f) S. Searles, H. R. Hays, and E. F. Lutz, *J. Org. Chem.*, 1962, **27**, 2832.
2. K. Isomura, G.-I. Ayabe, S. Hatano, and H. Taniguchi, *J. Chem. Soc., Chem. Commun.*, 1980, 1252.
3. J. S. Davies and T. T. Howarth, *Tetrahdron Lett.*, 1982, **23**, 3109.
4. M. L. Gilpin, J. B. Harbridge, T. T. Howarth, and T. J. King, *J. Chem. Soc., Chem. Commun.*, 1981, 929.

Chapter 10

1. See ref. 12 in Chapter 10.
2. W. H. Rastetter, *J. Am. Chem. Soc.*, 1976, **98**, 6350.
3. J. T. Sharp, R. H. Findlay, and P. B. Thorogood, *J. Chem. Soc., Perkin Trans. 1*, 1975, 102.
4. (a) T. Tsuchiya, M. Enkaku, and H. Sawanishi, *J. Chem. Soc., Chem. Commun.*, 1978, 568. (b) G. H. Field and L. H. Sternbach, *J. Org. Chem.*, 1968, **33**, 4438. (c) D. J. Anderson and A. Hassner, *J. Chem. Soc., Chem. Commun.*, 1974, 45.

Chapter 11

1. (i) (a) pyrazine (b) 1,4-diazine (c) 1,4-diazabenzene; (ii) (a) pyrazole (b) 1*H*-1,4-diazole (c) 1,2-diazacyclopenta-2,4-diene; (iii) (a) furazan (b) 1,2,5-oxadiazole; (c) 1-oxa-2,5-diazacyclopenta-2,4-diene; (iv) (a) morpholine (b) 1,4-oxazinane (c) 1-oxa-4-azacyclohexane.

2. (a) (b) (c) (d)

3. (a) (b) (c)

   (d)

4. (a) 1,2,4-triazolo[4,3-*a*]pyridine; (b) thiepino[4,5-*c*]furan; (c) pyrano[2,3-*b*]indole; (d) 3*H*-pyrano[2,3-*c*]isoquinoline.

Index